Random Testing of Digital Circuits

Random Testing of Digital Circuits
Theory and Applications

René David
Laboratoire d'Automatique de Grenoble
Institut National Polytechnique de Grenoble
Centre National de la Recherche Scientifique
Grenoble, France

MARCEL DEKKER, INC.   NEW YORK · BASEL · HONG KONG

Library of Congress Cataloging-in-Publication Data

David, René.
   Random testing of digital circuits: theory & application / by René David; with a foreword by Thomas W. Williams.
      p. cm.
   Includes bibliographical references and index.
   ISBN 0-8247-0182-8 (alk. paper)
   1. Digital integrated circuits–Testing. I. Title.
TK7874.D365 1998
621.3815--dc21   98-2765
                                                                                       CIP

The author credits the cooperation of Mireille Jacomino and Pascale Thevenod.

The publisher offers discounts on this book when ordered in bulk quantities. For more information, write to Special Sales/Professional Marketing at the address below.

This book is printed on acid-free paper.

Copyright © 1998 by MARCEL DEKKER, INC. All Rights Reserved.

Neither this book nor any part may be reproduced or transmitted in any form or by any means, electronic or mechanical, including photocopying, microfilming, and recording, or by any information storage and retrieval system, without permission in writing from the publisher.

MARCEL DEKKER, INC.
270 Madison Avenue, New York, New York 10016
http://www.dekker.com

Current printing (last digit):
10 9 8 7 6 5 4 3 2 1

PRINTED IN THE UNITED STATES OF AMERICA

Foreword

The introduction of complex integrated networks and the ability to separate good networks from defective ones has consistently been a difficult problem. With the introduction of Self-Testing Techniques, for example Built-In-Logic-Block-Observation technique (BILBO), Random Patterns have played a central role in helping to solve this problem. As a result of the research in this area, a better understanding of this fundamental concept of Self-Testing has led to many designs, from large mainframe computers to the smallest of networks with utility in manufacturing and in the field. This has resulted in considerable savings in manufacturing costs and repair costs. Because of the high demand for Self-Testing, research and development in Random Patterns is absolutely required to keep pace with these ever increasing complex networks.

The author of this book has been and continues to be one of the prominent researchers in the area of random pattern testing for more than two decades. He has brought a refreshing new outlook to the field which has resulted in a number of very innovative approaches to help solve some of these fundamental problems. The key to his innovative approach is strongly based in a solid mathematics foundation. With this approach he has helped solve a number of problems in self testing applicable to structures such as BILBO. Moreover, he and his coworkers have gone further to bring some insight into an area in which few others would even have ventured, that is, random pattern testing of complex sequential machines such as μProcessors. Today this body of work has found application not only in testing of hardware but is now even being applied to the very important area of software testing. All this is accomplished with a solid mathematical base which is very easily in the reach of practicing scientists and engineers.

This book is based on the author's work in the area of random pattern testing. With the analytical tools developed in this book you will be taken through the frontier of random pattern testing from the fundamentals up to very useful concepts that can be applied in today's designs. I am pleased to have the opportunity to write this foreword for one of the researchers for whom I have always had the highest respect, and I am sure that after studying this work the reader will have similar feelings.

Professor Thomas W. Williams
University of Hannover and IBM

Preface

Since people have been able to design and manufacture logic circuits of increasing complexity, the problem of determining whether a circuit is behaving correctly, according to the specifications, has steadily grown in importance. This problem, i.e., testing digital circuits, was seriously tackled during the sixties. It was stated in the following way: given a fault (or a set of faults), find an input sequence producing a faulty output sequence if this fault (or any fault in the set of faults) is present in the circuit. This approach is known as deterministic testing. The famous D-algorithm, published in 1966, is typical of that decade.

During the academic year 1969-70, students in the Laboratoire d'Automatique de Grenoble, directed by P. Deschizeaux, built a small random tester and performed some *random testing* trials. This approach consists of applying a random input sequence to both the circuit under test and a reference circuit and comparing the outputs in order to decide whether or not the circuit is faulty. A study aiming at a Ph.D. was started in 1970. The author of this book was involved in advising this study, and then decided to continue research in this area, even though, for several years, research on random testing was considered by some people to be an academic amusement. Approximately at the same time, and independently, J.-C. Rault at Thomson/CSF, Paris, E. J. McCluskey at Stanford University, and P. Agrawal and V. D. Agrawal at Automation Technology Company, Champaign, started to work on similar subjects. These teams stopped this activity around 1975-78. The author's team was alone in tackling random testing analysis for circuits as complex as microprocessors in the late seventies; for several years, studies on random testing seemed to be more or less abandoned by other researchers. Around 1980, this research was recommenced, at IBM Company, in particular by P. H. Bardell in Poughkeepsie and T. W. Williams in Boulder, at Stanford University, and in some other places. In the author's team, some part of the research activity has been devoted to random testing without discontinuity up to the present time.

Nowadays, random testing of digital circuits is used extensively for built-in self-test: test access problems of modern VLSI circuits are solved by the introduction of on-chip test generators and response compressors (called signature analyzers). This technique is known as built-in self-test (BIST). Test generation cost can be greatly reduced through the use of generators such as a linear feedback shift register (LFSR), which produces pseudorandom test sequences whose behavior is close to purely random test sequences.

The fitness of a random test sequence, for a circuit and a required level of confidence in the test result, must be assessed qualitatively and quantitatively. This book aims at providing tools for this estimation.

When writing this book, the author was guided by two main goals. The first one is to start from the basics of digital circuit testing, independent from the kind of test, deterministic or random, and to reach an accurate understanding of random testing: theory and applications. The second goal is to present a *didactic tutorial* which is easy to understand due to many simple examples and detailed figures. Roughly speaking, *three main parts* can be identified in the book.

Chapters 1, 2, 3, and partially 4, present the *basic concepts related to testing of digital circuits*. This part is independent of the kind of test, either deterministic or random. Its content was used by the author for an introductory course on testing of digital circuits at post-graduate level. It contains what should be known by a student before tackling the bibliography, articles and communications, relating to a Master's or Ph.D. thesis on testing of digital circuits.

After the performance measurements on random testing presented in a section of Chapter 4, Chapter 5 is devoted to the *basic principles* of this kind of test. The concepts presented in this chapter are qualitative, corresponding to a kind of *philosophy of random testing*. Analysis of the required test length for some level of confidence is explained in Chapters 6 to 9 for various kinds of circuits: combinational, sequential, memories, and microprocessors.

Chapters 10 to 13 present *various aspects related to random testing*. Generation of random test sequences and experimental results are presented in Chapters 10 and 11, respectively. Then, signature analysis and design for random testability, two important subjects as far as BIST is concerned, are presented in Chapters 12 and 13.

In addition to the thirteen chapters, thirteen *appendices* provide details on some specific points. These points are not dealt with in the chapters for the following reason: although they may be useful for an accurate understanding of some points, they are not absolutely necessary for understanding the content of chapters (for simplified reading of appendices, there is some slight redundancy).

There is a *continuity* between most of the successive chapters; for example Chapter 2 presents notations used to define fault models in Chapter 3, and some examples in Chapter 4 are based on both. However, a reader who has some basic knowledge of the test principles may read any chapter without reading the previous ones; he is helped by *cross references* to sections, definitions, properties, etc., where some useful notions are discussed. Cross references are often between chapters and appendices or from a chapter to a previous one; however, there are a few references to future chapters, when a notion or property will be illustrated later while this illustration is not absolutely necessary to understand the part presently read.

For simplified reading, only few bibliographical references are given in the body of the chapters. Each chapter ends with "Notes and References".

All the main *notations* are given at the beginning of the book. At the end, a complete *index* allows a reader to easily locate where a particular concept is defined or discussed.

Acknowledgements

This book is based on research which was conducted in our laboratory for almost three decades, hence involving a lot of people. The author wishes to express his gratitude to Pierre Deschizeaux and to all the Master and Ph.D. students who have contributed by their scientific work and helped in many, many ways. I cannot mention all of them, but I would like to point out some key contributions. Roberto Tellez-Giron was the first Ph.D. student on the topic; he designed and built the random test machine allowing the first experimental comparison of deterministic and random testings. Xavier Fédi designed and built the random tester for Motorola *6800* microprocessors. This machine, which was patented [DaFé 83], allowed experiments which were important for our understanding of what a random tester could and should be. Pascale Thévenod and Mireille Jacomino were my collaborators for about one decade each. Their thorough understanding of random testing was an invaluable help in writing this book. We jointly decided of the contents of the book and, chapter after chapter, they provided technical criticism and made en excellent contribution to discussions which have allowed considerable improvements. This book would not be what it is without them.

This book has also benefited from collaboration with researchers of other teams with whom I had the opportunity to work. The time I spent at Stanford University and Waterloo University was made possible thanks to Edward McCluskey and Paul Bardell on the one hand and to John Brzozowski on the other. I have also had the chance to work and write papers with Bernard Courtois, Jianwen Huang, Helmut Jürgensen, Jean-Luc Rainard, and Kenneth Wagner. Other contributions are cited in the body of the book.

The work of our team, on random testing, was supported by the *Institut de Recherche en Informatique et Automatique* (IRIA) and the *Projet Pilote SURF* (for *sur*eté de *f*onctionnement). It was also supported by, or conducted in cooperation with, a number of industrial partners: *Commissariat à l'Energie Atomique* (CEA), *Centre National d'Etudes des Télécommunications* (CNET), *company Efcis*, *International Business Machines* (IBM), *Laboratoire Central des Industries Electriques* (LCIE), and *company Thomson-CSF*.

I owe a special acknowledgement to Jacques Pulou and John Brzozowski who spent a lot of time reading the manuscript, writing their criticisms, suggesting improvements, and discussing technical points with me. Jacques' extensive bibliographical knowledge and John's eye for detail added to the extensive scientific knowledge of both and have been responsible for inestimable improvements to the manuscript.

I was encouraged and guided in writing this book by several people, particularly Mihalis Nicolaïdis and Yervant Zorian.

I would like to emphasize the invaluable and enthusiastic support of Thomas Williams who has shown his continuous interest in our work and who wrote the foreword to this book.

At Marcel Dekker, Maria Allegra and Brian Black have proved efficient and understanding editors. While I was writing the book, I was in contact with several publishers; I was provided with comments which have been helpful for improving the manuscript by Vishwani Agrawal at Kluwer and Peter Mitchell at John Wiley & Sons (who forwarded me comments of anonymous reviewers).

Although it was very time consuming, I was allowed and encouraged to write this book by the Director of the *Laboratoire d'Automatique de Grenoble* and researchers of the team *Systèmes Logiques et Discrets*. The technical team of the laboratory was very helpful to solve software problems during the preparation of the manuscript, Florence Pouget made an excellent type-written copy of the text, and Julia Summerton corrected a lot of mistakes and clumsy turns of phrase in my initial draft.

René David

Contents

Foreword *Thomas W. Williams*   iii

Preface   v

Notation   xvii

1 Random Testing and Built-In Self-Test   **1**

   1.1 TESTING OF DIGITAL CIRCUITS   1
      1.1.1 Testing needs
      1.1.2 Scan design
      1.1.3 Some fundamental difficulties leading to BIST

   1.2 BUILT-IN SELF-TEST   7
      1.2.1 On-line BIST
      1.2.2 Off-line BIST

   1.3 SOME QUESTIONS   12

   NOTES and REFERENCES   14

2 Models for Digital Circuits and Fault Models   **15**

   2.1 NOTATIONS AND MODELS FOR DIGITAL CIRCUITS   15
      2.1.1 Combinational circuits
      2.1.2 Sequential circuits
          2.1.2.1 Synchronous sequential machine
          2.1.2.2 Asynchronous sequential machine
          2.1.2.3 Formal models
      2.1.3 Sequences of vectors
          2.1.3.1 Input language and output language
          2.1.3.2 Equivalent machines

   2.2 BASIC FAULT MODELS   26
      2.2.1 Levels of modelling
      2.2.2 Defects, faults and errors
      2.2.3 Basic fault models
          2.2.3.1 Classical models
          2.2.3.2 Specific models
          2.2.3.3 Functional models

	2.3	FUNCTIONAL AND STRUCTURAL TESTING	35

- 2.3 FUNCTIONAL AND STRUCTURAL TESTING — 35
  - 2.3.1 Functional testing
  - 2.3.2 Structural testing
  - 2.3.3 Exhaustive testing
- NOTES and REFERENCES — 39

3 Basic Concepts and Test Generation Methods — 41

- 3.1 COMBINATIONAL CIRCUITS — 41
  - 3.1.1 Comparison between faulty and fault-free circuits
  - 3.1.2 Propagation through sensitized paths
    - *3.1.2.1 Illustration of the principle*
    - *3.1.2.2 About reconverging fanout and backtracking*
  - 3.1.3 Algebraic method
  - 3.1.4 Relations between faults
    - *3.1.4.1 Fault collapsing*
    - *3.1.4.2 Checkpoints*
    - *3.1.4.3 Redundancy and masking*
    - *3.1.4.4 Fault coverage*
- 3.2 SEQUENTIAL CIRCUITS — 57
  - 3.2.1 Initialization and evasive faults
    - *3.2.1.1 Initialization*
    - *3.2.1.2 Compatible initial states and evasive faults*
  - 3.2.2 Comparison between faulty and fault-free circuits: Observers
    - *3.2.2.1 The initial state of the faulty circuit is known*
    - *3.2.2.2 The initial state of the faulty circuit is not known*
  - 3.2.3 Propagation
  - 3.2.4 Relations between faults
    - *3.2.4.1 Fault collapsing*
    - *3.2.4.2 Fault detectability*
    - *3.2.4.3 Generation of a test sequence for a set of faults*
- NOTES and REFERENCES — 80

4 Performance Measurements for a Test Sequence — 83

- 4.1 DEFECT LEVEL — 83
- 4.2 FAULT MODELS AND FAULT COVERAGE — 86
- 4.3 RELATION BETWEEN FAULT COVERAGE AND DEFECT LEVEL — 88
- 4.4 MEASUREMENTS OF THE CONFIDENCE LEVEL FOR RANDOM TESTING — 90
  - 4.4.1 Measurements derived from deterministic concepts
  - 4.4.2 Measurements specific to random testing
  - 4.4.3 Some properties
    - *4.4.3.1 Relations between measurements*
    - *4.4.3.2 Relations between faults in the context of random testing*
- NOTES and REFERENCES — 105

5 Basic Principles of Random Testing — 107

5.1 PRINCIPLE OF RANDOM TESTING — 107
5.2 ALL THE SPECIFIED CASES AND ONLY THE SPECIFIED CASES SHOULD BE TESTED — 109
  5.2.1 Example of combinational circuit
  5.2.2 Example of asynchronous sequential circuit
  5.2.3 Example of synchronous sequential circuit
  5.2.4 Example of partly synchronous and partly asynchronous circuit
  5.2.5 Conclusion on specification

5.3 ABOUT DETECTION POWER OF RANDOM TESTING — 117
  5.3.1 Markov chains associated with the observer
   5.3.1.1 The initial state of the faulty machine is known
   5.3.1.2 The initial state of the faulty machine is unknown
   5.3.1.3 Generalized distribution
  5.3.2 Limits of random testing
  5.3.3 Limits of deterministic testing
  5.3.4 Conclusion on detection power

5.4 FIRST DETECTION AND MEMORY EFFECT — 129
  5.4.1 Combinational fault
  5.4.2 Sequential fault

NOTES and REFERENCES — 133

6 Random Test Length for Combinational Circuits — 135

6.1 RANDOM TEST LENGTH — 135
  6.1.1 Random test length for a fault
   6.1.1.1 Detection probability
   6.1.1.2 Random test length
  6.1.2 Random test length for a set of faults
   6.1.2.1 Expected fault coverage
   6.1.2.2 Worst case: minimum testing probability

6.2 COMPUTATION OF THE DETECTION PROBABILITY — 147
  6.2.1 Detection probability of a fault
   6.2.1.1 Detection function
   6.2.1.2 Extended Cutting Algorithm
  6.2.2 Detection probabilities for a set of faults
   6.2.2.1 Controllability, observability, and activity
   6.2.2.2 Simulation
   6.2.2.3 Lower bounds on detectability

6.3 NUMERICAL RESULTS — 162
NOTES and REFERENCES — 164

xii   Contents

7 Random Test Length for Sequential Circuits   167

7.1 ASYNCHRONOUS AND SYNCHRONOUS TESTS   167

7.2 TEST LENGTH FOR A FAULT   169
7.2.1 Example of exact calculation
7.2.2 Accurate approximation for a large test length
7.2.3 Obtaining the test length $L(\varepsilon)$

7.3 APPROXIMATE METHODS   177
7.3.1 Minimal detecting transition sequence (MDTS)
7.3.1.1 Detecting transition sequence
7.3.1.2 Detection set
7.3.2 Single transition faults
7.3.2.1 Approximate random test length
7.3.2.2 Topics not taken into account

7.4. NUMERICAL RESULTS   188

NOTES and REFERENCES   190

8 Random Test Length for RAMs   193

8.1 MODELS   193
8.1.1 Models of RAMs
8.1.2 Fault models
8.1.2.1 Faults in the decoder and read/write logic
8.1.2.2 Faults in the memory cell array

8.2 TEST LENGTH FOR SINGLE FAULTS   197
8.2.1 Stuck-at fault of a memory cell
8.2.1.1 Influence of the distribution ψ
8.2.1.2 Influence of initialization
8.2.1.3 Influence of the number of cells
8.2.1.4 Influence of the confidence level
8.2.1.5 Concluding remarks
8.2.2 Faults in address decoding and read/write logic
8.2.3 Faults in the memory cell array
8.2.3.1 Length coefficient for toggling fault
8.2.3.2 Other faults

8.3 EXTENSION TO OTHER MODELS   209
8.3.1 Word-oriented memory
8.3.1.1 All the cells involved belong to different words
8.3.1.2 PSFs such that some involved cells belong to the same word
8.3.1.3 Coupling of bits in a single word
8.3.2 Multiple faults

8.4 POWER OF RANDOM TESTING FOR RAMs   213
8.4.1 Linearity of test length as a function of the number of cells
8.4.2 Comparison with deterministic testing
8.4.3 Example of application to a batch of RAMs

NOTES and REFERENCES   220

9 Random Test Length for Microprocessors — 221

9.1 FUNCTIONAL MODELS — 221
9.1.1 Functional model of a microprocessor
9.1.2 Fault models
- 9.1.2.1 Faults in the registers
- 9.1.2.2 Faults in the operators
- 9.1.2.3 Faults in the register decoding function
- 9.1.2.4 Faults in instruction decoding and control function

9.2 MARKOV CHAINS AND MDTS — 228
9.2.1 First example
9.2.2 Second example

9.3 TEST LENGTH FOR FAULTS IN THE DATA PROCESSING SECTION — 233
9.3.1 Faults in the registers
9.3.2 Faults in the operators

9.4 TEST LENGTH FOR FAULTS IN THE CONTROL SECTION — 239

9.5 TEST LENGTH FOR A MICROPROCESSOR — 241
9.5.1 Example microprocessor
- 9.5.1.1 Faults in the registers
- 9.5.1.2 Faults in the operators
- 9.5.1.3 Fault in the control section
- 9.5.1.4 Fault in the whole microprocessor

9.5.2 Microprocessor Motorola 6800
- 9.5.2.1 Basic results
- 9.5.2.2 Further results

NOTES and REFERENCES — 247

10 Generation of Random Test Sequences — 249

10.1 NEEDS — 249
10.1.1 Set of vectors for combinational faults
- 10.1.1.1 Equal likelihood of all the input vectors
- 10.1.1.2 Constant but not equally likely distribution

10.1.2 Set of subsequences for sequential faults
- 10.1.2.1 Synchronous test
- 10.1.2.2 Asynchronous test (adjacent vectors)
- 10.1.2.3 Generalized distribution

10.2 SOFTWARE GENERATION — 254
10.2.1 Constant distribution
10.2.2 Generalized distribution
10.2.3 Comments on software generation

10.3 HARDWARE GENERATION — 257
10.3.1 Basic properties of LFSRs and M-sequences
10.3.2 Constant Distribution
- 10.3.2.1 Equally likely distribution
- 10.3.2.2 Weighted test vectors

10.3.3 Sequence of adjacent vectors

xiv   Contents

    10.3.4 Generalized distribution. Example for the Motorola *6800* microprocessor.
        10.3.4.1 General description of the random test machine
        10.3.4.2 Principle of the input sequence generation
        10.3.4.3 Hardware test pattern generator

  NOTES and REFERENCES   276

11 Experimental Results   277

  11.1 TTL CIRCUITS   277
    11.1.1 Batch of 61 circuits reference *483 E*
    11.1.2 Memory effect in sequential circuits

  11.2 LSI CMOS CIRCUITS   281

  11.3 MOTOROLA *6800* MICROPROCESSOR   283
    11.3.1 Experimental results for a set of 60 microprocessors
    11.3.2 Experiments versus theory

  11.4 OTHER EXPERIMENTAL RESULTS   290
    11.4.1 Experiments by W. Luciw: Intel *8080* microprocessor
    11.4.2 Experiments by A. Laviron et al.: Motorola *6800* microprocessor
    11.4.3 Experiments by R. Velazco et al.: Motorola *6800* microprocessor
    11.4.4 Experiments by D.A. Wood et al.: multiprocessor cache controller

  NOTES and REFERENCES   296

12 Signature Analysis   299

  12.1 GENERAL FEATURES   299
    12.1.1 Aliasing and non-revelation
    12.1.2 General property
    12.1.3 Choice of k

  12.2 SINGLE INPUT SIGNATURE ANALYSERS   306
    12.2.1 Counting methods
    12.2.2 Signature by linear feedback shift register
    12.2.3 Properties of SISR
    12.2.4 Cost of signature analysis

  12.3 MULTIPLE INPUT SIGNATURE ANALYSERS   322
    12.3.1 Space dependent and time dependent errors
    12.3.2 MISR if $m \leq k$
    12.3.3 MISR if $m > k$
    12.3.4 SISR for periodic errors

  NOTES and REFERENCES   334

13 Design For Random Testability — 337
  13.1 DESIGN FOR TESTABILITY IN GENERAL — 337
  13.2 EXTENDED SPECIFICATION — 338
  13.3 DECORRELATION BY EXOR GATES — 342
  13.4 FACTORIZATION IN COMBINATIONAL FUNCTIONS — 347
  NOTES and REFERENCES — 351

Postface — 353

Appendices — 357

  A Random Pattern Sources — 357
  B Calculation of a Probability of Complete Fault Coverage — 361
  C Finite Markov Chains — 363
  D Black-Box Fault Model — 365
  E Exact Calculation of Activities — 371
  F Comparing Asynchronous and Synchronous Test — 375
  G Proofs of Properties 7.1, 7.2, and 12.3 — 379
  H Microprocessor Motorola *6800* — 385
  I Pseudorandom Testing — 389
  J Random Testing of Delay Faults — 393
  K Subsequences of Required Lengths — 397
  L Diagnosis from Random Testing — 401
  M Conjectures on Multiple Faults — 407

Exercises — 411

Solutions to exercises — 425

Bibliography — 447

Index — 463

Notation

General

CUT	Circuit under test.
BIST	Built-in self-test.
LFSR	Linear feedback shift register.
$\Pr[\alpha]$	Probability of event α.
$\Pr[\alpha \mid \beta]$	Conditional probability of event α given event β.
$E[\gamma]$	Mathematical expectation of the random variable γ.
$\sigma[\gamma]$	Standard deviation of the random variable γ.
$G \setminus H$	Set of elements in G which are not in H, i.e., $G \cap \overline{H}$.
V^{T}	Matrix transposed from V.

Automata and Sequences

n	Number of primary inputs of a circuit, of cells in a RAM.
$x_i; x_j$	Input variable; input vector (meaning of index j in Section 2.1.1).
X	Set of input vectors.
$z_i; z_j$	Output variable; output vector (index j: see Section 2.1.1).
Z	Set of output vectors.
$q_i; q_0$	Internal state; initial state.
Q	Set of internal states (up to Chapter 9).
$\delta; \mu$	Transition function; output function.
CK	Clock.
$\$$	Unspecified entity: may have any value or does not exist.
A	Automaton (up to Chapter 8).
$M = (A, q_0)$	Machine i.e., initialized automaton.
$M' = (A', q'_0)$	Faulty machine.
$M' = (A', Q'_0)$	Faulty machine with unknown initial state ($q'_0 \in Q'_0$).
$\Omega(M, M')$	Observer for M and M'.
$X^*; Z^*$	Monoids built over the set of input vectors; of output vectors.
$\mathcal{P}(B)$	Set of prefixes of the regular expression B.
λ	Word of length 0.
$I; O$	Input language ($I \subset X^*$); output language ($O \subset Z^*$).
S	Input sequence or test sequence ($S \in I$).
R	Output sequence or response ($R \in O$).

Faults and Measurements

D; $\overline{D}$	Erroneous Boolean value 0 instead of 1; 1 instead of 0.
f or f_i	A fault.
$f_1 \equiv f_2$	f_1 and f_2 are equivalent.
$f_1 \cong f_2$	f_1 and f_2 are equidectable.
w/α	Fault: line w stuck-at α.
$[f_1 \& f_2]$	Double fault.
T_f	Detection function (or set of vectors) detecting the combinational fault f.
F or F_j	Set of faults.
$\mathcal{M}_i$; F_i	Fault model; set of single faults corresponding to $\mathcal{M}_i$.
F_i^*	Set of multiple faults whose components are in F_i.
F_u	Universal fault set.
F_{nt}	Set of non-target faults.
Y	Production yield.
DL	Defect level.
P_u	Faulty circuit coverage.
P	Fault coverage.
P_w	Weighted fault coverage.
P_m	Minimum testing probability.
P_c	Probability of complete fault coverage.
P_{wm}	Weighted minimum testing probability.

Probabilities and Sequence Lengths

ψ or ψ_i	Distribution of random input vectors (constant or generalized).
ψ_0	Equally likely distribution.
$\psi(x_j)$	Probability of input vector x_j (constant distribution).
$P(f, S)$	Probability of testing fault f by the input sequence S.
$P(f, \psi, L)$	$P(f, S)$ where ψ and L are the distribution and the length of S.
or $P(f, L)$ or $P(f)$	When ψ is implicit or when ψ and L are implicit.
k_f	Detectability of a combinational fault.
p_f	Detection probability ($= 1 - \lambda_2$ for a sequential fault).
$k_{\min}$; $p_{\min}$	k_f of the worst case fault (combinational); p_f for this fault and ψ_0.
$p_{av,f}$	Average detection probability (sequential fault).
MDTS	Minimal detecting transition sequence.
D_f	Detection set associated with the sequential fault f.
l_ω ; l_τ	Length to first detection; between two successive detections.
$\pi(l)$	State probability at time l (Markov process).
$\pi_i(l)$; $\pi_\omega(l)$	Probability of state i; that the fault is detected (state ω).
λ_2	Greatest eigenvalue of a stochastic transition matrix (except $\lambda_1 = 1$).
η	Vector $(0, ..., 0, 1)^T$ in Chapter 7, $(1, 0, ..., 0)^T$ in Chapter 12.
L	Test length.

$1 - \varepsilon$	Confidence level: may be P, P_u, P_w, P_m, P_c, or P_{wm} for a set of faults, or $P(f)$ for a fault.
ε	Detection uncertainty.
$\varepsilon(L)$	Detection uncertainty for test length L.
$L(\varepsilon); L_m(\varepsilon)$	Test length for detection uncertainty $\leq \varepsilon$; if $\varepsilon = 1 - P_m$.

Controllability and Observability

$C(\alpha, w)$	α-controllability of line w.
$O(w)$	Observability of line w.
$A(y, G_i)$	Activity of input combination y to gate G_i.

RAMs

n	Number of cells.
$\uparrow i \Rightarrow \uparrow j$	Idempotent coupling.
$\uparrow i \Rightarrow \updownarrow j$	Toggling.
PSF	Pattern sensitive fault.
$r^i; w^i_\alpha$	Read cell i; write α in cell i (input vectors of a RAM).
$V(f)$	Number of involved cells (fault f in a RAM).
H	Length coefficient L/n (for RAMs).

Generators and Signature Analysers

$Q(t); Q_i(t)$	Content of an LFSR at time t; content of the ith stage.
σ	Number of shiftings at a single clock pulse (LFSR).
SISR	Single input signature register.
MISR	Multiple input signature register.
A or A_i; A_0	A signature; fault-free signature (Chapter 12).
$\Pr[A = A_0]$	Probability of non-revelation.
$\Pr[R = R_0]$	Probability of non-detection.
AL	Probability of aliasing.
E_j	Error sequence (single output) or error matrix (multiple output).

Random Testing of Digital Circuits

1

Random Testing and Built-In Self-Test

This introductory chapter presents the testing needs of digital circuits, the difficulties of testing, which increase with the complexity of VLSI (very large scale integrated) circuitry, and the main ways to tackle them (Section 1.1). Built-in self-test *(BIST) is an approach allowing some of the most important problems to be solved. The main features of BIST are presented in Section 1.2. Pseudorandom generators are widely used as build-in test pattern generators. Some fundamental questions arise about random testing (which can also be used without BIST); they are stated in Section 1.3, then tackled in the rest of the book.*

In this first chapter, some terms might be unknown to a reader who has no knowledge of the test field. The more specific terms are explained. The very basic notations and concepts will be introduced in Chapter 2 and subsequent chapters.

1.1 TESTING OF DIGITAL CIRCUITS

First, the needs for digital testing are presented in Section 1.1.1. The difficulty of generating test patterns has led to scan design, presented in Section 1.1.2. Some fundamental difficulties (summarized in Section 1.1.3) remain. They lead to using built-in self-test (BIST).

1.1.1 Testing needs

An integrated circuit (i.e., a chip) may behave incorrectly for various reasons: error in the design (electrical or logical), defect because the manufacturing process is imperfect, aging of the components. Verification that there is no error in the design is not in the scope of this book (nevertheless, Section 11.4.4 presents an example where random behavior simulation was used to uncover errors in the design). Usually, *hardware testing* means verification of the correct behavior, given there is no design error.

The *manufacturer* has to test the manufactured circuits. He can first test them on the *wafer* (i.e., a large silicon slice in which hundreds of circuits may be integrated); the test may be very simple in order to avoid packaging of circuits obviously defective. The manufacturer should also test the circuits in their packages before shipping them to the users. Several kinds of tests can be performed: *parametric tests* (verification of electrical levels, voltages and intensities); *dynamic tests* (rising and falling times, delays); *logic tests* (logic function, Boolean or sequential). In order to eliminate short-lived parts, one can test the circuits with stress (*burn-in*): high temperature, temperature cycling, vibration etc. With the complexity of circuits existing today, the cost of testing is a significant part of the cost of the product.

The *quality level* is defined as the fraction of good circuits among the circuits that pass all the tests and are shipped. It is more usual to consider values of the *defect level* (*DL*) which is the complement of the quality level (the defect level depends on both the production yield and the "quality" of test, as will be seen in Chapter 4). The defect level is commonly given in parts per million or *ppm*. The defect level for complex chips was estimated to be about 200 *ppm* in [In 87] (corresponding to 0.02% of faulty circuits). The Semiconductor Research Corporation aims at a defect level of roughly one *ppm* for 10^7-transistor chips [SRC 85].

The *user* may have several testing environments. Some testing is done on receipt of a shipment of circuits (*incoming inspection*). Next, the *production test* is performed when the circuit is combined with a *printed circuit board* (or simply *board*) containing other circuitry (test of the whole board, or *in-situ* testing of a circuit if it can be electrically isolated). The next test relates to the *system* (in which the board has been implemented). The final user test environment is *field testing*.

The cost of testing increases rapidly (by a factor of about 10) at each step: wafer testing, package testing, incoming inspection, board testing, system testing, and field testing. Hence, it is important to detect that a circuit is faulty as soon as possible.

Assume a user implements a board containing 100 circuits. If every circuit has probability 0.99 of being fault-free, the board has probability $0.99^{100} \approx 0.37$ of being fault-free (this is the probability that all the circuits on the board are fault-free, assuming that no fault is added by the connections): two thirds of the boards are faulty! Now, if every circuit has a 0.9998 probability of being fault-free (i.e., defect level = 200 *ppm*), 2% of the boards are faulty since $0.9998^{100} \approx 0.98$. This example illustrates the importance of incoming inspection: the user saves board testing which is more costly. The user does not have the same information as the manufacturer on the internal structure of the circuits; he can perform a functional test, verifying the specifications published by the manufacturer.

Figure 1.1 represents a general testing scheme. A *test sequence* is the *input sequence* sent to the circuit under test (CUT). The *response*, i.e., the *output sequence* of the CUT, is observed by the test evaluator. The controller synchronizes the different elements, may initialize them if necessary, etc; in the sequel, the controller is not always represented, it is implicit.

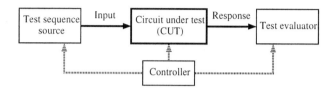

Figure 1.1 General testing scheme.

The *test sequence* may be either *deterministic or random* (or mixed if necessary, see [AgAg 72], [Go 78], [Ab et al. 86] for example). If it is deterministic, the test sequence is based on *fault hypotheses*, which may be either structural or functional (for a manufacturer's test, because the user may not know the internal structure). The test sequence may be *prestored* (deterministic or random) or *generated in real-time* (random).

The *test evaluator* may be a comparison of the whole response of the CUT with the expected response; this fault-free response may be either *prestored* or *obtained in real-time* from a reference circuit or model receiving the same input sequence (deterministic or random). The test evaluator may also be based on a *signature analyser* which maps the response (containing up to hundreds of millions of bits) into a few dozen bits: the signature of the CUT is compared with the fault-free signature. There is some randomness in the signature analysis since there is a non-zero probability that a faulty response produces a fault-free signature: Chapter 12 is devoted to this problem. The test evaluator may also measure the consumption of electric power of the circuit: this is known as I_{DDQ} testing. This kind of testing is not in the scope of this book; references are proposed to interested readers at the end of this chapter.

Let us return to the *generation of the test sequence, which is the most crucial problem*, and consider specifically *logic faults*. A deterministic test pattern generator (TPG) aims at testing all the faults in a prescribed set; it may be automated (ATPG), but even for the single stuck-at fault model (a simple model which does not cover all the possible defects), the automated test generation is very long and costly for large circuits. On the other hand, random testing has the disadvantage of requiring longer test sequence (the length of the test sequence will appear to be a key point) and the advantage of being easier to develop.

Chapter 1

Because of the difficulty of generating test sequences for large circuits[1], design methods aiming at reducing this difficulty have been developed; their generic name is *design for testability* (DFT). An important DFT method, the scan design, is presented in the next section.

1.1.2 Scan design

Full scan design is a technique of design for testability whose aim is to transform a sequential circuit in order to test it like a combinational circuit. In *partial scan*, the system remains sequential but the number of non-controllable state variables is reduced. The principle is illustrated in Figure 1.2; there are variants of this principle.

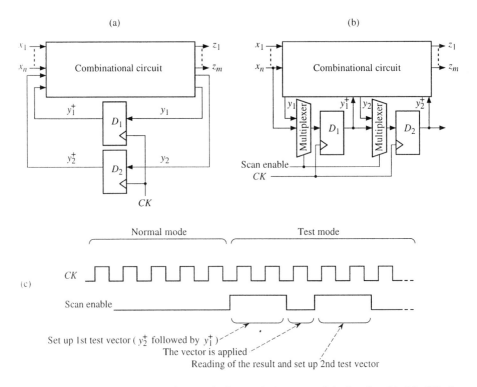

Figure 1.2 Illustration of scan design. (a) A sequential circuit. (b) Modified circuit with a scan path. (c) Use of clocks in normal and test modes.

---

[1] It has been observed that the computer run time required to perform test generation and fault simulation is approximately proportional to the number of logic gates to the power 3 [Go 80].

Figure 1.2a represents an n-input, m-output, 2-internal variable, sequential circuit[2]. It is made up of an $(n + 2)$-input, $(m + 2)$-output, combinational circuit, and of two D flip-flops[3] storing the internal state (the output y_i at time t becomes the input y_i^+ at $t+1$). Figure 1.2b represents the same circuit where a multiplexer precedes each D flip-flop. The multiplexer is controlled by the Boolean variable *scan enable*; when *scan enable* = 0, the variable y_i is the input of the flip-flop D_i; when *scan enable* = 1, the flip-flops are connected as a shift register (*scan path*), whose input is x_n.

Figure 1.2c illustrates the behavior of the circuit in Figure 1.2b. In *normal mode*, *scan enable* remains 0 and the circuit behaves as in Figure 1.2a. In *test mode*, the behavior is as follows. 1) When the variable *scan enable* = 1, during a series of s clock pulses (if s is the number of internal variables, $s = 2$ in our example), the values required in the flip-flops are set up by shifting from the input x_n (the required y_2^+ followed by the required y_1^+). 2) Then *scan enable* = 0 for one clock pulse and the desired $(n + s)$-input test vector is applied; it is made up of the n boolean variables present on $x_1, \ldots, x_n$ and the s outputs of the flip-flops. 3) The outputs $z_1, \ldots, z_m$ can immediately be observed, and a series of s clock pulses when *scan enable* = 1, shifts the s other outputs, stored in the flip-flops, to the output of the last flip-flop. During this series of s pulses, the next vector of values required in the flip-flops is set up. And so on.

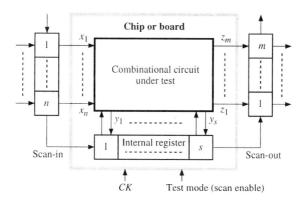

Figure 1.3 Boundary Scan.

In the sequel, we will explicitly distinguish between the interior of a circuit and the exterior of a circuit. Basically, we will consider that *a circuit* (i.e., all interior parts) corresponds to a chip; however, it could be a board or a subsystem. As illustrated in Figure 1.3, *boundary scan* allows the circuit to be

---

[2] Section 2.1.2.
[3] Figure 2.6.

tested via registers logically, and often physically, adjacent to the input/output pins of the chip. The template in Figure 1.3 allows the circuit to function normally: all the registers are used without any shifting. It also allows the internal logic to be tested: input data are applied from the input register and the internal register to the circuit, and the corresponding responses are latched in the output register; the results in the internal register can be shifted out and verified.

Scan design (including boundary scan) is useful because it allows good controllability and good observability of the internal memory elements of the circuit. In addition, test pattern generation is simplified, since it is performed for a combinational circuit (in case of full scan). However, scan design does not give a satisfactory answer to all the test problems, as is explained in the following section.

1.1.3 Some fundamental difficulties leading to BIST

The testing of large and complex integrated circuits presents some fundamental difficulties which naturally lead to the idea of built-in self-test, i.e., a design including some test device on the chip. A BIST may be used with or without scan design.

1) If the circuit under test and the *external tester* are based on the same technology, *a prestored test sequence cannot be applied at the nominal speed of the CUT*: the tester is slower than the tested circuit (hence some defects producing errors only at high speed cannot be detected). On the other hand, if a test sequence is *hardware generated on the chip* itself, it *may be applied at the nominal speed* of the circuit.

2) Scan design improves controllability and observability. However, it has some disadvantages: even if it is built in the circuit, since an input vector is set by shifting its components, the necessary time between two successive vectors *cannot correspond to the nominal speed*, and the *test time is long*.

3) The *internal complexity increases more quickly than the number of possible external pins*. Assume for example that we are able to produce (with a reasonable yield) square chips whose side dimension is d. The number of components may be proportional to the surface; hence, this number increases proportionally to d^2 when d increases (in addition, each component becomes smaller from year to year, then the number of components is still greater); on the other hand the number of possible pins is proportional to the perimeter[4], i.e., to d. Because of this limitation it becomes more difficult to test a circuit from the limited number of pins. If the test sequence generator and the test evaluator are built in the chip, controllability and observability may be improved by additional inputs and outputs to the nominal circuit, respectively, without using extra pins (since the

---

[4] If the size of the chip was defined to allow a larger number of pins, its surface would be partially unused.

corresponding wires come from the sequence generator or go to the test evaluator).

An advantage of BIST is that the test device is the same throughout the circuit life. From a test on the wafer to a field test, the same test device is used both by the manufacturer and the user.

Other advantages (particularly avoidance of test pattern generation, often cited in the literature) derive from the fact that pseudorandom tests are used in BIST. They will be explained at the end of Section 1.2.

1.2 BUILT-IN SELF-TEST

Various names have been used over twenty years for the ability of a circuit to test itself. Today, the name *built-in self-test* (*BIST*), merging built-in test (BIT) with self-test, is widely used. Two main categories may be considered: *on-line BIST* and *off-line BIST*.

1.2.1 On-line BIST

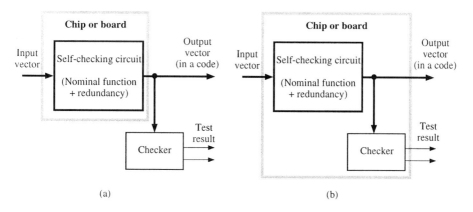

Figure 1.4 On-line BIST (a). The checker is not in the circuit. (b) The checker is in the circuit.

Figure 1.4 illustrates on-line BIST. The circuit is self-checking: informally, the circuit implements the *nominal* function (i.e., the basic function that the circuit is expected to perform) but also contains a redundant part in order to verify the correct behavior. The output vector is encoded, for example by a

k-out-of-m code (i.e., m outputs out of which exactly k have the value 1). Informally, the basic concepts are as follows. *Fault-secureness*: if there is a fault in the circuit, either the output is correct or it is not a valid code word. *Self-testing*: for any fault in the circuit there is an input vector in X such that the corresponding output is outside of the code. *Totally self-checking*: both fault-secure and self-testing. Variants of these concepts have been defined; some references are given in "Notes and References" at the end of the chapter.

The checker verifies that the output is in the code. It has two complementary outputs; if there is no error, the outputs are 01 or 10 (called double-rail code): a single output would not be able to detect a stuck-at of this output at the value corresponding to a fault-free behavior. The checker is itself self-testing and also *code-disjoint* (i.e., if its input vector is not in its code, its output is not in its code). In Figure 1.4a, the chip contains some redundancy but does not include the checker. In Figure 1.4b, the checker is built in the circuit (requiring two additional output pins).

Remark 1.1 A particular case of the scheme in Figure 1.4 is as follows. The self-checking circuit is made up of a duplicated nominal circuit: the first one performs the nominal function and the second one corresponds to the redundancy. If all the outputs of the second one are complemented, the output vector corresponds to a double-rail code.

1.2.2 Off-line BIST

A basic scheme of off-line BIST is presented in Figure 1.5. In normal mode, the n primary inputs are applied through the multiplexor to the *nominal circuit* (i.e., the circuit performing the function which is expected from the chip) and the m primary outputs are obtained. In test mode, the n inputs of the nominal circuit are produced by the built-in test pattern generator and the response is observed by the built-in test evaluator. If necessary, some inputs may be added to the circuit to improve controllability, and some outputs may be added to improve observability (Chapter 13): these additional inputs and outputs *do not need extra pins* as is shown in Figure 1.5. In this scheme, *only two extra pins* are required: one for normal mode/test mode, and the other for the test result i.e., pass/fail.

A test sequence generated off-line and prestored on the chip is in practice not realistic for big circuits (because of excessive memory requirements). Hence, the test sequence is generated in real time by the test pattern generator (TPG). This generation must be algorithmic. There are basically two main kinds of hardware test generation. The first one consists of producing successively all the input vectors; if the circuit under test is combinational, all the faults such that the circuit remains combinational can be detected (the test is then qualified as *exhaustive*[5]). This generation is too long if the number of inputs is large (the test length should be about 10^{12} vectors for a 40-input circuit). In addition, this generation may not be convenient for a sequential circuit under test. The second kind of generation produces a string of pseudorandom test vectors; *this kind of generation*, usually performed by a *linear feedback shift register*[6] (LFSR), *is widely used*. Variants to these main kinds of generation exists, such as pseudo-exhaustive testing (exhaustive testing of subcircuits) or generation of subsequences of two patterns (for testing specific faults).

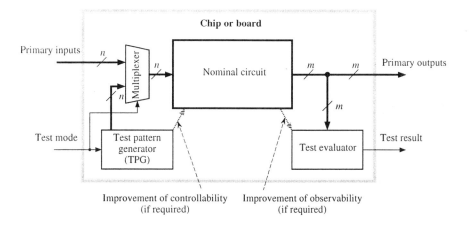

Figure 1.5 Off-line BIST: basic scheme[7].

In order to avoid storage of a long response of the nominal circuit, the test evaluator may be based on a *signature analyser*.

In Figure 1.5, the nominal circuit is tested as a whole system. This circuit may also be split into several CUTs corresponding to large blocks such as arithmetic and logic units (ALU), random access memories (RAM), programmable logic arrays (PLA); a BIST technique is applied to each of these blocks; all the blocks may be tested either in turn (the same TPG and the same signature analyser can be used for all the blocks) or concurrently (a TPG and a test evaluator can be dedicated to every block). There are thus various ways to structure the BIST. In the sequel, several BIST templates are presented.

In order to specify which primary outputs are faulty (this may be useful for location of a fault) the test evaluator may be built as shown in Figure 1.6.

---
[5] Section 2.3.3.
[6] Chapter 10.
[7] This figure is reprinted from [Th *et al.* 87] (© 1987 IEEE).

A single input signature analyser[8] is used in turn for every primary output. Assume for example that $m = 4$: there are 4 primary outputs, namely z_1, z_2, z_3, and z_4. The multiplexer is controlled by two address lines, a_1 and a_2 (since $\lceil \log_2 4 = 2 \rceil$). When $a_1 a_2 = 00, 01, 10,$ or 11, then the output of the multiplexer is z_1, z_2, z_3, or z_4, respectively; the four corresponding signatures can be obtained.

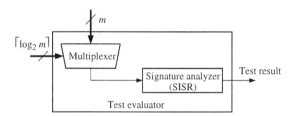

Figure 1.6 Test evaluator in which a single input signature analyser is used in turn for every primary output.

The BIST template in Figure 1.7 draws inspiration from the boundary scan scheme presented in Figure 1.3. The n-bit input register is replaced by an h-bit ($h \geq n$) linear feedback shift register corresponding to a pseudorandom pattern generator. The m-bit output register is replaced by a m-input signature analyser[9] of length k (k is not necessarily greater than or equal to m, see Chapter 12). These two registers, as well as the internal register, are built-in on the chips as illustrated in Figure 1.7.

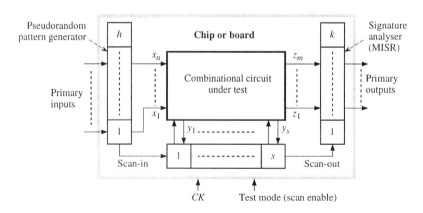

Figure 1.7 Built-in self-test boundary-scan template.

---

[8] Usually called SISR, short for *single input signature register* since the signature is usually performed by a linear feedback shift register (Chapter 12).
[9] Usually called MISR, short for *multiple input signature register*.

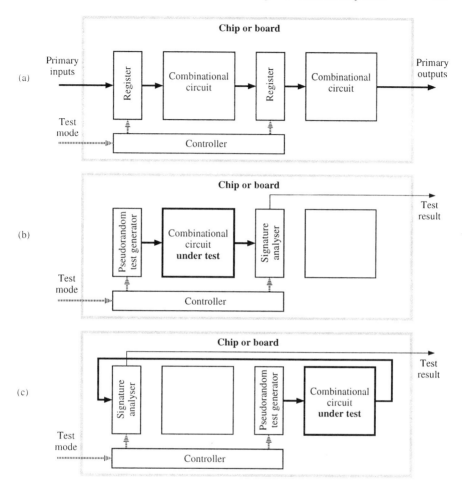

Figure 1.8 Illustration of the BILBO principle. (a) Normal working. (b) Test of the first combinational circuit. (c) Test of the second combinational circuit.

The BILBO (built-in logic block observer) architecture consists of partitioning a circuit into a set of registers and blocks of combinational circuits; the normal registers are replaced by BILBO registers able to perform several functions (specified by a controller). This is illustrated by an example in Figure 1.8. This example is made up of two registers and two combinational circuits. Three kinds of behavior can be specified for each register: normal operation, pseudorandom pattern generator, or signature analyser[10]. In Figure 1.8a, the

---

[10] This is only an example. For other applications, a BILBO register can be transformed into a scan path.

system is working normally: the system is fed by primary inputs and produces primary outputs. In Figure 1.8b the first combinational circuit is under test; for this purpose, the first register behaves as a pseudorandom test pattern generator (thanks to connections specified by the controller) and the second register as a signature analyser. In Figure 1.8c, the roles of the registers are reversed in order to test the second combinational circuit.

Summary of BIST properties

Although on-line BIST and off-line BIST can be defined, *the term BIST is often used instead of off-line BIST*. This meaning is used in the sequel.

BIST presents some *disadvantages*. The circuit is more complex: it may be slower and less reliable. The circuitry built in for the test purpose increases the surface of the chip (about 10 to 15 percent for circuits smaller than 10^5-transistor chips, only few percents for biggest circuits), and the number of pins (minimum 2, see Figure 1.5). Additional design time and cost are needed.

On the other hand, BIST presents several *advantages*. Some of these are the direct consequence of *building in*: the test device is the same throughout the life of the circuit, for the manufacturer and for the user (who usually has no knowledge of the internal structure); the nominal circuit may be tested at its nominal speed (hence detection of performance-related faults is possible); controllability and observability of the nominal circuit can be improved without additional pins. Some other advantages of BIST are due to the *pseudorandom test pattern generation*: development of pseudorandom test generation is much shorter than a deterministic test pattern generation based on fault hypothesis; because of randomness, faults which are not in the hypothesis[11] may have a high probability of detection; external deterministic testers are complex and expensive.

All these comments are more qualitative than quantitative. Some questions arise which are listed in the next section.

1.3 SOME QUESTIONS

The following features are explained in the above sections: BIST is a way of solving some of the major problems of digital circuit testing, and leads naturally to the use of pseudorandom pattern generators and signature analysis; pseudorandom test pattern generation may also be used as an alternative to deterministic test pattern generation (or they can be mixed) if an *external tester* is used.

To some extent, a pseudorandom test sequence behaves like a random test sequence; hence, in the sequel, the word random will often be used instead of

---

[11] The current fault models are imperfect; furthermore, new fault types arise from very large chips with smaller geometry.

pseudorandom and most theoretical concepts may be based on the assumption that the test sequence is *random*[12].

From these features, some theoretical and practical questions arise:

1) What is the "quality" of a random test sequence? Before answering this question it is necessary to define and compare the performance measurements which can be used for this purpose. This is the objective of Chapter 4.

2) The author would like to make the reader realize that *random testing should not be performed with a method chosen at random* (paraphrasing D.E. Knuth, about generation of random numbers [Kn 69]). What are the basic (theoretical) principles ensuring "good" random testing (*qualitative* aspect)? They are explained in Chapter 5.

3) Since there is a non-zero probability that a random test does not detect that a faulty circuit is defective, what is the test length required to reduce this probability under some fixed level (*quantitative* aspect)? Estimations of this test length for a combinational circuit, for a sequential circuit, for a RAM, or for a microprocessor, are presented in Chapters 6 to 9.

4) In practice, how can random testing be implemented in order to follow the theoretical principles presented in Chapter 5? Chapter 10 answers this question.

5) Now, beyond the theoretical aspect, what is the experimental efficiency of random testing for real circuits and real defects? Various experimental results are presented in Chapter 11.

6) In the BIST context, signature analysis is used. What are the basic principles of signature analysis and what is the probability that a faulty response to a random test sequence falls into a fault-free signature? These questions are answered in Chapter 12.

7) How can we design circuits such that the random tester is easy to build and the required test length is not too long? This problem is tackled in Chapter 13.

❑

The reader may have noted that Chapters 2 and 3 are not cited in this list of questions. These are two introductory chapters. Chapter 2 introduces the models of digital circuits used in the book and the commonly used fault models. Chapter 3 presents basic concepts and generation methods related to testing of digital circuits: most of the concepts presented are well established, a few are new.

---

[12] The pseudorandomness is taken into account for some specific cases, especially for the generation of the test sequences (Chapter 10). Advantage can also be taken of the fact that a pseudorandom sequence is deterministic, i.e., always the same: 1) signature analysis is possible (Chapter 12); and 2) a simulation can be performed in order to verify that a pseudorandom sequence tests the faults in a list (Section 3.2.4.3).

NOTES and REFERENCES

Many papers have been published on testing of digital circuits. Several books are entirely or partially devoted to this topic, for example [Fu 85], [Mc 86], [BaMcSa 87], and [AbBrFr 90].

The quality level of integrated circuits is treated in [SRC 85], [In 87], [McBu 88]. In [KrGa 93], the authors show that future quality level requirements for embedded memories will require built-in random testing.

I_{DDQ} testing, i.e., test by measurement of the consumption of the circuit is studied, for example, in [Ha et al. 89], [JaRaDa 89], [ChLi 92], [So et al. 92], [GuHa 93], [RoFi 97]. This kind of testing, in which no propagation is required for observing a faulty value, is outside of the scope of this book. However, random patterns may be used for I_{DDQ} testing [YoTaNa 94].

The scan-path technique was first presented in [FuWaAr 75]. The Level-Sensitive Scan Design (LSSD) was first described in [EiWi 78]. The Joint Test Action Group has presented a proposal for standardized boundary-scan techniques [BeMa 87] [JTAG 88]. Tools exist, for example [Sy 97], [Me 97]. Merging built-in self-test with boundary scan was proposed in [BaMc 82], [Ko 82], [Le 84], [GlBr 89], [NaKaLe 91], [Th et al. 91]. These topics are presented in the books cited above.

The basic ideas related to on-line BIST were introduced in [CaSc 68]. The totally self-checking (TSC) property for functional blocks and for checkers is formally defined in [An 71]. The *TSC goal* is that the first faulty output is outside the code. The largest class of functional circuits ensuring the TSC goal corresponds to *strongly fault secure* (SFS) circuits. This notion was introduced and defined as far as combinational circuits are concerned in [SmMe 78]. The definition of SFS circuits including both combinational and sequential circuits is given in [DaThe 78]. The largest class of checkers ensuring the TSC goal corresponds to *strongly code disjoint* (SCD) circuits, introduced in [NiJaCo 84]. The TSC goal is based on the fact that the *output vector* should be a code word. The sequentially self-checking (SSC) concept generalizes this notion: the output sequence should be in a language [ViDa 80].

Various proposals for implementing self-checking circuits are presented in the literature, for example [HaBo 84] [FuAb 84], [NaKa 85], [Ni 85], [NiCo 85], [Jh 93].

An approach mixing on-line and off-line BIST was proposed in [Ni 89]. BIST for mixed signal analog/digital circuits was also considered [CoReCa 92].

The basis of off-line BIST and signature analysis are found in [Be et al. 75] who suggested the test evaluator in Figure 1.6. The BILBO principle was presented in [KöMuZw 79 & 80]. A tutorial on BIST can be found in [AgKiSa 93].

Various references concerning (pseudo)random testing and signature analysis will be given in the other chapters.

2

Models for Digital Circuits and Fault Models

This chapter presents the models used in the book for representation of both combinational and sequential circuits, in Section 2.1. This is followed by a presentation of the basic fault models in Section 2.2 (when required, more specific models will be presented). The two basic ways to consider testing, i.e., functional and structural testing, are presented and commented on in Section 2.3.

2.1 NOTATIONS AND MODELS FOR DIGITAL CIRCUITS

It is assumed that the reader is familar with the content of this section. This reminder presents the notation used in the book: for combinational circuits in Section 2.1.1 and for sequential circuits in Section 2.1.2; notations related to sequences of input and output vectors are presented in Section 2.1.3.

The **complement** of a Boolean variable a will be written as a' (the notation $\bar{a}$ is also used).

2.1.1 Combinational circuits

Thin characters are used to denote input and output variables while bold characters are used to denote input and output vectors.

Let A be a combinational circuit and let its **input variables** be denoted by x_1, x_2, ..., x_n. The **input vector** is $x = (x_1, x_2, ..., x_n)$. The vector x takes a value from the set $X = \{x_0, x_1, ..., x_{2^n-1}\}$ corresponding to all the possible **input states**.

16   Chapter 2

If $z_1, z_2, \ldots, z_m$ are the **output variables**, then the **output vector** is $z = (z_1, z_2, \ldots, z_m)$, which takes a value in the set $Z = \{z_0, z_1, \ldots, z_{2^m-1}\}$ of possible **output states**.

Figure 2.1a represents a 4-input, 2-output combinational circuit. The set of *input states* is $X = \{x_0, \ldots, x_{15}\}$. The *indices j* of x_j could be arbitrary. However, it may be convenient to choose them so that x_j is the binary vector representing integer j, i.e., j is equal to $\sum_{i=1}^{n} x_i \cdot 2^{n-i}$. For example, x_5 corresponds to $x_1 x_2 x_3 x_4 = 0101$. Similarly, the set of output states is $Z = \{z_0, \ldots, z_3\}$.

A way to specify a combinational circuit consists of writing out the **truth table**, which is a *functional model*.

Figure 2.1b provides an example: an *output state* is associated with every input state. For example $z(x_1) = z_2$ since the output vector is 10 when the input vector is 0001.

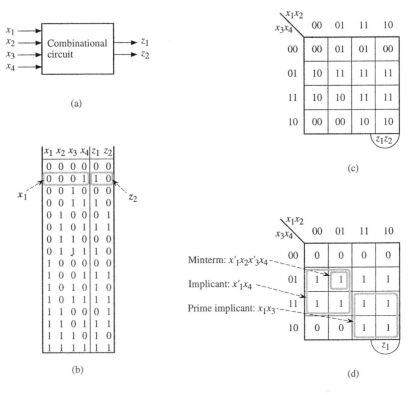

Figure 2.1 Combinational circuit. (a) Inputs and outputs. (b) Truth table. (c) Karnaugh map. (d) Illustration of implicants.

A Karnaugh map is a modified form of the truth table (Figure 2.1c). Each cell corresponds to an input vector and adjacent cells correspond to adjacent vectors. For example the cell defined by the first column and the second row corresponds to the input vector 0001. The corresponding output vector, 10, is written in this cell[1].

A variable x_i or x'_i is called a **literal**. A **product of literals** is also called a **cube**. Figure 2.1d is the Karnaugh map of the output z_1. This output has the value 1 when $x'_1 x_2 x'_3 x_4 = 1$. The cube $x'_1 x_2 x'_3 x_4$ is called a **minterm**; it is a product containing all the inputs, either in true or in complemented form. The cube $x'_1 x_4$ covers several minterms; this is called an **implicant**. The implicant $x_1 x_3$ is a **prime implicant**. That means that deleting any literal in it gives a cube that is no longer an implicant; informally there is no implicant that covers $x_1 x_3$ plus other minterms.

2.1.2 Sequential circuits

The behavior of a sequential circuit is modeled by a **sequential machine** or **finite-state machine** (or **finite automaton**). This is a *functional model*. The sets of input states and ouput states are defined as for a combinational circuit. In addition, a set of **internal states** $Q = \{q_0, q_1, ..., q_s\}$ is introduced.

2.1.2.1 Synchronous sequential machine

Figure 2.2a represents a synchronous sequential machine with two inputs, x_1 and x_2, and one output z. Input *CK* is a special input, a clock. The internal state and output state can only change on a rising edge of clock *CK* (the inputs are supposed not to change when *CK* rises). Since this applies to all transitions, this condition is implicit and is not represented on the state table and state diagram.

Figure 2.2b represents a machine with three *internal states* known as q_0, q_1 and q_2, or $Q = \{q_0, q_1, q_2\}$. Each column of this state table corresponds to an *input state*. Each row of the table corresponds to an *internal state*. Each cell of the table corresponds to a **total state**. Usually, **state** is short for internal state. In the cell (q_0, x_1), for example, the state q_2 is written. This means that, if the **present state** (*PS*) is q_0 and the input state is x_1 when the rising of *CK* occurs, the **next state** (*NS*) will be q_2. The right end column corresponds to the output states associated with the internal states, i.e., $z = 0$ in states q_0 and q_1, and $z = 1$ in q_2. Such a sequential machine where the output state depends only on the internal state is called a **Moore machine**.

---

[1] In the half circle under the Karnaugh map, "$z_1 z_2$" is written. This means that the content in each entry corresponds to $z_1 z_2$. Similarly, the content in each entry of Figure 2.2b corresponds to the next state *NS*.

Chapter 2

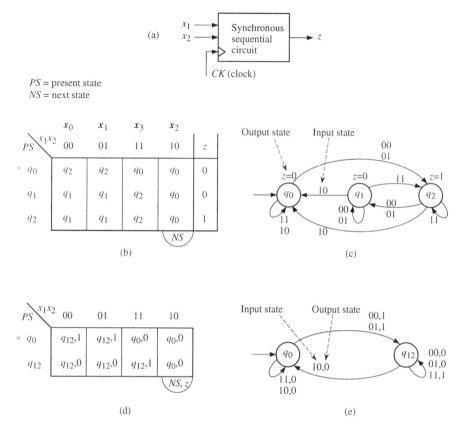

Figure 2.2 Synchronous sequential machine. (a) Block diagram. (b) and (c) State table and state diagram of a Moore machine. (d) and (e) State table and state diagram of a Mealy machine.

Figure 2.2c is a state diagram that contains exactly the same information as the state table in Figure 2.2b. An arrow from an internal state to another one is labelled by the input state(s) producing the transition between these internal states. For example, the arrow $q_0 \to q_2$ is labelled by both 00 and 01. This means that if the present state is q_0 and if the input state is either 00 or 01 when the rising of CK occurs, the next state will be q_2. The output states are written near the corresponding states. The **initial state** is represented by an asterisk in the state table and by an incoming short arrow in the state diagram.

Figures 2.2d and e represent an equivalent **Mealy machine** (equivalence will be formally defined in Section 2.1.3; roughly speaking, they have the same input/output behavior). States q_1 and q_2 have been merged together in a state q_{12}. In this case, the *output state* depends not only on the internal state but also on the input state; it then *depends on the total state* (this is the definition of a Mealy machine). On the state table, this is shown by the representation of the next output state after the next internal state. For example, with the cell (q_0, 01) is associated the next internal state q_{12} and the next output state 1.

The state diagram in Figure 2.2e contains exactly the same information. The output states are no longer associated with a internal states as in Figure 2.2c, but with the transitions. For example, the arrow $q_0 \rightarrow q_{12}$ is labelled by both 00,1 and 01,1. This means that a transition from q_0 to q_{12} can be produced either by the input state 00, and the next ouptut state will be 1, or by the input state 01, and the next output state will also be 1.

2.1.2.2 *Asynchronous sequential machine*

An asynchronous sequential machine is a machine without a clock. The internal and output states can only change on a change of the input state. Figure 2.3a represents an asynchronous machine with two inputs, x_1 and x_2, and one output, z.

Figure 2.3b is a **state table** (or **flow table**) representing a *Moore* machine with three internal states. Assume the input state is 11; then the internal state is q_1. This state does not change as long as the input state does not change: it is stable. This **stable state** is represented by q_1 circled in column 11. If input x_2 changes to 0, then we proceed to state q_2: on the state table, we move to column 10 (on the same row) where q_2 appears without a circle (this represents an **unstable state**). This unstable state q_2 indicates the stable state which will be reached: q_2 which corresponds to the third row.

The state diagram in Figure 2.3c contains the same information as the state table in Figure 2.3b. The state q_1 in Figure 2.3c corresponds to the *internal state* (union of three *total* stable *states*) which have the same name in the state table. The transition from q_1 to q_2 is represented by an arrow labeled by the input state causing this transition[2], i.e., 10.

It is usually assumed that, after a change in one input has occured, there is no other change in any input until the circuit enters a stable state. Such a mode of operation is referred to as the **fundamental mode**[3]. If the input state is 10 (internal state q_2), the next input state will be either 00 or 11. Since the next input state cannot be 01, the next state corresponding to the total state (q_2, 01) remains **unspecified**. Since fundamental mode is assumed, then, for example, a transition from (q_0, 00) to (q_1, 11) is not allowed in this model.

---

[2] For an asynchronous machine, the time is continuous. A self-loop labeled 10 from q_2 to itself does not have the meaning of a transition; we avoid self-loops in this model.

[3] Since the time is continuous, if x_1 and x_2 are two independent (not correlated) input variables, the probability that a change of x_1 occurs at the same time as a change of x_2 is 0. This is a mathematical model. In practice, the *implementation* should take into account that two input variables could change *almost* at the same time [Un 69][BrSe 95].

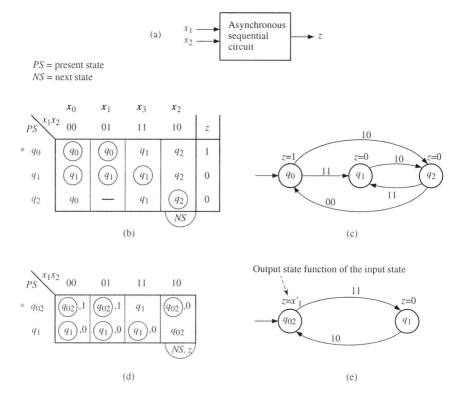

Figure 2.3 Asynchronous sequential machine. (a) Block diagram. (b) and (c) State table and state diagram of a Moore machine. (d) and (e) State table and state diagram of a Mealy machine.

Figures 2.3d and e represent an equivalent *Mealy* machine. States q_0 and q_2 have been merged together into a state q_{02}. The Mealy state table in Figure 2.3d looks like a Mealy state table of a synchronous machine, except that the output state is not necessarily specified for unstable states. If the present state is written on the left side of the table, then circling the entries is redundant (and vice-versa). Now, the state diagram of an asynchronous Mealy machine (Figure 2.3e) is different from a synchronous Mealy machine (Figure 2.2e). In the asynchronous case, an output is defined as follows: associated with each internal state is a *Boolean function of the input state*. For example, in Figure 2.3e, the output functions associated with the states q_{02} and q_1 are $z = x'_1$ and the trivial function $z = 0$, respectively.

Remark 2.1 When a sequential circuit has only one state, the output function is always the same. It is therefore clear that a combinational circuit is a particular case of a sequential circuit with a single internal state.

However, we would like to emphasize this fact because, unless otherwise specified, a **digital circuit** will denote either a sequential circuit or (as a particular case) a combinational circuit, and an *automaton* may denote a machine with a single internal state.

2.1.2.3 Formal models

A **deterministic finite-state automaton** is a quintuple

$$A = (Q, X, Z, \delta, \mu), \tag{2.1}$$

where Q, X, and Z are finite sets of internal, input, and output states.

The function δ is the **transition function** (or **next state function**). It is a mapping from $Q \times X$ into Q which gives the next state associated with the total state. For example $\delta(q_0, x_1) = q_2$ in Figure 2.2b and $\delta(q_1, x_2) = q_2$ in Figure 2.3b.

The function μ is the **output function**. It is a mapping from Q into Z in a Moore automaton and a mapping from $Q \times X$ into Z in a Mealy automaton. For example $\mu(q_2) = 1$ in Figure 2.2b and $\mu(q_0, x_3) = 0$ in Figure 2.2d.

We shall use the notation $ to represent an *unspecified* value[4]. For example $\delta(q_2, x_1) = \$$ in Figure 2.3b (the corresponding total state cannot occur in fundamental mode); and $\mu(q_{02}, x_3) = \$$ in Figure 2.3d (the output was 1 before the transition and will be 0 after the transition: both values can be admitted during the unstable state).

Henceforth, the term **automaton** will be used as a short-hand notation for deterministic finite state automaton. An **initialized automaton** is a pair consisting of an automaton and an initial state; it can be represented as a sextuple

$$(A, q_0) = (Q, X, Z, \delta, \mu, q_0), \tag{2.2}$$

where (Q, X, Z, δ, μ) is an automaton as previously defined and $q_0 \in Q$ is the initial state. Usually, *machine* and *automaton* are used synonymously; from now on, **machine** will mean an *initialized automaton*.

Remark 2.2 A truth table (abstract specification for a combinational circuit) defines an automaton where $|Q| = 1$. This automaton is always initialized[5].

Remark 2.3 Consider the particular case of an initialized Moore machine such that $Z = \{0, 1\}$. Let $Q_F \subset Q$ denote the set of states such that the output is 1. Since the output 0 is associated with all other states, this machine can be modeled as

---

[4] A function or an output is not specified either if it *may have an unspecified value* or if the corresponding total state *is impossible*. We use the same symbol for simplicity. The exact meaning is understood from the context.

[5] $|\alpha|$ denotes the cardinality of the set α.

$$A = (Q, X, \delta, q_0, Q_F). \tag{2.3}$$

Such a machine is known as a deterministic finite **acceptor**, i.e., a sequence of input vectors is *accepted* if and only if it leads the automaton from q_0 to any state in Q_F. The set Q_F is the set of **final** or **accepting** states.

2.1.3 Sequences of vectors

A combinational or sequential circuit is a discrete event dynamic system, i.e., its state changes at discrete times. Then, time will be denoted by $t = 0,1,2,...$, where $t = 0$ is the origin of time.

Usually, for an asynchronous system, time $t = i$ is defined by the ith change of input state. For a synchronous system, time $t = i$ is defined by the ith rising of the clock. For a combinational circuit, either definition may be used, depending on the practical application. In this book, where the inputs are generated by a tester, the synchronous one is more convenient.

Let M be a machine (modeling either a sequential or a combinational circuit). Let $x(t) = (x_1(t), x_2(t), ..., x_n(t))$, $q(t)$, and $z(t) = (z_1(t), z_2(t), ..., z_m(t))$ denote the input state, internal state, and output state at time t, respectively.

$$S = x(1)x(2)...x(m) \tag{2.4}$$

is the **input sequence**, or **test sequence** when the circuit is under test. The **length** of S is m.

$$R = z(1)z(2)...z(m) \tag{2.5}$$

is the **output sequence**, or **response** to S, which can be written as $R(q_0, S)$ or simply $R(S)$ if the circuit is combinational or if q_0 is implicit.

2.1.3.1 Input language and output language

We assume that the reader is familiar with the algebra of regular expressions for which we review the most essential operations. A **language** is a set of sequences built from an alphabet. Let $D = \{a, b, ..., u\}$ be an alphabet; D may be the set of inputs, X, the set of outputs, Z, or any other set. A language can be represented by a regular expression. The three operations in regular expressions are union, concatenation, and iteration.

Let $A = cabb$. This is a sequence built from the alphabet D. It corresponds to the successive occurrences of c, then a, then b twice. For example if $D = X$, $a = x_1$, $b = x_2$, $c = x_3$, the sequence A corresponds to $x_3x_1x_2x_2$. Let us say that A occurs if the sequence *cabb* occurs (i.e., the input sequence $x_3x_1x_2x_2$ is applied).

Union is represented by +. If A and B are two sequences (or regular expressions), then $C = A + B$ means that C occurs if either A or B or both occur.

For example $C = x_3x_1x_2x_2 + x_2x_4$ occurs if either the input sequence $x_3x_1x_2x_2$ OR the input sequence x_2x_4 is applied (the "OR" is inclusive, i.e., both sequences may be applied). It is clear that $A + B = B + A$.

Concatenation is represented as a product: $A \cdot B = AB = (A)(B)$ means A followed by B. For example $E = x_1x_2 \cdot x_4x_3 = x_1x_2x_4x_3$. If A and B represent several sequences, AB occurs if a sequence in A followed by a sequence in B occur. This corresponds to distributivity. For example $(x_1 + x_4x_3)(x_2 + x_3) = x_1x_2 + x_1x_3 + x_4x_3x_2 + x_4x_3x_3$.

Iteration is represented by an asterix: B^* represents the repetition of B any finite number of times, including zero times. This can be written as $B^* = \lambda + B + BB + \ldots = \lambda + B + B^2 + \ldots$, where λ is the sequence of length 0. This sequence has the following property: $\lambda B = B\lambda = B$. For example, $x_2x_3x_4 + x_1x_2x_3x_4 = (\lambda + x_1)x_2x_3x_4$.

$D^* = (a + b + \ldots + u)^*$ represents all the sequences that can be built from D. The set of sequences D^* is a monoid whose neutral element is λ. We denote $D^+ = D^* \setminus \{\lambda\}$ the set of all sequences in D^* except the null sequence λ.

Let α, β, and γ be three sequences such that $\alpha = \beta\gamma$. The sequence β is called a *prefix* of α (γ may be any sequence; β is a proper prefix of α if $\gamma \neq \lambda$). The sequence γ is called a *suffix* of α (β may be any sequence; γ is a proper suffix of α if $\beta \neq \lambda$). We denote by $\mathcal{P}(B)$ the set of all the prefixes of the regular expression B. For example, $\mathcal{P}(ab)^* = \lambda + a + ab + aba + abab + \ldots$.

The sequence β is a **subsequence** of $\alpha\beta\gamma$ (a prefix and a suffix are particular cases of subsequences).

The functions δ and μ defined in Section 2.1.2.3 can be extended. The *transition function* can be extended to a mapping from $Q \times X^*$ into Q.

$$\delta(q, S) = \begin{cases} q, & \text{if } S = \lambda \\ \delta(\delta(q, S_1), x), & \text{if } S = S_1x \text{ with } x \in X, S_1 \in X^*. \end{cases} \quad (2.6)$$

The *output function* can be extended to a mapping from $Q \times X^*$ onto Z for a *Moore machine*:

$$\mu(q, S) = \mu(\delta(q, S)); \quad (2.7)$$

and to a mapping from $Q \times X^+$ into Z for a *Mealy machine*:

$$\mu(q, S) = \mu(\delta(q, S_1), x), \quad \text{if } S = S_1x, \text{ with } x \in X, S_1 \in X^*. \quad (2.8)$$

From these definitions, one can write, for example, the response to S for a Mealy machine (Equation 2.5) as:

$$R(q_0, S) = \mu(q_0, x(1)) \, \mu(q_0, x(1)x(2)) \ldots \mu(q_0, S). \quad (2.9)$$

Consider the synchronous circuit in Figure 2.4a. The possible input states are $x_0 = 00$, $x_1 = 01$, $x_2 = 10$ and $x_3 = 11$. The **input alphabet** is $X = \{x_0, x_1, x_2,$

x_3}, and the **input language** is $X^* = (x_0 + x_1 + x_2 + x_3)^*$. That means that any sequence of input states can be applied. Similarly the **output alphabet** is $Z = \{z_0, z_1, z_2, z_3\}$, where the vector $z_0 = 00$ corresponds to the output values $z_1 = 0$ and $z_2 = 0$ for example. Since the specification is not given, we do not know if all **output sequences** in Z^* can be obtained. Some $R_j \in Z^*$ is a possible output sequence if there is an input sequence S_i such that R_j is the response to S_i. Let O be the set of possible output sequences. The set O is called the **output language**.

Figure 2.4 (a) Any 2-input, 2-output synchronous sequential circuit. (b) SR flip-flop.

Consider the circuit of Figure 2.4b. This is an SR flip-flop (Figure 2.6b in Section 2.2.1). The behavior of such a device is not defined if the inputs are $x_1 x_2 = 11$ (since x_1 is the Set input and x_2 is the Reset input). In that case the **specified input language**, or **input language** for short, is $I = (x_0 + x_1 + x_2)^*$ which is a proper subset of X^*. For such a flip-flop the two outputs always have complementary values, and the output states $z_1 z_2 = 00$ or 11 cannot be obtained. Hence the **output language** is $O = (z_1 + z_2)^*$ which is a proper subset of Z^*.

Remark 2.4 If S_i is a sequence in the input language I, then all prefixes of S_i are in I. If R_j is a sequence in the output language O, then all prefixes of R_j are in O. Thus *an input or output language may be represented as a set of prefixes $\mathcal{P}(B)$ for some language B.*
❑

The output language is a function of the input language. For example, assume that the input language for the circuit of Figure 2.4b is $I = \mathcal{P}(x_1 x_0 + x_2 x_0)^*$, i.e., after each set or reset there is exactly one input vector x_0 that does not change the output. In this case, the output language is $O = \mathcal{P}(z_1 z_1 + z_2 z_2)^*$.

In all examples for which the input language is not explicitly specified, we assume that $I = X^*$.

2.1.3.2 Equivalent machines

Definition 2.1 Two *machines* $M = (Q, X, Z, \delta, \mu, q_0)$ and $M' = (Q', X, Z, \delta', \mu', q'_0)$ are **equivalent** for some input language I if, for any input sequence $S_i \in I$, the response of M, $R(q_0, S_i)$, is identical to the response of M', $R'(q'_0, S_i)$. This is denoted by $M \equiv M'$.

◻

A non-specified Boolean value $ is **compatible** with both 0 and 1. Definition 2.1 can be extended to the case where $R(S_i)$ and $R'(S_i)$ are compatible. This is referred to as **pseudo-equivalence**.

Finally, two states q_1 and q_2 of the same automaton are equivalent ($q_1 \equiv q_2$) if $R(q_1, S) = R(q_2, S)$ for any input sequence S (they are pseudo-equivalent if the responses are compatible).

Property 2.1 For any Moore machine, one can construct an equivalent Mealy machine, for[6] $t > 0$, and vice versa.

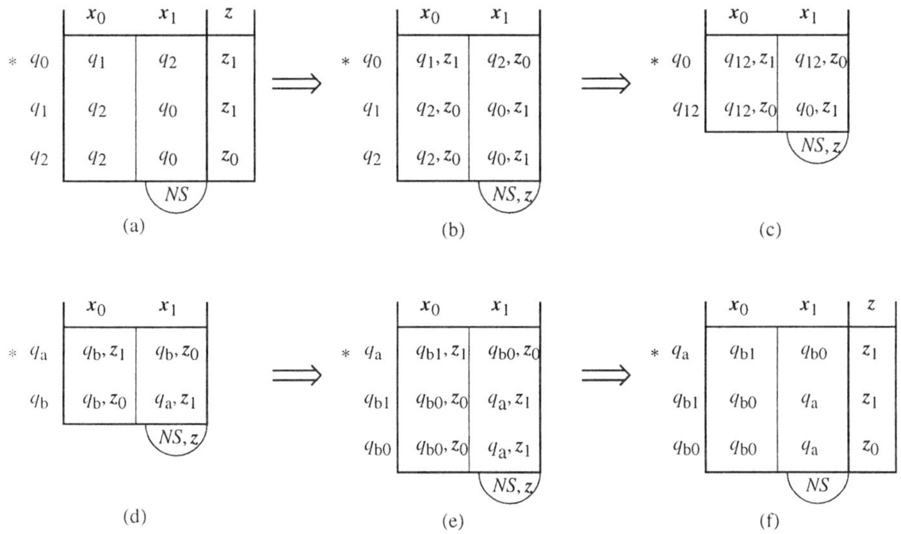

	x_0	x_1	z
* q_0	q_1	q_2	z_1
q_1	q_2	q_0	z_1
q_2	q_2	q_0	z_0

(a) NS

⇒

	x_0	x_1
* q_0	q_1, z_1	q_2, z_0
q_1	q_2, z_0	q_0, z_1
q_2	q_2, z_0	q_0, z_1

(b) NS, z

⇒

	x_0	x_1
* q_0	q_{12}, z_1	q_{12}, z_0
q_{12}	q_{12}, z_0	q_0, z_1

(c) NS, z

	x_0	x_1
* q_a	q_b, z_1	q_b, z_0
q_b	q_b, z_0	q_a, z_1

(d) NS, z

⇒

	x_0	x_1
* q_a	q_{b1}, z_1	q_{b0}, z_0
q_{b1}	q_{b0}, z_0	q_a, z_1
q_{b0}	q_{b0}, z_0	q_a, z_1

(e) NS, z

⇒

	x_0	x_1	z
* q_a	q_{b1}	q_{b0}	z_1
q_{b1}	q_{b0}	q_a	z_1
q_{b0}	q_{b0}	q_a	z_0

(f) NS

Figure 2.5 (a) (b) (c) Transformation of a Moore machine into a Mealy one. (d) (e) (f) Transformation of a Mealy machine into a Moore one.

This property is illustrated in Figure 2.5. The Moore machine in Figure 2.5a is transformed into the Mealy machine in Figure 2.5b: the output state associated with a present state in the first one is associated with the next states having the same label in the second one. Some equivalent states in Figure 2.5b can be merged to obtain Figure 2.5c. In our example, q_1 and q_2 are merged into q_{12}. The transformation from a Mealy machine to a Moore one is illustrated in Figures 2.5d to f. In Figure 2.5d both output states z_0 and z_1 can be associated with the

---

[6] For a Mealy machine, the output is not defined at $t = 0$.

state q_b, depending on the total state. Figure 2.5e is a new Mealy machine such that a single output vector is associated with some next state labels; this is obtained by splitting the state q_b into q_{b1} and q_{b0}. Then the Moore machine in Figure 2.5f can be obtained from Figure 2.5e. Note that the output state for a Mealy machine is unspecified at time $t = 0$. Then, if the initial state of a Mealy machine is split, any subsequent state of the Moore machine may be the initial one.

As it is possible to pass from one model to the other, we may choose the *Moore or Mealy type* to suit our purpose.

2.2 BASIC FAULT MODELS

A fault model depends on the level of modeling. Hence, the level of modeling are presented in Section 2.2.1 and an introductory presentation of fault models is given in Section 2.1.2.

2.2.1 Levels of modeling

Different levels of modeling are required during the design process of a digital system. Some *abstract* or **functional models** are used to specify a device to be built. Hardware Design Languages (HDLs) are text-based models of a system. Register Transfer Languages (RTLs) are used at the register and instruction set level.

Automata, represented by a state table or diagram, are not used in the design process, except for small parts, because they would be too large. However, they can be conceptually defined at any level.

A **structural model** may be obtained from interconnection of memory elements and gates, and, possibly, more complex primitive elements. If the memory elements (latches and flip-flops) are expanded to their gate-level model, then the whole structural model is a **gate-level model**. Now, if the gates are developed in a lower level we obtain a **transistor-level model**. Lower levels exist, e.g., technological layout level.

Some **register-level models** will be presented later. Let us introduce lower level models which will be used extensively in the book.

Figure 2.6 presents some basic memory elements. These elements usually have two complementary outputs denoted by Q and $\overline{Q}$. In this book, Q' denotes the complement of the value Q and $\overline{Q}$ the second output of the memory element (hence if this element is fault free $\overline{Q} = Q'$). The D latch and D flip-flop (Figure 2.6a) are such that the output Q does not change as long as $CK = 0$, and assumes the value of the input D when the clock CK rises from 0 to 1, i.e., when the *event* noted as ↑CK occurs. The difference between a latch and a flip-flop is illustrated in Figure 2.7. Figure 2.7a illustrates the usual case: the clock changes more often than D. For both a D latch and a D flip-flop the output Q is similar to the input D with some *delay*. This delay is due to waiting for the ↑CK event for changing. (In addition there is a small internal delay Δ corresponding to the propagation time in the memory element.) Figure 2.7b shows that the D **flip-flop** is **edge triggered** by the clock CK, i.e., its output Q can change only when ↑CK occurs. On the other hand, the output of a D **latch** follows the input value as long as $CK = 1$ and keeps the same value for the time when $CK = 0$.

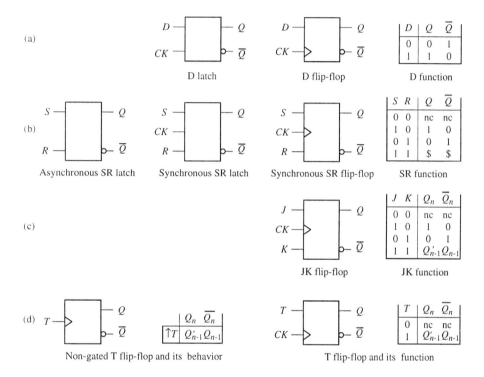

Figure 2.6 Some basic memories: latches and flip-flops (nc means *no change*).

The SR memory elements are shown in Figure 2.6b. The synchronous SR flip-flop output changes only when ↑CK occurs. If the **set** input $S = 1$, (and $R = 0$), the output assumes the value 1. If the **reset** input $R = 1$ (and $S = 0$), the output assumes the value 0. If $S = R = 0$, there is no change (noted nc in the SR function table). If $S = R = 1$, the behavior is not specified. The state of the asynchronous SR latch changes when SR changes from 00 to 10, if $Q = 0$, and when SR changes from 00 to 01, if $Q = 1$. The synchronous SR latch behaves

28   Chapter 2

like an asynchronous SR latch when $CK = 0$ and keeps the same output value as long as $CK = 1$.

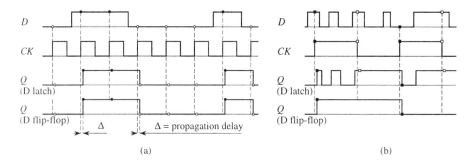

Figure 2.7 Illustration of the difference between a latch and a flip-flop.

The function of a JK flip-flop is virtually similar to the function of a SR flip-flop, except that it is completely specified. In Figure 2.6e, Q_{n-1} and Q_n denote the output state before and after occurrence of an event $\uparrow CK$, respectively. If $JK = 11$ when $\uparrow CK$ occurs, then the output state changes. The output of a non-gated T flip-flop changes each time T changes from 0 to 1 (it is unclocked). The output of a (clocked) T flip-flop (called *trigger*) changes if and only if $T = 1$ when $\uparrow CK$ occurs.

Figure 2.8a represents the primitive gates; Figure 2.8b is a gate-level model of a D latch.

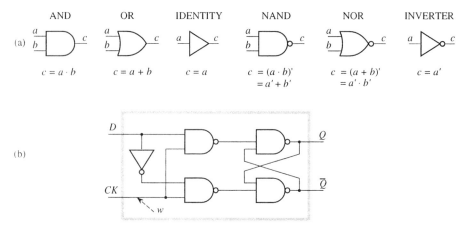

Figure 2.8 (a) Primitive gates. (b) Gate-level model of a D latch.

Remark 2.5

a) In a latch, the clock is sometimes noted **Enable**.

b) Assume a latch enables when $CK = 0$, instead of enabling when $CK = 1$ (if an inverter were placed at point w in Figure 2.8b, for example); or a flip-flop is triggered by the *trailing-edge*, i.e., by the event $\downarrow CK$ (clock changing from 1 to 0), instead of the *leading-edge*. This corresponds to complementing the clocking signal, with respect to the basic memory elements. The convention is then to use an inversion circle on the clocking input. This is consistent with the use of inversion circles in Figures 2.6 and 2.8a.

❑

A model level lower than the gate level is the *transistor level*. This model depends on the technology. The set of **bipolar** logic families includes the famous TTL, transistor-transistor logic. This technology is extensively used for SSI circuit (small scale integration) and MSI circuit (medium scale integration). For LSI circuit (large scale integration, i.e., more than one hundred gates in a single circuit) and VLSI circuit (very large scale integration), the **MOS** (metal-oxide semiconductor) families are extensively used, and particularly the CMOS logic (complementary MOS) which is illustrated in Figure 2.9.

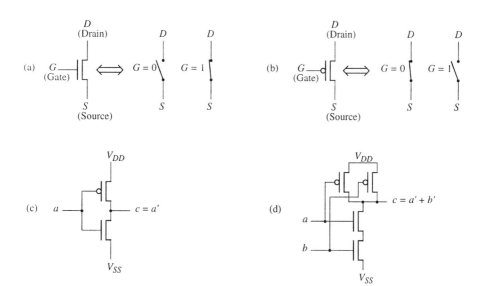

Figure 2.9 (a) NMOS transistor. (b) PMOS transistor. (c) CMOS inverter. (d) CMOS NAND gate.

Figure 2.9a illustrates the behavior of an NMOS transistor (i.e., *n*-channel transistor). By definition, the voltage of the *drain* D is always greater than the voltage of the *source S*. Roughly speaking this transistor behaves like a switch. If the voltage of the *gate G* with respect to *S* is about 0 volts (logical level of 0), the drain and source are open circuited. When the voltage $V_G - V_S$ reaches some threshold voltage (logical level of 1), the transistor turns "on" and current can flow from *S* to *D*. Figure 2.9b illustrates a PMOS transistor (*p*-channel). The behavior is logically complementary: the switch is "on" if and only if the voltage of the gate is lower than some threshold value (logical level of 0).

A CMOS gate is made up of two complementary networks: an *N*-network between V_{SS} (ground, logical 0) and the gate output; a *P*-network between V_{DD} (positive voltage, logical 1) and the gate output. Figure 2.9c and d present an inverter and a NAND gate, respectively. In both cases, as in any gate, there is always a path from V_{SS} to the output *c* or from *c* to V_{DD}, but not both.

The model level lower than the transistor level is the **layout level**. An example will be given in the next section.

2.2.2 Defects, faults and errors

Figure 2.10a shows a possible layout for an CMOS NAND gate[7]. Each crossing polysilicon-diffusion creates a transistor (*n*-channel or *p*-channel depending on the kind of diffusion). The lower level of connection corresponds to diffusions (the connection is interrupted by the transistors for some values of their gate voltages). The intermediate and upper levels correspond respectively to polysilicon and metal. There are some points of contact between the metal and either level.

A **defect** is any *physical imperfection occurring* in a circuit (manufacturing defect or physical failure). Figure 2.10a illustrates a defect: the chip is covered by some undesirable metal area which produces an electrical flow between two metal connections; the chip does not behave as it ought to. The faulty behavior may be modeled by a fault.

A **fault** *is a model of the faulty behavior, depending on the abstraction level*. The defect in Figure 2.10a can be modeled by a bridging fault in Figure 2.10a and b. The corresponding fault is the stuck-at-1 of the output *c* in Figure 2.10c (gate level), because the line *c* always has the voltage corresponding to V_{DD}. Note that both fault models (bridging and stuck-at) could be considered in the transistor level model of Figure 2.10b.

Many other defects can affect the circuit shown in Figure 2.10a, for example an open interconnection or a bad diffusion. Clearly some of these cannot be modeled at the higher level of abstraction which is the gate level (for example, a bad diffusion may lead to a behavior such that a voltage is too high to be considered as a 0 Boolean value and too low to be considered as a 1 Boolean value).

---

[7] This is a basis but nowadays more complex designs, using several metal layouts, exist.

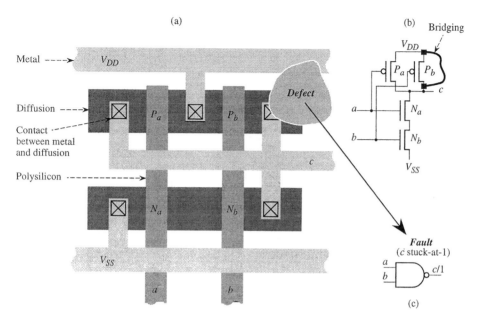

Figure 2.10 Illustration of the notions of defect and fault. (a) Layout level. (b) Transistor level. (c) Gate level.

An **error** is a *faulty value* somewhere in the circuit. Consider the faulty gate in Figure 2.10c. If $a = b = 1$, there is an error on line c, since the value is 1 when it should be 0. On the other hand, if $a = b = 0$, there is no error since the actual value of c is the expected value for a fault-free gate.

When there is an error on c, the faulty value may propagate errors to other parts of the circuit. When there is an *error on a primary*[8] *output* of the circuit, there is a **detection** of the fault. When there is *detection* of a fault, one only knows that there is a defect in the circuit. *Identification* of the fault present in the circuit is referred to as fault **diagnosis**.

2.2.3 Basic fault models

A fault is either **permanent** (always active after occurrence), or **intermittent** (sometimes *active* and sometimes *inactive*; due to the temperature of a component for example), or **transient** (occurs once; electromagnetic perturbation, for example). A **logical** fault replaces the fault free logic function by another logic function. There is another type of fault called a **delay** fault. It

---

[8] **Primary** inputs and outputs are directly accessible from outside (the words input and output can be also used for gates or subcircuits).

occurs when, although the logic function is correct, the output change occurs too late after the change of the input.

In this section, we consider only *permanent logical* faults.

2.2.3.1 Classical models

The classical fault models correspond to faults affecting the interconnections at the gate level. Typical faults affecting interconnections are shorts and opens. If a **short** is formed by connecting an equipotential line w to the positive supply (e.g. in Figure 2.10), the fault model is w *stuck-at*-1, which may be written as w s-a-1 or $w/1$. If w is connected to the ground, the fault is w *stuck-at*-0 (w s-a-0 or $w/0$). See Figure 2.11a.

Now, if there is an **open** due to a broken connection, the downstream part of the broken equipotential remains at constant voltage level. This may also be modeled by a stuck-at-0 or a stuck-at-1 fault, depending on the technology, and is illustrated in Figure 2.11b (could also be intermittent depending on the neighborhood).

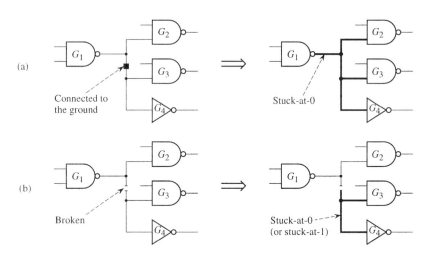

Figure 2.11 Examples of stuck-at faults (the stuck-at parts are shown in heavy lines). (a) Short. (b) Open.

The **stuck-at fault** is really the basic fault model which is the foundation of test theory (if the circuit is feedback-free, the fault-free and the faulty circuit are combinational: this is called a **combinational fault**).

When a short is between two signal lines, w_1 and w_2, both lines have the same voltage level. This defect usually creates a new logic function. The logical fault is called a **bridging fault**. Depending on the technology, one can distinguish OR bridging faults (logic 1 overrides logic 0) from AND bridging faults (logic 0 overrides logic 1).

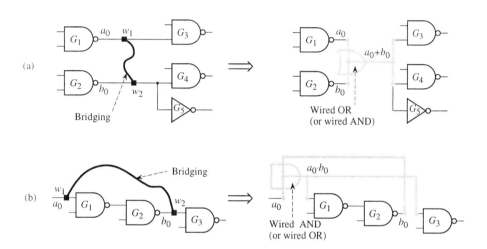

Figure 2.12 (a) Nonfeedback bridging fault. (b) Feedback bridging fault.

Figure 2.12a presents a bridging fault between signal lines w_1 and w_2, assuming that logic 1 overrides logic 0. Both w_1 and w_2 have the same value. Let a_0 and b_0 be the logic values of w_1 and w_2 if the circuit was fault-free, and a_f and b_f the logic values when the OR bridging fault is present. Due to the wired OR made by the bridging, $a_f = b_f = a_0 + b_0$. Figure 2.12b shows an AND bridging fault. Then, $a_f = b_f = a_0 \cdot b_0$. In addition, in this case, there is a path between w_1 and w_2 (in the fault-free circuit, through gates G_1 and G_2). This is a **feedback bridging fault** which creates a loop, i.e., an internal state. If the circuit was combinational it *becomes sequential* when affected by this fault. A feedback bridging fault is called a **sequential fault**[9], while a nonfeedback bridging fault is a *combinational fault*.

There are **single** and **multiple faults**. Let us consider the stuck-at faults affecting only inputs and outputs of gates. In this context, the single defect presented in Figure 2.11b is equivalent to a double fault, i.e., stuck-at of an input of G_3 and an input of G_4. More generally, a multiple stuck-at fault contains both stuck-at-0 and stuck-at-1 faults at various sites. A multiple bridging fault may correspond to several sites too, but a short between more than two signal lines is also a multiple bridging fault.

---

[9] A formal definition is given in Section 3.2.3.

2.2.3.2 Specific models

Specific fault models can be used for particular technologies or for particular structures. For example, for CMOS technology the stuck-open and the stuck-on faults are usually considered.

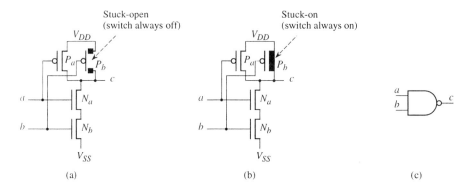

Figure 2.13 Fault models specific to CMOS technology. (a) Stuck-open. (b) Stuck-on. (c) Gate level model of the fault-free circuit.

Figure 2.13a illustrates the stuck-open fault model. The switch corresponding to transistor P_b is always off, for any value of b. Let us show that this fault cannot be modeled by a stuck-at at the gate level, considering that the transistors are modeled by switches (as shown in Figure 2.9, Section 2.2.1).

First case: $ab = 11$ then 10. When $a = b = 1$, both N_a and N_b are *on* while P_a and P_b are *off* (P_b is permanently *off* but just now it is the correct state). Then $c = 0$. When b changes to 0, N_b becomes *off* and P_b remains *off* while it should become *on*. Then c is neither connected to V_{SS} (because N_b is *off*) nor connected to V_{DD} (because P_a is normally *off* and P_b is incorrectly *off*). It follows that c remains at the past value, i.e., $c = 0$, instead of changing to the value 1: an error occurs.

Second case: $ab = 00$ then 10. When $a = b = 0$, both N_a and N_b are *off* while P_a is *on*; then $c = 1$ which is the correct value. When a changes to 1, c becomes disconnected from both V_{SS} and V_{DD} as in the first case. Then c remains at the past value, i.e., $c = 1$, which is correct.

We have seen that, when $ab = 10$, the value of c may be either faulty or correct, depending on the previous value of the vector ab. Then this fault is a *sequential fault*. It cannot be modeled by a stuck-at fault at the gate level[10] since such a fault is a combinational fault.

---

[10] It can be modeled by a stuck-at fault at the transistor level: line feeding the gate terminal of P_b stuck-at 1.

Figure 2.13b illustrates the stuck-on fault model. Transistor P_b is always *on*. Assume $ab = 11$. Then N_a and N_b are normally *on* while P_b is incorrectly *on*. It follows that there is a path from V_{SS} to V_{DD} through N_b, N_a and P_b. The voltage on line c depends on the resistances in these three transistors. If the resistance in the stuck-on transistor is very low with regards to the resistance in N_a and N_b, this can be modeled by $c/1$. Otherwise the level may be between the logic 0 and the logic 1. This is not a logical fault (it could be transformed into a logical fault, depending on the downstream gate which could see this level as a logic one).

Specific fault models can also be considered for particular structures or behaviors. For example, in RAMs (random access memories) one can consider faults where the activity of one memory cell may influence the behavior of another memory cell; examples will be given in Chapter 8. Faults in microprocessors are presented in Chapter 9. Delay faults, which are not logic faults, are considered in Appendix J.

2.2.3.3 *Functional models*

A *functional fault model* is characterized by the faulty performance that occurs instead of the fault-free function. *This model depends on the abstraction level*. At a low level, a functional fault model may be a faulty truth table for a combinational circuit or a faulty state table for a sequential circuit. At the register transfer level the following faults, for example, can be considered. Fault f_1: instead of writing information in register r_1, the information is written in both r_1 and r_2. Fault f_2: instruction I_3 is never executed. This kind of functional model will be used in Chapter 9.

2.3 FUNCTIONAL AND STRUCTURAL TESTING

Functional testing, structural testing, and exhaustive testing are presented respectively in Sections 2.3.1, 2.3.2, and 2.3.3.

2.3.1 Functional testing

Informally, functional testing consists in verifying that the system corresponds to the *functional specifications*. There are two main posibilities, depending on whether a fault model is considered. We give the following example. Figure 2.14a represents a combinational circuit producing an addition $Z = A + B$, where $0 \leq A \leq 3$, $0 \leq B \leq 3$ and $0 \leq Z \leq 6$. These numbers are binary encoded by the Boolean variables $a_1 a_2$ for A, $b_1 b_2$ for B and $z_1 z_2 z_3$ for Z.

Without a fault model, one can test the correct operation of the system by trying a few examples: for example, for $A = 3$ and $B = 2$ we expect $Z = 5$. That must be a non-sophisticated way of checking that your pocket calculator is OK.

When *a functional fault model is considered*, one can search for a set of operations able to test the faults defined by the model. For example, assume the fault is the following one: $Z = A$ is obtained instead of $Z = A + B$. This fault can be tested by any value of A and any value of B except $B = 0$, for example $A = 0$ and $B = 1$.

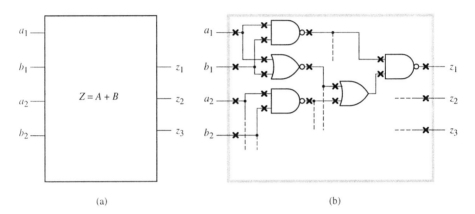

Figure 2.14 Two-bit adder. (a) Functional representation. (b) Part of the internal structure.

2.3.2 Structural testing

To derive a structural test for a digital circuit, a structural description of the circuit and a fault model at this level of description are required. For example, the 2-bit adder in Figure 2.14a can be represented at gate level as illustrated in Figure 2.14b. The crosses in heavy lines in Figure 2.14b (partially) represent all the gate inputs and outputs plus the primary inputs. If the single stuck-at fault model of these lines is considered, a test sequence must contain at least one vector detecting every stuck-at-0 and stuck-at-1 of the considered lines. Consider the input vector $a_1 b_1 a_2 b_2 = 1111$. The corresponding output vector should be $z_1 z_2 z_3 = 110$; then this input vector tests $z_1/0$, $z_2/0$, $z_3/1$ and other faults in the circuit. In most cases, all the single stuck-at faults can be tested by a test sequence in which some (or even many) possible input vectors are not present. In return one must know and take into account the structure.

In this example, the circuit is combinational and remains combinational for any fault in the fault model. Then each vector tests a set of faults independently from the preceding vectors, and this test sequence can be characterized by the **test set**, i.e., the set of vectors appearing at least once in the test sequence.

2.3.3 Exhaustive testing

A particular case of functional testing is called *exhaustive*. Usually, this expression is only applied to *combinational circuits* but it can also be used, with a different definition, to *sequential circuits*. This point will be emphasized at the end of this section.

Exhaustive testing of a combinational circuit consists in applying a test sequence containing *all* the possible input vectors, in any order. For the example in Figure 2.14, exhaustive testing consists in successively applying the 16 possible input vectors. Such a test is exhaustive in that it *fully checks the fault-free behavior* (as far as the circuit remains combinational). It does not require a fault model and it is able to detect many faults which are not in the classical models.

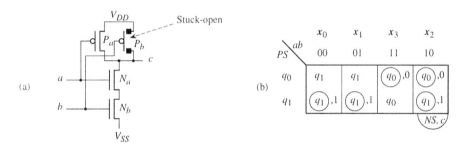

Figure 2.15 (a) Sequential fault. (b) Corresponding state table.

There is an obvious limitation to exhaustive testing: it becomes infeasible for large circuits.

Another limitation is that the implicit fault model contains all the faults such that *the circuit remains combinational*, but the test of a sequential fault is not guaranteed. This is illustrated in Figure 2.15. Figure 2.15a is a combinational circuit (NAND gate) with a stuck-open which is a sequential fault (explained in Section 2.2.3.2). Figure 2.15b shows the corresponding state table, in which q_0 and q_1 correspond to $c = 0$ and $c = 1$, respectively. Both test sequences $S_1 = x_0 x_1 x_3 x_2$ and $S_2 = x_0 x_2 x_3 x_1$, correspond to exhaustive tests. However, S_1 detects the fault (because x_2 is applied after x_3 and the output c keeps the value 0) while S_2 does not (because x_2 is applied after x_0 and the output c keeps the value 1, which is the fault-free value).

Now, what about *exhaustive testing of a sequential circuit*? It is clear that a single application of all the input vectors to a sequential circuit is not sufficient. As a matter of fact, there is at least one input vector such that the output obtained for this input vector depends on the input sequence applied before this input vector. However, one can find a test sequence which *completely checks the fault-free behavior* of a sequential circuit. This is called a **checking sequence**. Such a sequence for an s-state automaton A: 1) brings the automaton into a known state; 2) verifies that the s states can be reached; 3) verifies all the entries of the state table. References are given at the end of the chapter. We illustrate the main concepts by a simple example.

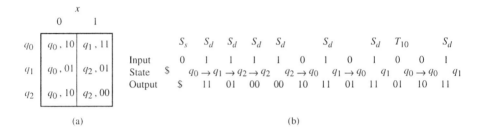

Figure 2.16 (a) Automaton A. (b) Checking experiment for automaton A.

When a sequential circuit is powered up, it reaches a **starting state** which may be unknown. A **synchronizing sequence** is an input sequence which, from any starting state, brings the circuit into a known **initial state**. For the automaton A in Figure 2.16a, $S_s = 0$ is a synchronizing sequence (practically all the real life circuits have a synchronizing sequence).

A **distinguishing sequence** S_d is such that the observation of the output sequence during application of the input sequence S_d informs us of the state of the automaton before applying S_d. For the automaton A in Figure 2.16a, $S_d = 1$ is a distinguishing sequence. Its application allows observation of either z_3 or z_1 or z_0 (where z_i denotes the output vector corresponding to the binary coding of i) if the internal state before S_d was q_0, q_1 or q_2, respectively. Not every machine has a distinguishing sequence, even if it is reduced (if it is not reduced, there is certainly no distinguishing sequence since there are equivalent states). An automaton with a distinguishing sequence can be obtained by adding one output to an automaton without distinguishing sequence; this is obtained thanks to a pertinent assignment of the values 0 and 1 to this output z_{m+1} in the various states [Ko 78].

Now, a checking sequence can be built[11] as shown in Figure 2.16b. From an unknown state denoted by $, application of $S_s = 0$ leads to q_0. Application of the distinguishing sequence $S_d = 1$ verifies that the state q_0 was reached (observing the output z_3). The state is now q_1 and this is verified by application of S_d. The state is now q_2 and this is verified by application of S_d. Just now, i.e., after application of the input sequence $S_4 = 0111$, we have verified that the three states can be reached. We still have to verify the transitions. Note that S_4 has verified the transitions $q_0 \to q_1$ and $q_1 \to q_2$ (the transitions verified are represented by arrows in Figure 2.16b). At the end of S_4, the transition $q_2 \to q_2$ has been made but not yet verified: application of S_d verifies that q_2 has been reached. And so on. The transition denoted by T_{10} is a sequence for moving from q_1 to q_0, in order to verify $q_0 \to q_0$.

We have obtained $S_{12} = 011110101001$. This is a checking sequence because automaton A in Figure 2.16a is the[12] *only* 1-input, 2-output *machine* with no more than 3 internal states which produces the output sequence $R(S_{12}) = \$z_3 z_1 z_0 z_0 z_2 z_2 z_3 z_1 z_3 z_1 z_2 z_3$ (the first output vector is not significant). Thus, this checking experiment is able to test *any fault such that the number of states has not increased*. Let us now define an exhaustive test in the general case.

Definition 2.2 An **exhaustive test** is a sequence testing all the faults such that the *number of states of the faulty circuit is less than or equal to* the number of states of the fault-free one.

❏

From this definition, *a checking experiment corresponds to an exhaustive test* for a sequential circuit. If a fault f transforms the automaton A in Figure 2.16a into a faulty automaton A_f having 4 or more states (this can happen with some bridging faults or stuck-open faults), it is not certain that S_{12} tests the fault f.

NOTES and REFERENCES

For sequential machines and regular expressions, the reader may refer to [Ko 78].

Details on the various levels of modeling can be found in [AbBrFr 90]. The reference [Mc 86] contains information about logic design and integrated technologies. Implementation of CMOS technology can be found in [NSC 81], [PiStZa 90].

The stuck-at fault model was introduced by Eldred [El 59]; it represents many physical faults [Ti et al. 83]. Bridging faults are studied in [Me 74], stuck-open

---

[11] If the automaton has a distinguishing sequence of length m, one can construct a checking sequence of length polynomial in m and the size of the machine [He 64].
[12] Corresponding to a class of equivalent machines obtained by renaming the states.

and stuck-on are introduced in [Wa 78] and a functional fault model for microprocessors is introduced in [TaAb 80].

Since exhaustive testing is not practicable for large circuits, pseudo-exhaustive testing has been considered for combinational circuits [Mc 84]: it consists in exhaustively testing subcircuits depending on subsets of the set of primary inputs.

The notion of checking sequence was introduced in [Mo 56] and [He 64] was an influencial paper; this topic is detailed in [FrMe 71] and [Ko 78].

3

Basic Concepts and Test Generation Methods

Here, we consider deterministic testing of combinational and sequential circuits. The basic methods for combinational circuits[1] assume combinational faults (i.e., faults such that the circuit remains combinational). If a sequential fault affects a combinational circuit, the methods for sequential circuits are relevant to it.

This chapter provides the basis for answering the following type of questions: 1) given a fault, find a test vector (combinational circuit) or a test sequence (sequential circuit) to detect it; 2) given a set of faults, find a test set (combinational) or a test sequence (sequential) to detect every fault in the set. For a set of faults, the problem of relations between faults arises.

The two basic principles are the comparison between faulty and fault-free circuits, and propagation of an error through a sensitized path. They are illustrated by simple examples. Algorithms based on these methods do exist and are used in the Automated Test Pattern Generator (ATPG).

Section 3.1 and Section 3.2 are respectively dedicated to combinational and sequential circuits.

3.1 COMBINATIONAL CIRCUITS

The comparison and propagation approaches are illustrated in Sections 3.1.1 and 3.1.2. An algebraic approach in presented in Section 3.1.3. Various relations among faults, including the concept of fault coverage are presented in Section 3.1.4

---

[1] *"Combinational circuits"* implicitly means without feedback. Combinational circuits with feedback exist, but they are quite marginal and practically not used.

3.1.1 Comparison between faulty and fault-free circuits

Figure 3.1a represents a 3-input, 1-output combinational circuit. Let us denote z_0 the output function of the fault-free circuit. Then $z_0 = x_1 + x_2x_3$. Let f denote the stuck-at fault w/0 and z_f the corresponding output function. Then $z_f = x_1$. Figure 3.1b presents the truth tables of both z_0 and z_f. Note that they are different for the input vector $x_1x_2x_3 = 011$, since the output should be 1 and is actually 0 for this input vector (the faulty value is presented in bold). This input vector is then a **test vector** (or **detecting pattern**) for the fault f.

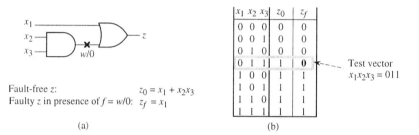

Fault-free z: $z_0 = x_1 + x_2x_3$
Faulty z in presence of $f = w/0$: $z_f = x_1$

(a)   (b)

Test vector $x_1x_2x_3 = 011$

Figure 3.1 (a) Combinational circuit with a fault. (b) Comparison of the faulty truth table with the fault-free one.

3.1.2 Propagation through sensitized paths

After the presentation of the principle in Section 3.1.2.1, reconverging fanout and backtracking are evoked in Section 3.1.2.2.

3.1.2.1 *Illustration of the principle*

A test vector x_i for a fault f has two basic properties. First, it **activates** the fault f, i.e., it **provokes** *an error*, which is a fault effect. Second, it **propagates** *the error to a primary output* in order to detect the faulty behavior. In a sensitization method, three steps can be considered. They are illustrated with the example in Figure 3.2.

Let us first define a **path**. This is a connected sequence of lines and gates between two points; for example $w_2w_8G_3w_3G_5z$ is a path between the output of G_2 and the primary output in Figure 3.2. *Given its extremities*, a path can be characterized[2] by a *string of gates*; for example, the preceding path may be

---
[2] There are pathological cases where a string of gates may correspond to several paths: if the same signal is feeding two or more inputs of a gate.

denoted by G_3G_5, as it is between the output of G_2 and the primary output (note that $w_8G_3w_3G_5z$ corresponds to the same string of gates, G_3G_5, but it does not start at the same spot).

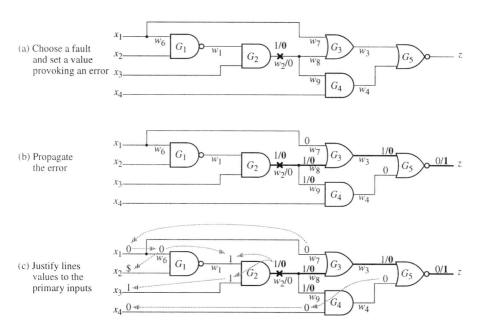

Figure 3.2 Path sensitization. (a) Choice of a fault. (b) Propagation. (c) Justification.

First step. Choose a fault and set a value provoking an error. In Figure 3.2a, the fault $w_2/0$ is considered. Setting $w_2 = 1$ (for the fault-free circuit) *provokes an error* on the line w_2. This is noted by the pair fault-free/**faulty** values, with the second one in bold.

Second step. Propagate the error to a primary output. There are 2 paths between w_2 and the output z: G_3G_5 and G_4G_5. Then, one can choose to propagate either through G_3G_5, or through G_4G_5 or through both. We have chosen path G_3G_5 (see Figure 3.2b). Propagation through the OR gate G_3 implies that $w_7 = 0$ (because $w_7 = 1$ would imply $w_3 = 1$ for any value of w_8). Then $w_7 + w_8 = 1$ for the fault-free circuit and 0 for the faulty one. Similarly, propagation through the NOR gate G_5 implies $w_4 = 0$. If these values $w_4 = w_7 = 0$ are met, the output z would be 1 for the faulty circuit, instead of 0 for the fault-free one.

Third step. Justify the line values (this means find primary input values implying the chosen values; this is obtained by upstream propagation). We have to justify $w_2 = 1$, $w_4 = 0$ and $w_7 = 0$. See Figure 3.2c, where the arrows in doted lines represent implications. The value $w_2 = 1$ implies that the inputs of the AND gate G_2 be $w_1 = x_3 = 1$. Since G_1 is an NAND gate $w_1 = 1$ can be obtained if $x_1x_2 = 00$ or 01 or 10. We do not choose for the moment. The value $w_4 = 0$ implies $x_4 = 0$. The value $w_7 = 0$ implies $x_1 = 0$, and then x_2 may have any value.

Then we have obtained two test vectors for the fault $w_2/0$: $x_1x_2x_3x_4 = 0010$ and 0110.

		OR	AND	INVERTER
1/0 ⟺ D	1- input error	$D + 0 = D$ $D + 1 = 1$ $\overline{D} + 0 = \overline{D}$ $\overline{D} + 1 = 1$	$D \cdot 0 = 0$ $D \cdot 1 = D$ $\overline{D} \cdot 0 = 0$ $\overline{D} \cdot 1 = \overline{D}$	$D' = \overline{D}$ $(\overline{D})' = D$
0/1 ⟺ $\overline{D}$	2- input error	$D + D = D$ $\overline{D} + \overline{D} = \overline{D}$ $D + \overline{D} = 1$	$D \cdot D = D$ $\overline{D} \cdot \overline{D} = \overline{D}$ $D \cdot \overline{D} = 0$	—
(a)		(b)		

Figure 3.3 (a) Meaning of symbols D and $\overline{D}$. (b) Basic logic operations.

The pairs 1/0 and 0/1 can be denoted by the symbols D and $\overline{D}$, respectively. Figure 3.3b presents the operations performed by the primitive gates (Figure 2.8 in Section 2.2.1). It appears that a "basically OR" gate (i.e., OR or NOR) allows propagation of a single error (i.e., D or $\overline{D}$), only if the other inputs of the gate are 0. Similarly, a "basically AND" gate (i.e., AND or NAND) propagates an error if the other inputs are 1. When two (or more) inputs of a gate present an error, the propagation of an error to the gate output is possible only if all the errors have the same "sign" (all D or all $\overline{D}$). The sign of the error is changed through an inverter.

Let us now return to the fault $w_2/0$. In Figure 3.2 the propagation through path G_3G_5 was performed. Let us now consider other propagations. This is illustrated in Figure 3.4, using the symbols D and $\overline{D}$. In Figure 3.4a an attempt of propagation through the path G_4G_5 is shown. That would require the value $w_3 = 0$. However, there is no value of w_7 able to produce $w_3 = 0$ (according to Figure 3.3b). Therefore propagation via the path G_4G_5 is impossible. Figure 3.4b shows the propagation through both paths G_3G_5 and G_4G_5. This leads to a solution similar to Figure 3.2c except that $x_4 = 1$, instead of 0.

Hence we have obtained 4 possible test vectors for the fault $w_2/0$ (2 in Figure 3.2b and 2 in Figure 3.4b), which can be represented by $x_1x_2x_3x_4 = 0\$1\$$.

The D-Algorithm [Ro 66] is the best known method of propagation through sensitized paths. For any detectable combinational fault it can obtain at least one test vector, even if that requires concurrent propagation through several paths.

Basic Concepts and Test Generation Methods 45

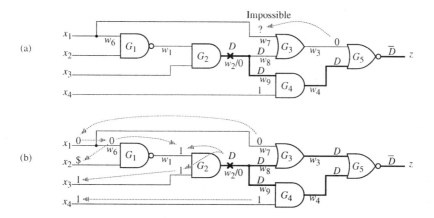

Figure 3.4 (a) Propagation through $G_4 G_5$. (b) Propagation through both paths.

Remark 3.1 The three steps presented above constitue a general technique for various fault models. For a multiple stuck-at fault, the first step presents a choice of one or several faulty spots, and the other two are exactly the same.

3.1.2.2 About reconverging fanout and backtracking

We now introduce some concepts that will also be used in subsequent chapters.

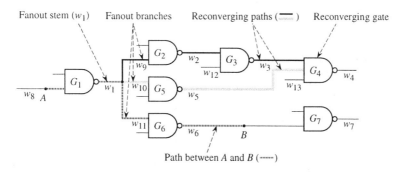

Figure 3.5 Illustration of the notions of path, fanout, and reconverging path.

A signal propagating from one source to several destinations is said to have **fanout**. The source, which may be either a primary input or gate output, is the **fanout stem**. In Figure 3.5, w_1 is a fanout stem. The **fanout branches** are the lines corresponding to destinations (gate inputs or primary outputs). The lines w_9,

w_{10} and w_{11} are the fanout branches in Figure 3.5. Two (or more) paths are said to be **reconverging paths** if 1) their origins are fanout branches corresponding to the same fanout stem, and 2) their ends are inputs of the same gate. In Figure 3.5, the paths $w_9G_2w_2G_3w_3$ and $w_{10}G_5w_5$ are paths reconverging to the gate G_4, called the **reconverging gate**. The fanout following G_1 is then a **reconverging fanout**. A circuit in which no signal has a fanout is said to be **fanout-free**; it is a **tree** circuit.

In circuits with fanout, during the sensitization process, **inconsistencies** may occur. One of these may be caused by *propagation through reconverging paths*. Assume that there is an error, denoted by D, on the line w_1 in Figure 3.5. Let us propagate that error to both inputs of G_4 by the two reconverging paths. Then, the value of w_3 is D because there are two inverters on the first path (G_2 and G_3 are inverting gates), and the value of w_5 is $\overline{D}$ since there is only one inverting gate on the second path. It follows that $w_4 = 1$ for both the fault-free and the faulty circuits, according to Figure 3.3b.

The number of inverting gates of a path may be odd or even; this is called the **parity** of the path. An error can be propagated to a primary output via two reconverging paths, only if they have the same parity (as in Figure 3.4b).

The *justification* step is another cause of inconsistency. Assume an error D on w_{13} is to be propagated through G_4 in Figure 3.5. This requires $w_3 = w_5 = 1$. One can try to justify $w_5 = 1$ by $w_{10} = 0$, and to justify $w_3 = 1$ by $w_2 = 0$. But $w_2 = 0$ implies $w_9 = 1$ which is inconsistent with $w_{10} = 0$ since $w_9 = w_{10}$. This contradiction is a consequence of the choices which have been made. A **backtracking** strategy allows a return to *past decisions* in order to remove them. For example, $w_3 = 1$ can be justified by $w_{12} = 0$ instead of $w_2 = 0$.

Property 3.1 In a fanout-free circuit, each justification is independent from the others. Then justification cannot lead to inconsistency.

❏

We present now some vocabulary which is used in other chapters. The part of the circuit through which the logical information propagates from primary inputs to a given connection is called the **upstream** or **cone of influence** (or *input cone* or *upstream cone*) of this connection. For example, in Figure 3.5, the upstream cone of w_2 consists of w_2, the gate G_2, w_9 and the other input of G_2, w_1, the gate G_1, and the inputs of this gate. Similarly, the part of the circuit through which the logical information propagates from a given connection to primary outputs is called the **downstream** of this connection (or *output cone* or *downstream cone*).

3.1.3 Algebraic method

The idea of the algebraic method is illustrated in Figure 3.6. When it is fault-free, the function z corresponding to Figure 3.6a is $z_0 = x_1 + x_2x_3$. This expression is a **sum of products**, where each *product* (or *implicant*, according to Section 2.1.1) contains some **literals** (a *literal* is a Boolean variable, complemented or not). This function is represented by the Karnaugh map in Figure 3.6b. Implicant x_1 covers four cells of the Karnaugh map corresponding to input vectors $\boldsymbol{x}_4$, $\boldsymbol{x}_5$, $\boldsymbol{x}_6$ and $\boldsymbol{x}_7$ and implicant $x_2 x_3$ covers the cells associated with input vectors $\boldsymbol{x}_3$ and $\boldsymbol{x}_7$ (by abuse of language, one can say that an implicant "covers some input vectors").

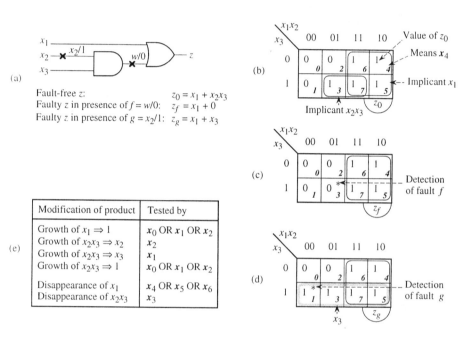

Figure 3.6 Illustration of the algebraic method.

Consider the fault $f = w/0$. The faulty function is $z_f = x_1$; the implicant $x_2 x_3$ becomes 0 and then "*disappears*". Function z_f is represented in Figure 3.6c. It is clear that the fault f can be tested by the input vector covered by $x_2 x_3$ but not covered by x_1, i.e., $\boldsymbol{x}_3 = x'_1 x_2 x_3$.

Consider now the fault $g = x_2/1$. The faulty function is $z_g = x_1 + x_3$; the implicant $x_2 x_3$ becomes x_3 under this fault and then it "*grows*", i.e., it covers a larger number of vectors. Function z_g is represented in Figure 3.6d. It is clear that fault g can be tested by the vector covered by implicant x_3 which is not in the fault-free function, i.e., $\boldsymbol{x}_1 = x'_1 x'_2 x_3$.

Property 3.2 and its dual one 3.2', which leads to a test set for all stuck-at faults in a combinational circuit are based on the observation that any Boolean expression can always be expanded into both a sum-of-product expression and a product-of-sums expression using distributivity. For example :

$$x_1(x_2 + x_3) + x_3 x'_4 = x_1 x_2 + x_1 x_3 + x_3 x'_4;$$

$$x_1(x_2 + x_3) + x_3 x'_4 = (x_1 + x_3)(x_1 + x'_4)(x_2 + x_3 + x_3)(x_2 + x_3 + x'_4).$$

Property 3.2 Suppose C is a combinational circuit without feedback and a *sum of products* z_0 represents the function implemented by the circuit (after expansion). For any single or multiple stuck-at fault in C, any product term in z_0 can either *remain unchanged* or *disappear* or *grow* by losing one or several literals (no product can be created, nor can a product decrease by the addition of literals).

Property 3.2'. Similar to Property 3.2 with the words *sum* and *product* interchanged.

❑

The proof of these properties can be found in [BoHo 71] and [ThDa 75].

In the case of Property 3.2' a sum of literals covers input vectors corresponding to 0 values of the output (instead of 1 values in Property 3.2).

Application of Property 3.2 to the circuit in Figure 3.6a is presented in Figure 3.6e. Note that the product $x_2 x_3$ could lose either one literal (becoming x_2 or x_3) or 2 literals (becoming 1). If the loss of each literal has been tested (vectors x_1 and x_2), the loss of both has also been tested (one vector among x_0, x_1, or x_2).

From Figure 3.6e, one can obtain a *test set* $T = \{x_1, x_2, x_3, x_4\}$ (x_4 is an arbitrary choice from x_4, x_5, and x_6 as can be seen in Figure 3.6b). This set of test vectors *detects all the single and multiple stuck-at faults in the circuit*. This method is "global" while the methods presented in Sections 3.1.1 and 3.1.2 are applied to a chosen fault.

Remark 3.2

a) If there is a redundant product term in the function z_0, its disappearance cannot be tested. In that case a complete test set can be obtained by testing: 1) all the growths of the products of the sum of products, and 2) all the growths of the sums of the product of sums.

b) The development in a sum of products may produce terms such as $A x_1 x'_1$. They must be kept because they can grow to become $A x_1$ and $A x'_1$. The expression where the terms $A x_i x'_i$ have not been deletted is called the **complete sum of products** (the complete product of sums is defined dually).

3.1.4 Relations between faults

The concepts of fault collapsing, checkpoints, redundancy and masking, and fault coverage are presented in this section.

3.1.4.1 Fault collapsing

The faulty outputs associated with the 10 single stuck-at faults of the circuit in Figure 3.7a are presented in Figure 3.7b. Let T_f denote the **set of test vectors for the fault** f. Figure 3.7b shows that the columns corresponding to faults $x_1/1$, $w/1$ and $z/1$ are identical. These faults have the same set of test vectors, namely $T_{x_1/1} = T_{w/1} = T_{z/1} = \{x_0, x_1, x_2\}$. Since these three faults have the same effect on the function performed by the circuit, they are said to be *equivalent*.

Definition 3.1 Two *combinational faults* f and g are **equivalent** if $z_f(x) = z_g(x)$ for any input vector x in X. This is denoted by $f \equiv g$.

❑

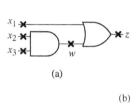

(a)

(b)

Inputs		Output in presence of faults								
$x_1\ x_2\ x_3$	z_0	$x_1/0$	$x_2/0$	$x_3/0$	$w/0$	$z/0$	$x_1/1$	$x_2/1$	$x_3/1$	$w/1$	$z/1$
x_0 0 0 0	0	0	0	0	0	0	1	0	0	1	1
x_1 0 0 1	0	0	0	0	0	0	1	1	0	1	1
x_2 0 1 0	0	0	0	0	0	0	1	0	1	1	1
x_3 0 1 1	1	1	0	0	0	0	1	1	1	1	1
x_4 1 0 0	1	0	1	1	1	0	1	1	1	1	1
x_5 1 0 1	1	0	1	1	1	0	1	1	1	1	1
x_6 1 1 0	1	0	1	1	1	0	1	1	1	1	1
x_7 1 1 1	1	1	1	1	1	0	1	1	1	1	1

Equivalence fault collapsing

(c)

Inputs		Output in presence of faults								
$x_1\ x_2\ x_3$	z_0	$x_1/0$	$x_2/0$	$x_3/0$	$w/0$	$z/0$	$x_1/1$	$x_2/1$	$x_3/1$	$w/1$	$z/1$
x_0 0 0 0	0	0	0	0	0	0	1	0	0	1	1
x_1 0 0 1	0	0	0	0	0	0	1	1	0	1	1
x_2 0 1 0	0	0	0	0	0	0	1	0	1	1	1
x_3 0 1 1	1	1	0	0	0	0	1	1	1	1	1
x_4 1 0 0	1	0	1	1	1	0	1	1	1	1	1
x_5 1 0 1	1	0	1	1	1	0	1	1	1	1	1
x_6 1 1 0	1	0	1	1	1	0	1	1	1	1	1
x_7 1 1 1	1	1	1	1	1	0	1	1	1	1	1

Dominance fault collapsing

(d)

Inputs		Output in presence of faults								
$x_1\ x_2\ x_3$	z_0	$x_1/0$	$x_2/0$	$x_3/0$	$w/0$	$z/0$	$x_1/1$	$x_2/1$	$x_3/1$	$w/1$	$z/1$
x_0 0 0 0	0	0	0	0	0	0	1	0	0	1	1
x_1 0 0 1	0	0	0	0	0	0	1	1	0	1	1
x_2 0 1 0	0	0	0	0	0	0	1	0	1	1	1
x_3 0 1 1	1	1	0	0	0	0	1	1	1	1	1
x_4 1 0 0	1	0	1	1	1	0	1	1	1	1	1
x_5 1 0 1	1	0	1	1	1	0	1	1	1	1	1
x_6 1 1 0	1	0	1	1	1	0	1	1	1	1	1
x_7 1 1 1	1	1	1	1	1	0	1	1	1	1	1

Figure 3.7 Comparison of faults.

The three faults $x_1/1$, $w/1$ and $z/1$ form an equivalence class. For test analysis, it is sufficient to consider only one fault representative of the class, for example $x_1/1$. Similarly the set of faults $\{x_2/0, x_3/0, w/0\}$ corresponds to a class of equivalent faults which can be represented by $x_2/0$. Replacing all the faults of a class by a representative fault is called **equivalence fault collapsing**.

The set of single stuck-at faults for the circuit in Figure 3.7a contains 10 faults:

$$F_1 = \{x_1/0, x_2/0, x_3/0, w/0, z/0, x_1/1, x_2/1, x_3/1, w/1, z/1\}.$$

After *equivalence fault collapsing*, the set F_2 is obtained containing only 6 faults (Figure 3.7c):

$$F_2 = \{x_1/0, x_2/0, z/0, x_1/1, x_2/1, x_3/1\}.$$

Now, one can observe in Figure 3.7c that $T_{x_2/0} = \{x_3\}$ is included in $T_{z/0} = \{x_3, x_4, x_5, x_6, x_7\}$. It follows that if $x_2/0$ is tested (by application of the input vector x_3), then $z/0$ is tested too. It is said that fault $z/0$ *dominates* $x_2/0$. The fault $z/0$ *dominates* $x_1/0$ too. Similarly $x_1/1$ *dominates* both $x_2/1$ and $x_3/1$ (see Figure 3.7c).

Definition 3.2 A combinational fault f **dominates** a fault g if $T_f \supset T_g$.
□

Since a fault f dominating a fault g is certainly tested when fault g is tested, it may be removed from the set of faults to be considered for test generation. This type of reduction is called **dominance fault collapsing**. If this reduction is applied to F_2, the set F_3 is obtained containing only 4 faults (Figure 3.7d):

$$F_3 = \{x_1/0, x_2/0, x_2/1, x_3/1\}.$$

This means that a test set able to test all the faults in F_3 is able to test all the faults in F_1 too. For example, x_3 which tests $x_2/0$ in F_3, tests also $x_3/0$ and $w/0$ which are equivalent to $x_2/0$ and $z/0$ which dominates $x_2/0$.

Definition 3.3 Let f and g be two combinational faults. They are **equidetectable** if $T_f = T_g$, i.e., if f dominates g and vice-versa. This is denoted by $f \cong g$.
□

Figure 3.8 Two faults which are equidetectable but not equivalent.

Note that two *equidetectable* faults are not necessarily equivalent if the circuit has several outputs (they are equivalent if the circuit has a single output). Consider for example the two faults f and g illustrated in Figure 3.8. These faults are equidetectable since $T_f = T_g = \{x_3\}$. However, they are not equivalent since the output vectors are different when x_3 is applied.

Remark 3.3

a) If detection of faults is considered, both equivalence and dominance fault collapsing can be used. For diagnosis of the fault present in a circuit, only equivalence fault collapsing must be used (there is no way to distinguish between two equivalent faults while two equidetectable faults may be distinguished).

b) Consider an n-input AND gate (the result for other primitive gates is obtained by duality and complementation); the equivalence classes correspond to every input and output stuck-at-1 plus one input (or output) stuck-at-0 (i.e., $n + 2$ classes of equivalence). Now, the output stuck-at-1 dominates every input stuck-at-1. Then, for any n-input primitive gate, there are $2(n + 1)$ single stuck-at faults corresponding to $n + 2$ classes of equivalence; after dominance fault collapsing, $n + 1$ faults remain to be considered.

❑

One can observe that, for the example in Figure 3.7, only faults on the primary inputs remain after collapsing (see F_3). The notion of checkpoints (next section) allows a generalization of this result.

Definitions 3.1 to 3.3 assume implicitly that any input vector in X can be applied to the circuit. However, for a test set $T \subset X$ (for example if the behavior is specified only for input vectors in T), they can be modified as follows.

Definitions 3.1' to 3.3'

Definition 3.1' is obtained from Definition 3.1 by replacing *equivalent* by **equivalent under a test set** T, and X by T.

Definition 3.2' is obtained from Definition 3.2 by replacing *dominates* by **dominates under a test set** T, T_f by $T_f \cap T$, and T_g by $T_g \cap T$.

Definition 3.3' is obtained from Definition 3.3 by replacing *equidetectable* by **equidetectable under a test set** T, and *dominates* by *dominates under* T. (In this case: $T_f \cap T = T_g \cap T$.)

3.1.4.2 Checkpoints

Fault collapsing means that testing of the stuck-at faults in a combinational circuit can be reduced to testing of faults at so called checkpoints, defined below.

Definition 3.4 The **checkpoints** in a combinational circuit are the *primary inputs* and the *fanout branches*.

❑

For the circuit in Figure 3.9 the checkpoints are the lines marked by a cross, i.e., primary inputs x_1 to x_5 and fanout branches w_1 to w_6.

Property 3.3 In a combinational circuit C, any test set that detects *all the single stuck-at faults on the checkpoints* of C detects all the single stuck-at faults in C.

Example 3.1 The circuit in Figure 3.9 contains 17 lines: 5 primary inputs, 6 gate outputs (including 2 fanout stems and 2 primary outputs) and 6 fanout branches. Then it has 34 single stuck-at faults, and only 22 checkpoints single stuck-at faults.

Moreover, one can observe, according to Remark 3.3b, that some of the checkpoints faults are equivalent: $x_1/0 \equiv w_1/0$, $w_2/0 \equiv w_3/0$, $x_3/0 \equiv x_4/0$ and $w_4/0 \equiv x_5/0$. Thus 18 single stuck-at faults are still to be considered.

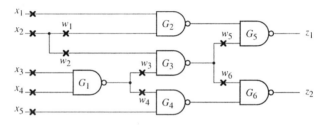

Figure 3.9 Checkpoints.

Property 3.4 In a combinational circuit C, any test set that detects *all the multiple stuck-at faults on the checkpoints* of C detects all the multiple stuck-at faults in C.

❏

The circuit in Figure 3.9 has $3^{17} - 1$ multiple stuck-at faults (each line may be either fault-free or stuck-at-0 or stuck-at-1; out of these 3^{17} cases, one corresponds to the fault-free circuit). One can observe that the multiple faults involving x_2 are taken into account by the multiple faults involving both w_1 and w_2. This observation leads to the following remark.

Remark 3.4 *Property 3.4 remains true if the primary inputs with fanout are deleted from the checkpoint list.*

3.1.4.3 Redundancy and masking

A circuit C is **redundant** if some line in C can be set to a constant logical value without changing its output function. Note that a redundancy does not imply incorrect design. In fact, redundancy is intentionally used by designers to solve some particular problems.

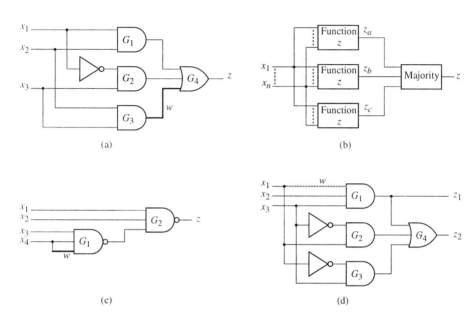

Figure 3.10 Illustration of some redundancies[3].

Figure 3.10a illustrates an intentional redundancy to avoid a static hazard. As a matter of fact, $x_1x_2 + x'_1x_3 + x_2x_3 = x_1x_2 + x'_1x_3$; however, the product $x_2x_3 = 1$ (line w) ensures the continuity of the value $z = 1$ if $x_2 = x_3 = 1$ when the logical value of x_1 changes.

Another example of intentional redundancy is triple modular redundancy, which is a basic technique used in fault tolerant design (Figure 3.10b). Three copies of the function z are made, namely z_a, z_b and z_c. Then a voter produces the majority function $z = z_a z_b + z_b z_c + z_c z_a$. In that case, z_a, for example, may have any value without changing the value of z if $z_b = z_c$.

Figure 3.10c illustrates the use of a 3-input gate G_1, when a 2-input gate is required. This may be convenient because G_2 is a 3-input gate; several 3-input gates can be contained in the same package. The line w can also be connected to the 1 level.

The circuit in Figure 3.10d presents a different kind of redundancy which arises because the circuit has several outputs. Since the function $x_1x_2x_3$ is needed (primary output z_1), it is used for designing z_2. However, if we consider only

---

[3] Figure 3.10d is reprinted from [KeMe 94] (© 1994 IEEE).

function z_2, the circuit is redundant since $x_1x_2x_3$ is not a prime implicant for z_2; it could be replaced (for z_2 only) by either x_2x_3 (corresponding to $w = 1$) or by x_1x_2.

Definition 3.5 A combinational fault f is **undetectable** (or **redundant**) if there is no input vector detecting this fault, i.e., $z_f(x) = z_0(x)$ for any input vector x in X.

Definition 3.5' A combinational fault f is **undetectable** (or **redundant**) for a test set[4] T if there is no input vector x in T such that $z_f(x) \neq z_0(x)$.
❑

In Figure 3.10a and c the faults $w/1$ are *undetectable*. In Figure 3.10b many faults are undetectable if only z is observable. For this system, testing can be carried out if z_a, z_b and z_c are observed during the test, i.e., one can test the 4-output circuit with $z = (z_a, z_b, z_c, z)$. All the stuck-at faults in Figure 3.10d are detectable. The fault $w/1$, for example, cannot be detected by observation of z_2 but it produces a faulty z_1. However, this *kind of redundancy* is a problem for *other kinds of faults* such as delay faults[5].

In summary, various *problems arise from redundancy*: in general, if a fault cannot be detected, a lot of computing time may be necessary to prove this redundancy; in addition, this fault may mask another one, as explained in the sequel.

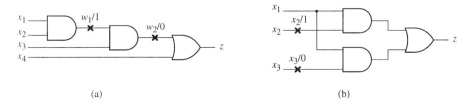

(a)   (b)

Figure 3.11 Fault masking. (a) $w_2/0$ masks $w_1/1$. (b) $x_2/1$ masks $x_3/0$.

The problem of *fault masking* concerns the presence of a *multiple fault*. Consider the single stuck-at fault $w_1/1$ in Figure 3.11a. Three input vectors can detect this fault, namely $x_1x_2x_3x_4 = 0110$ or 1010 or 0010. They all produce a faulty value $z = 1$ instead of 0. Now, if the double fault [$w_1/1$ & $w_2/0$] is present, none of the three input vectors detecting $w_1/1$ can detect this double fault. It is

---

[4] It is usual to define the properties related to combinational faults assuming that all the input vectors in X can be applied to the circuit in normal operation. Definitions 3.1 to 3.3 and 3.5 correspond to this option. However, the output could be specified only for a subset T of X. In that case, Definitions 3.1' to 3.3' and 3.5' can be used. They are more general since X is a particular case of T. For sequential circuits (Section 3.2.4) the properties are directly given for an input language I, which is a subset of X^*.

[5] As explained in [KeMe 94] from which this example is taken.

said that $w_2/0$ *masks* $w_1/1$. This is clear in this example since: 1) w_2 is on the only path between w_1 and z (then $[w_1/1 \,\&\, w_2/0] \equiv w_2/0$); and 2) $w_2/0$ can produce only faulty values $z = 0$ while $w_1/1$ alone would produce faulty values $z = 1$.

Figure 3.11b is an example of masking where the two faults are not on the same path to the primary output. Fault $x_3/0$ can be detected only by the input vector $x_1 x_2 x_3 = 101$. Now, this vector does not detect the double fault $[x_3/0 \,\&\, x_2/1]$, i.e., $x_2/1$ *masks* $x_3/0$.

Definition 3.6 Let T_f be the set of all input vectors detecting a fault f. Fault g **masks** fault f if there is no input vector in T_f detecting the multiple fault $[f \,\&\, g]$.

3.1.4.4 Fault coverage

Informally, **fault coverage** is the ratio between the number of faults detected by the test sequence and the total number of faults in the assumed fault set F. Fault coverage will be defined formally in the sequel. In order to have a single definition for both combinational and sequential circuits, let us consider now the most general case: given a circuit C and a set F of faults under consideration, fault coverage is a measurement of the *effectiveness of a test sequence S*.

Remark 3.5
a) When the fault set is defined by a fault model (for example single stuck-at faults), all the possible faults can be enumerated.
b) If all the faults in F are combinational, a test set T may be used instead of a test sequence; T is the set of input vectors that appear at least once in S. ❑

Now, what about the *redundant faults*? We can either add them to the detected faults or subtract them from the total number of faults [SeAgFa 90]. This leads to two possible definitions of fault coverage.

Definition 3.7 The **fault coverage** is defined either as

a) $P_{(a)} = \dfrac{\text{number of detected faults} + \text{number of redundant faults}}{\text{total number of faults}}$, or as

b) $P = \dfrac{\text{number of detected faults}}{\text{total number of faults} - \text{number of redundant faults}}$. ❑

The contexts where each definition can be used will be commented on at the end of this section. Let us first look at the meaning of fault coverage when there is no redundant fault; in this case Definition 3.7a and 3.7b are *equivalent*, i.e., $P_{(a)} = P$.

Let us return to the example in Figure 3.7, and the set of faults which were obtained in Section 3.1.4.1: F_1 before fault collapsing, F_2 after equivalence fault collapsing and F_3 after dominance fault collapsing. They are recalled here:

$$F_1 = \{x_1/0, x_2/0, x_3/0, w/0, z/0, x_1/1, x_2/1, x_3/1, w/1, z/1\};$$

$$F_2 = \{x_1/0, x_2/0, z/0, x_1/1, x_2/1, x_3/1\};$$

$$F_3 = \{x_1/0, x_2/0, x_2/1, x_3/1\}.$$

Let us denote $P(F_i, T)$ the fault coverage of the set F_i by the test set T. From the principle of *fault collapsing* (Section 3.1.4.1) we obtain:

$$P(F_1, T) = 100\% \Leftrightarrow P(F_2, T) = 100\% \Leftrightarrow P(F_3, T) = 100\%.$$

However, fault coverage is not the same for all the F_i's when it is not 100%. Consider the two test sets $T_1 = \{x_1, x_2, x_3\}$ and $T_2 = \{x_1, x_2, x_4\}$. Both have the same fault coverage with respect to F_3: $P(F_3, T_1) = P(F_3, T_2) = 75\%$, since each one does not detect one of the four faults in F_3 ($x_1/0$ for T_1 and $x_2/0$ for T_2). However, T_1 does not detect one fault in F_1 ($x_1/0$) while T_2 does not detect three faults in F_1 ($x_2/0, x_3/0$, and $w/0$). Then,

$$P(F_1, T_1) = 90\% > P(F_3, T_1)$$

and

$$P(F_1, T_2) = 70\% < P(F_3, T_2).$$

In this case $P(F_2, T_1) = P(F_2, T_2) = 83\%$, but the values could be different in another case.

The most *significant fault coverage* is $P(F_1, T)$ because all the faults are present with the same weight in F_1, while, after collapsing, the "weight" of a fault depends on the number of faults in the equivalence class and, possibly, of the faults dominating it. However, the more *usual fault coverage* concerns a set of faults after collapsing, because it is much easier to use (Properties 3.3 and 3.4 in Section 3.1.4.2 are used for this purpose).

Definition 3.7a of fault coverage is more pertinent when the set of faults before collapsing is considered: as a matter of fact, every fault is considered individually; since a circuit with a redundant fault may be considered as "correctly tested", fault coverage increases with the number of redundant faults.

On the other hand, *Definition 3.7b is more convenient if the set of faults after collapsing is considered*: as a matter of fact, the faults are not considered individually but they are gathered into classes; in that case, all the redundant faults are in the same class as the fault-free circuit.

From now on, the fault coverage P in Definition 3.7b will be used unless otherwise specified.

Generation of a test sequence for a set of faults
This is presented in Section 3.2.4.3 for both combinational and sequential circuits.

3.2 SEQUENTIAL CIRCUITS

Initialization, an important feature as far as testing is concerned, and the unusual concept of evasive faults are presented in Section 3.2.1. The comparison and propagation approaches are illustrated in Sections 3.2.2 and 3.2.3. In Section 3.2.4, new concepts related to collapsing and detectability of sequential faults are presented.

3.2.1 Initialization and evasive faults

By definition of a sequential circuit, its future behavior depends on its present state. It follows that a test sequence must be applied to an *initialized circuit*, modeled by an initialized automaton[6] (A, q_0) (Section 2.1.2.3). When a circuit is powered up, it generally reaches an unknown **starting state**. This starting state may be different for each instance of the circuit; for the same instance, each power-up does not lead necessarilly to the same starting state. Before a test sequence (either deterministic or random) is applied to the circuit, a known state is reached by an **initialization sequence** (or *preset sequence*) during which the output sequence is not significant. After the initialization sequence, the circuit is in the **initial state**. This is illustrated in Figure 3.12.

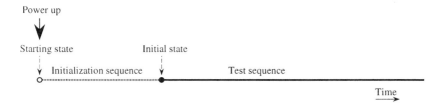

Figure 3.12 Starting state and initial state.

Fault-free circuits with different initial states can produce different responses. Therefore, the expected response of a fault-free circuit would not be unique if

---

[6] By abuse of language, if a fault corresponds to a faulty circuit which can be modeled by an automaton A', we shall speak of circuit A' or fault A'.

initialization were not performed. Initialization is therefore absolutely essential if the test consists in *comparing the response of the circuit under test to the fault-free response* (that is the usual way of performing deterministic and random testing). Practically all circuits can be initialized by one of the methods discussed in Section 3.2.1.1.

If initialization were not performed, several responses would be possible. In this case, one would have to decide if the *observed response belongs to some language* (that is more complicated and beyond our hypotheses). For example, if a frequency divider by 2 made up of a non-gated T flip-flop (Figure 2.6d in Section 2.2.1) were tested without initialization, the fault-free response would be a prefix of $(\lambda + 0 + 00 + 100)(1100)^*$.

3.2.1.1 *Initialization*

The initial state of the *fault-free* circuit is *determined solely* by the initialization sequence. On the other hand, the initial state of a *faulty* circuit is *not necessarily unique*, and it may be unknown even if the initialization sequence is known (examples will be given). However, some test methods do not take this fact into account, assuming that the initial states of a faulty circuit and of the fault-free circuit are the same.

Kinds of initialization

In some cases the initialization sequence is only an external reset which is not considered as an input of the automaton describing the circuit. If a power-up reset exists in the circuit, the input sequence of length zero, denoted by λ (Section 2.1.3.1), is a particular case of initialization sequence such that the initial state is the starting state (the circuit is self-initializing). Thus, there are 3 kinds of initialization:

First kind: power-up reset.

Second kind: external reset.

Third kind: synchronizing sequence.

For the first and second kinds, the initial state q_0 is unique and cannot be chosen by the user.

For the *third kind*, the user may choose the initial state. However, in this case, the *fault-free automaton must be strongly connected*, i.e., any "useful" state must be reachable from any other one; as a matter of fact 1) the state q_0 chosen to be the initial state must be reached from any starting state, and 2) any state which cannot be reached from q_0 is not "useful". Let us illustrate this notion with the example of Figure 2.16a (Section 2.3.3). This 3-state automaton is strongly connected. Since this automaton is already reduced, its implementation needs 2 internal variables; then, for any minimal-state implementation, there is a 4th state

q_3 which is not useful. If q_3 can be the starting state, then q_0 must be reached from q_3 too, for a third kind initialization (or $q_3 \equiv q_0$), but q_3 will never be reached after it has been left.

The first kind is an **implicit initialization**, while the second and third kinds are **explicit initializations**.

Remark 3.6

a) From the preceding comments on initialization, it follows that, if a machine is *not strongly connected*, the reset must be either a power-up or an external reset.

b) An automaton with an external reset may be represented as a strongly connected automaton with a synchronizing sequence by adding the reset line as an input line.

c) According to Section 2.3.3, a *synchronizing* sequence is an input sequence which, from any starting state, brings the circuit into a known initial state. A circuit without power-up reset or external reset, and having no synchronizing sequence, can exist; an example is the frequency divider by 2 made up of a non-gated T flip-flop evoked above (just before this section). Some authors say that such a circuit is not initializable. However, it can be initialized using a *homing sequence*. An input sequence S_h is said to be a **homing sequence** if the final state of the automaton (i.e., after S_h) can be determined uniquely from the automaton response to S_h regardless to the starting state (i.e., before S_h). A preset *homing sequence exists for every reduced[7] automaton* [Ko 78]. Let q_j be the state reached by the homing sequence. Since the machine is strongly connected (according to part a of this remark), a sequence S_j leading to a chosen initial state q_0 from the state q_j can be found. Hence, the circuit is initialized by the sequence $S_h \cdot S_j$ (S_j depends on the state reached after S_h has been applied).

From a practical point of view, initialization is more complicated since the *synchronization* sequence $S_h \cdot S_j$ depends on the starting state. However, from a theoretical point of view, this initialization will be considered as a third kind initialization.

Note that, if the faulty circuit does not produce one of the expected responses to the homing sequence S_h, the fault is immediately detected.

3.2.1.2 Compatible initial states and evasive faults

Let A be a fault-free automaton and A' be the faulty automaton of the fault f. Assuming an initialization has been performed to lead A into the initial state q_0, A' may be in a state q'_0 which is not always the same. Let Q'_0 be the set of

---

[7] If the automaton is not reduced, it contains equivalent states. In this case, a homing sequence determines the class of equivalent states which have been reached, and this is sufficient for our purpose.

possible initial states of the faulty automaton, i.e., $q'_0 \in Q'_0$. The initial states of the pair (q_0, q'_0) are said to be **compatible**.

Consider, for example, the automata A and A' given in Figure 3.13, and assume that the initialization is made by a synchronizing sequence (there is neither an external reset nor a power-up reset).

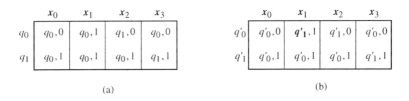

Figure 3.13 (a) Fault-free automaton A. (b) Faulty automaton A'.

Let $Sync(q_0)$ denote the set of synchronizing sequences leading the automaton A into the initial state q_0, from any starting state. All input sequences ending in x_0 or x_1 are in $Sync(q_0)$. There are infinitely many other sequences in $Sync(q_0)$, for example $x_1x_2x_2$. Some of the sequences in $Sync(q_0)$ synchronize automaton A' into q'_0, for example x_0. Other sequences in $Sync(q_0)$, for example $x_1x_2x_2$, do not synchronize A'; as a matter of fact if A' is started either in q'_0 or q'_1, the initial state after the application of the input sequence $x_1x_2x_2$ is either q'_1 or q'_0, respectively. The set $Sync(q_1)$ is similarly defined. For example x_1x_2 is in $Sync(q_1)$. Similarly, there are input sequences in $Sync(q_1)$, which do not synchronize the automaton A'. Then, in our example, if nothing is known about the initialization sequence, all pairs (q_0, q'_0), (q_0, q'_1), (q_1, q'_0), and (q_1, q'_1) correspond to *compatible* initial states.

Remark 3.7 If some *information about initialization* is known, the set of compatible initial states can be reduced. Given the initial state is q_0, the set of compatible initial states is reduced to $\{(q_0, q'_0), (q_0, q'_1)\}$.

Now, given the synchronizing sequence S_s, the set of compatible initial states may be still reduced. For example if $S_s = x_0$, there is only one pair of compatible initial states, (q_0, q'_0), since x_0 is also synchronizing for automaton A'.
□

We will now define the notion of evasive fault. Informally, a fault is evasive if there is a finite input sequence such that, if it is applied just after initialization, the fault will never be detected.

Definition 3.8
a) The automaton A' is said to be **evasive for the compatible pair** (q_0, q'_0) if there is a subautomaton $\hat{A}'$ of A' and an input sequence S such that the following two conditions are met.
   1) The responses of the fault-free and faulty automata are the same for S, i.e., $R(q_0, S) = R'(q'_0, S)$.
   2) The subautomaton $\hat{A}'$, reached from q'_0 by the input sequence S, behaves like the fault-free circuit for ever after S has been applied.

Otherwise, A' is called **non-evasive for the pair** (q_0, q'_0).

b) A' is called **non-evasive** if it is non-evasive for every pair of compatible initial states.

❑

According to Remark 3.7, a fault may be defined as **non-evasive given an initial state** of the fault-free machine, or **non-evasive given an initialization sequence**.

For a non-evasive fault, after every input sequence such that the responses are the same for A and A', it is still possible to make A' behave differently from A, and thus to detect the fault.

Example 3.2 A rather artificial example[8] of an evasive fault in a random access memory would be the following. A cell i is *stuck-at-1 until the first 1 is written* into it; after that, it behaves like any normal cell. The initialization sequence consists in writing 0 in every cell. In this case, if the fault has not yet been detected, writing 1 in cell i will destroy the possibility of detecting it.

Example 3.3 Figure 3.14a presents a specified 3-state machine, (A_1, q_0), and a possible implementation. As a matter of fact, since 2 internal variables are necessary to encode 3 states, the implemented machine needs at least 4 internal states. The specified machine is strongly connected (but the implemented machine is not) and x_0 is a synchronizing sequence. If the starting state is q_3, the initial state is q_0 after applying x_0.

Consider the fault (A'_1, q'_3) in Figure 3.14b: the initial state is q'_3 if the starting state is q'_3 and the initialization sequence is x_0. If the first test vector is x_0 (i.e., after the synchronizing sequence x_0), the fault is detected since $\mu(q_0, x_0) = 0$ and $\mu'(q'_3, x_0) = 1$; if it is x_1 the fault will never be detected.

Example 3.4 Figure 3.14c represents a 4-state machine, (A_2, q_0), which is not strongly connected: q_0 and q_1 are states which will never be reached after q_2 or q_3 has been reached. According to Remark 3.6 in Section 3.2.1.1, there is no synchronizing sequence; then initialization must be either a power-up reset or an external reset.

---

[8] This example is taken from [DaBrJü 93].

Fault (A'_2, q'_1) in Figure 3.14d corresponds to a faulty reset (the next state function and the output functions are fault-free). If the first test vector is x_0 the fault is detected because $\mu'(q'_1, x_0) \neq \mu(q_0, x_0)$; if it is x_1 the fault will never be detected since $\mu'(q'_1, x_1) = \mu(q_0, x_1)$ and the "same" state, $q'_2 \equiv q_2$, is reached from both states.

□

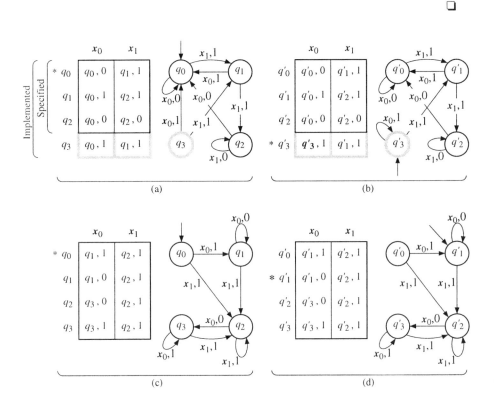

Figure 3.14 Evasive faults. (a) Fault-free machine (A_1, q_0) with an useless state. (b) Faulty machine (A'_1, q'_3). (c) Fault-free machine (A_2, q_0). (d) Faulty machine (A'_2, q'_1).

In Example 3.2, the faulty cell has 3 states instead of 2: the two normal states plus an initial state such that there is a 1 instead of 0 in the cell. Furthermore the additional state can be reached only when the circuit is powered up. Note that *an evasive fault always corresponds to a faulty initialization into a transient state*: in Examples 3.2 and 3.3, the initial states of the faulty machines do not exist in the specifications; in Example 3.4, the initial state of the faulty machine exists in the fault-free machine, but it is not the correct initial state.

3.2.2 Comparison between faulty and fault-free circuits: Observers

Two cases are successively considered: the initial state of the faulty circuit is known or is not known.

3.2.2.1 The initial state of the faulty circuit is known

Consider a fault-free initialized automaton (A, q_0) and a faulty automaton $A' = (Q', X, Z', \delta', \mu')$, corresponding to a faulty circuit.

Note that, in this fault model, *only the set X of input vectors is common to both the fault-free and the faulty circuit*. The set of internal states, the set of output states, the transition function and the output function may be different. As will be seen in the next section, the initial state may also be different and even unknown for the faulty circuit. This fault model corresponds to *all the permanent logic faults* (set F_u in Section 4.2).

Assume that, after the initialization sequence, A' has reached the initial state q'_0. To distinguish between the machines $M = (A, q_0)$ and $M' = (A', q'_0)$, one must apply input sequence S for which the responses of both circuits differ. Denote these responses by $R(M, S)$ and $R'(M', S)$. An observer can make a complete comparison of both machines. The *observer* for M and M' is an *acceptor* (Remark 2.3 in Section 2.1.2.3) in which a current state is a pair (q, q') of states of both machines; the only *final state*, denoted by ω, is the state reached when the difference between M and M' has been observed (i.e., the fault has been detected).

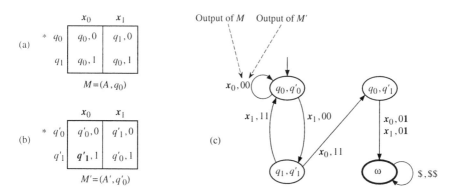

Figure 3.15 (a) Fault-free machine M. (b) Faulty machine M'. (c) Observer for the pair of machines (M, M').[9]

---

[9] This figure is reprinted from [Da 97] (© 1997 IEEE).

An observer is illustrated in Figure 3.15. The fault-free machine $M = (A, q_0)$ and the faulty machine $M' = (A', q'_0)$ are represented in Figures 3.15a and b. Figure 3.15c corresponds to the observer for the pair (M, M'). The initial state (q_0, q'_0) of the observer is composed of both initial states. If the input vector x_0 is applied to both machines, M remains in q_0 and M' remains in q'_0, while both outputs are 0. This is represented by the self loop labelled x_0, 00 in Figure 3.15c. If the input vector x_1 is applied from the state (q_0, q'_0), the state (q_1, q'_1) is reached and both outputs are 0, and so on. If the input vector x_0 is applied from the state (q_0, q'_1) then the fault-free machine produces a 0 output and the faulty one produces a 1 output. The corresponding arc is labelled x_0, 01. Since M' and M have delivered different outputs, the difference between both machines is detected: the absorbing state ω is reached. After this state has been reached, we are no longer interested in the outputs produced for any input vector. This is noted as \$, \$\$ and will be simply noted \$ in the next figures. Let us note that, given an initial state, some states (q, q') may be not reachable.

From the observer in Figure 3.15c, one can see that the shortest input sequence detecting the difference between M and M' is $S_1 = x_1 x_0 x_0$ (or $S_2 = x_1 x_0 x_1$). As a matter of fact, $R(M, S_1) = 010$ and $R(M', S_1) = 011$. One can also deduce that any input sequence S in the language $(x_0 + x_1 x_1)^* x_1 x_0 (x_0 + x_1)(x_0 + x_1)^*$ is a sequence detecting the fault since this language corresponds to all the input sequences leading from the initial state (q_0, q'_0) to the final state ω.

A formal definition of an observer is given in the sequel. However, in practice, the concept and the construction of an observer are relatively simple.

Definition 3.9 The **observer** for M and M', assuming that the circuit under test is M', is a finite acceptor $\Omega(M, M') = \Omega = (B, X, \gamma, b_0, \omega)$, with $B = (Q \times Q') \cup \{\omega\}$, $b_0 = (q_0, q'_0)$, and with γ defined as follows

$$\gamma(b, x) = \begin{cases} (\delta(q,x), \delta'(q',x)), & \text{if } b=(q,q') \text{ and } \mu(q,x)=\mu'(q',x) \\ \omega, & \text{otherwise,} \end{cases}$$

for all $b \in B$ and x such that there is $Sx \in I$ such that $\gamma(b_0, S) = b$.

□

Intuitively, the observer Ω is a reduced representation of the automaton in Figure 3.16. The state of the *observer before reduction* in this figure is (q, q', z), since the comparator is combinational and, then, has a single state. The observer previously defined is such that any state $(q, q', 0)$ is represented simply by (q, q') and all states $(q, q', 1)$ are merged and represented by ω. In Figure 3.16 the outputs of A and A' have been omitted as outputs of the observer.

Let us emphasize that the observer is a machine allowing to calculate *all* the test sequences detecting a fault.

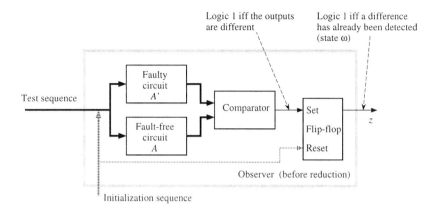

Figure 3.16 Illustration of the observer[10].

3.2.2.2 The initial state of the faulty circuit is not known

Figure 3.15a shows that the input sequence $S_s = x_0$ is a synchronizing sequence for automaton A since from any *starting state* the *initial state* q_0 is reached. On the other hand S_s is not a synchronizing sequence for automaton A' as can be seen in Figure 3.15b (furthermore there is no synchronizing sequence for A'). It follows that: if the starting state of A' is unknown, the initial state of A' reached by the "synchronizing" sequence S_s is unknown. Then the machine which is tested may be either $M' = (A', q'_0)$ (Figure 3.15b; observer in Figure 3.15c) or $M'' = (A', q'_1)$ (Figure 3.17c; observer in Figure 3.17d).

An observer can be constructed for both machines M' AND M'', simultaneously compared to M. This observer is presented in Figure 3.17e. The initial state[11] is $(q_0; q'_0, q'_1)$. The components of this state are the initial states of M, M' and M'', in this order. If the input vector x_0 is applied from the initial state, the outputs of the three machines are 0,0 and 1 in the same order. Then it is observed that M'' is different from M and the state reached is $(q_0; q'_0, \omega)$ where ω means that M'' has already been detected as faulty. From this state, only the outputs associated with M and M' will be compared; the output associated with M'' will be replaced by $. The state ω is reached when both machines under consideration have been detected as faulty.

From Figure 3.17e, one can observe that the shortest input sequences leading from $(q_0; q'_0, q'_1)$ to ω are $S_1 = x_1 x_0 x_0$ and $S_2 = x_1 x_0 x_1$. Each of these sequences is able to detect that automaton A' is faulty whatever its initial state.

---

[10] This figure is reprinted from [DaFuCo 89] (© 1989 IEEE).
[11] For increase clarity, we use a semi-colon to separate the state of the fault-free machine from the others, whenever the observer concerns a fault-free machine and *more than one faulty machine*.

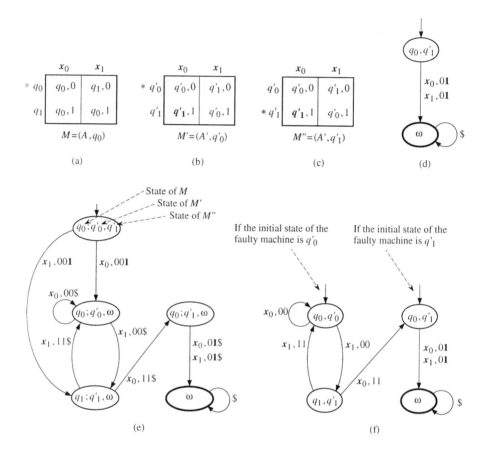

Figure 3.17 (a)(b)(c) Machines M, M', and M''. (d) Observer for M'': $\Omega(M, M'')$. (e) Observer for M' AND M'': $\Omega(M; M', M'')$. (f) Observer for M' OR M'': $\Omega(M, M'$ OR $M'')$.[12]

Machines M' and M'' in Figure 3.17 correspond to the same automaton A': only the initial state is different. Then, both observers for machine M' and for machine M'' (Figures 3.15c and 3.17d) can be merged into a single observer for M' OR M''. In this case (Figure 3.17f), the initial state depends on the automaton which is "observed".

Remark 3.8
a) The construction of an observer may be generalized to the comparison of any number of faulty automata to the faulty-free one. In Figure 3.17e the observer corresponds to two machines corresponding to the same automaton A'

---

[12] Figures 3.17a,b,c, and f are reprinted from [Da 97] (© 1997 IEEE).

with two initial states q'_0 and q'_1. An observer could also consider machines corresponding to different automata A'', A'''.

b) When *all the faulty machines* correspond to the *same automaton* (and only in this case), two kinds of observer can be built; we will call **AND-observer** an observer for M' AND M'' AND ... (example in Figure 3.17e); we will call **OR-observer** an observer for M' OR M'' OR ... (example in Figure 3.17f). Note that the observer to be used depends on the problem to be solved.

c) Roughly speaking, an observer is a model of parallel fault simulation. It applies to both sequential and combinational circuits as a particular case. This is illustrated in Figure 3.18.

❑

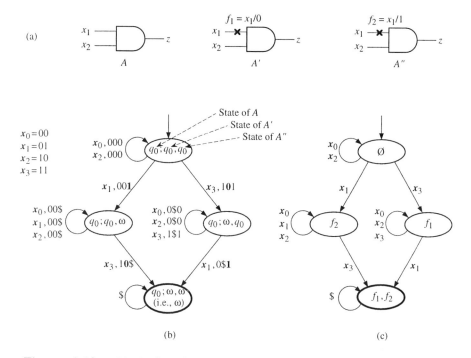

Figure 3.18 (a) A fault-free combinational circuit and two faulty ones. (b) Observer for the two faulty circuits: $\Omega(A; A', A'')$. (c) Simplified representation of the observer for a combinational circuit and combinational faults.

In Figure 3.18a, A corresponds to a combinational circuit consisting of a single gate, A' and A'' correspond to two faults $f_1 = x_1/0$ and $f_2 = x_1/1$, respectively. Figure 3.18b is the observer for A' AND A''. Since the three circuits are combinational, they have a single state noted q_0 for every circuit. However, the observer has several states. If x_1 is applied from the initial state, A'' produces a faulty output 1. Then the state $(q_0; q_0, \omega)$ is reached. From this state only A' has still to be detected as different from A. One can observe a symmetry in the observer; the input vectors x_1 and x_3 may be applied in any order since the faults are combinational ones. Since the circuit and the faults concerned are combinational, there is always a single internal state q_0. In that case, one can represent a state of the observer by the set of faults which have been detected in this state without loss of information. Figure 3.18c is an example of this simplified representation (in which the outputs have been omitted).

3.2.3 Propagation

An **iterative circuit** is a digital structure composed of a cascade of identical circuits called cells. It is well known that there is a similarity between a sequential circuit and a combinational iterative circuit: every finite output sequence that can be produced *sequentially* by a sequential circuit can also be produced *spacially* by a combinational iterative circuit. This property can be used to transform a sequential circuit into a combinational iterative one and then to apply some methods known for combinational circuits.

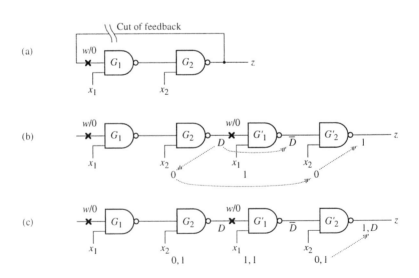

Figure 3.19 (a) Sequential circuit. (b) Iterative combinational circuit: an error is provoked and partially propagated. (c) Propagation is completed.

An example is presented in Figure 3.19. The 2-input, 1-output sequential circuit in Figure 3.19a is transformed into a combinational one by cutting the feedback. Then it is transformed into a combinational iterative circuit by cascading several copies of the basic combinational circuit, as shown in Figure 3.19b. The first copy is made of gates G_1 and G_2. The second one is made of gates G'_1 and G'_2. The feedback from G_2 output to G_1 input in the sequential circuit is replaced by a connection from G_2 output to G'_1 input in the iterative circuit. There are several copies of the inputs x_1 and x_2 in the iterative circuit. A single copy of the output has been represented for clarity.

With a single fault in the sequential circuit, w/0 in our example, a multiple fault is associated in the iterative circuit, as shown in Figure 3.19b.

Now, since we have obtained a combinational circuit, one can try to apply the propagation through sensitized paths which have been presented in Section 3.1.2. See Figure 3.19b. In order to provoke an error at the output of gate G_2 (value D), the value 0 must be obtained on the input x_2 of G_2 (note that one input of G_1 is not controllable). The error can be propagated to the output of G'_1 by the input $x_1 = 1$. Now, $x_2 = 0$ on gate G_2 implies $x_2 = 0$ on gate G'_2 too; then the error cannot be propagated to z. This means that the input vector $x(1) = x_2 = 10$ provokes an error and a propagation to G'_2 input. If the next input vector is $x(2) = x_3 = 11$, then the error is propagated to z as illustrated in Figure 3.19c. Hence we have obtained a test sequence for the fault w/0, namely $S = x_2 x_3$.

The test sequence corresponding to a spacial propagation in Figures 3.19b and c, corresponds to a sequential (i.e., temporal) propagation in Figure 3.19a.

3.2.4 Relations between faults

3.2.4.1 Fault collapsing

Fault collapsing in combinational circuits was considered in Section 3.1.4.1. It is clear that some faults affecting a component can be compared whether the circuit is combinational or sequential. Consider, for example, an AND gate; all the stuck-at-0 of either an input or an output are *equivalent* and the output stuck-at-1 *dominates* any input stuck-at-1 whatever circuit the gate is a component of.

However, these are particular cases of more general properties which will be presented further on. Let us first define a sequential fault.

Definition 3.10 A **sequential fault** is either:
1) any fault affecting a *sequential* circuit

or

2) a fault affecting a *combinational* circuit and such that the faulty circuit is not combinational.

❑

From this definition, a fault affecting a sequential circuit and such that the faulty circuit is combinational is a particular case of sequential fault.

Let (A, q_0) be a fault-free machine whose input language is I. In the general case a sequential fault in the corresponding circuit may be represented by (A', Q'_0) where $Q'_0 = \{q'_0, q'_1, \ldots\}$ is the set of possible initial states given that the initialization (leading to q_0 in the fault-free circuit) has been performed. In other words, the set of compatible initial states is $\{(q_0, q'_0), (q_0, q'_1), \ldots\}$. In the important particular case where Q'_0 contains a single state, the fault is represented by (A', q'_0). The model (A', q'_0) is used either if the initial q'_0 is *unique* or if it is *assumed to be known* when it is not unique.

We shall now use the notation $R(q_0, S)$ (introduced in Section 2.1.3), representing the output sequence of machine (A, q_0). The advantage of using the output sequence $R(q_0, S)$, instead of the output vector $\mu(q_0, S)$, is that if $R'(q'_0, S) \neq R(q_0, S)$ for some input sequence S, then $R'(q'_0, S_1) \neq R(q_0, S_1)$ for any S_1 such that $S_1 = S \cdot S_2$: this means that if a fault is tested by a sequence S, then it is tested by any sequence $S \cdot S_2$.

Definition 3.11 Two *sequential faults* (A', q'_0) and (A'', q''_0) are **equivalent** (denoted by $(A', q'_0) \equiv (A'', q''_0)$), for some input language I, if

$$R'(q'_0, S) = R''(q''_0, S), \quad \text{for any } S \in I.$$

❑

If a test sequence detects a fault several times, the *first detection* is the most important. When the first detection occurs, the state ω of the observer is reached. This feature leads to the following definition.

Definition 3.12 A test sequence S is **just detecting** for a fault $f = (A', q'_0)$, if the fault f is tested by S but is not tested by any proper prefix of S. Formally:

$$R'(q'_0, S_1) = R(q_0, S_1), \quad \text{for any } S_1 \in \mathcal{P}(S) \setminus S,$$

and

$$R'(q'_0, S) \neq R(q_0, S).$$

In this case, the fault f is **just detectable** by S.

❑

Definitions of dominating and equidetectable sequential faults are given in the sequel. Informally, a fault f dominates a fault g if any sequence testing g tests also f. The faults f and g are equidetectable if they are tested by the same set of input sequences.

Definition 3.13 A *sequential fault* $f = (A', q'_0)$ **dominates** a sequential fault $g = (A'', q''_0)$, for some input language I, if the following property is met: every sequence in I which is *just detecting* for the fault g, tests the fault f.

Definition 3.14 The *sequential faults* $f = (A', q'_0)$ and $g = (A'', q''_0)$ are **equidetectable** (denoted by $f \cong g$) for some input language I, if f and g are *just* *detectable by the same set of sequences* in I. In other words, f dominates the fault g and vice versa.

□

Informally, if two faults are equidetectable, their observers are quite similar (in fact isomorphic): 1) they have the "same" set of states (the names are different but there is a 1-to-1 mapping from one set to the other); the initial state is the same; all the transitions between states of the observer are similar and labelled by the same input vectors.

This is illustrated in Figure 3.20. The faults (A', q'_0), (A, q''_0), and (A''', q'''_0) correspond to the fault-free next state function and faulty output functions. They are equidetectable ("same" observers). From the states tables, it appears that all the detections (not only the first one) occur exactly at the same times for (A', q'_0) and (A'', q''_0); let us say that they are *strongly equidetectable* (however, they are not equivalent). Comparing faults (A', q'_0) and (A''', q'''_0), one can observe that the first detection occurs always at the same time for both (at the first input vector x_1 for any test sequence). However, all the detections after the first one do not occur at the same times: for the input sequence $x_1 x_1$ there is one detection for (A', q'_0) and two detections for (A''', q'''_0). Let us say that these faults are *weakly equidetectable*.

Definition 3.14 can be applied to combinational faults as a particular case. For combinational faults, weakly equidetectable faults do not exist.

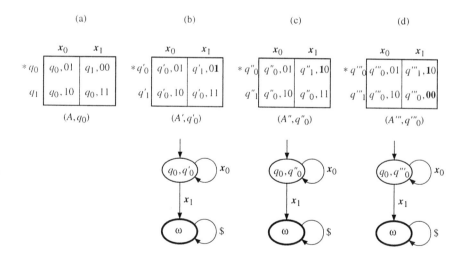

Figure 3.20 (a) A fault-free machine. (b) (c) (d) Three faulty machines (equidetectable) and the corresponding observers.

Consider now the faults (A', Q'_0) and (A'', Q''_0), i.e., the initial state is not always the same.

Definition 3.15 Two *sequential faults* (A', Q'_0) and (A'', Q''_0) are said to be **equivalent**, for some input language I, if:

for any $q'_i \in Q'_0$, there is $q''_j \in Q''_0$ such that $(A', q'_i) \equiv (A'', q''_j)$;

for any $q''_j \in Q''_0$, there is $q'_i \in Q'_0$ such that $(A', q'_i) \equiv (A'', q''_j)$.

❑

It is clear that Definition 3.11 corresponds to a particular case of Definition 3.15.

3.2.4.2 Fault detectability

The notion of detectable fault is quite simple for a combinational fault (Definition 3.5 in Section 3.1.4.3). Since the initial state of a faulty sequential circuit is not always unique, the notion of detectability is not so simple for a sequential fault. The definitions given in the sequel may be useful if one wants to show that a test sequence exists for any initial state of the faulty machine (or to find such a test sequence).

The definitions correspond to three cases: the fault-free initial state q_0 is unique and the faulty initial state q'_0 is unique (Definition 3.16); the fault-free initial state q_0 is unique and the faulty initial state is in the set Q'_0 (Definition 3.17); the fault-free initial state is not necessarily unique (this case could happen only if initialization is performed by a synchronizing sequence) and the faulty initial state is in the set Q'_k if the fault-free initial state is q_k (Definition 3.18, where the fault is denoted by $(A', \{Q'_k\})$).

Definition 3.16
a) The sequential fault (A', q'_0) is **detectable** for some input language I, if there is a test sequence S in I such that $R'(q'_0, S) \neq R(q_0, S)$.
b) The sequential fault (A', q'_0) is **undetectable**, or **redundant**, otherwise.

Remark 3.9 According to Definition 3.16, fault $M' = (A', q'_0)$ is detectable if and only if there is an oriented path from the initial state (q_0, q'_0) to the absorbing state ω in the observer for the pair of machines (M, M').

Example 3.5 Consider the fault-free machine $M = (A, q_0)$ in Figure 3.21a and the faulty machine $M' = (A', q'_0)$ in Figure 3.21b. Assume that the input language is $I = (x_0 + x_1)^*$.

As can be seen on the observer in Figure 3.21c, the fault $M' = (A', q'_0)$ is detectable for $I = (x_0 + x_1)^*$ since $R'(q'_0, x_1x_1) \neq R(q_0, x_1x_1)$ (according to Definition 3.16).

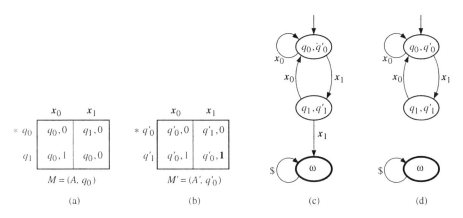

Figure 3.21 (a) Fault-free machine M. (b) Faulty machine M'. (c) Observer for the pair (M, M') given $I = (x_0 + x_1)^*$. (d) Observer for the pair (M, M') given $I = \mathcal{P}(x_0 + x_1 x_0)^*$.

Example 3.6 Similar to Example 3.5, except that the input language is $I = \mathcal{P}(x_0 + x_1 x_0)^*$.

As can be seen on the observer in Figure 3.21d, the fault $M' = (A', q'_0)$ is undetectable for $I = \mathcal{P}(x_0 + x_1 x_0)^*$ since $R'(q'_0, S) = R(q_0, S)$ for any S in I (according to Definition 3.16).

Common part for Examples 3.7 to 3.9, 3.11, and 3.12 Consider the fault-free and faulty machines in Figures 3.22a and b. It is assumed that initialization is made by a synchronizing sequence and the input language is $I = (x_0 + x_1)^*$. Observe that x_0 and x_1 are synchronizing sequences for A while A' is synchronized by x_0 but not by x_1. Note that the subautomaton $\hat{A}'$ of A' corresponding to the set of states $\{q'_0, q'_1\}$ is equivalent to A ($q'_0 \equiv q_0$ and $q'_1 \equiv q_1$).

Example 3.7 1) *The fault in the circuit is such that each time the faulty circuit is powered on, it reaches the starting state q'_2*; 2) *the synchronizing sequence is x_0.*

The initial states are q_0 and q'_0 (initialization α in Figure 3.22). As can be seen on the observer in Figure 3.22c, the fault will never be detected since there is no oriented path from state (q_0, q'_0) to state ω; according to Definition 3.16, the sequential fault (A', q'_0) is undetectable since $R'(q'_0, S) = R(q_0, S)$ for any sequence S. Note that the fault is evasive since the test sequence λ leads to the subautomaton $\hat{A}'$ equivalent to A without detection.

Example 3.8 Similar to Example 3.7, except that *the synchronizing sequence is x_1*.

The initial states are q_1 and q'_2 (initialization β in Figure 3.22). As can be seen on the observer in Figure 3.22d, the fault is detectable since $R'(q'_2, x_1) \neq R(q_1, x_1)$ (according to Definition 3.16), although it is evasive because the input sequence x_0 leads from the initial state (q_1, q'_2) to the subautomaton $\hat{A}'$ without detection.

❑

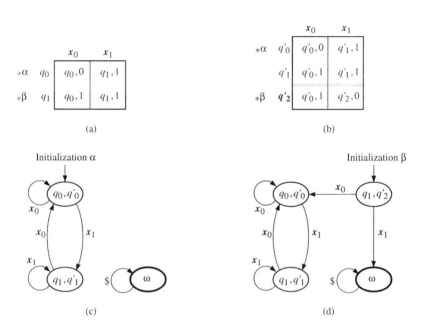

Figure 3.22 (a) Fault-free automaton A. (b) Faulty automaton A'. (c) Observer for initialization α. (d) Observer for initialization β.

Let us recall that, even if the fault-free circuit has a definite initial state q_0, the faulty one may have an unknown initial state in the set Q'_0 implied be q_0.

Definition 3.17
a) The sequential fault (A', Q'_0) is **detectable** for some input language I, if there is a test sequence S in I such that, for every q'_i in Q'_0, $R'(q'_i, S) \neq R(q_0, S)$ (i.e., S depends on Q'_0 but does not depend on q'_i).

b) The sequential fault (A', Q'_0) is **undetectable**, or **redundant**, if, for every sequence S in I and for every q'_i in Q'_0, $R'(q'_i, S) = R(q_0, S)$.

c) The sequential fault (A', Q'_0) is **partially detectable** if it is *neither detectable nor redundant*, i.e.:
   1) there is no test sequence in I fulfilling the condition in a); and

2) there is a test sequence S in I such that, $R'(q'_i, S) \neq R(q_0, S)$ for at least one q'_i in Q'_0, and $R'(q'_j, S) = R(q_0, S)$ for at least one q'_j in Q'_0.

Remark 3.10 According to Section 3.2.2.2 (Remark 3.8b), since Q'_0 contains more than one initial state, one can built an AND-observer and an OR-observer. Property 3.17 can be transposed as follows.

The sequential fault (A', Q'_0) is *detectable* if and only if there is an oriented path from the initial state to the absorbing state ω in the AND-observer.

The fault (A', Q'_0) is *partially detectable* if the two following conditions are met.

1) There is no oriented path from the initial state to ω in the AND-observer.

2) There is at least one initial state of the OR-observer such that there is an oriented path from this state to ω.

❑

Example 3.9 1) *The fault is such that, when the circuit is powered on, the starting state is sometimes q'_0, sometimes q'_1, and sometimes q'_2; 2) the synchronizing sequence is x_1.*

The initial states are q_1 for A, and $Q'_1 = \{q'_1, q'_2\}$ for A'. According to the AND-observer $\Omega((A,q_1); (A',q'_1), (A',q'_2))$ in Figure 3.23a, the fault is not *detectable*. However, it is *partially detectable* since there is an initial state in the OR-observer (state (q_1,q'_2) in Figure 3.23b) from which the fault can be detected. The same observation can be made on the AND-observer: the fault can be detected if the initial state of the faulty machine is q'_2.

❑

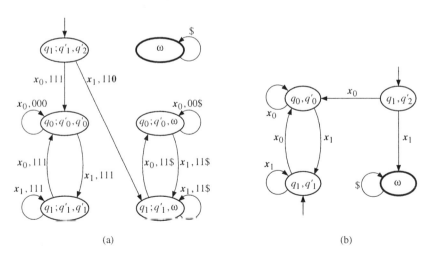

Figure 3.23 Illustration for Example 3.9. (a) AND-observer. (b) OR-observer.

In the preceding example, the fault cannot be detected if the initial state is (q_1, q'_1). The next example is such that, from any initial state in Q'_0 the fault can be detected; however the fault is partially detectable because there is no test sequence common to all initial states.

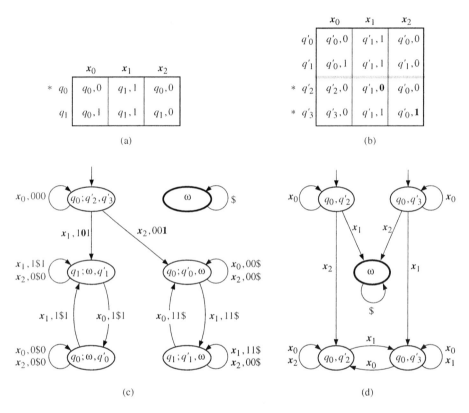

Figure 3.24 Illustration for Example 3.10. (a) Fault-free machine. (b) Faulty machine. (c) AND-observer. (d) OR-observer.

Example 3.10 The fault-free and faulty machines are given in Figures 3.24a and b (note that the subautomaton $\hat{A}'$ of A' corresponding to the set of states $\{q'_0, q'_1\}$ in Figure 3.24b is equivalent to A in Figure 3.24a.).

1) *The starting state of the faulty machine is sometimes q'_2 and sometimes q'_3;*
2) *the synchronizing sequence is x_0.*

It follows that the initial state of the faulty machine is $Q'_0 = \{q'_2, q'_3\}$. In the AND-observer of Figure 3.24c, there is no oriented path from the initial state to ω. From each initial state of the OR-observer in Figure 3.24d, there is an oriented path state to ω. Thus, the fault is *partially detectable* according to Definition 3.17.

As a matter of fact, any test sequence in $L_2 = x_0^* x_1 (x_0 + x_1 + x_2)^*$ detects the fault if the initial state is q'_2; any test sequence in $L_3 = x_0^* x_2 (x_0 + x_1 + x_2)^*$ detects the fault if the initial state is q'_3. However, $L_2 \cap L_3$ is empty.

❑

For a machine synchronized by a synchronizing sequence, the user may choose any initial state q_k (for the fault-free machine). As a matter of fact, the "useful" part of this machine must be strongly connected (Section 3.2.1.1); thus, if S_j is a synchronizing sequence to q_j, there is an input sequence S_{jk} such that $\delta(q_j, S_{jk}) = q_k$, then $S_j \cdot S_{jk}$ is a synchronizing sequence to q_k.

The next definition is related to such a machine *if the initial state is not yet known*, and to a fault such that the initial state of the faulty machine is in the subset Q'_k implied by q_k.

Definition 3.18 Take a strongly connected machine with a synchronizing sequence.

a) The sequential fault $(A', \{Q'_k\})$ of this machine is **detectable** for some input language I, if every (A', Q'_k) is detectable for I (i.e., the test sequence S depends on Q'_k which is implied by the initial state q_k of the fault-free machine).

b) The sequential fault $(A', \{Q'_k\})$ is **undetectable**, or **redundant**, for some input language I, if every (A', Q'_k) is *redundant* for I.

c) The sequential fault $(A', \{Q'_k\})$ is **partially detectable** if it is *neither detectable nor redundant*.

❑

For Examples 3.11 and 3.12, we return to the fault-free and faulty circuits of Figure 3.22.

Example 3.11 1) *The fault in the circuit is such that each time the faulty circuit is powered on, it reaches the starting state q'_2; 2) the synchronizing sequence which will be used is not known.*

Assume that the initial state of the fault-free machine is q_0. Then x_0 is a suffix of the initializing sequence (see Figure 3.22a) and thus the initial state of the faulty machine is q'_0 (see Figure 3.22b).

Assume now that the initial state of the fault-free machine is q_1. Then x_1 is a suffix of the initializing sequence (see Figure 3.22a) and, thus, the initial state of the faulty machine is either q'_1 or q'_2 (see Figure 3.22b); for example, if the synchronizing sequence is x_1 the initial state is q'_2, and if the synchronizing sequence is $x_0 x_1$ the initial state is q'_1.

Hence, since the synchronizing sequence is not known, the sequential fault can be denoted by $(A', \{Q'_0, Q'_1\})$ where $Q'_0 = \{q'_0\}$ and $Q'_1 = \{q'_1, q'_2\}$.

Initial state q_0: (A', q'_0) cannot be distinguished from (A, q_0) (Example 3.7).

Initial state q_1: $(A', \{q'_1, q'_2\})$ corresponds again to the AND-observer in Figure 3.23a and the OR-observer of Figure 3.23b (partially detectable).

Thus, the sequential fault $(A', \{Q'_0, Q'_1\})$ is partially detectable according to Definition 3.18.

Example 3.12 1) *The fault is such that, when the circuit is powered on, the starting state is sometimes q'_0, sometimes q'_1, and sometimes q'_2;* 2) *the synchronizing sequence which will be used is not known.*

This example includes all the cases in Example 3.11. Since the fault in Example 3.11 is partially detectable, this is also true for the present example, according to Definition 3.18.

Property 3.5

A sequential fault has *always the same initial state*, thus cannot be *partially detectable* (i.e., it is *either detectable or redundant* according to Definition 3.16) if one of the following conditions is true.

a) The initialization is of the first kind (power-up reset) and the starting/initial state of the faulty machine is always the same.

b) The initialization is of the second kind (external reset) and the reset always leads the faulty machine in the same initial state.

c) The initialization is of the third kind (synchronizing sequence) and: 1) the starting state of the faulty machine is always the same; 2) the synchronizing sequence is given.

d) The initialization is of the third kind, the fault-free machine is reduced, and $|Q'| \leq |Q|$.

e) The initialization is of the third kind and:

either 1) the synchronizing sequence is given and the fault is not evasive given this initialization sequence;

or 2) the initial state is given and the fault is not evasive given this initial state;

or 3) the fault is not evasive for any initial state.

f) The initialization is of the first or second kind and the fault is not evasive.

Proof

a) and b) There is *only one pair* of compatible initial states, (q_0, q'_0). According to Definition 3.16, if there is an oriented path in the observer from (q_0, q'_0) to ω, the fault is detectable. Otherwise it is redundant.

c) Let q'_s be the starting state, and S_s the synchronizing sequence. There is *only one pair* of compatible initial states (q_0, q'_0) where $q'_0 = \delta'(q'_s, S_s)$. Then the proof is similar.

d) Since there is a synchronizing sequence and $|Q'| \leq |Q|$, there is a checking sequence S able to detect if the automaton A' is equivalent to the automaton A (Section 2.3.3). Let us consider two cases. 1) If $|Q'| < |Q|$, then S detects the fault. 2) If $|Q'| = |Q|$, either S detects the fault or shows that A' is equivalent to A. In the latter case, any q_k in Q is equivalent to a state q'_k in Q'; then any synchronizing sequence leading the fault-free automaton into q_0 leads the faulty automaton into q'_0. Hence the fault is undetectable.

e)

1) Let q_0 be the initial state reached by the synchronizing sequence S_s. Let $Q'_0 = (q'_0, q'_1, \ldots, q'_t)$ be the set of possible initial states of the faulty machine, depending on the set of possible starting states.

If the fault is redundant, then (A', q'_i) is undetectable for any q'_i in Q'_0.

If the fault is not redundant, consider the observer $\Omega((A, q_0); (A', q'_0), (A', q'_1), \ldots, (A', q'_t))$. Assume that there is a state b_j of this observer which can be reached from the initial state $b_0 = (q_0; q'_0, q'_1, \ldots, q'_t)$, and having the following property: for at least one initial state q'_i, the fault has not been detected in b_j and cannot be detected from b_j (then there is no oriented path from b_j to ω). Then, according to Definition 3.8 (Section 3.2.1.2), the fault is evasive for the pair of compatible states (q_0, q'_i). This is a contradiction with the hypothesis, then the state ω may be reached from any state of the observer. Hence the fault is detectable since a sequence leading from b_0 to ω can be found.

2) The proof is similar. The state q_0 is given and the set Q'_0 may be larger than in the preceding case since the synchronizing sequence is not known.

3) The proof is similar to the case 2) for every initial state of the fault-free machine. (Note that, according to Definition 3.18, there is a test sequence for each fault-free initial state, but not necessarily the same test sequence for all.)

f) Similar to the proof of e).

❏

Informally, a fault may affect either the next state function or the output function or the reset function (or several of these functions). According to Properties 3.5a to c, the fault is always either detectable or redundant (but cannot be partially detectable) if the *reset function* is fault-free or, at least, *deterministic*.

According to Properties 3.5e and f, *a sufficient condition for a fault not to be partially detectable is to be non-evasive*.

3.2.4.3 Generation of a test sequence for a set of faults

A general scheme to obtain a test sequence is illustrated in Figure 3.25, for both combinational and sequential faults. It is assumed that a parallel simulation of faults can be performed. Let us comment on the various steps.

Step 1. This may be either an enumeration or a list of faults obtained from a fault model (e.g. single stuck-at). If some sequential faulty behavior may correspond to several initial states, one can consider that they correspond to different faults. For example one may have (A', q'_0) and (A', q''_0) in the set of faults.

Chapter 3

Step 2. Collapsing reduces the number of faults to simulate. However, if a fault f, dominating a fault g, remains in the list, the simulation will show that f is detected at the latest when g is detected.

Step 3. If all the faults are combinational, the order of the test vectors has no influence (Fig. 3.25a). We consider a test set instead of a test sequence.

Step 4. If the test generation is *fault-oriented*, a test for one specified fault, referred to as the **target fault**, is generated. If a *random generation* is considered, a test vector is randomly selected.

Step 5. This step finds the faults detected by the last vector.

Step 6. If the faults are combinational and if generation is random, there is no point keeping a useless vector x_i. For a sequential fault a vector which does not detect any fault is not necessarily useless since it may change the internal state.

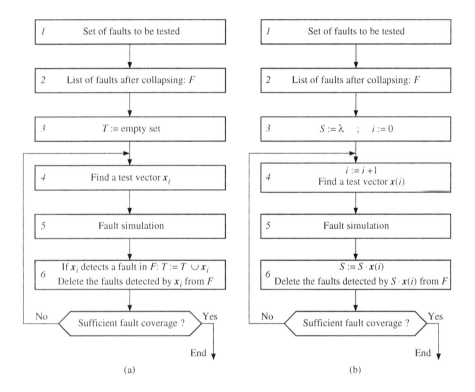

Figure 3.25 Generation of tests for a set of faults. (a) Test set of combinational faults. (b) Test sequence for sequential faults.

Remark 3.11 Many deterministic ATPGs generate randomly the first test vectors. The generation starts with random drawings in step 4. When several successive random vectors fail to detect new faults, the ATPG switches to a deterministic generation for finding tests for the hardest faults.

NOTES and REFERENCES

The notation D and $\overline{D}$ was introduced by J. P. Roth who defined the famous D-algorithm [Ro 66].

The main results relevant to the algebraic method can be found in [BoHo 71], and [ThDa 75] where relay logic and sequential circuits are also considered. Arguments for proofs of Properties 3.3 and 3.4 (Section 3.1.4.2) can be found in [AbBrFr 90] and [BoHo 71], respectively.

In [Ha 76b] a general type of redundancy is defined and properties of irredundant circuits are studied.

The discrepancy between predicted and observed failure rates, due to fault coverage of a collapsed set of faults, is described in [Wa 81].

The notion of evasive fault was introduced in [DaBrJü 92 & 93].

The general theory of observers was described in [BrJü 92]. The observer we have presented is a simplified version of the general construction, which is sufficient for the purpose of this book. The words "AND-observer" and "OR-observer" are introduced in this book.

One can find in [Ko 78] the similarities between sequential machines and iterative combinational circuits.

The notion of equidetectable faults for combinational circuits was introduced in [IsJaDa 93]. To the author's knowledge, all the definitions of Section 3.2.4.1 (sequential fault collapsing) are new.

The fact that the initial state of a faulty circuit may not be unique, i.e., $|Q'_0| > 1$, even if the initialization sequence is known, has been explicitly considered by several authors. The notion of "partially detectable" fault is introduced in [PoRe 93]. There are some similarities between our Section 3.2.4.2 and this paper. However, the definitions are different since we consider that there is always an initialization of the circuit before applying a test sequence: even if this initialization is inefficient for the faulty circuit, the fault-free circuit is in a known initial state and the set of possible initial states of the faulty one may be restricted by this "initialization" (this has been considered by the author for a long time [DaTh 80]; various initial states of a faulty RAM, and their probabilities given a synchronizing sequence, are considered in [DaBrJü 93]). A fault qualified "undetectable" in [AbPa 92], because the circuit cannot be correctly initialized, is detectable according to our definitions (Exercise 3.8).

4

Performance Measurements for a Test Sequence

The defect level *is a basic measurement corresponding to the proportion of faulty circuits in the circuits which have passed the test. It decreases as the* yield *increases. It decreases as the* performance of the test sequence *increases.*

Now, how can we obtain a good measurement of the performance of the test sequence?

*The usual measurement is the fault coverage of single stuck-at faults (introduced in Section 3.1.4). Some relations between this measurement and the defect level have been proposed. For random testing, the fault coverage can also be used (*expected *fault coverage). On the other hand, other measurements can be used in this context: the probability of detecting the worst case fault and the probability of having a 100% fault coverage.*

The defect level and related entities are presented in Section 4.1, and the concepts of fault model and fault coverage are given in Section 4.2. The relations proposed in the literature, between fault coverage and defect level, are presented in Section 4.3. In Section 4.4, the measurements of the confidence level for random testing are defined are compared.

4.1 DEFECT LEVEL

Some pertinent quantities are illustrated from the diagram in Figure 4.1.

Assume a manufacturer produces some kind of circuits. The number of very similar circuits is W (the magnitude of W is assumed to be very large). Among these W circuits, there are G **good** (fault-free) circuits and D **defective** (faulty) circuits, then

$$W = G + D. \tag{4.1}$$

Chapter 4

The production **yield** (or simply yield), denoted by Y, is the proportion of good circuits in the manufactured circuits, i.e.,

$$Y = \frac{G}{W}. \tag{4.2}$$

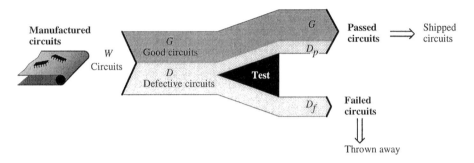

Figure 4.1 Principle of circuit testing[1].

The aim of the test is to separate the defective circuits from the good ones. As illustrated in Figure 4.1, this partition is imperfect. When a circuit has been tested, either the circuit has **failed** (i.e., a faulty response has been observed) or the circuit has **passed** (i.e., a fault-free response has been observed). We consider that all the G fault-free circuits pass the test (see Remark 4.1). Among the D faulty circuits, D_f fail and D_p pass the test. Then

$$D = D_f + D_p. \tag{4.3}$$

Remark 4.1 We consider that the observation of the response of the tested circuit is deterministic: one observes either the response itself or a compact deterministic function of it. This is usual for both deterministic and random testing. This implies that all the fault-free circuits pass the test (if the test device itself is not defective).

Few authors have considered an observation of statistical properties of the response [De et al. 76][Pa 76][Lo 77]. Assume, for example, a random input sequence with equally likely input vectors applied to a 2-input AND gate; this circuit passes the test if and only if the average output value is close to 0.25. With this kind of observation a good circuit could fail. This is not included in our hypotheses.

❏

---

[1] This figure is reprinted from [JaDa 89b] (© 1989 IEEE).

Performance Measurements for a Test Sequence

The proportion of faulty circuits in the passed circuits, i.e., in the shipped circuits, is the **defect level**, DL (The expression *field reject rate* is used by some authors):

$$DL = \frac{D_p}{G + D_p}. \tag{4.4}$$

It is clear that the defect level depends on the yield: if Y were equal to 1, then $D = 0$, hence $D_p = 0$ and $DL = 0$. It depends also on the *ability of the test sequence to detect the faulty circuits*. (We consider that the entire response of the circuit under test is observed; then, if the test sequence produces a faulty response, the circuit fails. Observation of a compact function of the response will be considered in Chapter 12.)

The proportion of faulty circuits which fail is called **faulty circuit coverage** (or *defect coverage*) and is denoted by P_u. Then (for $D > 0$):

$$P_u = \frac{D_f}{D}. \tag{4.5}$$

Property 4.1

$$DL = \frac{(1-Y)(1-P_u)}{Y + (1-Y)(1-P_u)}. \tag{4.6}$$

Proof From definition (4.4),

$$DL = \frac{D_p}{G + D_p}.$$

Dividing numerator and denominator by W yields:

$$DL = \frac{\frac{D_p}{W}}{\frac{G + D_p}{W}}. \tag{4.7}$$

Using (4.3) then (4.1), the numerator of (4.7) can be written as:

$$\frac{D_p}{W} = \frac{D - D_f}{W} = \left(\frac{D}{W}\right)\left(\frac{D - D_f}{D}\right) = \left(1 - \frac{G}{W}\right)\left(1 - \frac{D_f}{D}\right). \tag{4.8}$$

Since $\frac{G}{W} = Y$, from definition (4.2), and $\frac{D_f}{D} = P_u$, from definition (4.5), Equation (4.8) becomes $(1 - Y) \cdot (1 - P_u)$ which is then the numerator of (4.7).

The denominator of (4.7) can be written as

$$\frac{G+D_p}{W} = \frac{G}{W} + \text{Numerator of } (4.7) = Y + (1 - Y)(1 - P_u).$$

❑

From Equation (4.6) one can observe that:
1) *DL decreases when Y increases*, with the limit cases:
$DL = 1$ when $Y = 0$, and $DL = 0$ when $Y = 1$.
2) *DL decreases when P_u increases*, with the limit cases (for $Y > 0$):
$DL = 1 - Y$ when $P_u = 0$, and $DL = 0$ when $P_u = 1$.

Equation (4.6) is an *exact result*. However, the values of Y and P_u are never exactly known; they must be estimated. The yield can be estimated by experiments. The faulty circuit coverage estimation, from fault coverage, will be discussed in Section 4.3.

In practice, P_u is close to 1. For $P_u \approx 1$, the derivative $\frac{dDL}{dY} \approx \frac{P_u - 1}{Y^2}$; the estimation of *DL* is not very sensitive to an error on the estimation of Y since $P_u - 1$ is close to 0 (except in case of very bad manufacturing!).

On the other hand, the derivative $\frac{dDL}{dP_u} \approx \frac{Y-1}{Y}$; *the estimation of DL is appreciably sensitive to an error on the estimation of P_u* since Y is usually not close to 1; this is an important problem in the test context.

We can conclude this section as follows. The *faulty circuit coverage*, P_u, defined by Equation (4.5) is clearly a *very pertinent measurement of the performance of a test sequence*, as it is shown by Equation (4.6). However, Equation (4.5) is nothing but a definition.

4.2 FAULT MODELS AND FAULT COVERAGE

Given a kind of circuit, the **universal fault set** is an implicit fault set assuming that *any permanent logic fault* is possible. Let F_u denote this set. This implies that for any faulty circuit, the corresponding fault is in F_u.

The faulty circuit coverage (Section 4.1) of a test sequence S could be known if the following elements were known: 1) F_u; 2) for every f in F_u, the proportion of defective circuits affected by f; 3) for every f in F_u, the knowledge weither S does or does not detect this fault. Unfortunately, F_u is not known (hence the other elements are not known either). It follows that hypotheses about the faults must be made.

Notation 4.1
(a) Let $\mathcal{M}_i$ be a **fault model**.
(b) The *set of single faults* corresponding to $\mathcal{M}_i$ model is denoted by F_i.

(c) The *set of multiple faults* whose components are in F_i is denoted by F_i^* (A single fault is considered as a particular case of multiple fault, then $F_i \subset F_i^*$). ❏

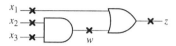

Figure 4.2 Example circuit.

For example, let $\mathcal{M}_{sa}$ denote the *stuck-at model*. For the circuit in Figure 3.7a, presented again in Figure 4.2, $F_{sa} = (x_1/0, x_2/0, x_3/0, w/0, z/0, x_1/1, x_2/1, x_3/1, w/1, z/1)$ (which is the set denotes as F_1 in Section 3.1.4.1.). The set F_{sa}^* includes all the faults in F_{sa} and all the multiple faults made from any subset of *compatible* faults in F_{sa}. For example $[x_1/0 \ \& \ x_2/0]$ and $[x_1/0 \ \& \ x_3/1 \ \& \ z/1]$ are in F_{sa}^*. But $[x_1/0 \ \& \ x_1/1]$ is not in F_{sa}^* since the components $x_1/0$ and $x_1/1$ cannot be present in the same circuit (they are *not compatible*).

For a CMOS type circuit, one may also consider the *stuck-open model*, $\mathcal{M}_{op}$, and the *stuck-on model*, $\mathcal{M}_{on}$. Then all the faults in F_{op}^* and in F_{on}^* are in F_u. A set of fault models can be considered as a fault model. Then the set of multiple faults $(F_{sa} \cup F_{op} \cup F_{on})^*$ is included in F_u.

Even if the simulator power were able to take into account as many faults as one wants, unmodeled faults would still remain because our imagination is limited. In practice, when studying the features of any method or test sequence, we restrict our attention to a set of faults, F, called the set of **faults under consideration**, the set of **prescribed faults** or the set of **targeted faults**. All the faults in F_u which are not in F are the **non-target faults**. Then[2]

$$F_{nt} = F_u \setminus F = F_u - F \tag{4.9}$$

where F_{nt} is the set of *non-target faults* corresponding to the set F of *faults under consideration*.

Now, given a set F of faults under consideration and a test sequence, S, the fault coverage, denoted by P, can be defined. According to Definition 3.7b in Section 3.1.4.4, P is the proportion of detectable faults in F which are detected by S. For most authors and for most simulators, $F = F_{sa}$, i.e., the set of faults under consideration is the set of single stuck-at faults.

Since there are equivalent faults, the set of faults corresponding to some models is often the set of faults *after equivalence fault collapsing* (Section

---

[2] The notation $A \setminus B$ represents the set of elements belonging to the set A which do not belong to the set B. When $B \subset A$, the notation $A - B$ may also be used.

3.1.4.1). For the circuit in Figure 4.2, the set of faults corresponding to $\mathcal{M}_{sa}$ after this collapsing should be $F'_{sa} = \{x_1/0, x_2/0, z/0, x_1/1, x_2/1, x_3/1\}$ (set denoted by F_2 in Section 3.1.4.1). As noted in Section 3.1.4.4, the fault coverage of a test sequence is not the same for F_{sa} and F'_{sa}. A *weighted fault coverage*, denoted by P_w, may be defined. It takes into account the proportion of circuits affected by every fault in the set of faults under consideration. For example, if the faults in F_{sa} are assumed to be equally likely, then, in F'_{sa}, the weight of $x_2/0$ should be 3 times the weight of $x_1/0$ (according to Figure 3.7b). These notions will be specified in Section 4.4.

4.3 RELATION BETWEEN FAULT COVERAGE AND DEFECT LEVEL

In this section, we consider that P is the fault coverage of the set of single stuck-at faults. Even if this measurement is not very good, it is considered to be a figure of merit from which some authors have attempted to estimate the defect level. On the basis of various hypotheses, some relations among *fault coverage*, P, the *production yield*, Y, and the *defect level*, DL, have been proposed. These relations are illustrated in Figure 4.3a (for $Y = 0.5$).

Every relation given in the sequel is based on a set of five hypotheses, here denoted by H0 to H4. The hypotheses H0 to H3 are common to the various approaches presented. Only hypothesis H4 is different for every approach.

H0. Let us consider a very large batch of similar circuits (they should be similar if they were fault-free). The ratio of circuits having a certain property to the number of circuits can be defined. Then, *a ratio for the set of circuits may be transformed into a probability to have this property, for each circuit*. Similarly, *the fault coverage P* is the ratio of faults tested by a test sequence to the total number of faults; it *may be transformed into a probability to be tested, for each fault*.

H1. The fault coverage, P, corresponds to a set F of faults and any fault in a circuit is assumed to be in F^*: the *non-target faults out of F^* are neglected*. (Usually $F = F_{sa}$, but similar results could be obtained for other fault models).

H2. The ratio of the number of circuits affected by a fault f to the total number of circuits is the same for every f in F.

H3. The fault $[f_1 \& f_2 \& ... \& f_k]$ whose multiplicity is k, is tested if at least one of its components is tested (i.e., multiple faults are assumed to be independent of one another). Thus, the probability of not detecting this multiple fault is $(1 - P)^k$.

H4. This hypothesis corresponds to the *distribution of the multiple faults* as a function of the multiplicity k.

❑

Various results have been obtained depending on hypothesis H4. These results are given in the sequel, without proof. The reader interested is referred to the corresponding papers.

In [Wa 78], it is assumed that the probability of occurence of a fault of multiplicity k is proportional to $(1 - Y)^k$ (*geometric distribution* function). The following equation is obtained:

$$DL = (1 - Y)(1 - P). \tag{4.10}$$

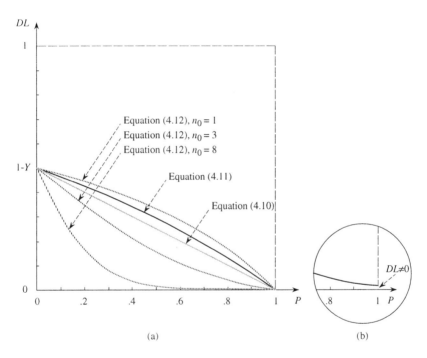

Figure 4.3 Relation among defect level, fault coverage and yield. (a) Theoretical forecasting. (b) Experimental results.

In [WiBr 81], the occurrence of a single fault in a circuit is independent of the presence of other faults in the circuit. Then the probability of occurrence of a multiple fault is the product of the probabilities of occurrence of its components. This hypothesis leads to a *binomial distribution* and to the approximate result:

$$DL = 1 - Y^{(1-P)}. \tag{4.11}$$

In [AgSeAg 82] a *Poisson distribution* is assumed. This assumption requires a new parameter, n_0, which is the average number of (single) faults in a faulty circuit (i.e., average multiplicity which may be not an integer value). The approximate result is obtained as follows:

$$DL = \frac{(1-P)(1-Y)e^{-(n_0-1)P}}{Y+(1-P)(1-Y)e^{-(n_0-1)P}}.\qquad(4.12)$$

Equation (4.11) is famous and has been used by many authors. Several authors have observed results resembling Equation (4.12) with $n_0 \geq 2$ (the value of n_0 is determined approximately from the observed results): in [Wa 78], $Y \approx 0.85$ and $n_0 \approx 2$ (for a small sequential circuit); in [AgSeAg 82], $Y \approx 0.07$, $n_0 \approx 8$ (for an LSI circuit); in [MaAi 93], $Y \approx 0.77$, $n_0 \approx 3$ (for an LSI sequential circuit). However, the defect level was not observed for values of P very close to 1 [AgSeAg 82], [MaAi 93]. Several authors have emphasized that, for physical circuits, the defect level is not 0 for $P = 1$ (Figure 4.3b). The reason is that, for any test sequence detecting all the faults in F^*, there are *non-target faults which are not detected* by this sequence, i.e., hypothesis H1 is not verified [Hu 93], [MaAi 93], [WaMeWi 95]. However, all Equations (4.10 to 4.12) are such that $DL \to 0$ when $P \to 1$, even if the corresponding derivatives $\frac{\mathrm{d}DL}{\mathrm{d}P}$ are different for $P = 1$.

Remark 4.2 Take the following two hypotheses: H1 and Poisson distribution with $n_0 = 1$. From H1, one deduces that $F^* \supset F_u$. Now, since $n_0 = 1$, it follows that every faulty circuit contains exactly one single fault which is in the set F. Then, F_u is restricted to F, and one obtains $P_u = P$. In this case, Equation (4.6) in Section 4.1 corresponds to Equation (4.12), where $P = P_u$ and $n_0 = 1$.

4.4 MEASUREMENTS OF THE CONFIDENCE LEVEL FOR RANDOM TESTING

Some measurements which are defined in terms of ratios in the context of deterministic testing, can be defined in terms of probabilities in the context of random testing. In addition, some measurements are used only in the context of random testing.

After introduction of some basic notations and concepts, measurements derived from deterministic concepts, then measurements specific to random testing are presented, in Sections 4.4.1 and 4.4.2 respectively. Relations among these measurements, and between faults in the context of random testing, are given in Section 4.4.3.

Let us introduce some notations. Let $F = \{f_1, ..., f_i, ..., f_r\}$ be a set of faults, and let f_0 depict the *fault-free circuit*.

Hypothesis 4.1 All the faults considered are mutually exclusive.

❏

For example if a circuit can be affected by $a/0$ or by $b/1$ or by both, then 3 faults are defined: $f_1 = a/0$ alone, $f_2 = b/1$ alone, $f_3 = [a/0 \& b/1]$.

Notation 4.2

a) $\Pr[f_i]$ is the *occurrence probability* of the fault f_i. Given a circuit, $\Pr[f_i]$ is the probability that this circuit is affected by f_i, $i \in \{0, 1, 2...\}$. (Note that this notation applies to the fault-free circuit, f_0, and to any target or non-target fault.)

b) $\Pr[f_i \mid F]$ is the *conditional occurrence probability* of f_i. It is the probability that a circuit is affected by f_i, $i \in \{1, 2, ...\}$, given it is affected by a fault in F.

c) $\Pr[F]$ is the probability that a given circuit is affected by a fault in F.

❏

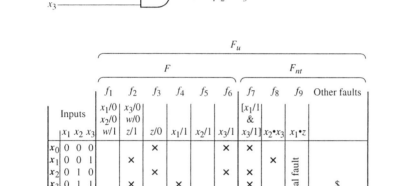

Figure 4.4 (a) A circuit. (b) Illustration of some notations and concepts with faults corresponding to this circuit.

Figure 4.4b is an example relative to some faults in the circuit in Figure 4.4a. The targeted faults are the single stuck-at faults i.e., $F = \{f_1, f_2, f_3, f_4, f_5, f_6\}$ after equivalent fault collapsing. The line of $\Pr[f_i]$ in Figure 4.4b is an *assumed example*. For example, $\Pr[f_1] = 0.05$ which means that f_1 is present in 5% of the circuits for a large number of circuits. In addition to the faults in F, non-target faults can occur. Fault f_7 is a double fault. Fault f_8 is a non-feedback bridging fault producing a wired AND between x_2 and x_3, and f_9 is a feedback bridging fault producing a wired AND between x_1 and z. The probability of being present in a circuit is 0.01 for f_7 and 0.04 for both f_8 and f_9. All the other non-target faults have a null probability.

From Hypothesis 4.1, one obtains for the example in Figure 4.4b:

$$\Pr[F] = \sum_{i:f_i \in F} \Pr[f_i] = 0.20,$$

and

$$\Pr[f_i \mid F] = \frac{\Pr[f_i]}{\Pr[F]}, \text{ for every } f_i \text{ in } F.$$

For example $\Pr[f_1 \mid F] = 0.05/0.20 = 0.25$.
Similarly,

$$\Pr[F_u] = \sum_{i:f_i \in F_u} \Pr[f_i] = 0.29,$$

and

$$\Pr[f_i \mid F_u] = \frac{\Pr[f_i]}{\Pr[F_u]}, \text{ for every } f_i \text{ in } F_u.$$

For example $\Pr[f_1 / F_u] = 0.05/0.29 = 0.172$.
The various values of $\Pr[f_i \mid F]$ and $\Pr[f_i \mid F_u]$ are given in Figure 4.4b.

When a test sequence (either deterministically or randomly generated) is known, one can determine by simulation if it detects a fault f_i. When a random test sequence is not known, there is some probability of it detecting the fault f_i. This probability depends on the distribution of the random input vectors and on the length, i.e., the number of input vectors.

Let $S = x(1)x(2) \ldots x(L)$ be a random test sequence whose length is L. Consider a **constant distribution**: the patterns $x(j)$ in S are *selected randomly with replacement* (i.e., a vector can be chosen several times, each time with the same probability); the probabilities of every possible pattern correspond to a distribution ψ. Assume, for example, that the circuit to be tested is a 2-input circuit. Then, there are four possible input vectors: x_0, x_1, x_2, and x_3. A constant distribution is a vector of probabilities $\psi = (\Pr[x_0], \Pr[x_1], \Pr[x_2], \Pr[x_3])$. For example $\psi_0 = (.25, .25, .25, .25)$ and $\psi_1 = (.3, .3, 0, .4)$ are two possible distributions since $\sum \Pr[x_i] = 1$. In ψ_0 all the input vectors are **equally likely** but not in ψ_1. From now on, ψ_0 will always represent the **equal likelihood distribution**.

Other distributions, here called **generalized distributions**, can be defined: the vector of probabilities may change after each random drawing of a test vector. This is considered in Appendix A: a constant distribution and a generalized distribution are generated from a *memoryless source* and a *Markov source*, respectively. **Distribution** means either constant distribution or generalized distribution.

Definition 4.1 Let $R(C_0, S)$ and $R(C_f, S)$ denote respectively the fault-free response and the response of the circuit affected by the fault f to the input sequence S. The **probability of testing** *fault f by the input sequence S*, is the probability that they are different, i.e.,

$$\Pr[R(C_0, S) \neq R(C_f, S)]. \qquad (4.13)$$

This probability is denoted by either

$P(f, S)$,

or $P(f, \psi, L)$, where ψ and L are the *distribution* and the *length* of S,

or $P(f, L)$, when ψ is implicit,

or $P(f)$, when ψ and L are implicit.

❑

In Figure 4.4b, the values of $P(f_i) = P(f_i, \psi_0, 20)$ are presented for faults f_1 to f_9 (i.e., all the input vectors are equally likely and the test length is 20). Note that f_9 is a sequential fault. The way to obtain these values will be explained in Chapter 6. However, let us calculate right now $P(f_1)$. Let S be a random test sequence; the input vectors are equally likely and $L = 20$. Since f_1 is detected only by the input vector x_7:

$$P(f_1) = \Pr[x_7 \text{ in } S] = 1 - \Pr[x_7 \text{ not in } S]. \qquad (4.14)$$

Let $x(i)$ be a random input vector. $\Pr[x(i) = x_7] = 1/8$ and $\Pr[x(i) \neq x_7] = 7/8$, since all the patterns are equally likely.

$$\Pr[x_7 \text{ not in } S] = \Pr[x(1) \neq x_7] \cdot \Pr[x(2) \neq x_7] \cdot \ldots \cdot \Pr[x(L) \neq x_7] = (7/8)^{20} \qquad (4.15)$$

Then, from (4.14) and (4.15), one obtains

$$P(f_1) = 1 - \left(\frac{7}{8}\right)^{20} = 0.931.$$

4.4.1 Measurements derived from deterministic concepts

It is clear that the *universal fault set*, F_u, has the same meaning for both deterministic and random testing.
The definitions given in Section 4.1 can be extended to the random context.

The **yield** (Equation (4.2) in Section 4.1) becomes

$$Y = \Pr[f_0] = 1 - \Pr[F_u]. \tag{4.16}$$

For the example in Figure 4.4, $Y = 1 - \Pr[F_u] = 1 - 0.29 = 0.71$.

The **faulty circuit coverage** (Equation (4.5) in Section 4.1) is the probability that a circuit *does not pass the test*, given it is faulty.

According to Hypothesis 4.1:

$$P_u = \sum_{i: f_i \in F_u} P(f_i) \cdot \Pr[f_i \mid F_u]. \tag{4.17}$$

For the example in Figure 4.4, one obtains $P_u = 0.931(0.172) + 1(0.172) + 1(0.034) + 0.931(0.104) + 0.931(0.104) + 1(0.104) + 1(0.034) + 0.997(0.138) + 0.702(0.138) = 0.932$.

The **defect level**, DL, is obtained by replacing the values of Y (4.16) and P_u (4.17) in Equation (4.6) (Section 4.1).

By definition, the entities Y, P_u, and DL are relevant to the *universal fault set* F_u. Now we shall consider notions which are relevant to a set of faults, F, which is a subset of F_u.

The **expected fault coverage**, or simply **fault coverage**, of a set F of faults containing $|F|$ faults is

$$P = \frac{1}{|F|} \sum_{i: f_i \in F} P(f_i), \tag{4.18}$$

which is the average detection probability of any fault in F.

Consider the example of Figure 4.4. One has $|F| = 6$ and $\sum_{i: f_i \in F} P(f_i) \approx 0.931 + 1 + 1 + 0.931 + 0.931 + 1 = 5.793$. Then $P \approx 5.793/6 = 0.965$.

The **weighted fault coverage** of a set F of faults is

$$P_w = \sum_{i: f_i \in F} P(f_i) \cdot \Pr[f_i \mid F]. \tag{4.19}$$

For the example in Figure 4.4 one obtains: $P_w \approx 0.931(0.25) + 1(0.25) + 1(0.05) + 0.931(0.15) + 0.931(0.15) + 1(0.15) = 0.962$.

The difference between P and P_w is that the probabilities of occurrence of the faults, $\Pr[f_i \mid F]$, are taken into account in the latter, while all have the same "weight" in P.

The difference between P_w and P_u is that only the targeted faults are taken into account in P_w, while all the faults of the universal set are taken into account in P_u.

Remark 4.3

a) For random testing, P_w is close to 1 if and only if P is close to 1.

b) On the other hand, P_u may be significantly less than 1 when P is close to 1 because the non-target faults are not taken into account in P. This is true for both deterministic and random testing.

4.4.2 Measurements specific to random testing

We now describe three figures of merit of random testing. Their application will be explained in Section 4.4.3.

A very commonly used measurement for random testing is the following one. The **minimum testing probability**, for a set F of faults, given ψ and L, is

$$P_m = \min_{i: f_i \in F} P(f_i). \tag{4.20}$$

This measurement corresponds to the **worst case** fault, i.e., the most difficult to detect fault. The expressions **hardest to detect** fault, **most difficult** fault, or **most resistant** fault, are also found in the literature.

For the set F in Figure 4.4b, $P_m = 0.931$. There are three hardest to detect faults, namely f_1, f_4, and f_5.

The **probability of complete fault coverage**, for a set of faults F is the probability that every fault is tested, i.e.,

$$P_c = \Pr[P = 1]. \tag{4.21}$$

In Figure 4.4b, one can observe that f_2 dominates f_5 and that f_3 dominates f_1. Then the complete fault coverage of F is obtained if and only if all the faults in $\{f_1, f_4, f_5, f_6\}$ are tested. Faults f_1, f_4, and f_5 are tested by x_7, x_3, and x_5, respectively. Fault f_6 is tested if at least one vector in the set $\{x_0, x_2, x_4\}$ is applied.

Then

$$P_c = \Pr[x_3 \text{ AND } x_5 \text{ AND } x_7 \text{ AND } (x_0 \text{ OR } x_2 \text{ OR } x_4) \text{ in } S]. \tag{4.22}$$

For the set F in Figure 4.4b, one obtains $P_c = 0.802$. Because the calculation is long, it is explained in Appendix B.

One can observe that $P_c < P_m$. This is clear from Equations (4.14) and (4.22). $P_m = P(f_1)$ is the probability that x_7 is in S (4.14), while P_c corresponds to the presence of x_7 *plus* other input vectors in S, according to (4.22).

A less usual measurement is the *weighted minimum testing probability*, denoted by P_{wm} since it is a combination of both P_w and P_m, defined from a partition of the set of faults under consideration.

A **partition** ρ of $F = \{f_1, ..., f_r\}$ is a collection of disjoint subsets of F (these subsets are called **blocks**) whose union is F, i.e.,

$$\rho = \{F_1, ..., F_s\}, \tag{4.23}$$

with $F_j \cap F_k = \emptyset$, for any $j \neq k$.

The identity partition, denoted by I, is the partition containing one block: $I = \{F\}$. The zero partition, denoted by O, is the partition such that each block contains one fault: $O = \{\{f_1\}, ..., \{f_r\}\}$. Let ρ_1 and ρ_2 be two partitions; the first one is thinner than the second one, denoted by $\rho_1 \leq \rho_2$, if any block of ρ_1 is contained in a block of ρ_2. One can easily show that any partition is ranged between the zero partition and the identity partition by the $\leq$ order relation:

$$O \leq \rho \leq I, \quad \text{for any } \rho. \tag{4.24}$$

Notation 4.3 For every subset F_j of F, corresponding to a block of the partition ρ, one can denote the following parameters.

a) $\Pr[F_j \mid F]$ is the probability that a circuit is affected by a fault in F_j, given that it is affected by a fault in F (*conditional occurrence probability* of F_j).

b) $$P_m(F_j) = \min_{i: f_i \in F_j} P(f_i) \tag{4.25}$$

(*minimum testing probability* for F_j).

c) $$P_w(F_j) = \sum_{i: f_i \in F_j} P(f_i) \cdot \Pr[f_i \mid F_j] \tag{4.26}$$

(*weighted fault coverage* of F_j).

□

Now, we can define $P_{wm}(\rho)$.

Let ρ be a partition of F, $\rho = \{F_1, ..., F_s\}$. The **weighted minimum testing probability** of F relative to ρ is the average of the minimum testing probability of the blocks of ρ:

$$P_{wm}(\rho) = \sum_{j: F_j \in \rho} P_m(F_j) \cdot \Pr[F_j \mid F]. \tag{4.27}$$

Consider for example $F = \{f_1, f_2, f_3, f_4, f_5, f_6\}$ in Figure 4.4b and $\rho = \{F_1, F_2\}$ such that $F_1 = \{f_1, f_4, f_5\}$ and $F_2 = \{f_2, f_3, f_6\}$. One can see in Figure 4.4b that: $P_m(F_1) = 0.931$ and $P_m(F_2) \approx 1$;

$$\Pr[F_1 \mid F] = \Pr[f_1 \mid F] + \Pr[f_4 \mid F] + \Pr[f_5 \mid F] = (0.05 + 0.03 + 0.03)/0.20 = 0.55;$$

$$\Pr[F_2 \mid F] = \Pr[f_2 \mid F] + \Pr[f_3 \mid F] + \Pr[f_6 \mid F] = (0.05 + 0.01 + 0.03)/0.20 = 0.45.$$

From the definition (4.27):

$$P_{wm}(\rho) = P_m(F_1) \cdot \Pr[F_1 \mid F] + P_m(F_2) \cdot \Pr[F_2 \mid F].$$

Then

$$P_{wm}(\rho) = 0.931 \times 0.55 + 1 \times 0.45 = 0.962.$$

One can observe that $P_m \leq P_{wm}(\rho) \leq P_w$ ($0.931 \leq 0.962 \leq 0.962$) and this property will be formally shown. In this arbitrary example, all the faults in a block have the same value $P(f_i)$; this is the reason why we obtain the limit case $P_{wm}(\rho) = P_w$.

When the measurement $P_{wm}(\rho)$ is used for a large circuit, the partition may be defined from various points of view. First example: consider a microprocessor with built-in test made of 4 parts: control section, data path, RAM, and test device. One can consider the partition $\rho = \{F_1, F_2, F_3, F_4\}$ where F_1 = faults in the control section, F_2 = faults in the data path, F_3 = faults in the RAM, F_4 = faults in the test device. Second example: the faults under consideration for a combinational CMOS circuit are the stuck-at faults and the stuck-open faults. One can consider $\rho = \{F_1, F_2\}$ when F_1 = stuck-at faults and F_2 = stuck-open faults. Many other ways to define a partition can be found. Some examples will be presented in Section 9.5.2.2 and Solution to Exercise 8.5.

Remark 4.4 Naturally, the measurements which are defined in a random test context can be defined, as particular cases, in the deterministic context. However, neither the minimum testing probability, P_m, nor the probability of complete fault coverage, P_c, are used in the deterministic context because they can only have two values: 0 or 1. Since both have the value 0 if and only if the fault coverage is less than 1, they are of no practical interest in the deterministic context. The measurement P_{wm}, based on P_m, suffers from the same weakness in the deterministic context; even if its value may be different from 0 if $P < 1$, it is excessively pessimistic.

4.4.3 Some properties

Section 4.4.3.1 presents the relations among the measurements which can be used to estimate the *confidence level* in a test. In Section 4.4.3.2 *relations among faults in the context of random testing* will be defined.

4.4.3.1 Relations among measurements

The **confidence level** in a test sequence will be denoted by $1 - \varepsilon$. The value of ε is called the **uncertainty level**[3]. For example, we shall often consider the values $\varepsilon = 0.001$, then $1 - \varepsilon = 0.999$. According to Section 4.4.1 and 4.4.2, the confidence level may be estimated by various measurements, namely P_u, P, P_w, P_m, P_c and $P_{wm}(\rho)$. The measurement P_u, the more pertinent, is unfortunately not available since it should take into account faults which are unknown. The relations among the five other measurements, which are defined from a set of targeted faults, are presented in Properties 4.3 and 4.5d.

---

[3] In the sequel, the *confidence level* and the *uncertainty level* are related to a *set of faults*. These notions can also be used for *a fault*.

Property 4.2 If all the faults in F have the *same occurrence probability* $\Pr[f_i \mid F] = \frac{1}{|F|}$, then the weighted fault coverage is equal to the fault coverage, i.e.,

$$P_w = P. \tag{4.28}$$

Proof By definition (4.19)

$$P_w = \sum_{i: f_i \in F} P(f_i) \cdot \Pr[f_i \mid F].$$

If $\Pr[f_i \mid F] = \frac{1}{|F|}$ for all faults f_i in F, then

$$P_w = \sum_{i: f_i \in F} P(f_i) \cdot \frac{1}{|F|} = \frac{1}{|F|} \cdot \sum_{i: f_i \in F} P(f_i) = P,$$

according to definition of P (4.18).

Property 4.3 For any circuit, for any set F of faults, and for any random test sequence (defined by the distribution ψ and the length L):
a) the *minimum testing probability* is less than or equal to both the *fault coverage* and the *weighted fault coverage*;
b) the *probability of complete fault coverage* is less than or equal to the *minimum testing probability*;
i.e.,

$$P_c \leq P_m \leq P, P_w. \tag{4.29}$$

Proof
1) $P_m \leq P_w$
By definition (4.19)

$$P_w = \sum_{i: f_i \in F} P(f_i) \cdot \Pr[f_i \mid F]. \tag{4.30}$$

By definition (4.20)

$$P_m = \min_{i: f_i \in F} P(f_i). \tag{4.31}$$

From (4.30) and (4.31)

$$P_w \geq \sum_{i: f_i \in F} P_m \cdot \Pr[f_i \mid F] = P_m \sum_{i: f_i \in F} \Pr[f_i \mid F]. \tag{4.32}$$

Since $\sum_{i: f_i \in F} \Pr[f_i \mid F] = 1$, Equation (4.32) becomes

$$P_w \geq P_m. \tag{4.33}$$

2) $P_m \leq P$

This is a particular case of $P_m \leq P_w$, since P is a particular case of P_w when all the faults have the same occurrence probability (Property 4.2).

3) $P_c \leq P_m$

Assume, without loss of generality, that $P_m = P(f_1)$. Let S be a random test sequence. From the definition of P_c (4.21), one can write:

$$P_c = \Pr[P = 1] = \Pr[S \text{ tests } f_1 \text{ AND } f_2 \text{ AND } \ldots \text{ AND } f_r]. \tag{4.34}$$

$$P_c = \Pr[S \text{ tests } f_1] \cdot \Pr[S \text{ tests } f_2 \text{ AND } \ldots \text{ AND } f_r \mid S \text{ tests } f_1]. \tag{4.35}$$

Since $\Pr[S \text{ tests } f_1] = P(f_1) = P_m$, and

$\Pr[S \text{ tests } f_2 \text{ AND } \ldots \text{ AND } f_r \mid S \text{ tests } f_1] \leq 1$,

one obtains from (4.35):

$$P_c \leq P_m. \tag{4.36}$$
□

For the example in Figure 4.4b, we have obtained $P = 0.965$ and $P_w = 0.962$ (Section 4.4.1), $P_m = 0.931$ and $P_c = 0.802$ (Section 4.4.2). These results provide an illustration of inequalities (4.29).

The Properties 4.4a and b, intuitively evident, link the measurements P_w and P_m relative to a set F to the corresponding measurements relative to the blocks of a partition of F.

Property 4.4 Let $\rho = \{F_1, \ldots, F_s\}$ be a partition on F.

a) $P_w = \sum_{j: F_j \in \rho} P_w(F_j) \cdot \Pr[F_j \mid F]. \tag{4.37}$

b) $P_m = \min_{j: F_j \in \rho} P_m(F_j). \tag{4.38}$

Proof

a) From the definitions of $P_w(F_j)$ and P_w (Equations 4.26 and 4.19, respectively), one obtains

$$\sum_{j: F_j \in \rho} P_w(F_j) \cdot \Pr[F_j \mid F] = \sum_{j: F_j \in \rho} \left(\sum_{i: f_i \in F_j} P(f_i) \cdot \Pr[f_i \mid F_j] \right) \cdot \Pr[F_j \mid F] \tag{4.39}$$

$$= \sum_{j: F_j \in \rho} \sum_{i: f_i \in F_j} P(f_i) \cdot \Pr[f_i \mid F_j] \cdot \Pr[F_j \mid F]$$

$$= \sum_{i: f_i \in F} P(f_i) \cdot \Pr[f_i \mid F] = P_w. \tag{4.40}$$

b) Obvious from the definitions (4.20) and (4.25).
□

Let us consider some properties relevant to the *weighted minimum testing probability*.

Property 4.5 For any circuit, for any set F of faults, and for any random test sequence (defined by the distribution ψ and the length L), one has the following properties[4].

a) For any partitions ρ_1 and ρ_2 on F such that $\rho_1 \geq \rho_2$,

$$P_{wm}(\rho_1) \leq P_{wm}(\rho_2). \tag{4.41}$$

b) $P_{wm}(O) = P_w.$ (4.42)

c) $P_{wm}(I) = P_m.$ (4.43)

d) For any partition ρ on F, Σ

$$P_m \leq P_{wm}(\rho) \leq P_w. \tag{4.44}$$

Proof

a) Let $\rho_1 = \{F_1, \ldots F_s\}$ and $\rho_2 = \{G_1, \ldots G_t\}$ be two partitions of F such that ρ_2 is thinner than ρ_1, i.e., $\rho_1 \geq \rho_2$. This means that for any $G_i \in \rho_2$, there is a $F_j \in \rho_1$ such that $G_i \subset F_j$. Then,

$$P_m(F_j) \leq P_m(G_i). \tag{4.45}$$

A block F_j contains one or several blocks in ρ_2, then

$$\Pr[F_j \mid F] = \sum_{i : G_i \subset F_j} \Pr[G_i \mid F]. \tag{4.46}$$

From (4.45), one can write

$$\sum_{i:G_i \subset F_j} P_m(G_i) \cdot \Pr[G_i \mid F] \geq \sum_{i:G_i \subset F_j} P_m(F_j) \cdot \Pr[G_i \mid F] = P_m(F_j) \cdot \sum_{i:G_i \subset F_j} \Pr[G_i / F] \tag{4.47}$$

From (4.47) and (4.46),

$$\sum_{i:G_i \subset F_j} P_m(G_i) \cdot \Pr[G_i \mid F] \geq P_m(F_j) \cdot \Pr[F_j \mid F]. \tag{4.48}$$

By definition (4.27),

$$P_{wm}(\rho_2) = \sum_{i:G_i \in \rho_2} P_m(G_i) \cdot \Pr[G_i \mid F]. \tag{4.49}$$

Since a block F_j contains a set of blocks G_i, Equation (4.49) may be rewritten as:

---

[4] The thinnest partition O and the identity partition I are defined in Section 4.4.2 (4.24).

$$P_{wm}(\rho_2) = \sum_{j:F_j \in \rho_1} \left(\sum_{i:G_i \subset F_j} P_m(G_i) \cdot \Pr[G_i \mid F] \right). \tag{4.50}$$

From (4.50) and (4.48)

$$P_{wm}(\rho_2) \geq \sum_{j:F_j \in \rho_1} \left(P_m(F_j) \cdot \Pr[F_j \mid F] \right). \tag{4.51}$$

The right part of (4.51) corresponds to the definition of $P_{wm}(\rho_1)$. Then

$$P_{wm}(\rho_2) \geq P_{wm}(\rho_1). \tag{4.52}$$

b) The partition O contains one fault in each block, then

$$P_{wm}(O) = \sum_{i:F_i \in F} P(f_i) \cdot \Pr[f_i \mid F] = P_w.$$

c) The partition I contains a single block corresponding to F, then

$$P_{wm}(I) = P_m(F) \cdot \Pr[F \mid F] = P_m.$$

d) Evident from Properties 4.5a, b and c, since any partition ρ is such that $O \leq \rho \leq I$.

$\square$

An example of weighted minimum testing probability was given in Section 4.4.2. Here is another one for the example in Figure 4.4b: $\rho = \{F_1, F_2\}$ with $F_1 = \{f_1, f_2, f_3, f_4, f_5\}$ and $f_2 = \{f_6\}$. One has: $P_m(F_1) = 0.931$ and $P_m(F_2) \approx 1$; $\Pr[F_1 \mid F] = 0.17/0.20 = 0.85$ and $\Pr[F_2 \mid F] = 0.03/0.20 = 0.15$. Then

$$P_{wm}(\rho) = 0.931 \times 0.85 + 1 \times 0.15 = 0.941.$$

This value is between $P_m = 0.931$ and $P_w = 0.962$ (Property 4.5d).

Remark 4.5 Comparison of the measurements
a) Given a targeted set of faults, F, the more pertinent measurement of the confidence level would be the weighted fault coverage, P_w, i.e., $1 - \varepsilon = P_w$, since P_w gives exactly the average probability that a circuit affected by a fault in F fails the test. Unfortunately, this measurement requires the knowledge of the occurrence probabilities of all the faults in F. Since these probabilities are usually not available, the (expected) *fault coverage*, P, is often used to estimate the confidence level. The value of P may be greater or lower than the value of P_w. For the example in Figure 4.4b, we have obtained $P = 0.965$ and $P_w = 0.962$ (Section 4.4.1); then $P > P_w$ in this example. However, for other values of $\Pr[f_i \mid F]$, the value of P_w could be greater than P. For example, if the conditional occurrence probabilities of f_1 and f_6 are permuted (i.e., $\Pr[f_1 \mid F] = 0.15$, $\Pr[f_6 \mid F] = 0.25$ and the other remains unchanged), one obtains $P_w = 0.969$, which is greater than P. Using P to estimate the confidence level $1 - \varepsilon$ corresponds

to an implicit assumption that all the faults in F have the same occurrence probability.

b) From Property 4.3a, the *minimum testing probability*, P_m, is a lower bound of both P_w and P. Then it provides a *conservative measurement* for the confidence level. It is very popular in the random test context since it is both conservative and much easier to obtain than P (only the most difficult fault has to be considered).

c) From Property 4.3b, the *probability of a complete fault coverage*, P_c, is a lower bound of P_m. Then it is *very conservative* (for the example in Figure 4.4b, we have $P_c = 0.802$, while $P_m = 0.931$). In addition, this measurement is very difficult to obtain since one must know the relations among the input vectors (or sequences for a sequential fault) allowing all the faults to be detected. (A calculation example is provided in Appendix B). Since this measurement is very pessimistic (more than P_m) and difficult to obtain (more than P_m), it is not a pertinent one. The intuitive idea of Property 4.3b is that: 1) *if the circuit is faulty*, then the probability of testing its fault is *at least* the *minimum testing probability*; 2) if the circuit is fault-free, there is no problem! No fault can be detected: the result 'fault-free circuit' is not **sure** as long as a complete fault coverage has not been performed (probability of being sure = probability of a complete fault coverage). However the result is **just** since this result is 'fault-free circuit' and the circuit is effectively fault-free (a result is just for a circuit if, when there is a box of 'passed circuits' and a box of 'failed circuit', according to Figure 4.1, the circuit is in the box corresponding effectively to its state.

d) The weighted minimum testing probability, $P_{wm}(\rho)$, is a measurement between P_m and P_w for both the estimation of the confidence level (Property 4.5d) and the difficulty to obtain them. If one can define a partition of faults and estimate the conditional occurrence probability of each block of faults (for example proportional to the number of devices of a part, or equal likelihood of the blocks), only $P_m(F_i)$ has to be estimated for every block F_i. An example will be presented in Section 9.5.2.2.

4.4.3.2 Relations between faults in the context of random testing

The relations between faults, *equivalence*, *dominance*, and *equidetectability*, (which are defined in Section 3.1.4.1 for combinational faults and in Section 3.2.4.1 for sequential faults), are independent of the test sequence. They can be used for both deterministic and random testing. As far as random testing is concerned, one has the following property.

Property 4.6 Let f and g be two faults (each one either combinational or sequential).

a) If fault f *dominates* fault g, then, according to the notation in Definition 4.1 (Section 4.4),

$$P(f, \psi, L) \geq P(g, \psi, L), \quad \text{for any } \psi \text{ and } L. \tag{4.53}$$

b) If the faults f and g are *equidetectable*, then

$$P(f, \psi, L) = P(g, \psi, L), \quad \text{for any } \psi \text{ and } L. \tag{4.54}$$

Proof The proof is straigthforward since:
a) from Definitions 3.2 and 3.13 (Sections 3.1.4.1 and 3.2.4.1), any test sequence detecting fault g detects also fault f;
b) from Definitions 3.3 and 3.14, any test sequence detecting fault g detects also fault f and vice-versa.

❑

In the context of *random testing*, additional relations can be used. Consider, for example, the faults f_1 and f_2 in Figure 4.4b. Fault f_1 is detected by the vector x_7 and f_2 is detected by any vector in the set $T_{f2} = \{x_1, x_3, x_5\}$. There is no relation of equivalence or dominance between these faults: given a test sequence S_1 which tests f_1, we have no information about the test of f_2 by S_1, and vice-versa. However, if a random test sequence with equal likelihood distribution, ψ_0, is applied, the probability that it tests f_1 is less than the probability that it tests f_2 because

$$|T_{f_1}| < |T_{f_2}| \Leftrightarrow P(f_1, \psi_0, L) < P(f_2, \psi_0, L), \quad \text{for any } L.$$

We will say that f_2 *R-dominates* f_1 (R stands for random) for the distribution ψ_0. Similarly,

$$|T_{f_1}| = |T_{f_4}| \Leftrightarrow P(f_1, \psi_0, L) = P(f_4, \psi_0, L), \quad \text{for any } L.$$

We will say that f_1 and f_4 are *R-equidetectable* for ψ_0.

Definition 4.2 A fault f **R-dominates** a fault g, *for the distribution ψ of random input vectors*, if

$$P(f, \psi, L) \geq P(g, \psi, L), \quad \text{for any } L. \tag{4.55}$$

❑

Definition 4.2 applies to both combinational and sequential faults. Figures 4.5a and b represent a fault-free sequential circuit and three faults f_1, f_2 and f_3. These three faults are such that only the output function is faulty.

Let $\psi_1 = (.7, .3)$, i.e., $\Pr[x_0] = 0.7$ and $\Pr[x_1] = 0.3$. It is clear, from the observers in Figures 4.5c and d, that f_1 R-dominates f_2 for the distribution ψ_1. As a matter of fact, both faults can be detected after q_0 has been reached in the fault-free circuit (hence q'_0 and q''_0 in the faulty ones); fault f_1 is detected if x_0 is applied (probability 0.7) and fault f_2 is detected if x_1 is applied (probability 0.3).

Definition 4.3 Two faults f and g are **R-equidetectable** *for the distribution ψ* of random input vectors, if

$$P(f, \psi, L) = P(g, \psi, L), \quad \text{for any } L, \tag{4.56}$$

i.e., f R-dominates g and g R-dominates f for ψ. ◻

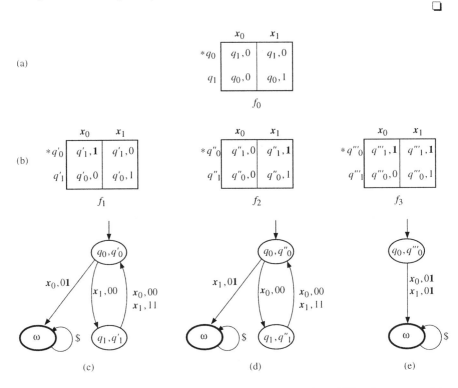

Figure 4.5 Illustration of R-dominance and R-equitestability for sequential faults. (a) A fault-free table. (b) Three faults. (c) Observer for f_1. (d) Observer for f_2. (e) Observer for f_3.

For the distribution $\psi_0 = (.5, .5)$, the faults f_1 and f_2 in Figure 4.5 are R-equidetectable.

The properties of R-dominance and R-equitestability are clearly function of the distribution ψ. Consider the faults f_1 and f_2 in Figure 4.4. Fault f_2 R-dominates f_1 for ψ_0. However, f_1 R-dominates f_2 for any distribution ψ such that $\Pr[x_7] \geq \Pr[x_1] + \Pr[x_2] + \Pr[x_3]$.

Now, there are some relations between the deterministic notions of *dominance* (see Definitions 3.2 and 3.13) and *equitestability* (see Definitions 3.3 and 3.14) and the notions of *R-dominance* and *R-equitestability*.

Property 4.7

a) *Fault f R-dominates fault g for any distribution* ψ if (and only if for combinational faults) *f dominates g*.

b) *Faults f and g are R-equidetectable for any distribution* ψ if (and only if for combinational faults) *f and g are equidetectable*.

Proof

a) *Sufficient condition for any faults.*

If f dominates g, any sequence testing g also tests f. Since this property is true for any input sequence, this is *a fortiori* true for any distribution.

Necessary condition for combinational faults.

If f R-dominates g for any distribution, then there is no input vector testing g which does not test f: as a matter of fact, if such a vector x_i exists, the fault f would not R-dominate g for the distribution such that $\Pr[x_i] = 1$. Hence $T_f \supset T_g$ which means that f dominates g.

b) The proof is similar since R-equitestability and equitestability correspond to reciprocal R-dominance and dominance, respectively.

☐

Fault f_3 in Figure 4.5 dominates both faults f_1 and f_2 in the same figure; f_3 R-dominates f_1 and f_2 for any distribution of the input vectors probabilities. This is an illustration of Property 4.7a.

Here is an abstract example showing that the sufficient condition in Property 4.7b is not a necessary condition for sequential faults. Let a sequential fault f be detected by the only input sequence $x_0 x_1$ and g be detected by the only input sequence $x_1 x_0$. These faults are not equidetectable according to Definition 3.14. However, they are R-equidetectable for any constant distribution ψ.

Remark 4.6 If two *faults* are *comparable* from a deterministic point of view, they are obviously comparable from a random testing point of view. In addition, even if two faults cannot be compared from a deterministic point of view, they can be compared from a random testing point of view since *one can always compare two probabilities*. That means that $P(f, \psi, L)$ and $P(g, \psi, L)$ can always be compared, given ψ and L.

NOTES and REFERENCES

The notion of *defect level* was introduced in [Wa 78] under the name *field reject rate*, then was made popular from [WiBr 81]. An extension to other faults than the stuck-at faults was considered in [KrGa 93].

The expression *universal fault model* is used in [AbBrFr 90] in order to define the set of all the faults that do not increase the number of states of the circuit (i.e., for a combinational circuit, it corresponds to all the combinational faults).

In the present book the *universal fault set* is larger since it includes all the permanent logic faults.

There is no "true" relation between fault coverage and defect level: there are several possible approximations depending on the hypothesis. This is the reason why various references are given in Section 4.3, while we usually avoid the references in the text. It is clear that the defect level is a function of both the quality of production (measured by the yield) and the quality of test (various measurements such as fault coverage); in the rest of the book only the quality of test is considered.

From the experimental results presented in [MaAi 93], it appears that the predictions are better for functional tests than for scan tests, which, according to the authors "produce significantly worse quality levels than predicted".

The *expected fault coverage* (or *fault coverage*) has been considered in many papers such as [AgSeAg 82], [MaYa 84], [WaChMc 87]. The *weighted fault coverage* was used with various names in [Sh 77], [SeAg 89], and [JaDa 89b]. In [CrJaDa 94], it is empirically shown (from random conditional occurrence probabilities) that the fault coverage is usually a good approximation of the weighted fault coverage for large circuits.

The *minimum testing probability* is the most popular measurement when random testing is concerned; it has been used for a long time [Ra 71] by many authors. The *probability of complete fault coverage* was implicitly used in [TeDa 74] when the *probability to test each fault* is considered. It is also used in [SaBa 84], where the *escape probability* is equal to 1 - P_c. The same notion is addressed in [WaChMc 87] under the name of *100% fault coverage*. In [DaBl 76] it was shown that the minimum testing probability was a more pertinent (less pessimistic) measurement than the probability of complete fault coverage (Property 4.3b in Section 4.4.3.1); in this paper, the minimum testing probability was called *detection quality* and the probability of complete fault coverage was called *testing quality*. The *weighted minimum testing probability* was introduced in [JaDa 89a & b] (it is based on the theory of partitions [HaSt 66]).

The fault relations in Section 4.4.3.2 are new to the best of our knowledge.

We have considered (Section 4.4) that a random test sequence could be obtained from drawings with a generalized distribution ψ. However, up to now, all the examples were constant distributions. Some authors have considered that one could generate several parts of the test sequence with several constant distributions [GrSt 93]. For example, the first part is obtained from a constant distribution ψ_1 such that the inputs have a high probability of being 0, and the second part is obtained from a constant distribution ψ_2 such that the inputs have a high probability of being 1; this random generation (whose aim is to shorten the test length) can be modeled by a generalized distribution (Figure A.1b in Appendix A, where the first three test vectors correpond to a 0.9 probability of being 0 for every input, and the other test vectors correpond to a 0.9 probability of being 1).

5

Basic Principles of Random Testing

The main principle of random testing consists in applying a random test sequence S to the circuit under test (CUT) and in verifying that the response R(S) is the expected one. Various implementations of this verification will be presented in Section 5.1.

The test sequence which is applied to the CUT is random, but it is not "anything": it must verify that the circuit behaves as it should be according to the specified behavior. All the specified cases should be tested. However, on the other hand, only the specified cases should be tested. This is explained and illustrated in Section 5.2.

An interesting property of random testing is that the probability of detecting a fault is an increasing function of the test length, tending to 1 for all the faults which are non-evasive. The conditions leading to this property, and some other theoretical results are presented in Section 5.3.

For a combinational fault, the probability that the jth vector of the input sequence produces a faulty output is independent from detection by the (j-1)th vector. This is not true for a sequential fault because there is some "memory effect". This is explained in Section 5.4.

5.1 PRINCIPLE OF RANDOM TESTING

Several possible implementations of random testing schemes are presented in Figure 5.1. The generic case is presented in Figure 5.1a. The test device applies a random input sequence, S, to the CUT, and the response of the circuit, $R(S)$, is observed in order to decide if the circuit passes the test or fails. But, how is the test device implemented ?

The most classical scheme is presented in Figure 5.1b. It corresponds exactly to the observer illustrated in Figure 3.16 (Section 3.2.2.1), in the case of random

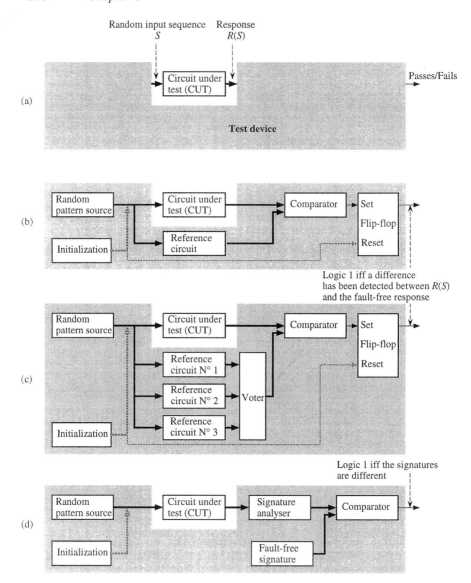

Figure 5.1 Various implementations of random testing. (a) Generic case. (b) Comparison with a fault-free circuit. (c) Comparison with several reference circuits. (d) Comparison of signatures.

test sequence. The test sequence is applied to both the CUT and a reference (fault-free) circuit. After initialization of both circuits (if they are sequential), a test sequence is applied from a *random pattern source*. This source may be either a *random pattern generator* producing the random sequence in real-time, or a *recorded random sequence* (which has been generated off line).

In order to avoid a bad decision (pass or fail) due to a failure of the reference circuit, several reference circuits can be used and the expected output of these reference circuits is decided by a voter. This is illustrated in Figure 5.1c.

The scheme in Figure 5.1d does not require a reference circuit. The response of the CUT, which may contain thousands or millions of bits, is compacted into a signature containing a few tens of bits. This signature is compared with the signature of a fault-free circuit previously recorded.

In order to compare the signatures, the random pattern source must always produce the same test sequence; this holds either if the sequence is recorded or if it is, in fact, a pseudorandom sequence, i.e., generated from a deterministic algorithm (see Chapter 10). The fault-free signature can be obtained either from a fault-free circuit, if it exists, or by simulation. This scheme is well adapted to built-in self-test: the test device may be integrated on the same chip as the circuit under test (see Chapter 1).

For all the implementations in Figure 5.1, the basic theoretical problem is the same: *what is the probability that R(S) is faulty* when a fault f_i affects the CUT, i.e., what is the value $P(f_i, \psi, L)$, given the fault f_i, the distribution ψ, and the length L? In Figures 5.1b and c, the CUT fails *if and only if* $R(S)$ is faulty. In Figure 5.1d, the CUT fails *only if* $R(S)$ is faulty; since the observation of $R(S)$ is not perfect (some information may be lost by compaction), the signatures could be similar for a fault-free and a faulty response $R(S)$. This is a problem due to signature analysis (see Chapter 12). The problem of random testing itself is the *study of the probability that R(S) is faulty, independently from the observation of this response*. From now on, we shall implicitly argue about implementation in Figure 5.1a, unless otherwise specified.

5.2 ALL THE SPECIFIED CASES AND ONLY THE SPECIFIED CASES SHOULD BE TESTED

A random generator for an m-input circuit is a system which, at times 1, 2, 3, ..., randomly draws an input vector from the set of 2^m possible vectors. At some time t, some input vectors may have a zero probability, depending on the vectors previously drawn (generalized distribution, Appendix A) or not (constant distribution). The distribution ψ (generalized or not) implies a set of input sequences which have a non-zero probability to be the test sequence randomly drawn.

Let I be the input language of the CUT (Section 2.1.3.1). An *ideal random test* must test all the cases for which the outputs are specified, and must not attempt to test any other case for which the output is not known. This is made precise in the following definition, assuming that the circuit (if sequential) has been previously initialized; otherwise a fault-free circuit could be classified as faulty since two fault-free circuits can produce two different responses if their initial states are not the same (Section 3.2).

After the definition of an ideal random test, examples related to a combinational circuit (Section 5.2.1), an asynchronous sequential circuit (Section 5.2.2), a synchronous sequential circuit (Section 5.2.3), a circuit partly synchronous and partly asynchronous (Section 5.2.4), are presented before a conclusion on specifications.

Definition 5.1 An **ideal random test** is such that for any test length L:

1) every sequence of length L belonging to the input language I has a non-zero probability of being a randomly drawn test sequence (for "*all the specified cases*");

2) any sequence which is not in the input language I has a zero probability of being a randomly drawn test sequence (for "*only the specified cases*").

❏

Let us note that this notion of *ideal random test* is qualitative but not quantitative.

5.2.1 Example of combinational circuit

Consider the 4-input, 4-output combinational circuit translating a binary code of a decimal digit into a reflected binary code, i.e., Gray code. The corresponding truth table and Karnaugh maps are presented in Figure 5.2a and b, respectively. It clearly appears that the output functions are incompletely specified since a decimal digit can have only 10 different values, while a 4 bit coding may have 16 different values.

In order to test *only the specified cases*, only the inputs vectors x_0 to x_9 (where the index i of x_i is the corresponding decimal value) must be applied. As a matter of fact, assume that the input vector x_{10} is applied during the test: there is no output specified for this input, then the circuit under test and the reference circuit could produce different outputs while both are fault-free. For example the output z_2 can be implemented as $z_2 = x_1 + x_2$ (Figures 5.2b and c) or as $z_2 = x_1 x'_3 + x_2$ (Figure 5.2d). The first one produces $z_2 = 1$ and the second one $z_2 = 0$ for the input pattern x_{10}, but both are correct.

Let us now restrict our attention to the 2-input, 1-output subcircuit implementing the function $z_2 = x_1 + x_2$. Figure 5.3a presents this function in CMOS technology (the complementation of x_1 and x_2 is not represented). The input vectors $x_0, x_1, x_2,$ and x_3 have the same effect on z_2, since this output does not depend on the inputs x_3 and x_4. Let us change the input alphabet: the new input alphabet is $\{y_0, y_1, y_2, y_3\}$, where $y_0, y_1, y_2,$ and y_3 correspond to $x_1 x_2 = 00, 01, 10,$ and 11, respectively. Then, y_0 corresponds to $x_0 + x_1 + x_2 + x_3$,

y_1 corresponds to $x_4+x_5+x_6+x_7$, y_2 corresponds to $x_8+x_9+x_{10}+x_{11}$, and y_3 corresponds to $x_{12}+x_{13}+x_{14}+x_{15}$.

If *only* the specified cases are tested, then the input y_3 is never applied since the patterns x_{10} to x_{15} are never applied. Then all the test sequences which can be

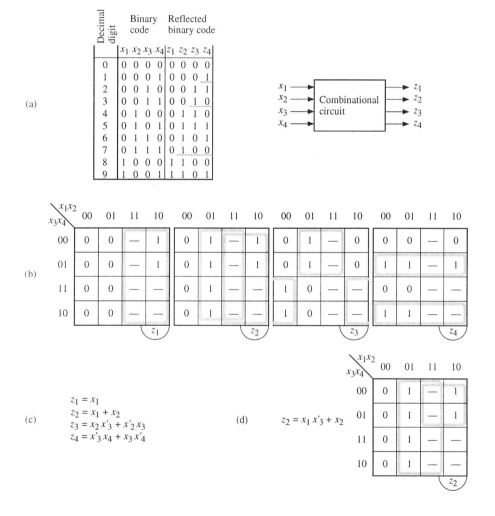

Figure 5.2 (a) Truth table. (b) Karnaugh-maps. (c) A possible set of equations. (d) Another equation for z_2.

applied are in the input language $I = (y_0 + y_1 + y_2)^*$. The distribution[1] represented in Figure 5.3b allows the production of any input string in this language (see Appendix A about generalized distributions). In this Figure 5.3b, the generator has a single state r_0, hence, the distribution ψ is constant, i.e.: in Figure 5.3a, input vector y_0 has a 0.4 probability to be applied; input vector y_1 has a 0.4 probability to be applied; input vector y_2 has a 0.2 probability to be applied; input vector y_3 has a zero probability to be applied.

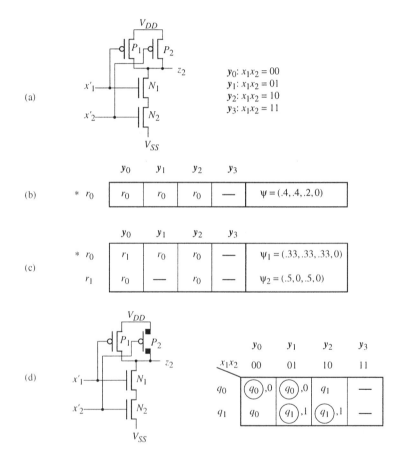

Figure 5.3 (a) Subcircuit implementing $z_2 = x_1 + x_2$. (b) A generalized distribution achieving an ideal random test. (c) A generalized distribution not testing all the specified cases. (d) Faulty circuit and the corresponding behavior.

---

[1] This distribution corresponds to the equal likelihood distribution of the input vectors x_0 to x_9.

Basic Principles of Random Testing

In normal operation of the code translator in Figure 5.2a, the input vectors can be applied in any order. Then for the subcircuit in Figure 5.3a, the patterns y_0, y_1, and y_2 can be applied in any order. Testing *all* the specified cases means that any test sequence in $I = (y_0 + y_1 + y_2)^*$ can be applied. Hence, according to Definition 5.1, the distribution represented in Figure 5.3b achieves an ideal random test.

On the other hand, the generalized distribution[2] represented in Figure 5.3c does not allow testing of *all* the specified cases since the input vector y_1 is *never applied after* the input vector y_0. As a matter of fact, in Figure 5.3c, the probability of chosing a vector depends on the past history of the test; this is modeled by a two-state sequential machine (states r_0 and r_1), and when state r_1 is reached, the input vector y_1 has a zero probability (constant distribution ψ_2). According to the state table of Figure 5.3c, the initial state is r_0; if y_0 is applied from this state, the state r_1 is reached and the next random drawing is made from the distribution ψ_2, where the probability of y_1 is zero.

Assume the transistor P_2 in Figure 5.3a is stuck open; the behavior of the faulty circuit is illustrated[3] in Figure 5.3d. This fault can be detected if and only if the subsequence $y_0 y_1$ occurs in the test sequence. Then it is not detected by any test sequence obtained from the distribution in Figure 5.3c.

5.2.2 Example of asynchronous sequential circuit

An asynchronous circuit working in *fundamental mode* (Section 2.1.2.2) is such that only one primary input changes at once. This classical assumption is illustrated in Figure 5.4a. This figure is a *primitive flow table* (a **primitive flow table** is such that there is exactly one stable total state in each row.). According to the classical methods [Ko 78], this machine, M_a, could be merged in a 4-line Moore machine, but, without merging, it clearly appears that all the multiple changes of input variable are unspecified because they do not occur.

Consider, for example, the state q_0 in Figure 5.4a which is stable for the input vector x_0. The next input change will correspond to a change of one of the input variables, i.e., a change of either x_1, or x_2, or x_3. If x_1 changes, then the next input state is x_4; if x_2 changes, the next input state is x_2; if x_3 changes, the next input state is x_1. Then, in normal operation, no input pattern in $\{x_3, x_5, x_6, x_7\}$ can be applied, just after the input pattern x_0, since these vectors are not **adjacent** to x_0, i.e., they differ from x_0 by more than one bit.

---

[2] A generalized distribution is such that the probabilities of the input vectors at time t depend on the vectors randomly drawn before t (Appendix A). The example is arbitrary; however, if we are not careful, we could build a generator missing generation of some subsequences (Section 10.3.2.1) as in this example.

[3] Caution: the order of the input vectors in the state table is (y_0, y_1, y_2, y_3) in order to be the same as in the representation of generalized distributions. This order is used throughout Section 5.2 while this order is usually (y_0, y_1, y_3, y_2).

114   Chapter 5

(a)

x_1	x_0	x_1	x_2	x_3	x_4	x_5	x_6	x_7	
	0				1				
x_2x_3	00	01	10	11	00	01	10	11	z
*	ⓠ₀	q_1	q_2	—	q_4	—	—	—	0
	q_0	ⓠ₁	—	q_3	—	q_5	—	—	1
	q_0	—	ⓠ₂	q_8	—	—	q_6	—	0
	—	q_9	q_2	ⓠ₃	—	—	—	q_7	1
	q_0	—	—	—	ⓠ₄	q_5	q_6	—	1
	—	q_1	—	—	q_4	ⓠ₅	—	q_7	0
	—	—	q_2	—	q_4	—	ⓠ₆	q_7	0
	—	—	—	q_8	—	q_5	q_6	ⓠ₇	0
	—	q_1	q_2	ⓠ₈	—	—	—	q_7	0
	q_0	ⓠ₉	—	q_3	—	q_5	—	—	0

(b)

	x_0	x_1	x_2	x_3	x_4	x_5	x_6	x_7	ψ^{r_i}
* r_0	—	r_1	r_2	—	r_4	—	—	—	$(0,.33,.33,0,.33,0,0,0)$
r_1	r_0	—	—	r_3	—	r_5	—	—	$(.33,0,0,.33,0,.33,0,0)$
r_2	r_0	—	—	r_3	—	—	r_6	—	$(.33,0,0,.33,0,0,.33,0)$
r_3	—	r_1	r_2	—	—	—	—	r_7	$(0,.33,.33,0,0,0,0,.33)$
r_4	r_0	—	—	—	—	r_5	r_6	—	$(.33,0,0,0,0,.33,.33,0)$
r_5	—	r_1	—	—	r_4	—	—	r_7	$(0,.33,0,0,.33,0,0,.33)$
r_6	—	—	r_2	—	r_4	—	—	r_7	$(0,0,.33,0,.33,0,0,.33)$
r_7	—	—	—	r_3	—	r_5	r_6	—	$(0,0,0,.33,0,.33,.33,0)$

Figure 5.4 (a) Asynchronous Moore Machine, M_a. (b) A generalized distribution, ψ_a, allowing testing of *all* the specified cases and *only* the specified cases of M_a (ψ^{r_i} is the constant distribution associated with the state r_i of the generator).

In order to test *only the specified cases*, the random drawing must take into account the normal operation of the circuit. The distribution machine[4] ψ_a, presented in Figure 5.4b, ensures that all the specified cases and only the specified cases of M_a are tested. Assume that the machines M_a and ψ_a are in states q_0 and r_0, respectively. According to Figure 5.4b, the next input pattern is randomly drawn among the set $\{x_1, x_2, x_4\}$ corresponding to the non-zero values of ψ^{r_0} (ψ^{r_0} denotes the constant distribution associated with the state r_0 of the machine ψ_a). Assume that x_1 is drawn. Then M_a and ψ_a reach the states q_1 and r_1, respectively. According to Figure 5.4b, the next input pattern is randomly drawn in the set $\{x_0, x_3, x_5\}$ corresponding to the non-zero values of ψ^{r_1}. These input vectors x_0, x_3, and x_5, correspond to the transitions which are specified from q_1 in M_a, and so on. One can observe that the number of states of ψ_a is the number of input states of M_a, i.e., 8. For example, if M_a is either in the state q_1 or in the state q_9 (input state x_1), ψ_a is in the state r_1; as a matter of fact, the input vectors for which transitions are specified are the same for both q_1 and q_9 in Figure 5.4a. The same comment applies to q_3 and q_8.

5.2.3 Example of synchronous sequential circuit

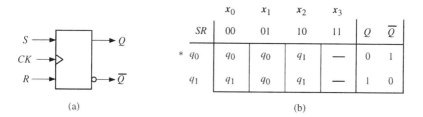

Figure 5.5 (a) Synchronous RS flip-flop. (b) Corresponding state table.

In many synchronous machines, the next state function is completely specified. In this case any input sequence can be applied. However, some synchronous machines can be incompletely specified, for example the RS flip-flop presented in Figure 5.5. The behavior of such a flip-flop is not defined for the input state $x_1 x_2 = 11$ (Section 2.1.3.1). Then all the specified cases and only the specified cases can be tested if the input sequence is randomly drawn in the language $\{x_0 + x_1 + x_2\}^*$. This can be performed, for example, by the constant distribution $\psi = (.33, .33, .33, 0)$.

---

[4] For a discussion on the implementation of this distribution machine, see Section 10.3.3.

5.2.4 Example of partly synchronous and partly asynchronous circuit

The circuit in Figure 5.6 behaves as a synchronous D flip-flop (Section 2.2.1) as long as $S = R = 0$. The input S and R are such that at most one of them must have the value 1. As soon as $S = 1$, the outputs Q and $\overline{Q}$ assume the values 1 and 0, respectively, *independent of the clock CK* and the value of D. Similarly, as soon as $R = 1$, the outputs Q and $\overline{Q}$ assume the values 0 and 1, respectively.

Then, D is a synchronous input while R and S are asynchronous ones. There is no way to represent the complete behavior of this circuit, except the representation of a 4-input asynchronous circuit taking the clock CK into account as an "ordinary" input. A way to test all the specified cases of this circuit is to test it as an asynchronous circuit. Naturally such a test is difficult to implement while the flip-flop is basically synchronous. (This topic will be adressed in Section 7.1 and Appendix F). However, some faults may remain undetected if not all the specified cases are tested. Assume the following test sequence made of two phases. During the first phase the clock remains at 0 and the S and R inputs are tested. Then, during the second phase, S and R remain at 0 and the synchronous behavior is tested from variations of CK and D.

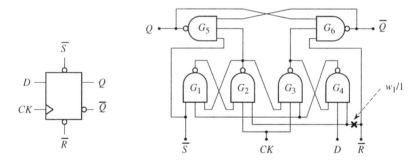

Figure 5.6 D synchronous flip-flop with asynchronous S and R inputs[5].

This is a functional test since it is designed without precise hypotheses about the faults which can affect the circuit. This test sequence may appear to be satisfactory. However, it does not detect the fault $w_1/1$ which is shown in Figure 5.6 (this fault represents a constant 1 on the rightmost inputs to gates G_2 and G_4). This fault can produce a faulty behavior only when $\overline{R} = 0$. It can be verified that the error is propagated only when the clock CK has the value 1.

---

[5] Figure 5.6b is reprinted from [HuDa 92] (and the example is taken from this paper), © (1992) page 201, with kind permission from Elsevier Sc. Ltd, The Boulevard, Langford Lane, Kidlington 0X5 1GB, UK.

5.2.5 Conclusion on specification

It is obvious that *all the specified cases* must be tested since a faulty behavior could appear in normal operation if this behavior has not been tested. A more specific result will be shown in the Section 5.3.2 (Property 5.1).

As previously illustrated by various examples, if one does not test *only the specified cases*, a fault-free circuit may be classed as faulty. Let us introduce the notion of extended specification: an **extended specification** *transforms an incompletely specified circuit into a completely specified one.*

Consider the combinational circuit in Figure 5.2. Its behavior is specified for only 10 input patterns. If this circuit is implemented according to equations in Figure 5.2c, this implementation implies a complete specification since for every input vector the output vector is now definite: $z(x_{10})$ = 1111, $z(x_{11})$ = 1110, $z(x_{12})$ = 1110, $z(x_{13})$ = 1111, $z(x_{14})$ = 1101, $z(x_{15})$ = 1100. This corresponds to an extended specification. If all the circuits in a batch are implemented in the same way, they will have exactly the same behavior for any input vector in $\{x_0, x_1, ..., x_{15}\}$ if they are fault-free.

The synchronous sequential circuit in Figure 5.5 is incompletely specified. If $\delta(q_0, x_3)$ and $\delta(q_1, x_3)$ are determined (possibly from an implementation), then we have an extended specification. While the **basic specification** corresponds to the input language $I = \{x_0 + x_1 + x_2\}^*$, the *extended specification* corresponds to the input language $X^* = \{x_0 + x_1 + x_2 + x_3\}^*$.

The 3-state automaton in Figure 3.14a (Section 3.2.1.2) seems to be "completely specified". However, its implementation requires a 4th state. Then, an extended specification of such an automaton is a 4-state automaton including it. It may be, for example, the 4-state automaton in Figure 3.14a. In a general case, an extended specification of a s-state machine is a completely specified[6] s'-state machine such that $s' = 2^{\lceil \log_2 s \rceil}$. This is important if initialization is performed by a synchronizing sequence.

For an extended specification, the input language is $I = X^*$, in which case the random pattern generator is easier to implement (see Chapter 13). Nevertheless, some probability remains that a circuit is faulty according to the extended specification, while it is fault-free according to the basic specification.

5.3 ABOUT DETECTION POWER OF RANDOM TESTING

When *deterministic testing* is used, a set of faults (structural or functional models) under consideration is *necessary* in order to build a test sequence able to

---

[6] The transition function of an asynchronous sequential automaton working in fundamental mode is completely specified if all the single changes of input variables are specified.

test these faults. In some cases, particularly when multiple faults are considered, it might not be possible to obtain a test sequence with a 100% fault coverage. In addition, in every case, *non-target* faults remain which are not necessarily tested even if a 100% fault coverage of targeted faults has been obtained.

When *random testing* is concerned, a test sequence can be generated independently of any fault model. A distribution ψ (constant or generalized) is the only data necessary to generate a test sequence of any length. Naturally the calculation of expected fault coverage requires specification of a set of faults F, since, by definition, the fault coverage is relative to a set of faults. In any case for any fault, targeted or not, the probability of detecting it is a non-decreasing function of the test length, L.

These features will be developed in this section. The general case, i.e., sequential circuits, is considered. It includes combinational circuits as a particular case.

Section 5.3.1 shows how a Markov chain can be associated with an observer. Some theoretical limits of random testing and deterministic testing are presented in Sections 5.3.2 and 5.3.3. Section 5.3.4 concludes on detection power.

5.3.1 Markov chains associated with the observer

Consider a fault f and a distribution ψ. We will show that the detection process can be modeled by a finite (homogeneous) Markov chain (the basic concepts related to these chains are presented in Appendix C). We shall consider first the usual case where the faulty machine has a single possible initial state and the distribution is constant. Then we shall consider the case where the initial state of the faulty machine is not unique. Finally the case of generalized distribution will be studied.

5.3.1.1 *The initial state of the faulty machine is known*

The Markov chain is constructed using the observer $\Omega = (B, X, \gamma, b_0, \{\omega\})$ (Definition 3.9 in Section 3.2.2.1) and the constant distribution ψ. The state set of the Markov chain is the set B of states of the observer. The entries $U_{b,b'}$ of the transition matrix U of the Markov chain are defined as

$$U_{b,b'} = \Pr[b' \text{ at time } l \mid b \text{ at time } l\text{-}1] = \sum_{x:\gamma(b,x)=b'} \psi[x]. \quad (5.1)$$

The observer in Figure 3.15 (Section 3.2.2.1), corresponding to the fault-free machine $M = (A, q_0)$ and the faulty one $M' = (A', q'_0)$ in the same figure, is represented again in Figure 5.7a, without the outputs. The corresponding finite (homogeneous) Markov chain is illustrated in Figure 5.7b for the constant distribution $\psi = (.4, .6)$. For example, the transition from the state (q_0, q'_0) to the state (q_1, q'_1) is achieved if the input vector x_1 is applied : then Pr $[(q_1, q'_1)$ at time $l \mid (q_0, q'_0)$ at time $l-1] = \psi(x_1) = 0.6$. From the Markov chain in Figure 5.7b, one obtains the transition matrix

$$U = \begin{bmatrix} 0.4 & 0.6 & 0 & 0 \\ 0.6 & 0 & 0.4 & 0 \\ 0 & 0 & 0 & 1 \\ 0 & 0 & 0 & 1 \end{bmatrix}.$$

Here the indices 1, ..., 4 of the rows and columns of U correspond to, and are identified with, the states of the observer as shown in Figure 5.7b.

The entry $U_{b,b'}$ is the probability of reaching the state b' when the next input is applied, given the present state is b. By construction, U is a stochastic matrix, i.e., $\sum_{b' \in B} U_{b,b'} = 1$, for any state b.

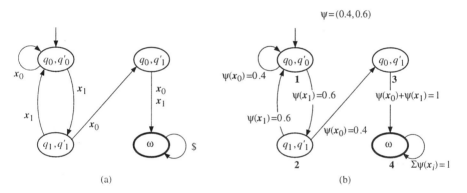

Figure 5.7 (a) Observer for the pair $((A, q_0), (A', q'_0))$. (b) Corresponding Markov chain for the distribution $\psi = (.4, .6)$.

Let $\pi(l)$ be a stochastic row vector denoting the *state probabilities*, i.e., the **probability vector**, at time l. Given $\pi(0)$, one has, for any positive integer l:

$$\pi(l) = \pi(l-1) \cdot U. \qquad (5.2)$$

Then, by iteration, for any non-negative integer l:

$$\pi(l) = \pi(0) \cdot U^l. \qquad (5.3)$$

The vector $\pi(0)$ corresponds to the initial state of the observer. For the example in Figure 5.7, one has

$$\pi(0) = (1 \ 0 \ 0 \ 0).$$

From this initial probability vector and the matrix U, one obtains[7] through Equation (5.3): $\pi(1) = (.4, .6, 0, 0)$, $\pi(10) = (.105, .076, .037, .782)$, $\pi(20) = (.017, .012, .006, .965)$.

Let $\pi_b(l)$ denote the probability of state b at time l. The probability $\pi_\omega(l)$ is the probability of the fault having been detected by time l. It is the last component of $\pi(l)$ since the index 4 corresponds to the state ω in Figure 5.7. The Markov chain is an **absorbing chain** with a single **absorbing state**, ω. This means that for the graph modeling the Markov chain: 1) from any other state b, there is an oriented path from b to ω, and 2) there is no arc leading from ω to another state.

From the above results, $\pi_\omega(1) = 0$, $\pi_\omega(10) = 0.782$, $\pi_\omega(20) = 0.965$. The problem to be addressed is usually as follows. Given a threshold ε, $0 < \varepsilon < 1$, what is the minimal length L of a random test such that the probability of testing the fault f, $P(f, \psi, L) = \pi_\omega(L)$, exceeds $1 - \varepsilon$? For the considered example, and for $\varepsilon = 10^{-3}$, one obtains $L = 40$, since $\pi_\omega(39) = 0.9989$ and $\pi_\omega(40) = 0.9991$.

Remark 5.1 For a *combinational fault* f, the observer (thus the Markov chain) is reduced to *two states*: the state (q_0, q'_0) and the absorbing state ω. The arc from (q_0, q'_0) to ω is labelled by the set T_f of input vectors detecting f; a self loop associated with (q_0, q'_0) contains all the input patterns which are not in T_f.

5.3.1.2 The initial state of the faulty machine is unknown

As it was pointed out in Section 3.2, the faulty circuit may not be in a known initial state after the initialization sequence has been applied. The faulty machine is denoted by (A', Q'_0), where Q'_0 is the set of possible initial states.

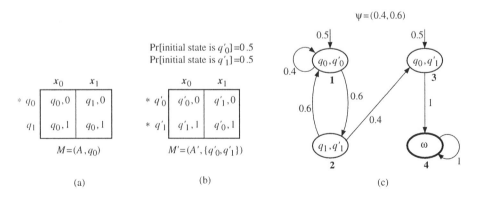

Figure 5.8 (a) Fault-free machine. (b) Faulty machine with two possible initial states. (c) Markov chain.

---

[7] In the text, we use commas to separate the components of a vector.

Figure 5.8 illustrates a case where $Q'_0 = \{q'_0, q'_1\}$: the input sequence $S_s = x_0$ is a synchronizing sequence for the automaton A (Figure 5.8a) but not for the automaton A' (Figure 5.8b). If we assume that each initial state of the faulty circuit has the same probability, i.e.,

Pr[initial state is q'_0] = Pr[initial state is q'_1] = 0.5,

one obtains the Markov chain in Figure 5.8c where the probability 0.5 is associated with each possible initial state of the OR-observer (Remark 3.8b, Section 3.2.2.2). The corresponding *initial probability vector* is

$\pi(0) = [.5\ 0\ .5\ 0]$.

In general, consider the fault-free automaton (A, q_0) and the fault f represented by the automaton (A', Q'_0). The number of possible initial states (q_0, q'_i) of the observer, where $q'_i \in Q'_0$, is $|Q'_0|$. Then the number of non-zero components in $\pi(0)$ is equal to $|Q'_0|$, and the values of these components, summing to 1, depend in general on the initialization sequence. From the Markov chain, and for a random test of length l, one obtains through Equation (5.3):

$$\pi_\omega(l) = \sum_{i: q'_i \in Q'_0} \Pr[f \text{ detected by } l \text{ vect.} | \text{ initial st. } q'_i] \times \Pr[\text{initial st. } q'_i],$$

which corresponds to the *average probability of testing* $P(f, \psi, l)$, weighted by the probability of the possible initial states.

Remark 5.2 If a Markov chain was built from an AND-observer (Remark 3.8b, example in Figure 3.17e; Section 3.2.2.2), the result would be very pessimistic. This kind of observer should be used if the probability of complete fault coverage (Section 4.4.2) was researched. But this measurement has proved to be too pessimistic (Section 4.4.3).

5.3.1.3 *Generalized distribution*

Figure 5.9a returns to the observer in Figure 5.7a where the non-absorbing states have been rechristened b_0, b_1, and b_2. Figure 5.9b represents a generalized distribution obtained from a Markov source (Appendix A). The corresponding Markov chain is presented in Figure 5.9c.

Since the Markov source of random patterns contains several states, a state of the Markov chain of the detection process is defined by a pair (state of observer, state of source). At the initial time, the observer is in the state b_0 and the source in the state r_0. Hence the initial state of the Markov chain is (b_0, r_0). A random drawing is performed with $\psi^{r_0} = (.4, .6)$. If the input pattern x_0 is drawn (probability 0.4), the observer and the source reach the states b_0 and r_1, respectively; then the state (b_0, r_1) of the Markov chain is reached with the probability 0.4. If the input pattern x_1 was drawn in the initial state (probability

0.6), the observer and the source would reach the states b_1 and r_0, respectively; then the state (b_1, r_0) of the Markov chain would be reached with the probability 0.6, and so on.

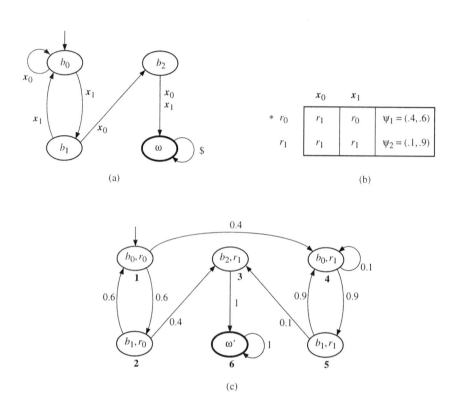

Figure 5.9 (a) Observer. (b) Generalized distribution. (c) Markov chain.

In the general case, let an observer be $\Omega = (B, X, \gamma, b_0, \omega)$ and a Markov source $\psi = (R, X, \Psi, \sigma, \mu, r_0)$. The Markov chain corresponding to the detection process is defined as follows.

The set of states of the Markov chain is $D \cup \{\omega'\}$, where $D = (B - \{\omega\}) \times R$ (some of these states may be not reachable). It is an absorbing chain, where ω' is the unique absorbing state corresponding to the set of states $\{\omega\} \times R$. Let $d = (b, r)$ and $d' = (b', r')$ be any two states in D. The entries of the transition matrix U are defined as

$$\begin{cases} U_{d,d'} = \sum_{\substack{x:\gamma(b,x)=b' \\ \text{and } \sigma(r,x)=r'}} \psi^r(x) \\ U_{d,\omega'} = \sum_{x:\gamma(b,x)=\omega} \psi^r(x) \\ U_{\omega',\omega'} = 1 \end{cases}$$

The initial state is (b_0, r_0). Naturally, as in Section 5.3.1.2, if the initial state of the faulty machine is not unique, the Markov chain may have several initial states, a fact which is taken into account in the initial probability vector $\pi(0)$.

Remark 5.3 For a *generalized distribution*, the Markov chain contains more states than the observer (see Figures 5.9a and c). Then, the Markov chain associated with a *combinational fault* and a generalized distribution contains more than two states; as a matter of fact, $|D| = |R|$ in this case.

5.3.2 Limits of random testing

Consider again the fault f and the distribution $\psi = (.4, .6)$ corresponding to the Figure 5.7. Figure 5.10 presents the probability of testing f as a function of l. It appears that this probability tends towards 1 when l tends towards infinity. This corresponds to a classical result about absorbing chains: if ω is the only *absorbing state*, then

$$\lim_{l \to \infty} \pi_\omega(l) = 1. \tag{5.4}$$

The fact that the Markov chain associated with a fault is an absorbing chain with a single absorbing state is practically always true. In order to specify exactly when, the notions of *compatible initial state* and *evasive fault* (defined in Section 3.2.1.2) are used.

If the random test is not ideal, an input sequence which is not in the specified cases may be applied and/or an input sequence in the specified cases may have a zero probability. The consequences are explained in Section 5.2, from examples. Property 5.1 is more precise; informally, this property means that: if there is an input sequence which has a zero probability to be applied, then there are faults which have a zero probability to be tested. On the other hand, Property 5.2 will show that any (non-evasive) fault can be detected with a probability as close to 1 as required is the random test is ideal. Thus, roughly speaking, an *ideal random test* (Definition 5.1 in Section 5.2) is necessary (Property 5.1) and sufficient (Property 5.2) for detecting all the targeted and non-targeted non-evasive faults.

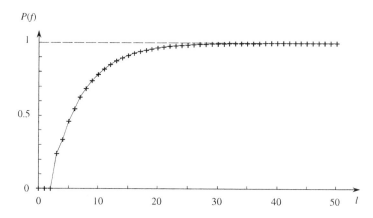

Figure 5.10 Probability of testing fault f given the distribution ψ (corresponding to Figure 5.7b), as a function of the test length, l.

Property 5.1 Consider the random test of a circuit C whose input language is I. Assume that, for a random test sequence of distribution ψ and length L, a sequence of length L in the input language I has a zero probability to be applied. Then, there is a fault of C, f in F_u (Section 4.2), whose probablity of testing is $P(f, \psi, L) = 0$.

Proof

Let C be defined by the machine $M = (Q, X, Z, \delta, \mu, q_0)$. Assume that the input sequence $S = x(1)x(2)...x(m)$, such that $m \leq L$ and $S \in I$, has the probability 0 to be applied.

First case: power-up or external reset. Let the fault f in F_u be modeled by the machine $M^{(f)} = (Q', X, Z, \delta', \mu', q'_0)$, where $Q' = Q \cup \{q'_0, q'_1, q'_2, ..., q'_m\}$, and defined as follows.

1) If $m = 1$,

$\delta'(q'_0, x) = \delta(q_0, x)$ for any $x \in I$,

$\mu'(q'_0, x(1)) \neq \mu(q_0, x(1))$, and $\mu'(q'_0, x) = \mu(q_0, x)$ for any $x \neq x(1)$, $x \in I$.

2) If $m > 1$,

for $i = 1, ..., m-1$:

$\delta'(q'_{i-1}, x(i)) = q'_i$, and $\delta'(q'_{i-1}, x) = \delta(q_0, x(1)...x(i-1)x)$
for $x \neq x(i)$ and $x(1)...x(i-1)x \in I$,

$\mu'(q'_{i-1}, x) = \mu(q_0, x(1)...x(i-1)x)$ for any $x(1)...x(i-1)x \in I$;

and for $i = m$:

$\delta'(q_{m-1}, x) = \delta(q_0, x(1)...x)$ for any $x(1)...x(m-1)x \in I$,

$\mu'(q'_{m-1}, x(m)) \neq \mu(q_0, x(1)...x(m))$, and $\mu'(q'_{m-1}, x) = \mu(q_0, x(1)...x(m-1)x)$
for any $x \neq x(m)$ and $x(1)...x(m-1)x \in I$.

By construction, $M^{t)}$ behaves exactly like M, except when the input sequence $S = x(1)x(2)...x(m)$ is applied from the initial states q'_0 and q_0, respectively. Since this sequence cannot occur, the fault $M^{t)}$ cannot be detected.

Second case: initialization sequence. The proof is almost similar. In the first case the reset device leads the faulty machine into the state q'_0 (by hypothesis). In the second case, given the initialization sequence S_s, it may be necessary to add some states $q'_{m+1}, q'_{m+2}, \ldots$ in order that $\delta'(q'_s, S_s) = q'_0$, where q'_s is the starting state; this is always possible and the machines M and M' reach respectively q_0 and q'_0 when the initialization sequence S_s is applied. ❑

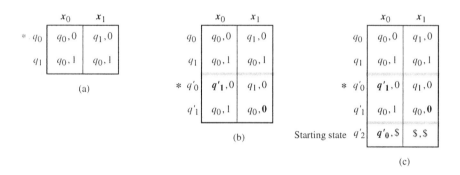

Figure 5.11 Illustration of the proof of Property 5.1, assuming the test input sequence $x_0 x_1$ has the probability 0. (a) Fault-free machine M. (b) Fault for the first case. (c) Fault for the second case if $S_s = x_0$.

The proof of Property 5.1, assuming the test input sequence $x_0 x_1$ has the probability 0, is illustrated in Figure 5.11. The faulty machine in Figure 5.11b has 4 states: the two states of the fault-free machine in Figure 5.11a plus q'_0 and q'_1 according to the construction in the proof. The reader may verify that: from the initial state q'_0, the test sequence $x_0 x_1$ could detect the fault, and no test sequence beginning with x_1 or $x_0 x_0$ can detect the fault. In Figure 5.11c, it is assumed that the fault-free machine is initialized by the synchronizing sequence x_0; the starting state q'_2, such that $\delta'(q'_2, x_0) = q'_0$, has been added.

The fault $M^{t)}$ in the proof is evasive by construction. However, in a general case, there are non-evasive faults which have a zero probability to be tested if the possible random test sequences correspond to a proper subset of the input language I. For example, let M be defined by Figure 5.12a. If the distribution (Figure 5.12c) produces the input language $\mathcal{P}(x_0 + x_1 x_0)^*$, then the fault M' in Figure 5.12b cannot be tested as illustrated by the observer and the Markov chain in Figures 5.12d and e.

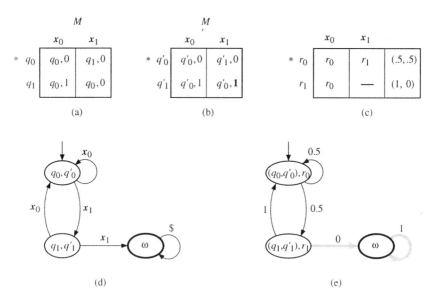

Figure 5.12 Example of fault and generalized distribution such that the fault cannot be detected (a) Machine M. (b) Fault M'. (c) Generalized distribution. (d) Observer. (e) Markov chain.

Let us now present the main property of this section. It is based on both the notions of ideal random test and non-evasive fault (Definitions 5.1 in Section 5.2 and 3.8 in Section 3.2.1.2, respectively).

Property 5.2 If an *ideal random test* is applied to a circuit C affected by a *non-evasive* fault f, then

$$\lim_{L \to \infty} P(f, \psi, L) = 1. \tag{5.5}$$

Proof. Let Ω be the observer associated with the fault f, and B_0 its possible initial states, corresponding to the compatible initial states of A and A'. Let b_0 be any initial state in B_0.

Since the fault f is non-evasive, from any state b_j of Ω reachable from b_0, there is at least one oriented path from b_j to ω (otherwise an input sequence leading from b_0 to b_j would meet the two conditions in Definition 3.8a). Then, state ω of the observer can be reached from any state, and cannot be left by construction.

Since the random test is ideal, every input sequence belonging to the input language of C has a non-zero probability to be applied. This implies that every arc of the Markov chain obtained from the observer and the distribution ψ has a probability which is not zero (i.e., the case illustrated in Figure 5.12e cannot occur). Then, this chain is an absorbing one and state ω is its only absorbing state. If a discrete time chain is absorbing, the probability of being in an absorbing state increases with the time and tends towards 1; this is a classical result. Since ω is the only absorbing state, Property 5.2 is obtained.

❑

The evasive fault in Example 3.2 (Section 3.2.1.2) is detected if and only if the faulty cell is read before a 1 is written in it. Then, if f denotes this fault,

$$P(f,\psi,L) = \frac{\Pr[\text{read the cell}]}{\Pr[\text{read the cell}] + \Pr[\text{write 1 in the cell}]}$$

$$\times \Pr[\text{cell has been read or 1- written at } L].$$

If all the inputs of the memory are equally likely, then $\Pr[\text{read the cell}] = 2 \times \Pr[\text{write 1 in the cell}]$ (see Section 8.2.1.1). It follows $\lim_{L \to \infty} P(f, \psi, L) = 2/3$. Note that the probability of testing this fault could be greater if several 'power down then power up then initialization' were carried out during the test experiment.

Remark 5.4 *Property 5.2 applies to any fault, even to non-target faults.* Then, according to this property, a random test can lead to a level of confidence as close to 1 as necessary.

5.3.3 Limits of deterministic testing

As already pointed out, a deterministic testing is based on a set of faults (structural or functional models). It follows that, even if a 100% fault coverage is obtained for a test sequence, non-target faults remain which may be not detected by this test sequence. This limitation, based on the notion of universal fault set, F_u, is expressed in the following property.

Property 5.3 *For any circuit C, and for any finite test sequence S (S is in the input language I), there is a non-evasive fault of C, in F_u, which is not detected by S.*

Proof. Let C be defined by the machine $M = (Q, X, Z, \delta, \mu, q_0)$ and let $S = x(1)x(2)...x(m)$. Let the fault f in F_u be modeled by the $(m+1)$-state machine $M' = (Q', X, Z, \delta', \mu', q'_0)$, defined as follows.

For any prefix S_i of S, $S_i = x(1)...x(i)$, $i = 1, ..., m$:

$\delta'(q'_0, S_i) = q'_i$,

$\mu'(q'_{i-1}, x(i)) = \mu(q_0, S_i)$,

and for some x in X such as $Sx \in I$:

$\mu'(q'_m, x) \neq \mu(q_0, Sx)$.

The state $\delta'(q'_m, x)$ may be any state. Up to now, machine M' is incompletely specified. It can easily be completed in order to be non-evasive.

The machine M' behaves like M for the input sequence S but not for the input sequence Sx.

5.3.4 Conclusion on detection power

For any finite test sequence one can find a fault in the universal fault set which is not tested by it. Then a 100 % fault coverage *for any specified set of faults* does not guarantee that all the faulty circuits can be detected since they can be affected by *unmodeled faults*.

By construction, *a deterministic test sequence has a bounded length*. On the other hand, *a random test sequence is not bounded*. In practice, it is necessarily finite; however, its length is not bounded since the test sequence can be defined from a generation algorithm (random drawings, see Chapter 10). Then, if necessary, one can apply a random test length allowing the test with a probability as close to 1 as required. The only limitation, for a permanent logic fault, is that it must be *non-evasive*.

According to Property 3.5 (Section 3.2.4.2), an evasive fault may be detectable. For every example of evasive fault in Section 3.2.1.2, there is an input sequence which can detect it: read the cell for Example 3.2, x_0 for Example 3.3, and x_0 for Example 3.4 (in Section 5.3.2). However, for random testing of an evasive fault the probability of testing tends to a limit which is less than 1; this limit was calculated in Section 5.3.2 for the Example 3.2.

For an evasive fault f whose initial state is not always the same (i.e., $|Q'_0| > 1$), $\lim_{L \to \infty} P(f, \psi, L)$ may be an average value, weighted by the probabilities of the possible initial states. While the hypotheses are not the same, this notion could be compared with that of *fault detection coefficient*, defined in the deterministic context for partially detectable faults [PoRe 93].

For a set of faults containing at least 2 detectable evasive faults, there may be no test sequence detecting all the faults in the set. In both cases (random and deterministic), the only way to have a efficient test consists in making several resets during a test experiment.

Fortunately, most of the faults occuring in practice are non-evasive.

5.4 FIRST DETECTION AND MEMORY EFFECT

In this section, we shall see that an average detection probability of a fault allows a good estimation of the required test length when the detection process is memoryless (combinational fault, Section 5.4.1) but that an accurate estimation requires an analysis of the Markov chain when there is a memory effect (sequential fault, Section 5.4.2).

Let (A, q_0) and (A', q'_0) be a fault-free and a faulty machine corresponding to fault f. Assume an infinite length random test sequence $S_\infty = x(1)x(2)...$ and let $S_l = x(1)...x(l)$ be a prefix of S_∞.

Definition 5.2 There is a **detection at time** l if the outputs are different at time l, i.e.,

$$\mu(q_0, S_l) \neq \mu'(q'_0, S_l).$$

❑

One can define two random variables, the length to first detection and the length between two successive detections.

Definition 5.3.
a) The **length to first detection**, denoted by l_ω, is such that

$$\mu(q_0, S_l) = \mu'(q'_0, S_l), \text{ for any } l < l_\omega,$$

and

$$\mu(q_0, S_{l_\omega}) \neq \mu'(q'_0, S_{l_\omega}).$$

b) The **length between** (successive) **detections**, denoted by l_τ, is such that: given

$$\mu(q_0, S_a) \neq \mu'(q'_0, S_a),$$

then

$$\mu(q_0, S_{a+b}) = \mu'(q'_0, S_{a+b}), \text{ for any } 0 < b < l_\tau,$$

and

$$\mu(q_0, S_{a+l_\tau}) \neq \mu'(q'_0, S_{a+l_\tau}).$$

❑

The event corresponding to *random variable* l_ω occurs at most once in every test sequence, while there may be many occurrences of the event corresponding to the random variable l_τ in the same test sequence. This is illustrated in Figure 5.13b.

5.4.1 Combinational fault

Consider a combinational circuit achieving the function z in Figure 5.13a, and the fault $f_1 = z/0$. A simulation of the detection process of this fault by a random test sequence is presented in Figure 5.13b; the distribution of input vectors is $\psi_0 = (.25, .25, .25, .25)$.

Let p_f denote the *probability of detecting fault f* by *any random test vector*. For our example, $p_{f_1} = \psi_0(x_1) + \psi_0(x_2) = 0.5$.

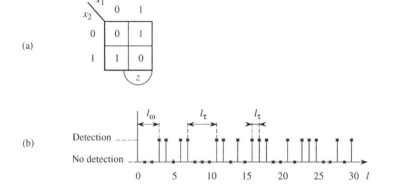

Figure 5.13 (a) Combinational function. (b) Simulation result for the fault $f_1 = z/0$ and the distribution ψ_0.

The length to first detection and the length between two successive detections have the same probability law (*binomial distribution*):

$$\Pr[l_\omega = i] = (1 - p_f)^{i-1} p_f, \quad i \in [1, \infty); \tag{5.6}$$

$$\Pr[l_\tau = i] = (1 - p_f)^{i-1} p_f, \quad i \in [1, \infty). \tag{5.7}$$

This is true because the process is *memoryless*. The probability of detection at any time l does not depend on the $l - 1$ previous test vectors:

$$\Pr[\text{detect. } f_1 \text{ at } l] = \Pr[\text{detect. } f_1 \text{ at } l \mid \text{detect. } f_1 \text{ at } l - i], \, i = 1,\ldots, l - 1.$$

The mean values[8] of l_ω and l_τ are

$$E[l_\omega] = E[l_\tau] = \frac{1}{p_f}, \tag{5.8}$$

and the standard deviations are

---

[8] For the proof, see Section 6.1.1.2.

Basic Principles of Random Testing   131

$$\sigma[l_\omega] = \sigma[l_\tau] = \sqrt{\frac{1-p_f}{p_f}}. \quad (5.9)$$

These are classical results of probability theory. The test length required for a combinational circuit will be developed in Section 6. For the time being, let us note that if a coverage $P(f_1) \geq 0.999$ is required, the required test length is $L = 10$, since $\Pr[l_\omega \leq 9] < 0.999$ and $\Pr[l_\omega \leq 10] > 0.999$.

5.4.2 Sequential fault

In the case of sequential fault a probability of detecting a fault f by any random test vector cannot be defined as for a combinational fault.

Consider the example of 1-input, 1-output synchronous sequential circuit presented in Figure 5.14a. The circuit is initialized by the synchronizing sequence $S_s = x_1 x_0$. Assume the fault f_2 such that only the output function is affected: $f_2 = z/0$. Then, S_s is also a synchronizing sequence for the faulty automaton; both fault-free and faulty circuits will always be in the same state after initialization. According to the automaton of Figure 5.14a, there is a detection each time one of the states q_2 or q_3 is reached (the output z should have the value 1 for all these transitions).

Assume a constant distribution $\psi_0 = (.5, .5)$ of the input vectors. The limit probabilities of the various states, when a long input sequence has been applied, could be obtained from a Markov chain (Section 7.3.2.1). For our example, one can observe that, except for the initialization, there is a symmetry between $(q_0, q_1, x_0, 0)$ on the one hand and $(q_2, q_3, x_1, 1)$ on the other hand: this means that if we replace q_0 by q_2 and q_2 by q_0, q_1 by q_3 and q_3 by q_1, x_0 by x_1 and x_1 by x_0, 0 by 1 and 1 by 0, we obtain the same graph. Since $\psi_0(x_0) = \psi_0(x_1)$, after some random test vectors have been applied the probability of being in state q_0 or q_1 is approximately the probability of being in state q_2 or q_3, respectively (at $l = 10$, one has $\pi_{q_0}(10) = 0.334$, $\pi_{q_1}(10) = 0.167$, $\pi_{q_2}(10) = 0.333$, $\pi_{q_3}(10) = 0.166$).

Let us now define the average detection probability.

Definition 5.4 The **average detection probability** of a fault f, given a distribution ψ is

$$p_{av,f}(\psi) = \lim_{l \to \infty} \frac{\text{number of detections}}{l}.$$

When ψ is implicit, $p_{av,f}(\psi)$ may be denoted by $p_{av,f}$.

□

For the considered example, $f_2 = z/0$ in the Figure 5.14a, one has $p_{av,f_2} = 0.5$ because of the symmetry.

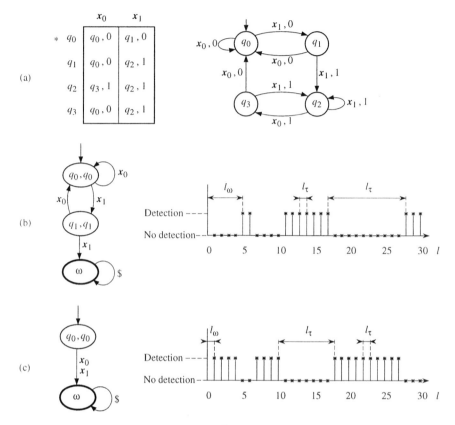

Figure 5.14 (a) Fault-free machine[9]. (b) Observer and a simulation for the fault $f_2 = z/0$. (c) Observer and a simulation for the fault $f_3 = z/1$.

From Definition 5.4, it is clear that $E[l_\tau] = \dfrac{1}{P_{av,f_2}} = 2$. However, neither l_ω nor l_τ correspond to a binomial distribution. In addition, and this is important, $E[l_\omega] \neq E[l_\tau]$.

Figure 5.14b presents the observer and a simulation for the example in Figure 5.14a and fault f_2. It clearly appears that the detections correspond to "bundles". This is because

$$\Pr[\text{detect. } f_2 \text{ at } l \mid \text{detect. } f_2 \text{ at } l-1] > \Pr[\text{detect. } f_2 \text{ at } l].$$

In other words there is a **memory effect**. As long as the automaton remains in the set of states $\{q_0, q_1\}$ there is no detection; as long as it remains in the set of

---

[9] Parts (a) and (b) of the figure are reprinted from [Da 97] (© 1997 IEEE).

states $\{q_2, q_3\}$ there is a detection at each time unit. From the observer in Figure 5.14b and the corresponding Markov chain, given ψ_0, the required test length is $L = 34$ for $P(f) \geq 0.999$.

Now, consider the fault $f_3 = z/1$ in the automaton of Figure 5.14a. The observer and a simulation are presented in Figure 5.14c (same random test sequence as in Figure 5.14b). In this case one obtains $L = 1$ for $P(f) \geq 0.999$, since the fault is always detected by the first input vector.

The conclusion is that for the same average detection probability, $p_{f_1} = p_{av, f_2} = p_{av, f_3} = 0.5$, for the same probability of testing, 0.999, the required test length is 10 for the *memoryless case*, while it may be 34 (for f_2) or 1 (for f_3) in case of *memory effect*. In the following chapters, it will be shown that neglecting the memory effect often leads to a length shorter than the necessary one (see Sections 7.3.1.2, 7.3.2.2, 11.3.1, and Exercise 11.4).

NOTES and REFERENCES

Some of the principles of random testing presented in this section are either well known or implicitly assumed by various authors. Some are formalized for the first time, for example the concept of ideal random test is new to the best of the author's knowledge.

Modeling the detection process of a fault in a sequential circuit by a Markov chain was introduced in [ShMc 76]. It has been shown to be a general approach and extensively used in [DaFuCo 89]. In [BrJü 92] this approach is formalized in relation with the theory of observers. The generalization to Markov sources of test patterns is introduced in this chapter.

The memory effect, whose consequence is that the average detection probability of a sequential fault does not allow calculation of an accurate estimation of the test length to first detection, has been observed, at least implicitly, by many authors. Experimental results pointing out the "bundles" of detection are reported in [DaTh 81]. The length to first detection was called the error latency in [ShMc 76].

For the properties of Markov chains, the reader may consult [KeSn 76], for example. A short review of these chains is presented in Appendix C.

The results of probability theory used in Section 5.4 can be found in any book about this topic, for example [Fe 68], [Pa 90].

6

Random Test Length for Combinational Circuits

As was shown in the preceding sections, a fault hypothesis is not needed to perform a random test. However, the estimation of a random test length requires some assumptions.

This chapter describes the calculation of the required test length for combinational circuits. This is a very important case since, in addition to purely combinational circuits it also covers parts of sequential circuits designed using the scan-path method. Recall that scan design (Section 1.1.2) transforms a sequential circuit into a combinational circuit plus scan-path registers.

Most of this chapter is devoted to combinational faults, i. e., assuming that both the fault-free and the faulty circuits are combinational. If the faulty circuit is sequential, then the techniques developed for sequential circuits (Chapter 7) must be used. However, it will appear (particularly in Chapters 7 and 9) that approximate methods for sequential faults use the results presented in Section 6.1 for combinational circuits.

It is assumed that the random input vectors correspond to a constant distribution, i.e., drawing at time t is independent from the input vectors previously drawn.

Calculation of random test length is presented in Section 6.1. Methods for calculating the detection probabilites are explained in Section 6.2. Numerical results for usual circuits are given in Section 6.3.

6.1 RANDOM TEST LENGTH

In order to simplify the presentation, most of the calculation are developed under the most usual hypotheses that follow: 1) the circuits are *entirely specified*, i.e., an output vector is specified for every input vector; 2) the *faults* under

136   Chapter 6

consideration are *combinational* and *permanent*; 3) the random test sequence is a string of *equally likely input vectors*.

However, each of these hypotheses can be removed. This will be shown: for incompletely specified circuit in Remark 6.2 (Section 6.1.1.2); for sequential fault in Remark 7.2 (Section 7.2.3); for intermittent fault in Notes and References at the end of chapter; and for weighted input vectors in Section 10.1.1.2.

The test length for a fault and for a set of faults are studied in Sections 6.1.1 and 6.1.2 respectively.

6.1.1 Random test length for a fault

6.1.1.1 Detection probability

Let C be a n-input combinational circuit and C_f a circuit affected by a combinational fault f. According to the notation and concepts presented in Chapters 2 and 3, the vector x_i is a test vector for f if

$$z_0(x_i) \neq z_f(x_i), \tag{6.1}$$

where z_0 and z_f are the output vectors of a fault-free circuit and the faulty circuit, respectively.

The set of input vectors that satisfy inequality (6.1) is the *set*, denoted by T_f, of *test vectors* for f:

$$T_f = \{x_i \mid z_0(x_i) \neq z_f(x_i)\}. \tag{6.2}$$

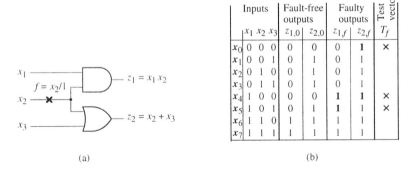

Figure 6.1 (a) Combinational circuit with a fault f. (b) Test vectors for f.

Random Test Length for Combinational Circuits 137

The fault $f = x_2/1$ in Figure 6.1a may be detected by any vector in the set $T_f = \{x_0, x_4, x_5\}$ as illustrated in Figure 6.1b. For x_0, there is an error on z_2; for x_4, there is an error on both outputs; for x_5, there is an error on z_1.

Detection of the fault f is a stochastic process. When a random input vector is applied to the circuit C_f, there is some probability that this vector is a test vector: this is the detection probability, formally defined as follows.

Definition 6.1 The **detection probability** of the fault f, denoted by p_f, is the probability that a random input vector x is a test vector for f, i.e.,

$$p_f = \Pr[x \in T_f \mid x \text{ is random}] \tag{6.3}$$

❏

For any constant distribution ψ (Section 4.4 and Appendix A), the detection probability may be written as:

$$p_f = \sum_{x_i \in T_f} \Pr[x = x_i \mid x \text{ is random}] = \sum_{x_i \in T_f} \psi(x_i). \tag{6.4}$$

If the input vectors are **equally likely**, then

$$\psi(x_i) = \psi_0(x_i) = \frac{1}{2^n}, \text{ for any } x_i. \tag{6.5}$$

Let k_f denote the number of test vectors for the fault f, i.e., $k_f = |T_f|$. If the input vectors are *equally likely*, then, given (6.5), Equation (6.4) becomes

$$p_f = \frac{k_f}{2^n}. \tag{6.6}$$

The number k_f is called the **detectability** of the fault f.

Return to the fault f in Figure 6.1. For random testing with a *constant distribution* ψ, the detection probability is $p_f = \psi(x_0) + \psi(x_4) + \psi(x_5)$. If the input vectors are *equally likely*, then, according to (6.6):

$$p_f = \frac{3}{2^3} = 0.375,$$

since $n = 3$ and $k_f = 3$.

6.1.1.2 Random test length

Assume that a test sequence $S = x(1)x(2) \ldots x(L)$ of length L is applied to the circuit C_f. According to Definition 4.1 in Section 4.4, the **probability of testing** the combinational fault f by the input sequence S, denoted by $P(f, S)$, is the probability that the sequence S contains at least one vector in T_f, i.e.,

138   Chapter 6

$$P(f, S) = \Pr[\text{there is an } x_i \text{ in } S \mid x_i \in T_f]. \qquad (6.7)$$

If the sequence S is random, the *probability of testing* fault f by $x(1)$ is the probability that this vector is a test vector, i.e.,

$$P(f, x(1)) = \Pr[x(1) \in T_f] = p_f.$$

Then, the *probability of not testing* fault f by $x(1)$ is

$$1 - P(f, x(1)) = \Pr[x(1) \notin T_f] = 1 - p_f.$$

Now, the *probability of not testing* fault f by the sequence $x(1)x(2)$ is

$$1 - P(f, x(1)x(2)) = \Pr[x(1) \notin T_f] \cdot \Pr[x(2) \notin T_f] = (1 - p_f)^2,$$

since the vectors $x(1)$ and $x(2)$ are independent and have the same detection probability.

Finally, the following generalization can be made:

$$1 - P(f, S) = 1 - P(f, x(1) \ldots x(L)) = (1 - p_f)^L,$$

and the following property obtained.

Property 6.1 The probability of testing the combinational fault f by a random sequence S of length L can be written:

$$P(f, S) = 1 - (1 - p_f)^L. \qquad (6.8)$$
❑

According to Section 4.4.3.1, $1 - P(f, S)$ corresponds to the uncertainty level ε. Thus, the uncertainty level corresponding to the test length L for the fault f is

$$\varepsilon(L) = (1 - p_f)^L \qquad (6.9)$$

Let $P(f, l)$ denote the probability of testing the fault f in Figure 6.1 by a random test of length l, given the distribution is ψ_0. The detection probability of this fault is $p_f = 0.375$. Then, according to (6.8),

$$P(f, l) = 1 - (1 - 0.375)^l. \qquad (6.10)$$

This is illustrated in Figure 6.2a, where it is clear that $P(f, l)$ tends to 1 when l approaches ∞ (this is true for any value of p_f).

The probability that the first detection occurs at the lth vector of the test sequence, i.e., that $l_\omega = l$, according to the notation in Section 5.4, is

$$\Pr[l_\omega = l] = P(f, l) - P(f, (l-1)) = (1 - p_f)^{l-1} \times p_f.$$

This is illustrated in Figure 6.2b, where it appears that $\Pr[l_\omega = l]$ decreases exponentially to 0. The origin of the exponential curve is at $p_f / (1 - p_f)$ for $l = 0$, and $\Pr[l_\omega = 1] = p_f$ (note that $p_f/(1-p_f)$ is close to p_f if p_f is close to 0).

(a)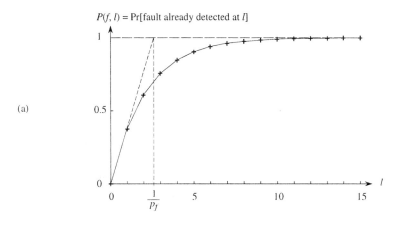

(b)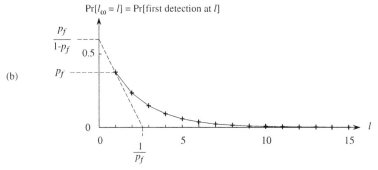

Figure 6.2 Test of fault f in Figure 6.1. (a) Probability that the fault has already been detected at the lth vector. (b) Probability that the first detection occurs at the lth vector.

Now, what about the average length to detection $E[l_\omega]$? A well known result is that, since l_ω and l are positive integers:

$$E[l_\omega] = \sum_{l \geq 0} \Pr[l_\omega > l]. \qquad (6.11)$$

Since $\Pr[l_\omega > l] = (1 - p_f)^l$ and $\sum_{l \geq 0} a^l = 1/(1-a)$, the following property is obtained from (6.11).

Property 6.2 The expected length to detection is the inverse of the detection probability:

$$E[l_\omega] = \frac{1}{p_f}. \qquad (6.12)$$

❑

For our example (fault f in Figure 6.1), the expected test length is $E[l_\omega] = 1/0.375 = 2.67$.

Let us now calculate the probability of testing the fault f for some values of l, using Equation (6.10): $P(f, 2) = 0.609$, $P(f, 3) = 0.756$, $P(f, 10) = 0.991$, $P(f, 20) = 0.9999$.

A question naturally arises: what is the *test length required to obtain some given level of confidence*? Let $1 - \varepsilon$ be the required *confidence level* (ε is the *uncertainty level*). Let us denote $L(\varepsilon)$ the minimal test length ensuring a confidence level at least equal to $1 - \varepsilon$. Since only one fault is considered, the confidence level is given by Equation 6.9, and the following equation must hold:

$$1 - (1 - p_f)^{L(\varepsilon)} \geq 1 - \varepsilon. \tag{6.13}$$

It follows that $(1 - p_f)^{L(\varepsilon)} \leq \varepsilon$, and $L(\varepsilon) \times \ln(1 - p_f) \leq \ln \varepsilon$. Since $(1 - p_f)$ and ε are less than 1, their logarithms are negative, then

$$L(\varepsilon) \geq \frac{\ln \varepsilon}{\ln(1 - p_f)}. \tag{6.14}$$

Since the test length is an integer, Property 6.3 is obtained from Equation (6.14).

Property 6.3 The length $L(\varepsilon)$ of a random sequence that ensures a *level of confidence* of $1 - \varepsilon$ is[1]

$$L(\varepsilon) = \left\lceil \frac{\ln \varepsilon}{\ln(1 - p_f)} \right\rceil \tag{6.15}$$

❑

Return to the fault f in Figure 6.1. For a confidence level equal to 0.999, the required random test length is

$$L(0.001) = \left\lceil \frac{\ln(0.001)}{\ln(1 - 0.375)} \right\rceil = \lceil 14.70 \rceil = 15.$$

Remark 6.1
a) From now on, we will write $L(\varepsilon) = \alpha$ instead of $L(\varepsilon) = \lceil \alpha \rceil$, in order to simplify the notation. Knowing that $L(\varepsilon)$ is a number of input vectors, this implicitly means that the least integer greater than or equal to α is the required value.

b) In this book we shall often consider a required level of confidence equal to 0.999. For this value, $\ln \varepsilon = \ln(0.001) = -6.91$. Since p_f is usually a value close

---

[1] $\lceil \alpha \rceil$ denotes the least integer greater than or equal to α.

to 0, one has $\ln(1 - p_f) \approx - p_f$. Then, for this value of ε, the test length can be roughly estimated using the following equation:

$$L(\varepsilon) \approx \frac{7}{p_f};\qquad(6.16)$$

(i.e., the required test length is rougthly 7 times the expected length to detection).

Remark 6.2 **Incompletely specified circuits**

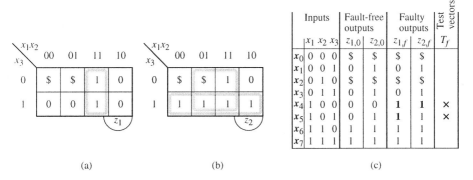

Figure 6.3 Incompletely specified circuit. (a) and (b) Karnaugh maps of the two outputs. (c) Test vectors for the fault $f = x_2/1$ in Figure 6.1a.

Assume that the circuit in Figure 6.1a is incompletely specified: in normal operation, the input vectors x_0 and x_2 never occur. This is illustrated in Figure 6.3. The vector x_0, which was a test vector in Figure 6.1b, is no longer a test vector since the outputs are not specified for this input vector. An ideal random test applies a test sequence in $(x_1+x_3+x_4+x_5+x_6+x_7)^*$. For a constant distribution ψ, the detection probability is $p_f = \psi(x_4) + \psi(x_5)$, since, according to Figure 6.3c, the test vectors are x_4 and x_5. For example, if the six possible input vectors are equally likely, i.e., $\psi(x_1) = \psi(x_3) = \psi(x_4) = \psi(x_5) = \psi(x_6) = \psi(x_7) = 1/6$, then $p_f = 0.33$. For an uncertainty level $\varepsilon = 0.001$, one obtains from Equation (6.15): $L(0.001) = 18$.

6.1.2 Random test length for a set of faults

6.1.2.1 Expected fault coverage

Let $F = \{f_1, f_2, ..., f_r\}$ be the set of faults under consideration, and consider a random test sequence S of length L. The expected fault coverage is expressed in the following property.

Chapter 6

Property 6.4 The *expected fault coverage* of the set of faults F by the test sequence S of length L can be written:

$$P = 1 - \frac{1}{r} \sum_{i=1}^{r} \left(1 - p_{f_i}\right)^L. \tag{6.17}$$

Proof According to Equation (4.18) in Section 4.4.1 and the notation in Definition 4.1 in Section 4.4, the *expected fault coverage* by the test sequence S is

$$P = \frac{1}{r} \sum_{i=1}^{r} P(f_i, S). \tag{6.18}$$

From Equation (6.8), giving $P(f_i, S)$, Equation (6.18) becomes:

$$P = \frac{1}{r} \sum_{i=1}^{r} \left(1 - \left(1 - p_{f_i}\right)^L\right). \tag{6.19}$$

Since $\sum_{i=1}^{r} \left(1 - \left(1 - p_{f_i}\right)^L\right) = r - \sum_{i=1}^{r} \left(1 - p_{f_i}\right)^L$, Equation (6.17) is obtained from (6.19).

□

If the *level of confidence* $1 - \varepsilon$ is defined as the expected fault coverage, then, given the *uncertainty level* ε, the required *random test length*, $L(\varepsilon)$, is the minimum value satisfying the following equation:

$$\frac{1}{r} \sum_{i=1}^{r} \left(1 - p_{f_i}\right)^{L(\varepsilon)} \leq \varepsilon. \tag{6.20}$$

Let us consider the circuit in Figure 6.4a. Since it contains 14 lines (x_1 to x_3, w_1 to w_9, z_1 and z_2), it may be affected by one out of 28 single stuck-at faults. The detectability of every fault is given in Figure 6.4b. The faults are gathered into classes of *equivalent* faults; for example $w_1/0 \equiv w_4/1$ (a class will be named like one of its elements). In addition, the detectability is given for the set of *equidetectable* classes. For example, $x_1/0 \cong w_1/0$. As a matter of fact: $x_1/0$ is detected on *both* outputs by the input vectors x_6 and x_7; $w_1/0$ is detected on z_1 by x_6 and x_7 too, while z_2 remains fault-free. According to Definition 3.3 in Section 3.1.4.1, $x_1/0 \cong w_1/0$ since $T_{x_1/0} = T_{w_1/0}$, but these faults are not equivalent.

Remark 6.3 According to Section 6.1.1.1, the measurement usually called "detectability", i.e., k_f, implicitly assumes that the input vectors are equally likely. The faults $x_1/0$ and $x_1/1$ have the same *detectability*: $k_{x_1/0} = k_{x_1/1} = 2$. However, $T_{x_1/0} = \{x_6, x_7\} \neq T_{x_1/1} = \{x_2, x_3\}$; then $p_{x_1/0}$ may be different from $p_{x_1/1}$ if the input vectors are not equally likely. On the other hand the faults we have defined as *equidetectable* have the same detection probability for any distribution ψ.

□

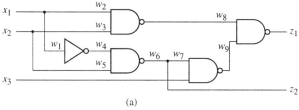

Classes of equivalent faults

$x_1/0$ $w_4/1$	$w_1/0$	$x_1/1$	$w_1/1$ $w_4/0$ $w_5/0$ $w_6/1$	$w_2/1$	$z_2/1$	$x_2/0$	$x_2/1$	$x_3/0$ $w_7/0$ $w_9/1$	$x_3/1$	$w_2/0$ $w_3/0$ $w_8/1$	$w_3/1$	$w_5/1$	$w_6/0$	$z_2/0$	$w_7/1$	$w_8/0$ $w_9/0$ $z_1/1$	$z_1/0$
2		2				3	3	2	2	1	1	2	6		1	4	4

Detectability

(b)

Figure 6.4 (a) Example of circuit. (b) Detectability of all the single stuck-at faults.

From Figure 6.4b, the detectability is: 1 for 5 faults, 2 for 15 faults, 3 for 2 faults, 4 for 4 faults, and 6 for 2 faults. For the distribution ψ_0, one obtains, from Equations (6.17) and (6.6):

$$P = 1 - \frac{1}{28}\left(5\left(1-\frac{1}{8}\right)^L + 15\left(1-\frac{2}{8}\right)^L + 2\left(1-\frac{3}{8}\right)^L + 4\left(1-\frac{4}{8}\right)^L + 2\left(1-\frac{6}{8}\right)^L\right) \tag{6.21}$$

This equation gives $P = 0.986$ for $L = 20$. If a level of confidence $P \geq 0.999$ is required, then $L(\varepsilon) = L(0.001) = 39$ is obtained from the following equation, directly obtained from Equation (6.20):

$$\frac{1}{28}\left(5\left(1-\frac{1}{8}\right)^{L(\varepsilon)} + 15\left(1-\frac{2}{8}\right)^{L(\varepsilon)} + 2\left(1-\frac{3}{8}\right)^{L(\varepsilon)} + 4\left(1-\frac{4}{8}\right)^{L(\varepsilon)} + 2\left(1-\frac{6}{8}\right)^{L(\varepsilon)}\right) \leq 0.001 = \varepsilon. \tag{6.22}$$

If the input vectors are equally likely, the "*testability*" of a circuit may be characterized by the detectability profile. The **detectability profile** of a circuit, given a set of faults under consideration, is a 2^n-component vector,

$$\pi = (\pi_1, \cdots, \pi_k, \cdots, \pi_{2^n}), \tag{6.23}$$

where each component π_k is the number of faults whose detectability is k. For the example in Figure 6.4, the detectability profile is $\pi = (5, 15, 2, 4, 0, 2, 0, 0)$. This is illustrated in Figure 6.5a.

144   Chapter 6

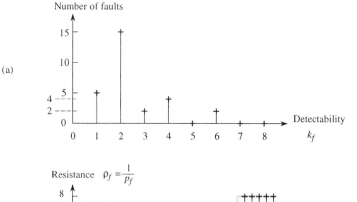

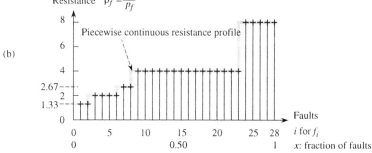

Figure 6.5 (a) Detectability profile. (b) Resistance profile.

Then, for the distribution ψ_0, Equations (6.17) and (6.20) become

$$P = 1 - \frac{1}{r}\sum_{k=1}^{2^n} \pi_k \left(1 - \frac{k}{2^n}\right)^L \quad \text{and} \quad \frac{1}{r}\sum_{k=1}^{2^n} \pi_k \left(1 - \frac{k}{2^n}\right)^{L(\varepsilon)} \leq \varepsilon. \qquad (6.24 \ \& \ 25)$$

The reader may verify that Equations (6.21) and (6.22) correspond to (6.24) and (6.25) applied to our example.

Since $L(0.001) = 39$, Equation (6.22) corresponds to

$$\frac{1}{28}\left(5\left(\frac{7}{8}\right)^{39} + 15\left(\frac{6}{8}\right)^{39} + 2\left(\frac{5}{8}\right)^{39} + 4\left(\frac{4}{8}\right)^{39} + 2\left(\frac{2}{8}\right)^{39}\right) < 0.001. \qquad (6.26)$$

$$\frac{1}{28}\left(5(5.5 \cdot 10^{-3}) + 15(1.3 \cdot 10^{-5}) + 2(1.1 \cdot 10^{-8}) + 4(1.8 \cdot 10^{-12}) + 2(3.3 \cdot 10^{-24})\right) < 0.001.$$

(6.26')

Some faults are **easy to detect** (for example, $z_2/0$ has detectability $k_{z_2/0} = 6$), while others are **hard to detect** or **random pattern resistant** (for example, $w_2/0$ has detectability $k_{w_2/0} = 1$). It is clear from Equation (6.26'), that the contribution to the detection uncertainty is mainly due to the hardest faults: for a fault f whose detectability is 1, $(1 - p_f)^{39} = 5.5 \cdot 10^{-3}$; for a fault f whose detectability is 2, $(1 - p_f)^{39} = 1.3 \cdot 10^{-5}$, which is two orders of magnitude lower. If k_{min} denotes the **detectability of the hardest fault**, practically all the faults f whose detectability are $k_f \geq 2 \cdot k_{min}$ have a negligible contribution to the detection uncertainty (however, their number is taken into account in the coefficient $\frac{1}{r}$). For our example:

$$\frac{1}{28}\left(5(5.5 \cdot 10^{-3}) + 15(1.3 \cdot 10^{-5}) + 2(1.1 \cdot 10^{-8}) + 4(1.8 \cdot 10^{-12}) + 2(3.3 \cdot 10^{-24})\right)$$

$$\approx \frac{1}{28}\left(5(5.5 \cdot 10^{-3})\right).$$

Let us call

$$\rho_f = \frac{1}{p_f} \qquad (6.27)$$

the **resistance** *of fault f to random testing* (according to Property 6.2 in Section 6.1.1.2), this is the *expected length to detection*). Let us simplify the notation in the following way:

$$p_i = p_{f_i} \quad \text{and} \quad \rho_i = \rho_{f_i}. \qquad (6.27')$$

The "*testability*" of a circuit may be characterized by the resistance profile. The **resistance profile** of a circuit, given a set of faults under consideration, is a r-component vector,

$$\rho = (\rho_1, ..., \rho_i, ..., \rho_r), \qquad (6.28)$$

where the faults are ordered into a non-decreasing order of their resistances, i.e., $\rho_i \leq \rho_{i+1}$. For the example in Figure 6.4, the resistance profile ρ is illustrated in Figure 6.5b. Given the resistance profile, the expected fault coverage is obtained from Equation (6.17), and the required random test length for some level of confidence is obtained from (6.20).

If all the input vectors are equally likely (constant distribution ψ_0), the resistance profile and the detectability profile contain exactly the same information as far as the calculation of the fault coverage P is concerned (the only difference is that the number of inputs, n, can be deduced from the detectability profile but not from the resistance profile). *The resistance profile may be used for any constant distribution ψ*, while the detectability profile is only useful for the equally likely distribution ψ_0.

Remark 6.4
a) Basically, the detectability profile and the resistance profile are defined for the set of faults under consideration. However, these profiles can also be defined for the set of *classes of equivalent faults*, i.e., after equivalence fault collapsing

(Section 3.1.4.1), or even for the set of *classes of equidetectable faults* (Definition 3.3 in Section 3.1.4.1).

b) The (discrete) resistance profile may be replaced by a *piecewise continuous resistance profile* as illustrated in Figure 6.5b: p_i is associated with the interval between faults $(i - 1)$ and i. If the abscissa $x \in [0, 1]$, representing a fraction of faults, replaces the number of faults $i \in \{1, ..., r\}$, then p_i becomes $\rho(x)$ and Equation (6.17) becomes

$$P = 1 - \int_0^1 \left(1 - \frac{1}{\rho(x)}\right)^L dx. \qquad (6.29)$$

6.1.2.2 Worst case: minimum testing probability

Obviously, it is difficult to obtain the detection or resistance profile for a large circuit. A common approach consists of computing the test length for the **most resistant** fault (also called *most difficult* fault, *worst case* fault, or *hardest* fault). Consider a set F of detectable combinational faults. Let

$$p_{\min} = \min_{f \in F} p_f. \qquad (6.30)$$

be the **minimum detection probability**, i.e., the detectability probability of the worst case fault of the set F.

For a random test sequence of length L, the following is deduced from (6.8):

$$\min_{f \in F} P(f) = 1 - (1 - \min_{f \in F} p_f)^L. \qquad (6.31)$$

Then, the *minimum testing probability* P_m, defined by Equation (4.20) in Section 4.4.2, is obtained from the most difficult fault:

$$P_m = 1 - (1 - p_{\min})^L. \qquad (6.32)$$

P_m is a lower bound of P (Property 4.3 in Section 4.4.3.1). Then, an upper bound of the required random test length, $L(\varepsilon)$, for some level of confidence $1 - \varepsilon$, may be obtained from the minimum detection probability. Let us denote $L_m(\varepsilon)$ the length of a random sequence that ensures a level of confidence of $1 - \varepsilon$ for the most difficult fault, i.e.,

$$L_m(\varepsilon) = \left\lceil \frac{\ln \varepsilon}{\ln(1 - p_{\min})} \right\rceil \quad \text{and} \quad (1 - p_{\min})^{L_m(\varepsilon)} \leq \varepsilon. \qquad (6.33 \text{ \& } 6.34)$$

It follows from (6.20), (6.30), and (6.33), that:

$$L_m(\varepsilon) \geq L(\varepsilon). \qquad (6.35)$$

If a test of length $L_m(\varepsilon)$ is applied, for any circuit affected by a fault in F, the testing probability is at least $1 - \varepsilon$ (according to Property 4.3b, and Remark 4.5c in Section 4.4.3.1).

If all the input vectors are *equally likely* (constant distribution ψ_0), then the value

$$p_{\min} = \frac{k_{\min}}{2^n} \tag{6.36}$$

may be used in (6.32), (6.33), and (6.34). For the example in Figure 6.4, $k_{\min} = 1$ and $p_{\min} = 1/2^3 = 0.125$. For $L = L(\varepsilon) = 39$, $P_m = 0.995$ is obtained from Equation (6.32). For $\varepsilon = 0.001$, $L_m(\varepsilon) = 52$ is obtained from Equation (6.34). Hence, if the test length is estimated for the worst case (52 test vectors), the estimation is pessimistic since 39 test vectors are sufficient. The ratio is $\frac{L_m(\varepsilon)}{L(\varepsilon)} = 1.33$. For larger circuits (ISCAS benchmark [BrFu 85]), this ratio has been found to be approximately between 2 and 25 [CrJaDa 94].

6.2 COMPUTATION OF THE DETECTION PROBABILITY

Given a fault f, its detection probability, p_f, is the probability that a random input vector detects this fault. All the available methods to obtain this probability may be used under the assumption of independent and equally likely input patterns. Some of them can take into account *probabilities of input lines* different from 0.5. Given *probabilities of input vectors* (particularly for incompletely specified circuits), the choice is more limited.

In this section, it is assumed that the distribution of input vectors, ψ, is constant.

Section 6.2.1 is devoted to the detection probability of a fault, and Section 6.2.2 to the detection probabilities of a set of faults.

6.2.1 Detection probability of a fault

6.2.1.1 Detection function

Let f be a fault in a single output circuit. The output functions for the fault-free and faulty circuits are denoted by z_0 and z_f respectively. The set of test vectors for the fault f (Section 3.1.4.1) is defined by the logical expression[2] (where $\oplus$ denote the exclusive-OR function);

$$T_f = z_0 \oplus z_f = z_0 z'_f + z'_0 z_f, \qquad (6.37)$$

which will be called the **detection function**. If the function z_0 is incompletely specified, $T_f = 0$ for all the input vectors for which z_0 is not specified. This is illustrated in Figure 6.6. The function z_0 in Figure 6.6a is specified for all the input vectors except x_0 and x_4. The faulty function z_f in Figure 6.6b is similar to z_0 except for the input vectors x_5 and x_6. Then $T_f = (x_5 + x_6) = x_1 x'_2 x_3 + x_1 x_2 x'_3$.

In the general case, let f be a fault in a n-input, m-output, circuit. The output z_j is denoted by $z_{j,0}$ and $z_{j,f}$ for the fault-free and faulty circuits, respectively. In this case the *detection function* is defined as:

$$T_f = \sum_{j=1}^{m} z_{j,0} \oplus z_{j,f}. \qquad (6.38)$$

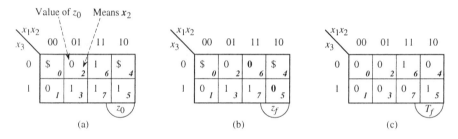

Figure 6.6 Incompletely specified circuit. (a) Fault-free function. (b) Faulty function. (c) Detection function.

Assume that the fault f in Figure 6.6 is tested by vectors randomly drawn with the constant distribution

$$\psi_1 = (0, .2, .2, .2, 0, .1, .2, .1), \qquad (6.39)$$

where $\psi_1(x_0) = \psi_1(x_4) = 0$ since the function is not specified for these two input vectors (Section 5.2.1). The detection probability is $p_f = \psi_1(x_5) + \psi_1(x_6) = 0.1 + 0.2 = 0.3$.

In this case, where the *probabilities of input vectors* are given, the only way to calculate p_f is to find the complete list of vectors in T_f because all the methods

---

[2] In Chapter 3, T_f was defined as a set of input vectors. For simplicity's sake, the same symbol is used for the function having the value 1 exactly for this set of input vectors, i.e., $T_f(x) = 1$ if and only if x is in the set T_f.

based on the probalilities of input lines cannot be applied[3]. This can be done by Boolean comparison of the fault-free and faulty outputs, by propagation through sensitized paths, or by an exhaustive simulation (specified input vectors only). It becomes unworkable when n is very large since the number of input vectors is exponential in n.

A necessary condition to avoid this difficulty is a complete specification of the circuit (we observe the *usefulness of extended specifications* defined in Section 5.2.5). In this case there is no input vector with a null probability; then an equal likelihood distribution ψ_0, or any other constant distribution obtained from probabilities of input lines, can be used without performing a test outside the specified cases.

We assume from now on that the circuits are *completely specified*. Figure 6.7 illustrates a way to obtain a lower bound and an upper bound on T_f. Consider the circuit whose output is z in the figure (*nominal circuit* made of gates G_i). The fault under consideration is $f = x_3/1$, and the gates H_j correspond to an *auxiliary circuit* which is built as explained below.

According to Section 3.1.2, an input vector is a test vector if 1) it provokes an error, and 2) it propagates the error to a primary output. The conditions for detecting the fault are ANDed by the gate H_5 (these conditions are defined on the fault-free circuit). The necessary and sufficient condition to provoke an error is that $x_3 = 0$, i.e., that the value 1 is present at the output of gate H_1. This error can be propagated to the primary output z either through path $G_1G_2G_4$ or through path $G_1G_3G_4$. The necessary and sufficient condition to propagate the error through gate G_1 (common to both paths) is that the other input of G_1 is 1. The necessary and sufficient conditions to propagate the error through G_2 is that the other input of this gate is 1 (and similarly for G_3). Now, the propagation conditions through G_4 are more complicated since it is a reconverging gate (Section 3.1.2.2): the values of input lines to G_4 are not independent. If propagation along the path $G_1G_2G_4$ is considered, it is *sufficient* that the other input[4] of G_3 is 1, i.e., $w_3 = 1$, to propagate through G_4. Note that this condition is not always necessary since, when $w_2 = D$, the propagation through G_4 is obtained also if $w_3 = D$ (Section 3.1.2.1); but we prefer to avoid a detailed study of the circuit. From the previous explanation, the auxiliary circuit in Figure 6.7a is obtained: the sufficient conditions of propagation through G_2G_4 and G_3G_4 are ANDed by gates H_2 and H_3, respectively. The OR gate H_4 means that propagation

---

[3] In Chapter 10, generation of random vectors given the *probabilities of input vectors* or given the *probabilities of input lines* will be considered. It will be shown that, in the general case, when the probabilties of input vectors are given, there is no way to find probabilities of input lines producing these probabilities. On the other hand, probabilities of input lines imply probabilities of input vectors (Section 10.1.1.2).
[4] This is because the parity is the same for both reconverging paths. If these parities are different, or if there are more than two reconverging paths, a sufficient condition can be obtained in a different way [DaWa 90].

150   Chapter 6

by one path is sufficient. The output T_f^- denotes a lower bound on T_f, i.e., $T_f^- \cdot T_f = T_f^-$, which can be written $T_f^- \leq T_f$.

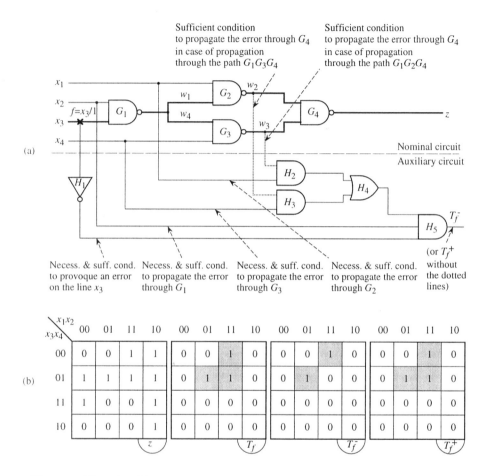

Figure 6.7 Auxiliary circuit providing bounds to the detection function. (a) Construction of the auxiliary circuit. (b) Corresponding Karnaugh maps.

Assume the dotted lines representing the sufficient conditions are deleted in Figure 6.7a. The circuit can be simplified: gates H_2 and H_3 are no longer useful since each one has a single input which is equal to its output. In the corresponding auxiliary circuit, only conditions which are *necessary* remain. Then the corresponding output, T_f^+, is an upper bound on T_f, i.e., $T_f^+ \cdot T_f = T_f$, which can be written $T_f^+ \geq T_f$.

The reader may verify that the functions obtained correspond to the Karnaugh maps presented in Figure 6.7b. It is clear that $T_f^- \leq T_f \leq T_f^+$.

There are several possible uses for T_f^- and T_f^+. For example, T_f^- may be a useful measure for automatic test generation. For random testing we are concerned with the following properties:

$$\Pr[T_f^- = 1] \leq p_f \leq \Pr[T_f^+ = 1], \tag{6.40}$$

$$\text{(Number of minterms in } T_f^-) \leq k_f \leq \text{(Number of minterms in } T_f^+). \tag{6.41}$$

For a conservative measure of the detection probability, T_f^- is more useful than T_f^+. In our example, $T_f^- < T_f$. However, there are many cases where $T_f^- = T_f$. A particularly important case is the following: *if there is no reconvergent fanout in the downstream cone of a line w*, then, for a stuck-at fault of this line, the lower bound is equal to the upper bound of the detection function; i.e.,

$$T_{w/0}^- = T_{w/0}^+ = T_{w/0} \quad \text{and} \quad T_{w/1}^- = T_{w/1}^+ = T_{w/1}. \tag{6.42}$$

This result applies to all the stuck-at faults of the checkpoints x_1, x_4, w_1, w_2, w_3, and w_4 for the nominal circuit in Figure 6.7a; only the checkpoints x_2 and x_3 have a reconvergent fanout in their downstreams.

Remark 6.5 The construction of the functions T_f^- and T_f^+ can be generalized to other faults such as multiple stuck-at faults and nonfeedback bridging faults.

❑

The next section presents an approximate method, based on the detection function, requiring probabilities of input lines, whose complexity is less than exponential in n.

6.2.1.2 *Extended Cutting Algorithm*

The aim of the Cutting Algorithm is to find the probability of an output signal being equal to logic value 1 [SaDiBa 84]. The Extended Cutting Algorithm (EC algorithm) is an improved version of this [DaWa 90]. The basic notions are presented with the help of an example: the function T_f^- in Figure 6.7a. This will be illustrated in Figure 6.9.

Let us first present the probability of a gate output being equal to 1, given the probabilities of the gate inputs being equal to 1, when these *probabilities are independent* from each other. These probabilities are given in the left-hand side of Figure 6.8 for AND, OR, and NOT gates. From these expressions, the probabilities for NAND and NOR gates can be deduced (right-hand side).

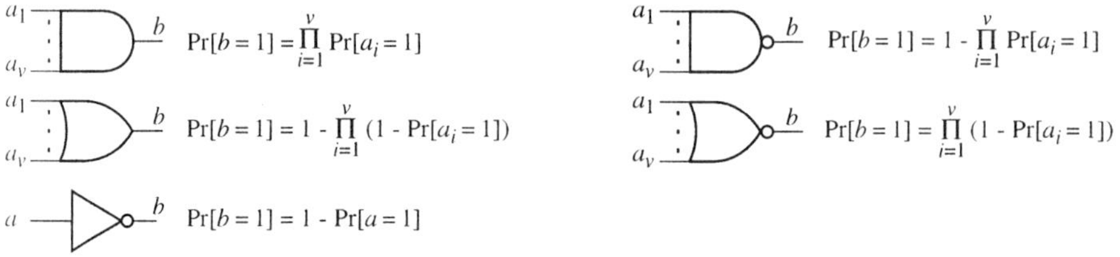

152   Chapter 6

$$\Pr[b=1] = \prod_{i=1}^{v} \Pr[a_i = 1]$$

$$\Pr[b=1] = 1 - \prod_{i=1}^{v} \Pr[a_i = 1]$$

$$\Pr[b=1] = 1 - \prod_{i=1}^{v} (1 - \Pr[a_i = 1])$$

$$\Pr[b=1] = \prod_{i=1}^{v} (1 - \Pr[a_i = 1])$$

$$\Pr[b=1] = 1 - \Pr[a=1]$$

Figure 6.8 Propagation of independent probabilities through primitive gates.

Consider the circuit whose output is T_f^- in Figure 6.9a and assume the probability of every input is 0.5 (and they are independent). According to Figure 6.8, $\Pr[w_1 = 1] = 1 - \Pr[x_2 = 1] \cdot \Pr[x_3 = 1] = 0.75$. One can also calculate the exact probabilities $\Pr[w_2 = 1] = \Pr[w_3 = 1] = 0.625$. However, we cannot calculate the exact probabilities of the outputs of gates H_2 to H_5, by the expressions in Figure 6.8, because their input probabilities are not independent. In order to avoid the reconvergence, *some lines of the circuit are cut*.

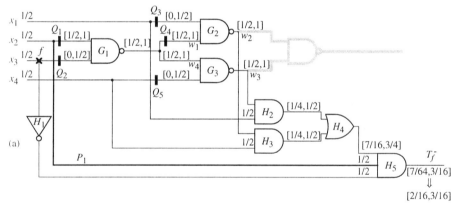

Figure 6.9 (a) Application of the EC algorithm. (b) Restricted-range assignments[5].

---

[5] Figure 6.9b is reprinted from [Jo 86] (© 1986 IEEE). Initially, eight cases were considered [SaDiBa 84]; in [Jo 86] it was observed that they can be gathered into only four cases.

Consider the input x_2. It is a fanout stem from which several reconverging paths start: $G_1G_2H_3H_4$, $G_1G_3H_2H_4$, and a direct path reconverge to the gate H_5. If the fanout branch Q_1 is cut, a single path remains between the stem x_2 and the input of H_5, noted P_1 in Figure 6.9a. Similarly, stems Q_2, Q_3, Q_4, and Q_5 can be cut and the circuit obtained is a tree: no reconvergent fanout is left. Bounds of the signal probability can be inserted at each cut point, thus guaranteeing that all the signal probability bounds computed by propagation on the tree enclose the true values. The probability range [0, 1] can always be used. In some cases, the restricted-range $[0, p]$ or $[p, 1]$ may be used at the cut Q_i instead of the full-range, where p is the actual probability of the line. For some cuts in the circuit, the value of p may be not exactly known; if p is known to be in a range $[l, u]$, the restricted range associated with the cut may be either $[0, u]$ or $[l, 1]$.

Consider for example the cut Q_1. The path from Q_1 to the output passes through G_1, G_3, H_2, H_4, and H_5. The "other reconverging branch" (other fanout branch from the same fanout stem as Q_1) is P_1. These two branches reconverge to H_5 which is an AND gate. The parity of P_1 is even. According to Figure 6.9b, since the reconverging gate is an AND gate and the parity of the other reconverging path is even, the restricted-range is $[l, 1]$. The exact signal probability at Q_1 is $1/2 = l = u$. It follows that the restricted range associated with Q_1 is $[1/2, 1]$.

The restricted-range provides tighter bounds than the full-range [0, 1]. The following formulas are used.

Tree formulas for propagating signal probability bounds[6]

Let l_j and u_j be a lower bound and an upper bound, respectively, of the line j signal probability. This is denoted by $[l_j, u_j]$.

    NOT gate.   Input probability range: $[l, u]$.
                       Output probability range: $[1 - u, 1 - l]$.   (6.43)

    AND gate.   Input probability ranges: $[l_j, u_j], j = 1, \ldots, v$.
                       Output probability range: $\left[\prod_{i=1}^{v} l_i, \prod_{i=1}^{v} u_i\right]$.   (6.44)

    OR gate.   Input probability ranges: $[l_j, u_j], j = 1, \ldots, v$.
                       Output probability range: $\left[1 - \prod_{i=1}^{v}(1 - l_i), 1 - \prod_{i=1}^{v}(1 - u_i)\right]$.   (6.45)

The formulas for NAND and NOR can be easily derived from the preceding ones.

❏

In Figure 6.9a, these formulas are applied, after the assignment of the restricted ranges to every cut[7]. The range [7/64, 3/16] is obtained at the output,

---

[6] Formulas (6.43) to (6.45) are reprinted from [SaDiBa 84] (© 1986 IEEE).

154   Chapter 6

which means that $7/64 \leq \Pr[T_f^- = 1] \leq 3/16$. The number of minterms in T_f^- is necessarily integer; since the 16 input vectors are equally likely, 7/64 may be replaced by $\dfrac{\lceil 7 \cdot 16/64 \rceil}{16} = \dfrac{\lceil 1.75 \rceil}{16} = \dfrac{2}{16}$. Hence the range is finally [2/16, 3/16], which means that the lower bound of the detectability, i.e., the number of minterms in T_f^-, is $2 \leq k_f^- \leq 3$ (the exact value is 2 as shown in Figure 6.7b).

6.2.2 Detection probabilities for a set of faults

Methods which can be used for a detailed study of one fault may become unworkable for a complete set of faults, which is usually the set of single stuck-at faults. The controllability and observability values related to the value of a line provide a way to estimate the detection probabilities of single stuck-at faults. Simulation is another method. Estimation of a lower bound of the minimum detection probability is a conservative way. Naturally, only the *detectable faults* have to be taken into account; this is usually implicit, particularly when a lower bound is considered.

6.2.2.1 Controllability, observability, and activity

The use of controllability and observability has been proposed in various methods. We shall now explain the main ideas with small examples. It is assumed that the circuits are completely specified. Consider the fault w/0, where w is some line in the circuit. The probability that a random input vector provokes an error on w is the probability that $w = 1$ in the fault-free circuit. This probability is called **1-controllability** and is denoted[8] by $C(1,w)$. The concept of **0-controllability** can be defined in a similar way.

$$C(1, w) = \Pr[w = 1] \quad \text{and} \quad C(0, w) = \Pr[w = 0], \tag{6.46}$$

$$C(0, w) + C(1, w) = 1. \tag{6.47}$$

If an error is present on w, i.e., $w = D$ in the example w/0, this error must be propagated to a primary output to be detected. The probability of this propagation is called the **observability**. If the circuit is a tree this observability is independent of the value on w, i.e., D or $\overline{D}$, and is easy to find; it is denoted

---

[7] In this example there is only one "other path" for each cut. When there are more than 2 fanout branches for a fanout stem, there are several "other paths". A way to affect the restricted-ranges when there are several cuts for a fanout stem is proposed in [DaWa 86 & 90]. It is also proposed that the only path which is not cut be the path with the minimum number of gates between the fanout stem and the reconverging gate.

[8] This notation is not the usual one. However, it is consistent with the notation in Appendix E for controllability of black-box input combinations.

by $O(w)$. When there are fanout stems the computation of observability is more difficult.

Consider the tree circuit in Figure 6.10 where the probabilities of the input lines are 0.5 and are independent from each other. The 1-controllabilities are calculated progressively from the primary inputs to the primary output; note that the 0-controllabilities are obtained by (6.47). For the primary inputs, $C(1, x_i) = \Pr[x_i = 1] = 0.5$. The 1-controllabilities $C(1, w)$, and then $C(1, z)$ are obtained from the propagation equations in Figure 6.8. Once the controllabilities are known, the observabilities are calculated in the reverse order. Since z is a primary output, $O(z) = 1$. The line w is observable if z is observable and $x_3 = 1$ (because G_2 is an AND gate); then $O(w) = O(z) \cdot C(1, x_3) = 0.5$. An input of AND or NAND (respectively OR or NOR) gate is observable if 1) its output is observable and 2) all the other inputs are 1 (respectively 0). These events are independent in a tree: their probabilities may be multiplied. This is illustrated in Figure 6.10.

Tree circuit	Controllabilities	Observabilities
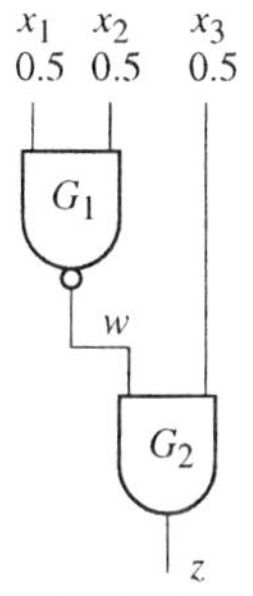 x_1 x_2 x_3 0.5 0.5 0.5 G_1 w G_2 z	$C(1,x_1) = 0.5$ $C(1,x_2) = 0.5$ $C(1,x_3) = 0.5$ $C(1,w) = 1 - C(1,x_1) \cdot C(1,x_2) = 0.75$ $C(1,z) = C(1,w) \cdot C(1,x_3) = 0.375$	$O(x_1) = O(w) \cdot C(1,x_2) = 0.25$ $O(x_2) = O(w) \cdot C(1,x_1) = 0.25$ $O(x_3) = O(z) \cdot C(1,w) = 0.75$ $O(w) = O(z) \cdot C(1,x_3) = 0.5$ $O(z) = 1$

Figure 6.10 Propagation of controllability and observability probabilities in a tree circuit.

Given a fault, its detection probability can be calculated from these controllabilities and observabilities. Consider for example the fault $w/1$:

$$p_{w/1} = C(0, w) \cdot O(w) = (1 - C(1, w)) \cdot O(w) = (1 - 0.75)(0.5) = 0.125.$$

The OR bridging (Section 2.2.3.1) of lines w and x_3, denoted by $w+x_3$, is detected if 1) the *combination* $(w, x_3) = (0, 1)$ or $(1, 0)$ is present and 2) the output of G_2 is observable. Then

$$p_{w+x_3} = \big(C(0, w) \cdot C(1, x_3) + C(1, w) \cdot C(0, x_3)\big) \cdot O(z) = 0.5.$$

Examples of black-box-faults are presented in Appendix D.

Chapter 6

One must now tackle the tricky problem of *circuits with fanout stems*. The concepts of controllability and observability are no longer sufficient for calculating the detection probabilities of various faults. Let us present the concept of *activity of an input pattern at a gate input* (or more generally at the input of a subcircuit called a black box, see Appendix D). The main notions are presented with the help of a simple example.

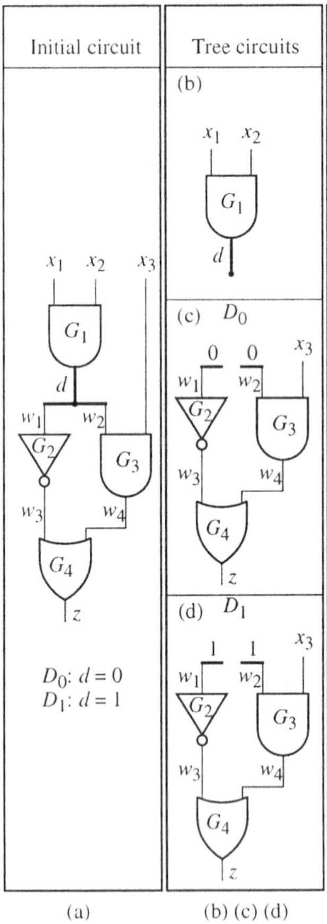

Initial circuit	Tree circuits	Conditional controllabilities	Conditional observabilities	Conditional activities
(a) x_1 x_2 x_3 G_1 d w_1 w_2 G_2 G_3 w_3 w_4 G_4 z $D_0: d=0$ $D_1: d=1$	(b) x_1 x_2 G_1 d	$C(1,x_1) = 0.5$ $C(1,x_2) = 0.5$ $C(1,d) = 0.25$	$O(x_1|D_0) = 0.25$ $O(x_1|D_1) = 0.25$ $O(x_2|D_0) = 0.25$ $O(x_2|D_1) = 0.25$ $O(d|D_0) = 0.5$ * $O(d|D_1) = 0.5$ *	$A(00, G_1|D_0) = 0.125$ $A(01, G_1|D_0) = 0.125$ $A(10, G_1|D_0) = 0.125$ $A(11, G_1|D_0) = 0.125$ $A(00, G_1|D_1) = 0.125$ $A(01, G_1|D_1) = 0.125$ $A(10, G_1|D_1) = 0.125$ $A(11, G_1|D_1) = 0.125$
	(c) D_0 0 0 x_3 w_1 w_2 G_2 G_3 w_3 w_4 G_4 z	$C(1,x_3|D_0) = 0.5$ $C(1,w_1|D_0) = 0$ $C(1,w_2|D_0) = 0$ $C(1,w_3|D_0) = 1$ $C(1,w_4|D_0) = 0$ $C(1,z|D_0) = 1$	$O(x_3|D_0) = 0$ $O(w_1|D_0) = 1$ $O(w_2|D_0) = 0$ $O(w_3|D_0) = 1$ $O(w_4|D_0) = 0$ $O(z|D_0) = 1$	$A(0, G_2|D_0) = 1$ $A(1, G_2|D_0) = 0$ $A(00, G_3|D_0) = 0$ $A(01, G_3|D_0) = 0$ $A(10, G_3|D_0) = 0$ $A(11, G_3|D_0) = 0$ $A(00, G_4|D_0) = 0$ $A(01, G_4|D_0) = 0$ $A(10, G_4|D_0) = 1$ $A(11, G_4|D_0) = 0$
	(d) D_1 1 1 x_3 w_1 w_2 G_2 G_3 w_3 w_4 G_4 z	$C(1,x_3|D_1) = 0.5$ $C(1,w_1|D_1) = 1$ $C(1,w_2|D_1) = 1$ $C(1,w_3|D_1) = 0$ $C(1,w_4|D_1) = 0.5$ $C(1,z|D_1) = 0.5$	$O(x_3|D_1) = 1$ $O(w_1|D_1) = 0.5$ $O(w_2|D_1) = 0.5$ $O(w_3|D_1) = 0.5$ $O(w_4|D_1) = 1$ $O(z|D_1) = 1$	$A(0, G_2|D_1) = 0$ $A(1, G_2|D_1) = 0.5$ $A(00, G_3|D_1) = 0$ $A(01, G_3|D_1) = 0$ $A(10, G_3|D_1) = 0.5$ $A(11, G_3|D_1) = 0.5$ $A(00, G_4|D_1) = 0.5$ $A(01, G_4|D_1) = 0.5$ $A(10, G_4|D_1) = 0$ $A(11, G_4|D_1) = 0$
(a)	(b) (c) (d)			

* These values cannot be calculated easily: see Appendix E.

Figure 6.11 Conditional controllabilities, observabilities, and activities, when the circuit is not a tree. Heavy lines refer to fanout points.

Consider the AND gate G_3 in the circuit in Figure 6.11a. Assume a fault on this gate which provokes an error if the input pattern $w_2 x_3 = 11$ is present at this gate input (this fault could be either $w_2/0$ or $x_3/0$ or $w_4/0$). This fault is detected if 1) the input pattern $w_2 x_3 = 11$ is present at the gate input, and 2) the error on w_4

is propagated to the primary output z. If both conditions are satisfied, the pattern $w_2x_3 = 11$ is said to be **active**. For example, if the input vector is $x_1x_2x_3 = 111$, then the pattern $w_2x_3 = 11$ is active since: 1) $w_2 = 1$ and $x_3 = 1$ (the pattern is present), and 2) $w_3 = 0$ (the error is propagated). The **activity** of the pattern is the *probability that it is active when a random input vector is applied to the circuit*; it may be denoted by $A(w_2x_3 = 11, G_3)$. If an order of inputs is defined for each gate, *the notation can be simplified*. For example, let the inputs of each gate be ordered *from left to right* in the examples we will consider. Then $A(w_2x_3 = 11, G_3)$ can be simply written $A(11, G_3)$. If a fault related to a gate may be detected by several input patterns at this gate input, the detection probability of this fault is the sum of the corresponding activities. For example, $f = w_4/1$ may be detected by $w_2x_3 = 00$ or 01 or 10; then $p_f = A(00, G_3) + A(01, G_3) + A(10, G_3)$. When all the activities are known, *the detection probability may be obtained for various kinds of faults*. Let us calculate the activities for all the input patterns of all gates of the circuit in Figure 6.11a.

In a general case, there are several fanout stems and D_i denotes a pattern of values at these stems. For example, if there are three fanout stems d_1, d_2, and d_3, D_0 represents $d_1d_2d_3 = 000$, D_1 represents $d_1d_2d_3 = 001$, etc (if the various d_j are not independent, some patterns D_i may never appear). In our example, where there is a single fanout stem, D_0 represents $d = 0$ and D_1 represents $d = 1$. Given a pattern D_i, each fanout stem has a definite value, 0 or 1; then the circuit may be split into a set of tree circuits whose inputs are primary inputs or constant values corresponding to the fanout stems. In our example, for D_0, the tree circuits in Figure 6.11b and c are obtained; for D_1, the first tree circuit is the same, and the second one is shown in Figure 6.11d. For every tree circuit the *conditional controllabilities* are calculated as explained above and illustrated in Figure 6.10. For example, from Figure 6.11c $C(1, w_1 \mid D_0)$ is obtained; this is the 1-controllability of w_1 given the pattern D_0. The conditional 0 and 1-controllabilities are complementary like the unconditional ones: $C(0, w_1 \mid D_0) + C(1, w_1 \mid D_0) = 1$. If all the inputs of a tree circuit are primary inputs (Figure 6.11b for example), $C(1, w \mid D_i) = C(1, w)$ for all patterns D_i and any w in this circuit. The *conditional observabilities* are also calculated from the output to the inputs of each tree circuit. When the output of the tree circuit is a primary output there is no difficulty (Figures 6.11 c and d). However, if it is a fanout stem its observability must be previously calculated; this is more difficult as explained in Appendix E.

When all the conditional controllabilities and observabilities are known, the *conditional activities* can easily be derived. Consider for example the gate G_3 with inputs w_2 and x_3, and output w_4. For example, for the pattern D_1 the following equation is obtained.

$$A(10, G_3 \mid D_1) = C(1, w_2 \mid D_1) \cdot C(0, x_3 \mid D_1) \cdot O(w_4 \mid D_1) = 0.5. \tag{6.48}$$

All the conditional activities are presented in Figure 6.11. We now have to calculate the activities, taking into account the probabilities of the various patterns which are not difficult to calculate. In our example $\Pr[D_0] = \Pr[d = 0] = 0.75$ and $\Pr[D_1] = 0.25$. Then, for example,

$$A(10, G_3) = A(10, G_3 | D_0) \cdot \Pr[D_0] + A(10, G_3 | D_1) \cdot \Pr[D_1]$$

$$= 0 \times 0.75 + 0.5 \times 0.25 = 0.125. \tag{6.49}$$

Once all the activities are known, fault detection probabilities can be determined. Assume that the fault $x_3/0$ is present in the circuit. An error is provoked if the pattern $w_2 x_3 = 10$ is present at the input of gate G_3. Then $p_{x_3/0} = A(10, G_3) = 0.125$.

Remark 6.6 The examples related to the cutting algorithm, controllability, observability, and activity (Figures 6.9 to 6.11), correspond to $\Pr[x_i = 1] = 0.5$. These methods can be used if other *probabilities of input lines* are given, not necessarily the same for all. Nevertheless, in the general case, these methods cannot be used if only the *probabilities of the input vectors* are given (as will be explained in Section 10.1.1.2, probabilities of input lines imply probabilities of the input vectors, but the reciprocal property is not true).

6.2.2.2 *Simulation*

Fault simulation consists of simulating a circuit in the presence of faults. Both the fault-free and the faulty models of the circuit are available in the host computer. In *parallel fault simulation*, the fault-free circuit and a fixed number of faulty circuits are simultaneously simulated; considerable simulation time is saved. The faults tested by the input sequence S can be determined by comparing the fault simulation results with those of the fault-free simulation of the circuit simulated with the same sequence S. If the sequence S contains all the possible input vectors (each vector appearing once), this is an exhaustive simulation; then the number of times each fault is detected is equal to its detectability.

When the number of primary inputs is too large, an exhaustive simulation is not realistic: for example, if $n = 100$ the number of input vectors is about $2^{100} \approx 10^{30}$. In this case, the detection probabilities of the faults may be estimated from the simulation of a very long random test sequence (whose length is nevertheless much lower than the length of an exhaustive test). Assume that the input vectors are equally likely and the number of vectors in the simulated sequence is N. If the fault f is detected by $N(f)$ input vectors among the N, then the detection probability of f can be estimated by

$$p_f \approx \frac{N(f)}{N}. \tag{6.50}$$

For large N, if $N(f)$ is large enough, the distribution of p_f can be approximated with a normal distribution whose standard deviation is [BaMcSa 87]

$$\sigma_f \approx \sqrt{\frac{N(f)}{N^2}\left(1 - \frac{N(f)}{N}\right)}. \qquad (6.51)$$

According to (6.51), the estimate of p_f includes a random error that approaches zero when N tends to infinity. If $N(f)$ is small the estimation of p_f is not accurate. However, the main problem occurs when $N(f) = 0$. This may correspond either to a redundant fault or to a very resistant fault. It is important to distinguish between the two cases by an accurate study of the fault: if it is redundant it is quite normal that it is not detected; if it is very resistant, the circuit may be modified to increase its detection probability (Chapter 13).

In Appendix D, a simulation to estimate the activities of input patterns of subcircuits called black-boxes is presented.

6.2.2.3 Lower bounds on detectability

The estimation of fault detectabilities for a set of faults is a time consuming process, even with approximate methods. Even the worst case detectability, k_{min}, may be difficult to obtain since the most difficult fault is not known *a priori*. A lower bound of k_{min} is then usually considered satisfactory if this lower bound implies a test length which is reasonable.

Let us recall that only *detectable combinational faults* are considered for the time being. Hence, at least one test vector exists for any fault in the circuit; a rather self-evident lower bound is then:

$$k_{min} \geq 1. \qquad (6.52)$$

Consider n-input, m-output circuits.

Notation 6.1
a) The number of input vectors detecting the fault f at the primary output z_j is called the detectability of f related to z_j, and is denoted by $k_{j,f}$.
b) For a set of faults F, $k_{j,min} = \min_f k_{j,f}$, such that $f \in F$ and $k_{j,f} \neq 0$.
c) The number of primary inputs upstream of the output z_j is denoted by n_j.

❑

It is clear that

$$k_f \geq \max_j k_{j,f}, \quad j = 1, \ldots, m, \qquad (6.53)$$

since the fault f which is detected by $k_{j,f}$ input vectors at output z_j may be detected, in addition, by other input vectors at other outputs. For example, in Figure 6.1 (Section 6.1.1.1), $k_{1,f} = 2$, $k_{2,f} = 2$, and $k_f = 3 \geq \max(2, 2)$.

Property 6.5 The detectability of the most resistant fault in the set of detectable multiple stuck-at faults of a m-output circuit is

$$k_{\min} \geq \min_j k_{j,\min}, \quad j = 1, ..., m, \tag{6.54}$$

$$k_{\min} \geq \min_j 2^{n-n_j} = 2^{n-\max_j n_j}, \quad j = 1, ..., m. \tag{6.55}$$

Proof A detectable fault, f, is detectable at least at one primary output z_j. Then $k_f \geq k_{j,\min}$ and inequality (6.54) follows.

The fault f is detected by at least one combination of the n_j values of the primary inputs upstream of z_j, since only stuck-at faults are considered. This combination is independent of the $n - n_j$ primary inputs which are not upstream of z_j; thus

$$k_{j,f} \geq 2^{n-n_j}. \tag{6.56}$$

From (6.53) and (6.56),

$$k_f \geq 2^{n-n_j}. \tag{6.57}$$

Since $k_{\min} = \min_f k_f$, (6.55) is obtained from (6.57).

□

Consider, for example, the circuit in Figure 6.12. As illustrated in heavy lines, the upstream cone of z_1 contains $n_1 = 7$ primary inputs. The output z_1 does not depend on $n - n_1 = 5$ primary inputs, and any fault f detectable at z_1 is detectable by at least $2^{n-n_1} = 32$ input vectors. Similarly, $2^{n-n_2} = 32$. Then, from (6.55), $k_{\min} \geq 32$.

Let us now state a result related to a class of irredundant functions called multiplexing functions. The proof is omitted since it is not important for our purposes; it can be found in [DaBl 76].

Definition 6.2 Let a function be represented by its complete sum-of-products (Remark 3.2 in Section 3.1.3). The function $z = \sum_a m_a$ is called a multiplexing function if, 1) all products are disjoint prime implicants and 2) each product m_a may not have a consensus with more than one other product with respect to each of its literals. In other words, the product term obtained by deleting any literal of m_a can have a non-zero intersection with no more than one other product.

□

Random Test Length for Combinational Circuits

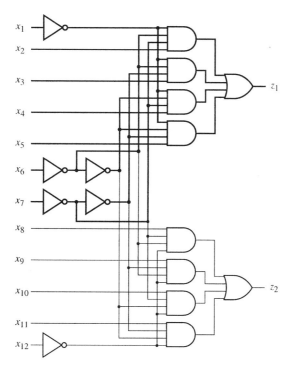

Figure 6.12 Dual 4 line to 1 line multiplexors[9]. The upstream cone of z_1 is shown in heavy lines.

For example, the function

$$z_1 = m_1 + m_2 + m_3 + m_4 = x'_1 x_2 x'_6 x'_7 + x'_1 x_3 x'_6 x_7 + x'_1 x_4 x_6 x'_7 + x'_1 x_5 x_6 x_7 \qquad (6.58)$$

in Figure 6.12 is a multiplexing function according to Definition 6.2, as is z_2.

Property 6.6 Let $z_j = \sum_a m_a$ be a multiplexing function. The detectability of the most resistant fault in the set of multiple stuck-at faults detectable at the output z_j is

$$k_{j,\min} \geq \min_a 2^{n-n_a-1} = 2^{n-\max_a n_a - 1}. \qquad (6.59)$$

❑

Return to the circuit in Figure 6.12. According to (6.58), $\max_a n_a = 4$. From (6.59), $k_{1,\min} \geq 2^{12-4-1} = 2^7 = 128$. Similarly $k_{2,\min} \geq 128$. Then, from (6.54), $k_{\min} \geq 128$. This means that any single or multiple stuck-at fault in the circuit of Figure 6.12 has a detectability which is greater than or equal to 128.

---

[9] This logic diagram is taken from [Se 75], reference SF.C 4153 E.

6.3 NUMERICAL RESULTS

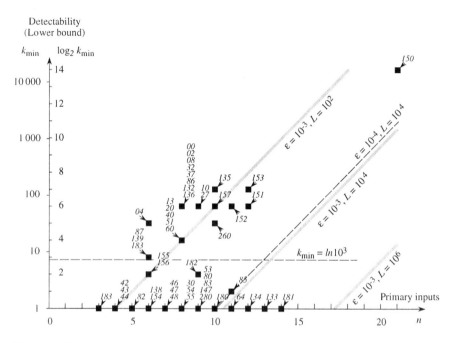

Figure 6.13 Upper bounds of test lengths for standard TTL integrated circuits[10] (SSI and MSI).

Figure 6.13 presents upper bounds of test lengths for a number of standard TTL packages (SSI and MSI circuits), given $\varepsilon = 10^{-3}$ (for most of the examples). Each package is represented by a number which is used by various manufacturers. For example, *181* represents a 14-input, 8-output, arithmetic and logic unit (ALU) which is referred to as *SN 54 181 M* (Texas Instruments), *SF.C 4 181 EM* (Sescosem), or *MIC 54 181 J* (I.T.T.). Each package is represented by a point $(n, \log_2 k_{min})$ in the plane; for example, circuit *260* has 10 inputs and $\log_2 k_{min}$ is 5. All the values for k_{min} are lower bounds which were roughly estimated using Properties 6.5 and 6.6 in Section 6.2.2.3: for the package *153*, a lower bound of k_{min} was calculated at the end of Section 6.2.2.3; for the 21-input, 1-output, multiplexer *150*, a lower bound $k_{min} \geq 2^{14}$ is found using Property 6.6; for the ALU no better lower bound that $k_{min} \geq 1$ (6.52) was found. The straight lines labelled with the values ε and L indicate circuits which are testable by a test

---

[10] This figure is reprinted from [DaBl 76] (© 1976 IEEE).

sequence of length L with probability $1 - \varepsilon$. They are obtained using Equations (6.33) and (6.36) in Section 6.1.2.2. For example, circuit *156* has 6 inputs and $\log_2 k_{min} = 2$, and is testable with a test sequence of length 100 with $\varepsilon = 10^{-3}$. Circuit *157* is testable by a sequence of the same length. The higher detectability of this circuit compensates for the higher number of inputs. The dotted straight line corresponding to $\varepsilon = 10^{-4}$ and $L = 10^4$ shows that the necessary length is not very sensitive to ε.

Let us note that: 1) with a random test sequence of about $L = 10^5$ (about 0.1 sec if the test frequency is 1 MHz), the fault coverage is at least 0.999 for all the circuits represented in Figure 6.13; for every circuit above the dotted line, corresponding to $k_{min} > \ln(1/\varepsilon)$, the random testing is shorter than exhaustive testing.

The lower bounds in Figure 6.13 may be far from the exact value of k_{min}. For the ALU, for example, the lower bound $k_{min} \geq 1$ was obtained while the true value is $k_{min} = 96$ for the set of single stuck-at faults. With this value, given $n = 14$, from Equations (6.36) and (6.33) in Section 6.1.2.2, $L_m(10^{-3}) = 1\ 176$ is obtained.

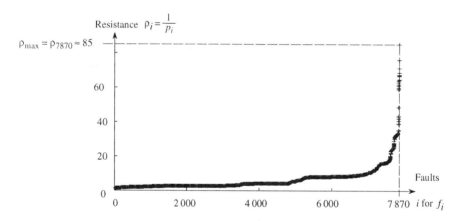

Figure 6.14 Resistance profile of the circuit *c6288* of the ISCAS benchmark[11].

Finally, let us present the resistance profile of a circuit of the ISCAS benchmark [BrFu 85]. The circuit *c6288* is a 32-input, 32-output, combinational circuit made of 2 416 gates. The number of black-box faults (Appendix D) is 7 870. The resistance profile in Figure 6.14 was obtained by simulation; thus the value of $\rho_{max} \sim 85$, corresponding to $p_{min} \approx 0.012$, is approximate. One can observe that the appearance of the resistance profile is *roughly* a *convex* graph such that: 1) the slope has a small value in the part correponding to easy faults

---

[11] This figure is reprinted from [CrJaDa 94] (© 1994 IEEE).

(left-hand side), i.e., there are many faults easy to detect (for any combinational circuit the easiest stuck-at fault is always such that $1 \leq \rho_{min} \leq 2$); the slope is much larger in the part correponding to hard faults (right-hand side). These features were observed for all the big circuits [CrJaDa 94], [PrDa 97].

NOTES and REFERENCES

Many authors have studied random testing of combinational circuits over a long period of time. The fundamentals were established in the early seventies: [Ra 71], [AgAg 72], [TeDa 74], [AgAg 75], [ShMc 75], [DaBl 76].

Several names were used for the same entity. The *detectability* of a fault, used in [MaYa 84], was called *detection surface* of the fault in [DaBl 76]; in this paper, the minimun value k_{min} was called *detection surface of the circuit*. The expression *random pattern resistant* was introduced in [EiLi 83].

The *detectability profile* and the *resistance profile* were introduced in [MaYa 84] and [CrJaDa 94], respectively. Another measurement, the *detection probability distribution*, is used in [SeAgFa 90]: it is a distribution $p(x)$ such that $p(x)dx$ is the fraction of detectable faults with probability of detection between x and $x + dx$. It is a normalized detectability profile if the vectors are equally likely; however, the detection probability distribution, like the resistance profile, may be used even if the vectors are not equally likely. In [PrDa 96], a continuous approximation of the resistance profile is presented: it leads to a 2-parameter measurement of the testability for a set of faults (one of these parameters is the resistance of the worst case fault. From this measurement, a rough estimation of the test length required for a fault coverage close to 1 can be obtained from a relatively short simulation of a random test sequence [PrDa 97].

A way to build *auxiliary circuits* associated with a fault in a nominal circuit, corresponding to a lower bound and an upper bound of the detection function as illustrated in Figure 6.7, can be found in [DaWa 86 & 90]. A very low bound considering propagation by a single path was used in [SaBa 84] (auxilliary gate), where the *Cutting Algorithm* is introduced. The Extended Cutting Algorithm [DaWa 86 & 90] introduces various improvements allowing tightest lower bounds of the detection probabilities of faults to be obtained. Given an auxiliary circuit, the probability that its output has the value 1 could be calculated by an *algebraic* method [PaMc 75a & 75b]; this method suffers from an exponential storage problem.

Various works have been devoted to calculation of controllability and observability. SCOAP [Go 79] introduced the first measures of testability. Normalized measures of controllability and observability were proposed in TMEAS [Gr 79] and CAMELOT [BeMaRo 81]. Unfortunately, these measures are not independent and their products are not always a good indication of the testability [Sa 83]. STAFAN [AgJa 85], PREDICT [SePaAg 85], and the method of Chakravarty and Hunt [ChHu 90], are more recent methods taking this fact into account; the last two methods use the notion of supergate. Our Section 6.2.2.1 and Appendix E are drawn mainly from the work of Simeu [Si 92]. There are some similarities with PREDICT; in case of multi-output circuit, Simeu's method gives the exact value while PREDICT gives a lower bound of the actual detectability. Naturally, the complexity of the circuits which can be studied by these methods is limited.

Parallel simulation was introduced in [Se 65]. This topic is detailed in [Fu 85], [AbBrFr 90]. The application with the black-box fault model is presented in Appendix D.

Only permanent faults have been considered in the body of this chapter. For an *intermittent* combinational fault, the test length must be divided by the proportion of time when the fault is active, to obtain the same probability of testing (because there is no memory effect and the periods of fault activity are assumed to be random) [DaTh 80b].

Methods for sequential circuits should be applied for sequential faults in combinational circuits (Remark 7.2 in Section 7.2.3). The method for obtaining an upper bound for an important class of sequential faults is presented in Remark E.1 in Appendix E.

7

Random Test Length for Sequential Circuits

According to Section 2.1.2, a distinction should be made between a synchronous circuit and an asynchronous circuit. The difference, as far as testing is concerned, is first presented (Section 7.1). Then synchronous sequential circuits are considered in the main part of the chapter; however, the methods presented could be adapted to asynchronous circuits.

A sequential fault may be analysed using a Markov chain model, as already stated in Section 5.3. The detection process is studied in more detail in this chapter, particularly the behavior for a long test sequence (Section 7.2). Then approximate methods are presented in Section 7.3.

Although this book concentrates on testing digital circuits, the exact and approximate methods presented in this chapter may be applied to other systems which are modeled by finite state machines, particularly certain types of program and communication protocols.

7.1 ASYNCHRONOUS AND SYNCHRONOUS TESTS

How should asynchronous and synchronous circuits be tested? This question draws inspiration from Section 5.2 which explains that all the specified cases and only the specified cases should be tested to perform a good test. The analysis in this section, which is not usual, is related to both *random and deterministic* testing. The aim of this section is to focus on the test of *all the specified cases*. In order to simplify the presentation, we assume that the circuits considered are completely specified.

Consider first an asynchronous circuit. According to Section 2.1.2.2, it is assumed that the circuit works in fundamental mode, i.e., that two input variables do not change simultaneously. Then the test of this circuit should be as illustrated in Figure 7.1 for a single output circuit. Periodically, a single input, randomly chosen, is changed, and the output state is tested (compared with the output state of a fault-free circuit according to Section 5.1). At t_1 a primary input changes, the output is compared with the fault-free one when it is stable, between t_1 and t_2. This method corresponds to comparison times represented by comparison C, as shown in Figure 7.1.

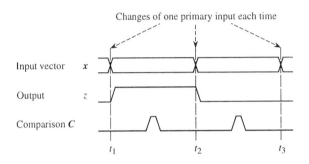

Figure 7.1 Illustration of an asynchronous test.

Consider now a synchronous circuit. The input vector is randomly changed periodically at times t_1, t_2, ..., according to Figure 7.2. Between two changes there is a clock pulse. Thus, three intervals of time are defined within a period: (t_i, t'_i), (t'_i, t''_i), and (t''_i, t_{i+1}). Figure 7.2 illustrates when the output changes for a latch type and flip-flop type circuit (see Section 2.2.1). For every case (i.e., latch type and both flip-flop types) the output is stable for three consecutive intervals, according to Figure 7.2. It follows that Comparison C_1 or C_2 or C_3 could be used.

If the output is faulty for three consecutive intervals, then any one of these comparisons will detect the fault. However, as shown in Appendix F, faults remain which can be detected only by a strict subset of the comparisons $\{C_1, C_2, C_3\}$; for example, if an error is present only when $CK = 1$, this fault can be detected only by Comparison C_2. These faults can be detected by the extended synchronous test illustrated in Figure 7.2. In addition, faults exist, detectable by an asynchronous test, which cannot be detected by the extended synchonous test. Appendix F is devoted to these problems.

An asynchronous test can be considered for small synchronous circuits (Section 7.4), but it is of course difficult to implement for a large circuit such as a microprocessor. From now on we shall return to the usual synchronous model of digital circuits.

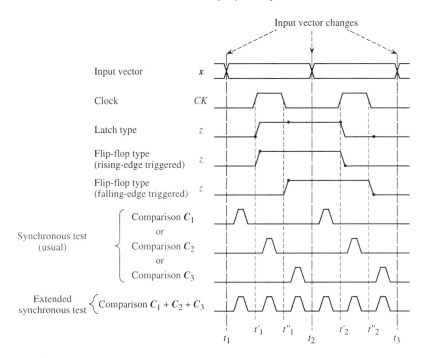

Figure 7.2 Illustration of a synchronous test.

7.2 TEST LENGTH FOR A FAULT

According to Section 5.3.1, the detection process of a fault in a sequential circuit can be modeled by a *finite Markov chain*. The basic concepts and notation related to these chains are presented in Appendix C. Section 7.2.1 gives an exact calculation of random test length, and an accurate approximation for large test length is shown in Section 7.2.2. Ways to obtain the test length required for a given detection uncertainty are presented in Section 7.2.3.

7.2.1 Example of exact calculation

Figure 7.3a represents a 2-input, 2-output, 4-state synchronous circuit whose state table is given in Figure 7.3b. If the fault $f = w/1$ is present in the circuit, the behavior corresponds to the state table in Figure 7.3c. Let us assume that the input vectors are equally likely (distribution ψ_0). According to Sections 3.2.2 and 5.3, the observer/Markov chain associated with the detection process of this fault is given in Figure 7.3d. The initial state of the observer is (q_0, q'_0) which can be reached by the synchronizing sequence x_3. The self-loop associated with this state is labelled both $x_0 + x_2 + x_3$ (observer) and 0.75 (Markov chain), and so on. The transition matrix of the Markov chain is

$$U = \begin{bmatrix} 0.75 & 0 & 0.25 & 0 & 0 \\ 0.75 & 0 & 0.25 & 0 & 0 \\ 0.5 & 0.25 & 0 & 0.25 & 0 \\ 0.5 & 0 & 0 & 0 & 0.5 \\ 0 & 0 & 0 & 0 & 1 \end{bmatrix}.$$

Time l is associated with the length l of the input sequence applied. According to Section 5.3.1.1, $\pi(l)$ is a stochastic row vector denoting the **probability vector** at time l. Given the initial state $\pi(0)$ of the observer, one has:

$$\pi(l) = \pi(l-1) \cdot U, \quad l > 0. \tag{7.1}$$

Then, by iteration:

$$\pi(l) = \pi(0) \cdot U^l, \quad l \geq 0. \tag{7.2}$$

The probability that the state ω has been reached if a random test sequence of length l has been applied (i.e., the probability of testing the fault) is denoted by $\pi_\omega(l)$. Then,

$$\pi_\omega(l) = \pi(l) \cdot \eta = \pi(0) \cdot U^l \cdot \eta, \tag{7.3}$$

where[1] $\eta = (0, ..., 0, 1)^T$ is an m-row vector (the absorbing state ω is the mth state of the matrix U).

For our example, $\pi(0) = (1, 0, 0, 0, 0)$ and $\eta = (0, 0, 0, 0, 1)^T$. For a level of confidence 0.999, a test length $L = 277$ is obtained since

$$\begin{bmatrix} 1 & 0 & 0 & 0 & 0 \end{bmatrix} \cdot \begin{bmatrix} 0.75 & 0 & 0.25 & 0 & 0 \\ 0.75 & 0 & 0.25 & 0 & 0 \\ 0.5 & 0.25 & 0 & 0.25 & 0 \\ 0.5 & 0 & 0 & 0 & 0.5 \\ 0 & 0 & 0 & 0 & 1 \end{bmatrix}^l \cdot \begin{bmatrix} 0 \\ 0 \\ 0 \\ 0 \\ 1 \end{bmatrix}$$

equals 0.99899 for $l = 276$ and 0.99902 for $l = 277$. Efficient methods of finding this length are discussed in Section 7.2.3.

---

[1] The matrix X^T denotes the transpose of the matrix X.

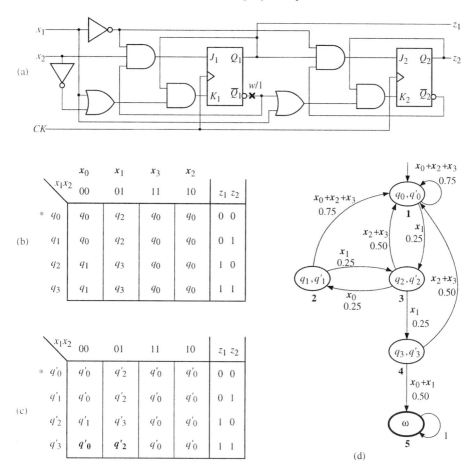

Figure 7.3 (a) Circuit under consideration. (b) State table of the fault-free circuit. (c) State table of the circuit affected by the fault $f = w/1$. (d) Observer/Markov chain corresponding to the detection of this fault.

7.2.2 Accurate approximation for a large test length

By construction, the Markov chain associated with an observer contains an absorbing state ω (Section 5.3.1). If the fault is non-evasive (Section 3.2.1.2), there is no ergodic set (see Appendix C) other than this absorbing state. This is illustrated by the examples in Figures 7.4a and b (an arrow in heavy lines may correspond to several transitions between states or several possible initial states).

For many faults, an error may be repeatedly caused, propagated, and *cancelled*, as long as it remains undetected. In this case, there is a single transient set as illustrated in Figure 7.4c; the corresponding transition matrix is shown in Figure 7.4d, where V corresponds to an irreducible sub-matrix. The Markov chains associated with all the faults studied in Chapter 8 have this structure; this means simply that as long as the fault has not been detected an error in the memory can be cancelled, before detection, by some input sequence (i.e., the subgraph of the Markov chain corresponding to submatrix V is strongly connected).

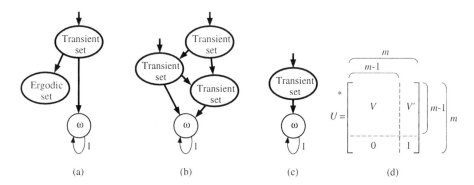

Figure 7.4 Examples of Markov chains[2]. (a) For an evasive fault. (b) For a non-evasive fault. (c) For commonly encountered faults. (d) Transition matrix corresponding to c.

Let us order the eigenvalues of the transition matrix U in Figure 7.4d in a non-increasing order of their absolute values: $\lambda_1, \lambda_2, ..., \lambda_m$. Since U is a stochastic matrix, $|\lambda_j| \leq 1$ for every j, and $\lambda_1 = 1$. Due to the shape of the matrix U: 1) $\lambda_1 = 1$ is the eigenvalue corresponding to the submatrix associated with the absorbing state; 2) $\lambda_2, ..., \lambda_m$ are the eigenvalues of the submatrix V and $|\lambda_j| < 1$ for $j \geq 2$, since V corresponds to a transient class.

Property 7.1 Let U be a transition matrix of the form in Figure 7.4d. If the submatrix V is irreducible and regular (i.e., there is a single acyclic transient class), then:

a) λ_2 is a single real eigenvalue; and

b) $\pi_\omega(l) \approx 1 + K_2 \cdot \lambda_2^l,$ for $l \to \infty,$ (7.4)

where K_2 is a constant value.

---

[2] This figure is reprinted from [Da 97] (© 1997 IEEE).

Proof See Appendix G.

□

For the example in Appendix C, presented again in Figure 7.5, one obtains $\lambda_1 = 1$, $\lambda_2 = 0.6505$, $\lambda_3 = -0.1253 + 0.3276i$, $\lambda_4 = -0.1253 - 0.3276i$, $K_1 = 1$, $K_2 = -1.4856$, $K_3 = 0.2428 - 0.0416i$, and $K_4 = 0.2428 + 0.0416i$. After some classical calculations one obtains

$$\pi_\omega(l) = 1 - 1.4856(0.6505)^l + 0.4926(0.3507)^l \cdot \cos(4.3471l + 0.1697), \quad (7.5)$$

$$\pi_\omega(l) \approx 1 - 1.4856(0.6505)^l, \quad (7.6)$$

hence,

$$\varepsilon(l) \approx 1.4856(0.6505)^l. \quad (7.6')$$

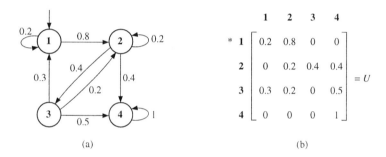

Figure 7.5 Absorbing Markov chain. (a) Graph representation. (b) Transition matrix.

This is illustrated in Figure 7.6. By definition, $\Pr[l_\omega = l] = \pi_\omega(l) - \pi_\omega(l-1)$. Then, from (7.4), $\Pr[l_\omega = l] \approx (1 + K_2 \cdot \lambda_2^l) - (1 + K_2 \cdot \lambda_2^{l-1})$, which can be rewritten as:

$$\Pr[l_\omega = l] \approx -K_2 \cdot \lambda_2^{l-1} \cdot (1 - \lambda_2). \quad (7.7)$$

Assume the initial state of the Markov chain, i, is unique (i.e., $\pi(0) = (0, ..., 0, 1, 0, ..., 0)$). Then, K_2 (which may be specified by $K_2(i)$) is obtained by[3]:

$$K_2 = K_2(i) = \left(\frac{z - \lambda_2}{\det(zI - U)} \Gamma_{i,m} \right)_{z = \lambda_2}, \quad (7.8)$$

---

[3] $(...)_{z=\lambda_2}$ corresponds to the value in $(...)$ when the value λ_2 is assigned to z.

Chapter 7

where $\Gamma_{i,m}$ is the co-factor of the matrix $(zI - U)^T$ corresponding to row i and column m. If the initial state is not unique, then K_2 is the weighted sum of cofactors:

$$K_2 = \pi(0) \cdot [K_2(1)\ K_2(2)\ ...\ K_2(m)]^T. \tag{7.9}$$

Thus, the following (usually very good) approximate value can be directly obtained from (7.4).

$$L(\varepsilon) \approx \frac{\ln \varepsilon - \ln(-K_2)}{\ln \lambda_2}. \tag{7.10}$$

Now, what about the value of K_2? An answer is given in the following property.

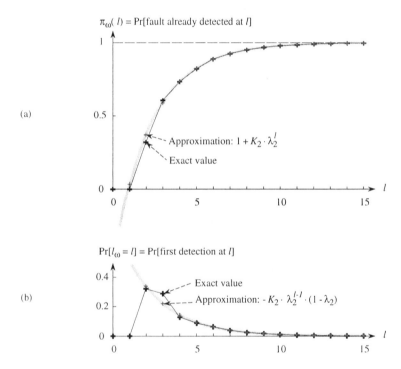

Figure 7.6 Illustration of the approximation[4] in Property 7.1.

Property 7.2 Given the conditions in Property 7.1, there is an initial probability vector $\pi(0)$ such that $K_2 = -1$.

---

[4] This figure is reprinted from [Da 97] (© 1997 IEEE).

Proof The proof is detailed in [MoDa 97]. The broad outline of the proof is given in Appendix G. This proof is based on the two following features: 1) there is at least one state i of the Markov chain such that $|K_2(i)| \leq 1$; 2) there is at least one state i of the Markov chain such that $|K_2(i)| \geq 1$.

☐

The value K_2 depends on the initial state of the Markov chain according to (7.8) where i is the initial state. If $|K_2| > 1$, the initial state is called **unfavourable** because the test length required for some detection uncertainty is greater than if $|K_2| = 1$; it is **favourable** if $|K_2| < 1$. For the example in Figure 7.5, $K_2(1) = -1.49$, $K_2(2) = -0.84$, and $K_2(2) = -0.94$. If the test length is calculated from (7.10), for $\varepsilon = 0.001$, assuming that $K_2 = -1$ provides an error which is about -9% if K_2 were -2 and about $+10\%$ if it were -0.5 (this error is calculated from the numerical values of the numerator in (7.10)). Then, when the initial state is not known, *the value $K_2 = -1$ can be used*. Accordingly, the approximate values in Property 7.3 can be used.

Property 7.3 For a large test length, the following expressions provide approximate values:

$$\pi_\omega(l) \approx 1 - \lambda_2^l, \qquad (7.11)$$

$$\varepsilon(L) \approx \lambda_2^L, \qquad (7.12)$$

$$L(\varepsilon) \approx \frac{\ln \varepsilon}{\ln \lambda_2}. \qquad (7.13)$$

Proof From Property 7.2 and (7.4) in Property 7.1, there is an initial state such that $\pi_\omega(l) \approx 1 - \lambda_2^l$. Thus, this expression provides an approximate value if initial state in unknown. Approximations (7.12) and (7.13) follow; (7.13) is derived from (7.10) where it is clear that the approximate value (7.13) becomes more accurate when ε decreases.

☐

Hence, *for a large test length, the detection of a sequential fault can be treated as the detection of a combinational fault* such that $p_f = (1 - \lambda_2)$. As a matter of fact, the following pairs of equations may be compared: (6.8) in Section 6.1.1.2 and (7.11); (6.9) and (7.12); (6.15) and (7.13).

Let us call

$$p_f = 1 - \lambda_2 \qquad (7.14)$$

the *equivalent detection probability* (or **e-detection probability**) of a sequential fault. Note that the *e*-detection probability is different from the average detection probability (Definition 5.4 in Section 5.4.2) which does not take into account the memory effect in a detection process (Section 5.4). For the fault f_2 in Section 5.4.2, whose detection process is illustrated in Figure 5.14b,

$p_{av,f_2} = 0.50$ while $p_{f_2} = 0.19$ (this illustrates that the average detection probability is only a rough estimate). Using $p_f = 0.19$ in Equation (6.15) (or equivalently $\lambda_2 = 0.81$ in (7.13)), leads to $L(0.001) \approx 33$, while the exact value is 34.

Remark 7.1 The approximation (7.4) is based on three conditions, namely that V is non-negative, regular, and irreducible. The first and second conditions are always verified: the matrix V is always non-negative, by definition, and an irreducible matrix such that $|\lambda_2| = |\lambda_3|$ with λ_2 real and λ_3 complex would be a particular case rarely found in practice [Ga 89] that it has no chance of being found in a detection process. The third condition may not be true: V could be reducible into several transient sets whose transition matrices are V_1, V_2, ... Let $\lambda_{2,i}$ be the largest eigenvalue of V_i. It could happen that $\max_i(\lambda_{2,i}) = \lambda_{2,1} = \lambda_{2,2}$, for example, and Property 7.1 would not be necessarily true. An abstract fault can be built in the universal set F_u presenting this feature; an example is given in Remark 8.4, Section 8.4.1 (the author has *never encountered an actual fault with this feature*). However, even in the particular case where Property 7.1 is not true, Property 7.3 provides an admissible approximation (a numerical example is presented in Remark 8.4).

7.2.3 Obtaining the test length $L(\varepsilon)$

Consider a fault f, whose detection process is modeled by a Markov chain (Section 5.3.1). Two problems may be considered.

First problem. Given a test length, L, what is the probability of testing $P(f, L) = \pi_\omega(L)$, or equivalently what is the uncertainty level $\varepsilon(L) = 1 - \pi_\omega(L)$.

Second problem. Given an uncertainty level ε, for which length $L(\varepsilon)$ is the probability of reaching the absorbing state at least $1 - \varepsilon$.

Solution of the first problem: $\varepsilon(L)$

The exact value of the uncertainty level $\varepsilon(L) = 1 - \pi_\omega(L)$ can easily be obtained from Equation (7.3); or from Equation (G.7) in Appendix G if the transition matrix has been solved (Equation (7.5) in our example). An approximate value can be obtained from (7.4) i.e., $\varepsilon(L) \approx - K_2 \cdot \lambda_2^L$, or from (7.12) i.e., $\varepsilon(L) \approx \lambda_2^L$.

Solutions of the second problem: $L(\varepsilon)$

The exact value of $L(\varepsilon)$ cannot be directly calculated. It can be obtained by iteration based on the solution of the first problem. A dichotomic approach when it is known that $l_1 \leq L(\varepsilon) \leq l_2$ is efficient. The following algorithm may be used; two input parameters are given, an arbitrary test length l_0, and $a > 1$ (% precedes a comment which is not executed in the program; other comments are given after the algorithm).

Algorithm 7.1

Step 1. $l_1 = l_0; l_2 = l_0$ % initialization
Step 2. **while** $\pi_\omega(l_1) > 1 - \varepsilon$, $l_1 = \lfloor l_1/a \rfloor$ **end** % decrease l_1 if necessary
Step 3. **while** $\pi_\omega(l_2) < 1 - \varepsilon$, $l_2 = \lceil l_2 \times a \rceil$ **end** % increase l_2 if necessary
Step 4. **while** $l_2 - l_1 > 1$
$\quad\quad l_{12} = \lfloor (l_1 + l_2)/2 \rfloor$
$\quad\quad$ **if** $\pi_\omega(l_{12}) \geq 1 - \varepsilon$ **then** $l_2 = l_{12}$ **else** $l_1 = l_{12}$
end
Step 5. $L(\varepsilon) = l_2$.

Comments on algorithm 7.1

If $L(\varepsilon)$ is approximatively known, then l_0 may be close to $L(\varepsilon)$ and the coefficient a may be close to 1, for example $a = 1.1$; otherwise l_0 may be any value, possibly 1, and a should be greater, for example $a = 2$. If the exact value of $L(\varepsilon)$ is not required, the iteration may be stopped when a satisfactory approximate value is obtained; *Step 4* may be replaced for example by: **while** $l_2 / l_1 > 1.001$ ....

Remark 7.2 A sequential fault in a combinational circuit is related to this chapter. Consider for example the sequential fault in Figure 2.15 (Section 2.3.3). The fault-free state table, the faulty state table, and the observer are represented in Figures 7.7a, b, and c, respectively (states q_i in Figure 2.15 are renamed q'_i here). The initial probability vector in Figure 7.7c is obtained if one assumes that the input vector is x_0 when the circuit is powered-up.

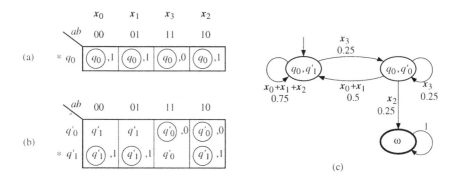

Figure 7.7 Sequential fault in a combinational circuit. (a) State table of the fault free circuit (NAND gate). (b) State table for the stuck-open in Figure 2.15. (c) Corresponding observer/Markov chain.

7.3 APPROXIMATE METHODS

The first approximate method, studied in Section 7.3.1, is based on the notion of detection set (set of minimum detecting transition sequences). It allows an approximate value of the average detection probability of a fault to be obtained. The systematic manner for obtaining the detection set is explained from the state tables of both the fault-free and the faulty circuits. However, when the concept has been clearly understood, a detection set can be obtained in a more intuitive manner, which will be used for microprocessor faults (Chapter 9).

The second approximate method is based on the single-state-transition fault model. This model appears to cover most faults in practical finite state machines [ChJo 92]. The method in Section 7.3.2 estimates the random test length if any single-state-transition fault is present in the circuit.

7.3.1 Minimal detecting transition sequence (MDTS)

The notion of detecting transition sequence is defined in Section 7.3.1.1. A detection set is a set of minimal detecting transition sequences; definitions and properties related to detection sets are given in Section 7.3.1.2.

This section is based on [ThDa 78b] and [DaTh 80a] where more details can be found.

7.3.1.1 Detecting transition sequence

Let us introduce intuitively the main concepts with the help of simple examples. Figure 7.8a represents a fault-free machine M. A transition sequence is illustrated, in heavy lines, on the corresponding state diagram of Figure 7.8b. This **transition sequence** is a string $J=(q_1 \rightarrow q_2)(q_2 \rightarrow q_0)(q_0 \rightarrow q_3)$ which is denoted without ambiguity by the state where the sequence starts and the string of input vectors producing the successive transitions, i.e.,

$$J = q_1 x_1 x_0 x_1.$$

Let us say that, during a testing experiment, a *transition sequence* $J = q_i S_j$ **is applied** to a machine if 1) the machine should be in the state q_i if it were fault-free and 2) the input sequence S_j is applied to the machine.

Let us consider a circuit under test whose behavior should correspond to machine M in Figure 7.8a, but corresponds to machine M' in Figure 7.8c. Assume the state should be q_2 at some time t in the test experiment. According to

Figure 7.8a, the last vector applied[5] was x_1 since q_2 can be reached only by this input vector, from either q_1 or q_2. Now, given the last input vector was x_1, the faulty machine is in a state of the set $\{q'_0, q'_2, q'_3, q'_4\}$ according to Figure 7.8c. The states in this set are said to be *i*-**compatible** (input-compatible) with q_2. The output associated with q_2 is 0. The outputs associated with q'_0, q'_2, q'_3, and q'_4 are 0, 0, 0, and 1, respectively. The states q'_0, q'_2, and q'_3 are said to be *o*-**compatible** (output-compatible) with q_2. Since q'_4 is not *o*-compatible with q_2, if M' is in the state q'_4 while it should be in state q_2, the fault is detected at time t. This is represented in Figure 7.9: to the state q_2, is assigned the set of *i*-compatible states $\{q'_0, q'_2, q'_3, q'_4\}$; the state q'_4 is deleted from this list because it is not *o*-compatible with q_2. That means that, at time t, if the fault has not yet been detected, the faulty machine is in one of the states q'_0, q'_2, or q'_3.

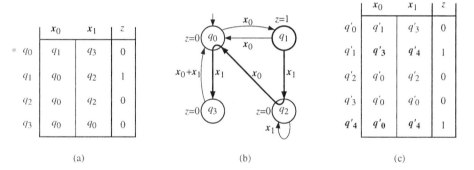

Figure 7.8 (a) Fault-free machine M. (b) Corresponding state diagram; a transition sequence in heavy lines. (c) Faulty machine M'.[6]

$$q_2; q'_0, q'_2, q'_3, q'_4 \xrightarrow{x_0} q_0; q'_0, q'_1 \xrightarrow{x_0} q_1; q'_1 \xrightarrow{x_1} q_2; q'_4$$

$$t \qquad\qquad t+1 \qquad\qquad t+2 \qquad\qquad t+3$$

Figure 7.9 The transition sequence $q_2 x_0 x_0 x_1$ is a detecting transition sequence.

Assume that the input sequence $x_0 x_0 x_1$ is applied just after time t. At $t+1$, machine M reaches the state q_0, while M' reaches q'_0 or q'_1. Since q'_1 is not *o*-compatible with q_0, the fault would be detected if the state q'_1 were reached.

---

[5] The "last vector applied" exists at any time $t \geq 0$ if initialization takes place by a synchronizing sequence, since this sequence contains at least one input vector, and for any time $t > 0$ otherwise since the first input vector is applied at $t = 1$ in this case.
[6] This figure is reprinted from [DaTh 80] (© 1980 IEEE).

Then, if the fault has not yet been detected at $t+1$, machine M' is in the state q'_0. Now, at time $t+2$, the states are respectively q_1 and q'_1 for M and M', and these states are o-compatible. Finally, if the fault was not yet detected at $t+2$, it is certainly detected at $t+3$ since the states which are reached, q_2 and q'_4, are not o-compatible. The conclusion is that, if the transition sequence $J_1 = q_2x_0x_0x_1$ is applied to the machine under test, the fault is detected; this is a **detecting transition sequence** (DTS).

Remark 7.3 The concept of DTS resembles the notion of distinguishing sequence (Section 2.3.3), in the sense that the input sequence S of a DTS, $J = qS$, makes a distinction between a state q of the fault-free machine M and those of the faulty machine M'. However, S is not associated with the machine M, but with the state q of M. A transition sequence (string of arcs on a state diagram) is of a different nature than an input sequence.

7.3.1.2 Detection set

Our aim is to find a set of detecting transition sequences for a fault f. For this purpose we need some algebra related to transition sequences.

The **concatenation of two transition sequences** $J_1 = q_1S_1$ and $J_2 = q_2S_2$ is defined if and only if $\delta(q_1, S_1) = q_2$, by $J_1 \cdot J_2 = q_1S_1 \cdot q_2S_2 = q_1S_1S_2$. For example, in Figure 7.8b, $q_1x_1x_0 \cdot q_0x_1 = q_1x_1x_0x_1$. The transition sequence S_i, in $J = q_iS_i$, may be null input sequence λ. Then, $J = q_i\lambda$ or $J = q_i$ is a particular case of transition sequence, the length of which is zero. Let $J_1 = q_iS_i$, $J_2 = q_i$, and $J_3 = q_j$ such that $q_j = \delta(q_i, S_i)$. Then $J_2 \cdot J_1 = J_1$ and $J_1 \cdot J_3 = J_1$. A single symbol v will denote without ambiguity a transition sequence such that $v \cdot J = J$ or $J \cdot v = J$.

A transition sequence J_1 is a subsequence of the transition sequence J_2 if there are J' and J'' such that $J_2 = J' \cdot J_1 \cdot J''$. The transition sequence J_1 is a *strict subsequence* of J_2 if either J' or J'' (or both) is different from v.

Definition 7.1 A transition sequence J is a **minimal detecting transition sequence** (MDTS) for the fault f if the following two conditions are met:
1) J is a DTS for f;
2) there is no strict subsequence of J which is a DTS for f.

❑

For example $J_1 = q_2x_0x_0x_1$, which have been found to be a DTS for the fault M' in Figure 7.8c is not an MDTS. As a matter of fact $J_1 = q_2x_0x_0 \cdot q_1x_1$ and q_1x_1 is also a DTS (the reader can easily verify this property from Figures 7.8a and c).

Definition 7.2 The **detection set** associated with a fault f is the set of MDTSs associated with the fault f. We shall denote this set $D_f = J_1 + J_2 + \ldots$.

❑

Since an MDTS may be represented by the concatenation of a state and an input sequence, the operations of regular expressions (Section 2.1.3.1) may be used for the detection sets. For example, $q_1S_1 + q_2S_1 = (q_1 + q_2)S_1$, $q_1S_1 + q_1S_2 = q_1(S_1 + S_2)$, $S_1^* = \lambda + S_1 + S_1S_1 + \ldots$.

A way to obtain the detection set D_f is illustrated in Figure 7.10. The fault f corresponds to the faulty machine M' in Figure 7.8 (the fault-free one is M in the same figure).

A state of the **compatibility machine** (Figure 7.10b) contains two parts separated by a semi-colon: the first part contains a state q_i of M, and the second part a set of states of M' which are both *i*-compatible and *o*-compatible with q_i. The second part of an *initial state* contains all the states q'_j which have the above property. The second part of a *final state* is empty. Other states are *current states*. Accordingly, the number of initial states is the number of states in M; they are obtained from the *compatibility table* (Figure 7.10a). For each of these states the next states corresponding to the various input vectors are obtained as illustrated in Figure 7.9. For example, from $(q_2; q'_0, q'_2, q'_3)$ the input vector x_0 leads to $(q_0; q'_0)$ which is a new state. The process continues until a final state is reached. When no initial state is also a final state, the final states can be gathered in a single state noted ω.

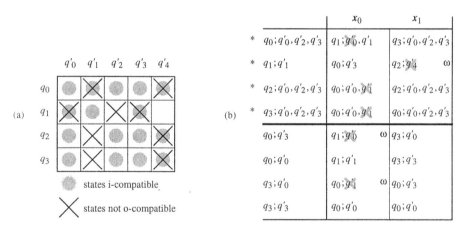

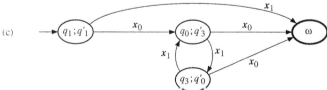

Figure 7.10 Searching for the detection set for M' in Figure 7.8c compared to M in Figure 7.8a. (a) Compatibility table. (b) Compatibility machine. (c) Detection graph.

The **detection graph** is now obtained from the compatibility machine as illustrated in Figure 7.10c. The final state ω represents all the final states in the compatibility machine. We add all the states from which this state can be reached, and the corresponding arcs; for example, ω can be reached from (q_0, q'_3) by the input vector x_0. The process continues and stops when an initial state is reached. For our example the detection graph in Figure 7.10c is obtained. From this graph the detection set

$$D_f = q_1(x_1 + x_0 x_1^* x_0) \tag{7.15}$$

is obtained.

Remark 7.4
a) The detection graph gives all the detecting transition sequences from an initial state to a final state and passing only through current states, i.e., all the DTS J such that 1) J is a DTS and 2) there is no DTS T_1 such that $T = T' \cdot T_1 \cdot T''$, with $T'' \neq v$. This means that we obtain all the MDTS but we can obtain in some cases some DTS which are not MDTS and which must be cancelled.

b) If M and M' are strongly connected and if there is no DTS (i.e., there is no final state), then M and M' are equivalent[7].

❏

The way to obtain an approximate value for the test length from the detection set is presented now. Let us say that a transition sequence **has just been applied**, at a given time in the testing experiment, if no more input vectors have been applied since then. For example, in Figure 7.9, $q_2 x_0 x_0 x_1$ has just been applied at $t+3$. In the sequel, the probabilities obviously depend on the distribution ψ. This is implicit.

Definition 7.3 The probability $\Pr[J]$ of a transition sequence J is the average probability that J has just been applied, i.e.,

$$\Pr[J] = \lim_{t \to \infty} \frac{\text{number of times } J \text{ was just applied up to } t}{t}.$$

The *probability* $\Pr[D_f]$ *of a detection set* is the average probability that any MDTS in D_f has just been applied.

Property 7.4 The probability of the detection set D_f is

$$\Pr[D_f] = \sum_{i : J_i \in D_f} \Pr[J_i]. \tag{7.16}$$

---

[7] More details on Remark 7.5a and proof of Remark 7.5b can be found in [DaTh 80a].

Proof Let us assume that an MDTS J_i in D_f has just been applied at time t. Hence no other MDTS J_j in D_f has just been applied at t. Indeed, either J_i should be a proper suffix of J_j or vice versa. Thus, either J_i should be a proper subsequence of J_j or vice versa, which contradicts the definition of an MDTS (Definition 7.1).

□

Let D_f and D_g be two detection sets. If for every J_i in D_f there is at least one J_j in D_g such that J_j is a subsequence of J_i, then D_g *dominates* D_f and $\Pr[D_f] \leq \Pr[D_g]$ for any distribution ψ.

Given a fault f, $\Pr[D_f]$ may be used as a rough approximation[8] of the e-detection probability (Section 7.2.2). Then, given a test length,

$$P(f, L) = 1 - \varepsilon(L) \approx 1 - (1 - \Pr[D_f])^L, \quad (7.17)$$

and, given an uncertainty level,

$$L(\varepsilon) \approx \frac{\ln \varepsilon}{\ln(1 - \Pr[D_f])}. \quad (7.18)$$

This approximation is usually not accurate since it does not take into account the *memory effect* (Section 5.4.2). However, it provides an order of magnitude of the required test length, and this is important for complex circuits such as microprocessors. Before giving examples, let us provide a property used to calculate $\Pr[D_f]$ given D_f. The general expression of D_f is

$$D_f = \sum_{i: q_i \in Q} q_i \cdot E_i, \quad (7.19)$$

where E_i is a regular expression whose alphabet is X. For some i, E_i may be the empty set of sequences denoted by ϕ. For example, in (7.15), $E_1 = x_1 + x_0 x_1^* x_0$, and $E_0 = E_2 = E_3 = \phi$. Let a **detection subset** D be a subset of a detection set (i.e., a set of MDTS), such that all sequences start with the same state q.

Property 7.5 Let $D = q \cdot E_1 E_2^* E_3$ be a detection subset where E_1, E_2 and E_3 are regular expressions. For a random test sequence, generated from a constant distribution

$$\Pr[D] = \Pr[q] \cdot \Pr[E_1 E_2^* E_3] = \Pr[q] \frac{\Pr[E_1] \cdot \Pr[E_3]}{1 - \Pr[E_2]}. \quad (7.20)$$

Proof
Case 1: E_1, E_2 and E_3 are input sequences. Since D is a set of MDTS, there is no problem of inclusion between the transition sequences qE_1E_3, $qE_1E_2E_3$,... . For

---

[8] $\Pr[D_f]$ is less than or equal to the average detection probability, as shown in [ThDa 78b].

a constant distribution, the successive input vectors are independent (Appendix A). Then,

$$\Pr[q\,E_1 E_2^* E_3] = \Pr[q] \cdot \Pr[E_1] \cdot (1 + \Pr[E_2] + \Pr[E_2^2] + \ldots) \cdot \Pr[E_3],$$

and

$$\Pr[E_2^n] = (\Pr[E_2])^n.$$

Hence,

$$(1 + \Pr[E_2] + \Pr[E_2^2] + \ldots) = \frac{1}{1 - \Pr[E_2]}.$$

Case 2: E_1, E_2, and E_3 are regular expressions. By definition, two transition sequences J_1, J_2 in D cannot be a subsequence of one another. Then, the result in *case 1* remains true.

□

Any detection set can be expressed by a sum of terms of the form $q\,E_1 E_2^* E_3$ where E_1, E_2, and E_3 are regular expressions. Then, Property 7.5 may be applied to any detection set.

For the examples given in the sequel, it is assumed that the equal likelihood distribution ψ_0 is used, and the uncertainty level $\varepsilon = 0.001$ is required.

First example. Consider the example in Figures 7.8 and 7.10 and assume the synchronizing sequence is $x_0 x_1 x_0$. The initial state of M is q_0, while the initial state of M' is q'_1 (probability 0.2 corresponding to starting state q'_1) or q'_0 (probability 0.8 corresponding to other starting states). The exact calculation from a Markov chain shows that the required test length is $L = 42$. For a long test sequence $\Pr[q_1] = 0.2$. To be presented as a sum of terms of the form $q\,E_1 E_2^* E_3$, (7.15) may be rewritten as $D_f = q_1 x_1 \lambda^* \lambda + q_1 x_0 x_1^* x_0$). From this expression and using Property 7.5:

$$\Pr[D_f] = 0.2 \left(\Pr[x_1] + \frac{\Pr[x_0] \cdot \Pr[x_0]}{1 - \Pr[x_1]} \right) = 0.2.$$

Then, from (7.18), $L(0.001) \approx 31$.

Second example. The circuit in Figure 7.3 (Section 7.2.1) is a 2-stage shift register with a synchronous reset. When a rising edge of *CK* occurs: 1) if $x_1 = 0$, $z_2 = Q_2$ takes the previous value of $z_1 = Q_1$, and $z_1 = Q_1$ takes the value of x_2; 2) if $x_1 = 1$, z_1 and z_2 are reset. The fault w/1 is detected if there is an error on the line w which is propagated to z_2 via the input K_2. These conditions are met if $\overline{Q_1} = 0$, $x_1 = 0$, and $z_2 = 1$ when the clock rises. This happens if the state is q_3 and the next input vector is x_0 or x_1. As a matter of fact, if the state is q_3, $z_1 = 1$ (which implies that $\overline{Q_1} = 0$) and $z_2 = 1$; in addition, $x_1 = 0$ for both input vectors x_0 and x_1.

Hence $D_{w/1} = q_3(x_0 + x_1)$. The values $\Pr[q_3] = 0.0625$, then $\Pr[D_{w/1}] = 0.03125$ can be obtained. From (7.18), it follows $L(0.001) \approx 218$ (to be compared to the exact value $L = 277$ obtained in Section 7.2.1).

7.3.2 Single transition faults

This section is based on a fault model which is very easy to use for the analysis of random testing when the behavior of the sequential circuit is specified by a state table or state diagram. In this fault model, called **single transition** fault model, a fault causes the destination state transition to be faulty. Formally, the fault-free automaton $A = (Q, X, Z, \delta, \mu)$ is transformed into the faulty one $A' = (Q, X, Z, \delta', \mu)$ where there is a single pair (q, x) such that $\delta'(q, x) \neq \delta(q, x)$. A **multiple transition** fault is such that there are several pairs (q, x) such that $\delta'(q, x) \neq \delta(q, x)$. In both cases the output function μ is correct.

In [ChJo 92], it is shown, using a Mealy model, that a test sequence that detects all the *multiple transition* faults detects all the irredundant physical faults in the machine (in particular all the faults corresponding to a faulty output function $\mu' \neq \mu$ are covered). This result is based on the explicit assumption that the number of states is the same in both machines, and the implicit assumption that the initial state is the same for both machines, i.e., $q'_0 = q_0$ according to our notations. It is also shown that a test sequence that detects all the single transition faults detects almost all multiple transition faults. These results show that this model is pertinent.

7.3.2.1 *Approximate random test length*

Let us introduce some notation and definitions. The machines M and M' denote respectively the *initialized automata* A and A'. The transition $\tau_i = (q_a, x, q_b)$ is such that $\delta(q_a, x) = q_b$. A fault $f = (\tau_i, \tau'_i)$ corresponds to a **faulty transition**[9] in M', $\tau'_i = (q_a, x, \delta'(q_a, x))$ such that $\delta'(q_a, x) \neq q_b$. There are $|Q| - 1$ various τ'_i which can be associated with the same transition τ_i. In the context of random testing, the **probability of a transition** is used:

$$\Pr[\tau_i] = \lim_{l \to \infty} \frac{\text{number of times } \tau_i \text{ occurs in } M}{l}. \qquad (7.21)$$

We assume that the fault $f = (\tau_i, \tau'_i)$ is *detected when the machine M passes through τ_i*. From this hypothesis, the $|Q| - 1$ faults corresponding to various τ'_i are in the same equivalence class, and it is assumed that $\mu(\delta(q_a, x)) \neq \mu(\delta'(q_a, x))$, which is not always true. Furthermore the memory effect is neglected. From

---

[9] Since the sets of states in M and M' are the same, a state q'_i in M' may be denoted by q_i when there is no ambiguity.

these hypotheses, a rough *approximation* is obtained; the phenomena not taken into account, hence the quality of this approximation, will be commented on at the end of this section.

According to the hypotheses, for a *strongly connected* automaton,

$$p_f \approx p_{av,f} \approx \Pr[\tau_i]. \qquad (7.22)$$

For a constant distribution ψ of input vectors, this probability is obtained by

$$\Pr[\tau_i] = \Pr[q_a] \cdot \psi(x). \qquad (7.23)$$

Let us illustrate this approach with an example. Consider the machine M in Figure 7.11a and assume a random test sequence with a constant distribution $\psi = (.4, .6)$, i.e., $\psi(x_0) = 0.4$ and $\psi(x_1) = 0.6$. With the state diagram in Figure 7.11a, a Markov chain, whose transition matrix is

$$U = \begin{bmatrix} 0.4 & 0.6 & 0 & 0 \\ 0.4 & 0 & 0.6 & 0 \\ 0.4 & 0 & 0 & 0.6 \\ 0 & 1 & 0 & 0 \end{bmatrix}, \qquad (7.24)$$

can be associated.

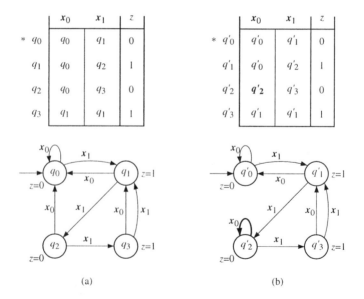

Figure 7.11 (a) Fault-free machine M. (b) Faulty machine M'.

The matrix U is irreducible and regular (equivalently the state diagram in Figure 7.11a is strongly connected and acyclic). Then, when l tends to infinity, the value of $\pi(0) \cdot U^l$ tends towards

$$\pi(\infty) = [0.3524 \quad 0.3304 \quad 0.1982 \quad 0.1190], \tag{7.25}$$

which is independent of $\pi(0)$ (Appendix C).

According to (7.23), the probability of every transition may be obtained from (7.25) and $\psi = (.4, .6)$. For example, $\Pr[(q_0, x_1, q_1)] = \pi_{q0}(\infty) \cdot \psi(x_1) = 0.3524 \times 0.6 = 0.2114$. It follows that the *expected fault coverage* may be approximated, according to (7.22), using Equation (6.17) in Section 6.1.2.1, where the number r of faults is the number of transitions, i.e., 8 in our example. If the uncertainty level ε is required, the calculation may be iterated according to Algorithm 7.1. For example, the length $L \approx 102$ is obtained for $\varepsilon = 0.001$.

Let us now consider the *worst case fault* in the set of transition faults. It corresponds to the transition $\tau_m = (q_3, x_0, q_1)$, whose probability is $\Pr[\tau_m] = 0.0476$. From Equation (6.33) in Section 6.1.2.2, given $p_{min} = 0.0476$, the test length $L \approx 142$ is obtained for $\varepsilon = 0.001$.

7.3.2.2 Topics not taken into account

The detection uncertainty $\varepsilon = 0.001$ is considered for all the examples in this section.

In the previous approximation, three topics are not taken into account. They are illustrated by a fault corresponding to the transition $\tau_1 = (q_2, x_0, q_0)$ in Figure 7.11a. For this fault, $\Pr[\tau_1] = \pi_{q2}(\infty) \cdot \psi(x_0) = 0.1982 \times 0.4 = 0.0793$; then, from (6.14) in Section 6.1.1.2, the test length $L \approx 84$ is obtained.

First topic. The *memory effect* is neglected. The fault is assumed to be detected when the transition τ_1 occurs. From this hypothesis, the actual value of the test length should be based on the *first time* that τ_1 occurs. This length L corresponds to $\pi_\omega \geq 0.999$ for the Markov chain in Figure 7.12a. This Markov chain is obtained directly from the state diagram in Figure 7.11a, given $\psi(x_0) = 0.4$ and $\psi(x_1) = 0.6$; the faulty transition τ_1 is replaced by a transition to the absorbing state ω. From this Markov chain the test length $L \approx 74$ is obtained[10]. For this example, ignoring the memory effect has led to a random test length of 84 instead of 74; but in other cases the length could be shorter (Section 5.4.2).

Second topic. It is assumed that the fault is *immediately detected* when the faulty transition occurs; this is not always true. The fault $f = (\tau_1, \tau'_1)$, where

---

[10] The test length is not very sensitive to changes in the initial state. If the initial state of the Markov chain in Figure 7.11a were **3** (very close to state **5** = ω) instead of **1**, $L = 69$ would be obtained instead of 74.

$\tau_1 = (q_2, x_0, q_0)$ and $\tau'_1 = (q'_2, x_0, q'_i)$ is such that $q'_i = q'_1$, or q'_2, or q'_3. If $q'_i = q'_1$ or q'_3 the fault is immediately detected since the outputs are different: $\mu(q'_1) \neq \mu(q'_0)$ and $\mu(q'_3) \neq \mu(q'_0)$. In the case $q'_i = q'_2$, the fault is illustrated in Figure 7.11b and the detection process is modeled by the Markov chain in Figure 7.12b. The test length obtained is 77 which is just a bit more than 74. If, after the faulty transition has occurred the circuit could reach a "good" state before detection, the test length could be significantly longer than the approximation. Otherwise, like in our example, the test length is not significantly affected.

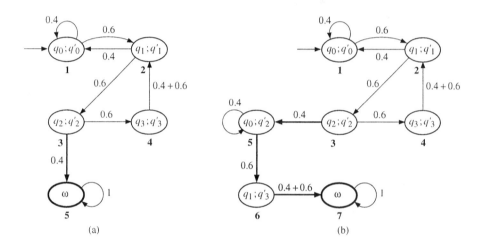

Figure 7.12 Markov chains for detection of the fault in Figure 7.11b. (a) Fault assumed to be detected immediately when the faulty transition occurs. (b) Exact model.

Third topic. The *initial state* is assumed to be the right one, but this is not always true. For example $x_0 x_0$ is a synchronizing sequence for the machine in Figure 7.11a. This sequence leads the machine in Figure 7.11b either to the state q'_0 (probability 0.75) or to the state q'_2 (probability 0.25). The detection process is modeled by the Markov chain in Figure 7.12b with $\pi(0) = (0.75, 0, 0, 0, 0.25, 0, 0)$. The test length obtained is 74, which is a bit less than 77, obtained from $\pi(0) = (1, 0, 0, 0, 0, 0, 0)$.

Conclusion on topics not taken into account. Usually, ignoring the memory effect is the most important cause of inaccuracy for the method presented in Section 7.3.2.1.

7.4. NUMERICAL RESULTS

The test lengths in Figure 7.13 have been obtained by the approximation using the detection sets (Section 7.3.1.2). For the synchronous test, the constant distribution ψ_0 is considered, and for the asynchronous test, each input has the same probability of change at each step (see Section 7.1). The uncertainty level $\varepsilon = 0.001$ is required for the worst case single stuck-at-fault[11].

For each *flip-flop* circuit, for example the SR flip-flop in the package *471L*, the detection set is obtained for every single stuck-at-fault. Each detection set dominating another one (Section 7.3.1.2) is deleted from the list since the worst case fault is considered. The probabilities of all states are calculated from a Markov chain (as in Section 7.3.2.1). Next, the probability of the detection sets and the corresponding test lengths are calculated according to Section 7.3.1.2.

Package	Content	**Asynchronous test**				**Synchronous test**				
		1 flip-flop		Package		1 flip-flop		Package		
		n	L	n	L	n	L	n	L	
471 L	1 SR flip-flop	9	32 000	9	32 000	8	1 300	8	1 300	Flip-flops
472	1 JK flip-flop	9	70 000	9	70 000	8	2 700	8	2 700	Flip-flops
472 H	1 JK flip-flop	9	16 000	9	16 000	8	2 700	8	2 700	Flip-flops
473 4107	2 JK flip-flops	4	8 500	8	17 000	3	220	6	220	Flip-flops
476	2 JK flip-flops	5	9 000	10	18 000	4	160	8	160	Flip-flops
478 E	2 JK flip-flops with 2 common inputs			8	14 000			7	160	Flip-flops
474	2 D flip-flops	4	1 100	8	2 200	3	80	6	80	Flip-flops
475	4 D latches with 1 clock for 2 latches	2	80	6	240	1	24	4	24	Registers
4174	6 SR } flip-flops with 2 common inputs	3	1 800	8	4 800	2	24	7	24	Registers
4175	4 SR } flip-flops with 2 common inputs			6	3 600			5	24	Registers
4164	8-bit shift register serial input, // output			4	$92 \cdot 10^6$			3	50	Registers
492	Divide-by -welve counter			4	15 000			2	110	Counters
473	4-bit binary counter			4	67 000			2	600	Counters

Figure 7.13 Test lengths for some SSI and MSI sequential circuits[12] ($\varepsilon = 0.001$, worst case stuck-at faults).

For *a register or a counter*, the circuit is broken down into subcircuits corresponding to flip-flops, and possibly some additional combinational parts. For example the reference *492* is broken down into four similar JK flip-flops and

[11] The results related to the asynchronous test were published in [ThDa 78b] and [DaTh 81]. Some results related to the synchronous test were given in [Ba 78] and the others were not published.

[12] See, for example, [Se 75].

a gate [Se 75]. The detection sets are calculated for a JK flip-flop; the detection sets of the three other flip-flops are similar since the flip-flops are similar. The only difference is that some inputs of a flip-flop, and possibly its clock, may not be primary inputs but correspond to outputs of other subcircuits. This fact must be taken into account when calculating the probabilities of the states and detection sets related to the considered flip-flop.

The lengths obtained are approximate, then they are rounded. The following observations can be made from the results in Figure 7.13.

1) Since 10^6 input vectors can be applied to these circuits in one second, all these circuits can be randomly tested in a few milliseconds, at least in the synchronous test. 2) The synchronous test is much shorter than the asynchronous one (however, a few faults cannot be detected by a synchronous test as shown in Appendix F). 3) When a package contains several independent or almost independent circuits, considering the whole package as a single circuit increases the asynchronous but not the synchronous test length. For example the package *476* contains two independent flip-flops. In the asynchronous test a single input changes at each step, then for one of the flip-flops there is no input change (the test length is then twice the length obtained for a single flip-flop). In the synchronous test the whole input vector is randomly drawn at each step, then the two flip-flops are tested concurrently.

The circuits considered in Figure 7.13 are relatively small. The methods for evaluating the test length for microprocessors (Chapter 9) is relevant to sequential circuits containing a large number of components.

NOTES and REFERENCES

As mentioned at the very beginning of this chapter, the methods presented can be used for various systems as soon as they are modeled by finite state machines [CoCz 96] [LeYa 96]; for example software and communication protocols [Ch 78] [AhSeVl 86] [Ho 91].

A synchronous test may be inefficient for testing some faults in a sequential circuit [Da *et al.* 77]. This observation is relevant to *both deterministic and random tests*. However, in practice, for large circuits only a synchronous test is realistic.

The more accurate way to estimate the test length related to a sequential fault is based on modeling of the detection process by a Markov chain. This idea was introduced in [ShMc 76]. The construction of the Markov chain may be based on the theory of observers [BrJü 92]. In this chapter, it is shown that, for a long test sequence, the test length is almost similar to that of a combinational fault such that $p_f = (1 - \lambda_2)$, where λ_2 is the greater eigenvalue (except $\lambda_1 = 1$) of the Markov chain.

The detection set, i.e., set of MDTS, provides a way of obtaining an approximate test length for a fault [ThDa 78b], [DaTh 80a]. It can be used without building the Markov chain and does not need the sequential circuit to be represented by a state table or diagram. It will be used for faults in microprocessors (Chapter 9). According to the main assumption of this approach (no memory effect) for an *intermittent* sequential fault, the test length must be divided by the proportion of time when the fault is active, to obtain the same probability of testing [DaTh 80b].

The assumption that an input sequence correctly tests the faults in a sequential circuit if all the transitions occur was considered a long time ago in the context of random testing [TeDa 74], [DaTe 79]. The functional model called *single transition fault* has gained some popularity in the context of deterministic testing for a few years, since the finite state machine model is used for modeling in a wide range of areas [ChJo 92], [KaSa 93], [CoCz 96]. The main limitation is the size of the sequential circuit since this fault model requires a finite state model of the circuit.

8

Random Test Length for RAMs

Random access memories (RAMs) are very widely used. There is no other kind of VLSI circuit with so many internal states: a one-megabit RAM has about $10^{315,000}$ internal states. Fortunately, the structure of a RAM is regular and its basic operation is fairly simple.

Due to the particular structure of a RAM, specific fault models for this kind of circuits have been derived. The main models and the corresponding test lengths are presented in Section 8.1 and in Section 8.2, respectively. The results are then extended to multiple faults and word-oriented memories in Section 8.3.

The power of random testing for RAMs is analysed and compared with deterministic testing in Section 8.4.

8.1 MODELS

Let us present first models of static RAMs (Section 8.1.1), then models of faults in these RAMs (Section 8.1.2).

8.1.1 Models of RAMs

Figure 8.1 presents a 1 024 word by 1 bit RAM.

The memory part is array-organized. In the example, the array is made up of 32 rows and 32 columns. The decoder (made up of a row decoder and a column decoder) addresses the cell which is specified by the address $i = a_0 a_1 ... a_9$. A read/write logic determines the behavior of the memory. The boolean value on the R/W line specifies whether the input x must be written in address i ($R/W = 0$) or the content in t must be read, i.e., sent to output z ($R/W = 1$). A chip-select input specifies whether the memory is to be used: a read or write operation on a memory can be performed only if $CS = 1$.

193

194 Chapter 8

In a more general case there may be several chip-select inputs. In a word oriented memory, we have b bits of input data and b bits of output data; each address corresponds to a b-bit word.

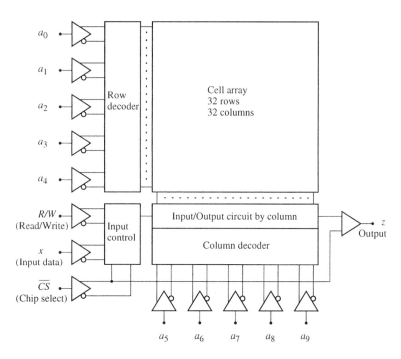

Figure 8.1 Bloc diagram of static RAM SF.F.80102.

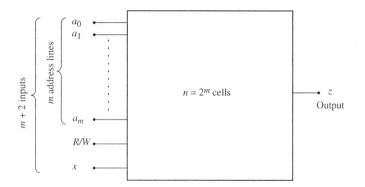

Figure 8.2 Simplified model of a RAM.

It will be shown, in Section 8.2.2., that all the faults in the decoder and the read/write logic are very easy to detect in comparison with the faults in the cell array. Then, the attention will be focused on the faults affecting the cell array. The simplified model in Figure 8.2 is used. A n word by 1 bit RAM is modeled by an $(m+2)$-input, 1-output circuit. The $m+2$ boolean inputs are m address lines, a read/write line, and a data input line. When the word-oriented RAMs are concerned, in Section 8.3.1, there are $b-1$ additional data input lines and $b-1$ additional output lines.

8.1.2 Fault Models

Two main categories of single faults are presented in this section: faults in the decoder and read/write logic on the one hand, and faults in the memory cell array (assuming word of 1 bit) on the other hand. The faults in a word oriented memory are similar except that coupling between bits of the same word can be present; they will be presented in Section 8.3.1. A multiple fault corresponds to the presence of two or more single faults in the same RAM.

8.1.2.1 Faults in the decoder and read/write logic

The stuck-at faults in the decoder and read/write logic result in behaviors which can be represented by the following models.

$\mathcal{M}_l$: an address line is stuck-at-α.

$\mathcal{M}_{rd}$: a row is always decoded. When selecting a cell of another row, we address two cells. This means that we read or write in two cells at the same time. When two cells are read, the output is the AND (or OR) function of both contents.

$\mathcal{M}_{ld}$: a column is always decoded (similar to $\mathcal{M}_{rd}$).

$\mathcal{M}_{cd}$: a cell is always decoded. If cell j is always decoded, both cells i and j are addressed when i should be addressed $(i \neq j)$.

$\mathcal{M}_{rw}$: the *R/W* line is stuck-at-0 or 1.

8.1.2.2 Faults in the memory cell array

The models of faults in the cell array have been progressively defined by many authors. The faults usually considered can be divided into six models belonging to three categories. The two models in each category are distinguished by *pattern sensitiveness*.

A pattern is given by a Boolean expression involving cell variables. For example, let k_1 and k_2 be two cells; the pattern with the content of k_1 being 0 and that of k_2 being 1 is denoted by $k'_1 k_2$.

Notation 8.1

a) If f is any fault and K is any pattern, then [f when $K = 1$] denotes the **pattern-sensitive fault** (PSF) of f being present when K is true.

b) $\uparrow i$: cell i changes from value 0 to value 1 ($\downarrow i$: cell i changes from value 1 to value 0)

c) $\updownarrow i$: cell i changes value, i.e., to value 1 if it was 0 or to 0 if it was 1. ❏

Category of **passive faults**.

$\mathcal{M}_{sa}$. **Stuck-at** fault: a cell i is stuck-at-α ($\alpha = 0$ or 1)

$\mathcal{M}_{saPSF}$. **Passive PSF**: a transition ($\uparrow$ or $\downarrow$) of the cell i cannot occur when writing the cell i when some pattern $K = 1$.

Category of active faults with inversion.

$\mathcal{M}_{to}$. **Toggling** (also called *inversion coupling*). A toggling fault is present from a cell i to a second cell j if, when cell i has value α, and α' is written into cell i, then the state of cell j is completed or "toggled" with respect to its previous value. This is denoted by $\uparrow i \Rightarrow \updownarrow j$, if $\alpha = 0$, or $\downarrow i \Rightarrow \updownarrow j$ if $\alpha = 1$.

$\mathcal{M}_{toPSF}$. **Pattern sensitive toggling** (PSF toggling), for example [$\uparrow i \Rightarrow \updownarrow j$, when $K = 1$], where K is independent of i and j.

Category of active faults with idempotence.

$\mathcal{M}_{ic}$. **Idempotent coupling**. An idempotent coupling is present from a cell to a second cell j if, when the cells contain a specific pair of binary values α and β, and α' is written into cell i, then cell j, as well as cell i, changes state. This is denoted by $\uparrow i \Rightarrow \uparrow j$ if $\alpha = \beta = 0$, $\downarrow i \Rightarrow \uparrow j$ if $\alpha = 1$ and $\beta = 0$, $\uparrow i \Rightarrow \downarrow j$ if $\alpha = 0$ and $\beta = 1$, and $\downarrow i \Rightarrow \downarrow j$ if $\alpha = \beta = 1$.

$\mathcal{M}_{icPSF}$. **Pattern sensitive idempotent coupling** (PSF idempotent coupling), for example [$\uparrow i \Rightarrow \uparrow j$ when $K = 1$], where K is independent of i and j. ❏

For PSFs, the cells involved in K may be either in the **neighborhood** of the cell which may have a faulty value (basic cell), or anywhere. The neighborhoods are usually a special pentomino or a rectangle of cells including the basic cell in the middle.

Other kind of faults can be considered. For example a transition fault such that a cell i of the memory fails to make the transition from 0 to 1; however, both states are possible for the cell, for example at power-up. Such a fault is similar to a stuck-at fault. It behaves as a stuck-at fault except if the starting state (at power-up) is 1 and the initial state is 1. As far as random testing is concerned, one can easily show that it is almost as easy to detect as a stuck-at fault.

Remark 8.1 In every model presented in Section 8.1.2, except $\mathcal{M}_{rw}$, all the faults are R-equidetectable for ψ_0 (Definition 4.3 in Section 4.4.3.2), given some mild conditions. Consider for example $\mathcal{M}_{sa}$. If the entire RAM is initialized at 0, then all the stuck-at-0 faults of cells are R-equidetectable, and all the stuck-at-1 faults of cells are R-equidetectable. If initialization of the RAM is made of random values or is unknown, then all the faults in $\mathcal{M}_{sa}$ are R-equidetectable.

8.2 TEST LENGTH FOR SINGLE FAULTS

The test length depends of various parameters which are analyzed in Section 8.2.1 using an example of fault (single stuck-at of a cell). Faults in the address decoding and read/write logic are considered in Section 8.2.2, and faults in the memory cell array in Section 8.2.3.

Let us first specify some useful notation. The behavior of a fault-free n-cell RAM can be specified by an automaton $A = (Q, X, Z, \delta, \mu)$ when $Q = \{0, 1\}^n$, $X = \{r^i, w_0^i, w_1^i \mid i = 1, ..., n\}$, and $Z = \{0, 1, \$\}$. The input symbols r^i, w_0^i, and w_1^i denote the actions of reading cell i, writing a 0 to cell i and writing a 1 to cell i, respectively. The state of the memory is defined by the values α_i which are stored in every cell i, i.e., $q = (\alpha_1, ..., \alpha_n)$. The transition function δ and the output function μ are defined by

$$\delta((\alpha_1, ..., \alpha_i, ..., \alpha_n), x) = \begin{cases} (\alpha_1, ..., \alpha_i, ..., \alpha_n), & \text{if } x = r^i \\ (\alpha_1, ..., 0, ..., \alpha_n), & \text{if } x = w_0^i \\ (\alpha_1, ..., 1, ..., \alpha_n), & \text{if } x = w_1^i \end{cases}$$

and

$$\mu((\alpha_1, ..., \alpha_i, ..., \alpha_n), x) = \begin{cases} \alpha_i, & \text{if } x = r^i \\ \$, & \text{otherwise.} \end{cases}$$

The output symbol $, used for unspecified values[1], denotes here "no output".

8.2.1 Stuck at fault of a memory cell

This is a *basic fault*. As we will show, it is *more difficult* to detect than the faults in the decoder and read/write logic, and it is the *easiest* to detect among the faults in the memory array.

Without loss of generality, let us consider the fault cell i stuck-at-0 (symmetrical results can be obtained for stuck-at-1). When the state of the memory should be $(\alpha_1, ..., \alpha_i, ..., \alpha_n)$, the state of the faulty memory is actually $(\alpha_1, ..., 0, ..., \alpha_n)$. Since the values α_j, $j \neq i$, have no influence on the testing process, a state of the observer denoted by $(\alpha_i, 0)$ may represent the merging of

---

[1] According to Note 2 in Chapter 2, the same symbol is used for both cases: *may have any value* or *cannot exist*. The exact meaning is understood from the context.

2^{n-1} states such that cell i should have value α_i. The corresponding observer is represented in Figure 8.3a: the states $(0,0)$ and $(1,0)$ correspond to $\alpha_i = 0$ and 1, respectively. The notation $a^{\neq i}$ corresponds to addressing any cell except i, i.e.,

$$a^{\neq i} = \bigcup_{j \neq i} \left\{ r^j + w_0^j + w_1^j \right\}. \tag{8.1}$$

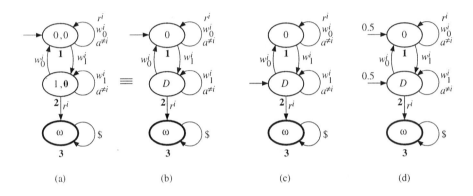

Figure 8.3 Observer for stuck-at-0 of cell i. (a) and (b) Initialization to 0. (c) Initialization to 1. (d) Unknown initialization.

Using the notation D introduced in Section 3.1.2.1 (D means "should be 1 but is actually 0"), the states of the observer can be denoted only by the state of cell i in the faulty memory. The states $(0, 0)$ and $(0, 1)$ in Figure 8.3a are denoted by (0) and (D), respectively, as shown in Figure 8.3b.

Figure 8.3b is obtained if the memory is initialized to 0 in every cell. The initial state is (0). The observer remains in this state if cell i is read, or if 0 is written into cell i, or if another cell is addressed; the state (D) is reached, i.e., there is an error in cell i, if 1 is (should be) written into cell i. From state (D) there are three possibilities: writing 0 into cell i leads back to (0); writing 1 into cell i or addressing another cell does not change the state; reading cell i leads to state ω, i.e., the fault has been detected.

Assume a constant distribution ψ of the input patterns. According to the indices of states shown in Figure 8.3, the transition matrix of the corresponding Markov chain is

$$U = \begin{bmatrix} 1-\psi(w_1^i) & \psi(w_1^i) & 0 \\ \psi(w_0^i) & 1-\psi(w_0^i)-\psi(r^i) & \psi(r^i) \\ 0 & 0 & 1 \end{bmatrix}. \tag{8.2}$$

Given initialization to 0, the initial probabilistic state is $\pi(0) = (1, 0, 0)$. Given initialization to 1 (Figure 8.3c), then $\pi(0) = (0, 1, 0)$; note that, due to symmetry, this gives the same result as initialization to 0 for a stuck-at-1 cell. If initialization is not known, the "**average initialization**" can be considered, then $\pi(0) = (.5, .5, 0)$ as illustrated in Figure 8.3d.

Let us now analyse the influence of various parameters on the test length: distribution ψ, initialization, number of cells n, and confidence level, $1 - \varepsilon$.

8.2.1.1 Influence of the distribution ψ

Assume a constant distribution ψ which corresponds to the probability distribution of the 2^{m-2} input patterns illustrated in Figure 8.2. Because of the symmetry between cells, it is clear that every cell should have the same probability of being addressed. Then:

$$\psi(r^i) + \psi(w_0^i) + \psi(w_1^i) = \frac{1}{n}, \quad \text{for } i = 1, ..., n. \tag{8.3}$$

Because of the symmetry between 0 and 1, it is clear that input x should have the probability $\Pr[x = 1] = 0.5$. Then:

$$\psi(w_0^i) = \psi(w_1^i), \quad \text{for } i = 1, ..., n. \tag{8.4}$$

Now, the only question concerns the probability of the R/W input. Is it interesting to have more read operations than write operations or vice versa? We now proceed to answer this question. Let

$$\Pr[R/W = 1] = \Pr[\text{reading}] = n \cdot \psi(r^i). \tag{8.5}$$

Definition 8.1 Given a test length, L, the **length coefficient** is the ratio $H = \frac{L}{n}$, i.e., the average number of accesses to every cell.

❑

Figure 8.4 shows the test length as a function of the probability of reading, $\Pr[R/W = 1]$, calculated for $n = 1\,024$ and $1 - \varepsilon = 0.999$. This is obtained from Equations (8.3) to (8.5) and the transition matrix U (8.2) for $\pi(0) = (1, 0, 0)$ and $\pi(0) = (0, 1, 0)$. From Figure 8.4, it is clear that

$$\Pr[R/W = 1] = 0.5 \tag{8.6}$$

is a very good value since it corresponds to the smallest value of H for initialization at 0 (worst case) or to a value close to the minimal value of H for initialization at 1. From now on this value will be used; then, from (8.3), (8.4), and (8.6)

$$\psi(r^i) = \frac{1}{2n} \quad , i = 1, \cdots, n; \tag{8.7}$$

$$\psi(w_0^i) = \psi(w_1^i) = \frac{1}{4n} \quad , i = 1, \cdots, n. \tag{8.8}$$

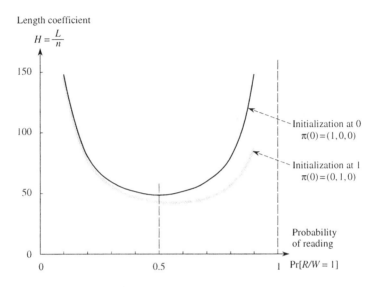

Figure 8.4 Influence of Pr[R/W = 1] on the length coefficient ($n = 1\,024$, $\varepsilon = 10^{-3}$).

8.2.1.2 Influence of initialization

The influence of initialization is also clear from Figure 8.4. If cell i is initialized at 0, its stuck-at-0 fault takes longer to detect than if the cell is initialized at 1. Let Pr[1 written in cell i] denote the probability that the value in cell i *should be* 1. The probability that there is a *detection* of the fault *at time l* (Definition 5.2 in Section 5.4) is proportional to this probability. Pr[1 written in cell i] is close to 0.5 when many random patterns have been applied: it increases from 0 to 0.5 if cell i is initialized at value 0; it decreases from 1 to 0.5 if cell i is initialized at value 1. This is illustrated in Figure 8.5a: the Markov chain represents the state of the fault-free cell i; the probabilistic state at time l is $v(l) = (v_1(l), v_2(l))$ and the curve in Figure 8.5a represents Pr[1 written in cell i] $= v_2(l)$ as a function of l for various initial probabilistic states. They correspond to $v(0) = (0, 1)$ for initialization at 1, $v(0) = (1, 0)$ for initialization at 0, and

$v(0) = (.5, .5)$ for average initialization. The probability that the fault has been detected at time l, $\pi_\omega(l)$, cannot be obtained directly from the probability of error, $v_2(l)$. However, one can *qualitatively* compare initialization at 1 and initialization at 0 of cell i. In the first case, at any time l, the probability of an error in cell i is greater than in the second case. It follows that, at any time l, the probability of a detection is greater in the first case than in the second case. Since this is true for any time $1, 2, \ldots l-1, l$, then the probability that the first detection has occurred by time l, $\pi_\omega(l)$ is greater for initialization at 1 (remind that the fault under consideration is the stuck-at-0 of the cell; if the fault was the stuck-at-1 of the cell, $\pi_\omega(l)$ would be lower for initialization at 1). This is illustrated in Figure 8.5b.

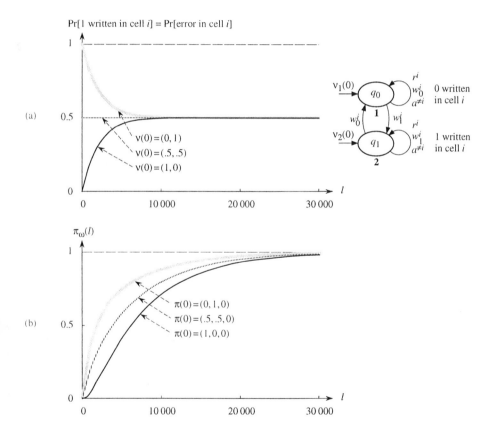

Figure 8.5 Stuck-at-0 of cell i in a 1 024-bit RAM. (a) Probability of an error in cell i. (b) Probability that the fault has been detected.

Initial state $\pi(0)$	$i = 0$ (1,0,0)	$i = 1$ (0,1,0)	Average (.5,.5,0)
L = test length	49 614	43 452	47 191
H = length coefficient	48.5	42.4	46.1
Variation relative to average initialization	+ 5.1 %	- 7.9 %	0 %

Figure 8.6 Influence of initialization on the length coefficient ($n = 1\,024$, $\varepsilon = 10^{-3}$).

Given Equations (8.7) and (8.8), the transition matrix (8.2) becomes

$$U = \begin{bmatrix} 1 - \dfrac{1}{4n} & \dfrac{1}{4n} & 0 \\ \dfrac{1}{4n} & 1 - \dfrac{3}{4n} & \dfrac{1}{2n} \\ 0 & 0 & 1 \end{bmatrix}. \qquad (8.9)$$

Assume a length L such that $\pi_\omega(L) \geq 0.999$ is sought (i.e., $\varepsilon = 10^{-3}$) for stuck-at-0 of cell i in a 1 024-bit RAM. If the initialization is $i = 0$, from the transition matrix (8.9) with $n = 1\,024$, we obtain $L = 49\,614$. Hence the length coefficient H (Definition 8.1 in Section 8.2.1.1) is $H = 49614/1\,024 = 48.5$. Similar results can be obtained if i is initialized at 1 or if initialization is unknown (average intialization). The test length, and hence the length coefficient, depends on initialization but not very much: few percent for $\varepsilon = 0.001$ as shown in Figure 8.6.

Remark 8.2
a) If a test is to be performed, the length coefficient relative to the worst case (i initialized to α for i stuck-at-α) or relative to the unknown initialization[2] (average) can be chosen.

b) The *variation relative to average initialization decreases when ε decreases*. As a matter of fact, when the test length is longer, the initial state has a smaller effect.

8.2.1.3 *Influence of the number of cells*

Figure 8.7 shows that the length coefficient is not exactly independent of n, but practically independent for large (usual) values of n. This means that the

---

[2] If the length corresponding to the unknown initialization is chosen, i.e., $L = 47\,191$ in our example, the actual detection uncertainty is $\varepsilon = 0.0014$ or 0.0006, depending on the actual initial state.

testing time increases linearly with n. This property is true for practically all the fault models in RAMs: this is a *very important property* which is addressed in Section 8.4.1.

n = RAM capacity	1 bit	32 bits	1 Kbit	1 Mbit
H = length coefficient	43	46.0	46.1	46.1

Figure 8.7 Influence of the number of cells ($\varepsilon = 10^{-3}$, average initialization).

8.2.1.4 Influence of the confidence level

As for any sequential circuit (Property 7.3 in Section 7.2.2) the test length is approximately proportional to $\ln \varepsilon$ when ε is small. It follows that the length coefficient is also proportional to $\ln \varepsilon$, i.e., proportional to the number of 9s in the confidence level $1 - \varepsilon$ if $\varepsilon = 10^{-i}$ for some integer i. This is illustrated in Figure 8.8.

ε = uncertainty level	10^{-1}	10^{-2}	10^{-3}	10^{-4}	10^{-5}	10^{-6}
H = length coefficient	15.0	30.4	46.1	61.8	77.5	93.3
Variation relative to $H(10^{-3}) \frac{\ln \varepsilon}{\ln 10^{-3}}$	-2.4 %	-1.2 %	0 %	+0.6 %	+0.9 %	+1.2 %

Figure 8.8 Influence of the uncertainty level ($n = 1\,024$, average initialization).

It has been noted, in Remark 8.2 (Section 8.2.1.2), that the variation relative to average initialization becomes smaller when ε decreases. For $\varepsilon = 10^{-6}$, the variation relative to average intialization is +2.5% and -3.9% for initialization of i at 0 and at 1, respectively (for i stuck-at-0).

8.2.1.5 Concluding remarks

We have extensively developed the case of a stuck-at fault of memory cell. The main reason is that most of the observations which have been made may be *generalized to other faults in RAMs*.

a) The relative difference between the length coefficient for some specified initialization and the length coefficient for average initialization is less than 10%

for other faults which have been studied [DaFuCo 89], for $\varepsilon = 10^{-3}$. For smaller ε the relative difference is still smaller.

b) The test length is practically proportional to the number of cells, n, for all usual faults.

c) The test length is approximately proportional to $\ln \varepsilon$ for the other faults too (the Markov chains for other faults have the property explained in Section 8.4.1). These properties are shown in Section 8.4.1 (Property 8.9).

8.2.2 Faults in address decoding and read/write logic

These faults correspond to the fault models $\mathcal{M}_l$, $\mathcal{M}_{rd}$, $\mathcal{M}_{ld}$, $\mathcal{M}_{cd}$, $\mathcal{M}_{rw}$, which are defined in Section 8.1.2.1.

Property 8.1 Any fault in $\{F_l \cup F_{rd} \cup F_{ld} \cup F_{cd} \cup F_{rw}\}$ is easier to detect by a random test than a fault in F_{sa}, i.e., a stuck fault of memory cell.

☐

Let us explain why with a single example, a fault in F_l, i.e., corresponding to stuck-at-α of an address line.

Assume, without loss of generality, that address line a_0 is stuck-at-0 (noted $a_0/0$). Let[3] $i^+ = i + \frac{n}{2}$ ($i = 1, 2, ..., \frac{n}{2}$). All the cells 1, 2, .., $\frac{n}{2}$ are normally addressed. The cells $1^+, 2^+, ..., \left(\frac{n}{2}\right)^+$ are never addressed since a_0 can never have the value 1, and the cell i is addressed instead of i^+. Let α_i denote the value in cell i ($\alpha_i \in \{0, 1, D, \overline{D}\}$), and α_{i^+} denote the value which should be in cell i^+. Let $b = (\alpha_i, \alpha_{i^+})$ denote the state of the pair (i, i^+): the possible states are $b_0 = (0, 0)$, $b_1 = (0, 1)$, $b_2 = (1, 0)$, $b_3 = (1, 1)$, $b_4 = (D, 0)$, and b_5 $(\overline{D}, 1)$. State b_0 is reached when the last value written in cell i and the last value written in cell i^+ are both 0 (it is the initial state if the entire memory is initialized at 0). State b_4 is reached if the last value written in cell i was 1, the last value written in cell i^+ was 0, and the writing in cell i^+ occurred after writing in cell i, and so on.

Consider detection of the fault when reading either i or i^+. The fault is detected when reading i (input vector r^i) if the state is either $b_4 = (D, 0)$ or $b_5 = (\overline{D}, 1)$ because the value in i is faulty. The fault is detected when reading i^+ (input vector r^{i^+}) if the state is either $b_1 = (0, 1)$ or $b_2 = (1, 0)$, because the value in i is read instead of the value in i^+. One can build an observer made up of the states $b_0, ..., b_5$, and ω as shown in Figure 8.9a (the notation $\neq$ associated with self-loops means 'input vectors other than those provoking a transition'). Since $\psi(w_0^i) = \psi(w_1^i) = \psi(w_0^{i^+}) = \psi(w_1^{i^+}) = 1/4n$, and $\psi(r^i) = \psi(r^{i^+}) = 1/2n$, the corresponding Markov chain can be reduced by standard methods to the 3-state

---

[3] The number of cells is even since $n = 2^m$ (Figure 8.2 in Section 8.1.1).

Markov chain in Figure 8.9b. Two states correspond to the merging of $\{b_0, b_3\}$ on one hand, and $\{b_1, b_2, b_4, b_5\}$ on the other hand; the third state is ω. From this Markov chain, and for $n = 1\,024$, one obtains a length coefficient $H_{ii^+} = 35.0$. This length coefficient has the same order of magnitude as the length coefficient for the stuck-at fault of cell i, and is lower than $H_{sa} = 46.1$. In addition, while the stuck at fault of cell i can be detected *only* when reading cell i, the fault $a_0/0$ may be detected by reading either cell 1 or 1^+, or 2 or 2^+, ..., or $\frac{n}{2}$ or $\left(\frac{n}{2}\right)^+$. Then the required test length is *much lower* than for a stuck-at cell.

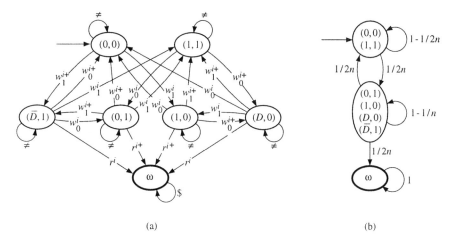

Figure 8.9 Fault "address line a_0 stuck-at-0". (a) Observer (detection by reading either cell i or cell i^+). (b) Reduced Markov chain.

After a long test sequence, the stationary probabilities of the states of the pair (i, i^+) are $\Pr[b_0] = \Pr[b_3] = 1/4$, and $\Pr[b_1] = \Pr[b_2] = \Pr[b_4] = \Pr[b_5] = 1/8$. This is true *for any* i. Let j be the cell addressed at any time l, for large l. The average detection probability (Definition 5.4 in Section 5.4.2) is

$$p_{av,a_0/0} = \Pr[\text{reading}] \left(\Pr[j < n/2] \left(\Pr[b_4] + \Pr[b_5] \right) \right.$$
$$\left. + \Pr[j \geq n/2] \left(\Pr[b_1] + \Pr[b_2] \right) \right), \tag{8.10}$$

$$p_{av,a_0/0} = \frac{1}{2}\left(\frac{1}{2} \times \left(\frac{1}{8}+\frac{1}{8}\right) + \frac{1}{2} \times \left(\frac{1}{8}+\frac{1}{8}\right)\right) = \frac{1}{8}. \tag{8.11}$$

This value may be compared to the average detection probability of a cell i stuck-at-α:

$$p_{av,i/\alpha} = \Pr[\text{reading}] \times \Pr[\text{cell } i \text{ addressed}] \times \Pr[\text{cell } i \text{ faulty}],$$

$$p_{av,i/\alpha} = \frac{1}{2} \times \frac{1}{n} \times \frac{1}{2} = \frac{1}{4n}. \tag{8.12}$$

Other proofs for every fault in F_{rd}, F_{ld}, F_{cd}, or F_{rw} are similar. For these faults there are a *lot of cells* whose reading allows detection, provided that some conditions are met [ThDa 78a]. For a fault in F_{sa}, there is *only one cell* whose reading allows detection.

8.2.3 Faults in the memory cell array

For any fault in the sets of faults defined in Section 8.1.2.2, we can model the detection process by a Markov chain and evaluate the test length L, then the test coefficient H, for some value ε. We shall explain only an example, the toggling fault $\uparrow i \Rightarrow \updownarrow j$, then present the results given in [DaFuCo 89].

8.2.3.1 Length coefficient for toggling fault

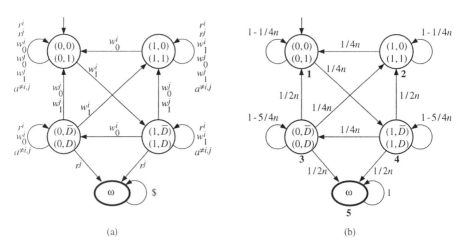

Figure 8.10. (a) Reduced observer for the toggling fault $\uparrow i \Rightarrow \updownarrow j$. (b) Corresponding Markov chain[4].

Let $f = (\uparrow i \Rightarrow \updownarrow j)$, where i and j are any two cells in the memory. Let $\alpha_i \in \{0,1\}$, and $\alpha_j \in \{0,1,D,\overline{D}\}$, denote the values in cells i and j. Let (α_i, α_j) denote the state of the pair (i, j). The observer for the fault f contains 9 states,

---
[4] Figure 8.10b is reprinted from [DaFuCo 89] (© 1989 IEEE).

namely (0, 0), (0, 1), (1, 0), (1, 1), (0, $\overline{D}$), (0, D), (1, $\overline{D}$), (1, D), and ω. By standard methods this observer can be reduced to 5 states: {(0, 0), (0, 1)}, {(1, 0), (1, 1)}, {(0, $\overline{D}$), (0, D)}, {(1, $\overline{D}$), (1, D)}, and ω, as illustrated[5] in Figure 8.10a. In this figure, the assumed initial state is $\alpha_i = 0$ and α_j fault-free (either 0 or 1). From this state, only w_1^i can change the state; after this writing, the state is such that $\alpha_i = 1$ and α_j is faulty (either $\overline{D}$ or D). From this state, three other states can be reached: r^j detects the fault; w_0^j and w_1^j cancel the error in cell j; and w_0^i changes the value of i but j remains faulty, and so on.

The Markov chain corresponding to Figure 8.10a is represented in Figure 8.10b. From this chain, for $\varepsilon = 10^{-3}$ and $n = 1\,024$, we obtain $L = 103\,481$; then $H = 103\,481 / 1\,024 \approx 101.1$. If we assume an average initialization, i.e., $\pi(0) = (.25, .25, .25, .25, 0)$, we obtain $H = 100.4$. Assume now that all the cells of the RAM are initialized at the same value, either 0 or 1. If all the cells are initialized at 0, the initial state is (0, 0). If all the cells are initialized at 1, the initial state is either (1, 1) or (1, D): it is (1, D) if cell i is initialized after cell j *and* the starting value of i is 0; else it is (1, 1). Then this assumption leads to $\pi(0) = (.5, .375, 0, .125, 0)$, and we obtain $H = 102.6$.

This analysis is an example. Other faults in the memory cell array can be analysed in a similar way.

8.2.3.2 Other faults

Let $V(f)$ be the number of *involved cells* for the fault f. This number includes the base cell, which may be faulty, and $V - 1$ influencing cells. For example $V(i$ stuck-at-0$) = 1$, for the pattern sensitive toggling $f = [\uparrow i \Rightarrow \updownarrow j$ when $k_1 k_2 k'_3 = 1]$, $V(f) = 5$. When a neigborhood is defined, V is the number of cells in it, but in a general case, the V cells may be anywhere in the RAM.

Figure 8.11 presents the rounded test lengths for the faults in the memory cell array, for $\varepsilon = 10^{-3}$ and assuming an average initialization. There are three categories of faults (defined in Section 8.1.2.2). *First category*: stuck-at cell if $V = 1$ and passive PSF if $V \geq 2$. *Second category*: toggling fault if $V = 2$ and PSF toggling if $V \geq 3$. *Third category*: idempotent coupling if $V = 2$ and PSF idempotent coupling if $V \geq 3$.

The rounded test lengths up to $V = 8$ are presented in Figure 8.11a. Figure 8.11b illustrates that, in every category, the length coefficient is approximately multiplied by 2 when V increases by 1 unit.

The notation for various sets of faults is introduced in Section 8.1.2.2. For example F_{saPSF} is the set of passive PSF. Let $F_{saPSF} = \{F_{saPSF,2} \cup F_{saPSF,3} \cup ...\}$ where $F_{saPSF,V}$ corresponds to the subset corresponding to V involved cells. Assume one wants to detect any fault f in $F_{saPSF,3}$ with a probability at least $P(f) = 1 - 10^{-3}$. Then

---

[5] $a^{\neq i,j}$ means: addressing any cell of the memory except i and j.

from Figure 8.11a a test length $L = 228n$ is required. It is observed that all the faults in $\{F_{sa} \cup F_{saPSF,2} \cup F_{saPSF,3} \cup F_{to} \cup F_{toPSF,3} \cup F_{ic}\}$ are detected with a probability of at least $1 - 10^{-3}$, because all these faults have a length coefficient, for $\varepsilon = 10^{-3}$, at the most equal to 228.

(a)

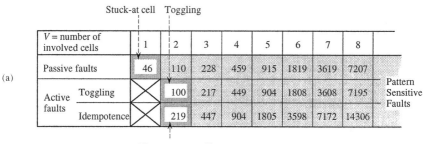

(b)

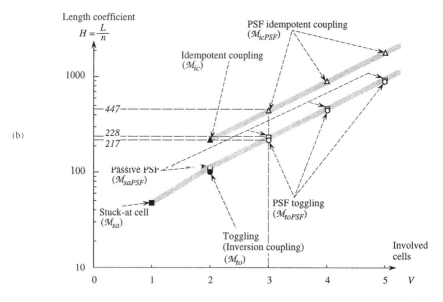

Figure 8.11 Length coefficients for faults in the memory cell array[6], for $\varepsilon = 10^{-3}$.

According to c) in Section 8.2.1.5, the test length is approximately proportional to $\ln \varepsilon$. Then, if a test length $228n$ is applied: a fault in $F_{saPSF,2}$ or in F_{to} is detected with a probability of about $1 - 10^{-6}$ (since 228 is about twice the

---

[6] This figure is reprinted from [DaFuCo 89] (© 1989 IEEE).

required length coefficient for $\varepsilon = 10^{-3}$); a fault in F_{sa} is detected with a probability of about $1 - 10^{-12}$ (228 about 4 times the required length coefficient); a fault in $F_{icPSF,3}$ is detected with a probability of $1 - 10^{-1.5} \approx 0.968$ (228 about half the required length coefficient for $\varepsilon = 10^{-3}$).

8.3 EXTENSION TO OTHER MODELS

Section 8.3.1 is dedicated to word-oriented memories and multiples faults are studied in Section 8.3.2.

8.3.1 Word-oriented memory

In a n-word by b-bit RAM, the model in Figure 8.2 (Section 8.1.1.1) is modified as follows: there are b data input lines, x_1, ..., x_b (a random input pattern is a $(m+b+1)$-bit vector); there are b output lines, z_1, ..., z_b. The total number of bits in the RAM is $n \cdot b$. The b bits of a word are written simultaneously, and read simultaneously.

Let (i,p), $i \in \{1,...,b\}$ and $p \in \{1,...,n\}$, denote the ith bit of the word p. Let $H_b(f)$ denote the length coefficient for the fault f in the considered memory (for some value of ε, for example 10^{-3}).

8.3.1.1 All the cells involved belong to different words

Consider a toggling fault $\uparrow(i, p) \Rightarrow \updownarrow(j, q)$. It has a meaning only if $p \neq q$, since all the bits in a word are written simultaneously. It follows that the number b has no influence on the test length. This is also true for a single stuck-at fault, for an idempotent coupling, and more generally for any fault such that every involved cell belongs to a different word. This leads to the following property.

Property 8.2
a) H_b(single cell stuck-at-α) = H_1(single cell stuck-at-α).
b) H_b(toggling) = H_1(toggling).
c) H_b(idempotent coupling) = H_1(idempotent coupling).
d) $H_b(f$ when $\alpha_{(i,p)} \cdot .... \cdot \alpha_{(j,q)} = 1) = H_1(f$ when $\alpha_p \cdot .... \cdot \alpha_q = 1)$,
if $p \neq ... \neq q$, and p, ..., q are different from the words involved in f ($\alpha_{(i,p)}$ and α_p are the value in the cell (i,p) and p, respectively).

❑

Property 8.2 applies to fault models which are not *PSFs* (parts a, b, and c), and to *PSFs* such that every involved cell belongs to different words (part d). Two kinds of faults have still to be considered: 1) *PSFs* such that some involved cells belong to the same word, and 2) coupling of bits in a single word, a new kind of fault since it cannot occur if $b = 1$.

8.3.1.2 PSFs such that some involved cells belong to the same word

From calculations based on Markov chains, the following observation has been made [Fu 86b, DaFuCo 89]. Given a kind of *PSF* and a number of involved cells, *the length coefficient increases very slightly when some involved cells belong to the same word* (a few percent).

Consider, for example, the following fault: $[\uparrow(i, p) \Rightarrow \updownarrow(j, q)$ when $\alpha_{(k1,r)} \cdot \alpha_{(k2,s)} = 1]$. The conditions $p \neq q$, r, s and $q \neq r$, s are required. If $r \neq s$, the test length is $H = 449$ (Property 8.2d); if $r = s$, the test length is $H = 458$.

Consider now the fault $[\uparrow(i, p) \Rightarrow \updownarrow(j, q)$ when $\alpha_{(k1,r)} \cdot \alpha_{(k2,s)} \cdot \alpha_{(k3,t)} = 1]$. The test lengths are $H = 904$ if r, s and t are three different words (Property 8.2d), $H = 913$ if $r = s \neq t$, and $H = 937$ if $r = s = t$.

8.3.1.3 Coupling of bits in a single word

Two or more bits in a single word could be coupled in such a way that they always have the same value. This is modeled by an OR-coupling or an AND-coupling. The 2-bit OR-coupling corresponds to[7] $(i, p)_f = (j, p)_f = (i, p)_0 + (j, p)_0$, $i \neq j$, where index 0 corresponds to the fault-free values. All these couplings are at least as easy to detect as a stuck-at fault of a memory cell. More precisely one has the following properties.

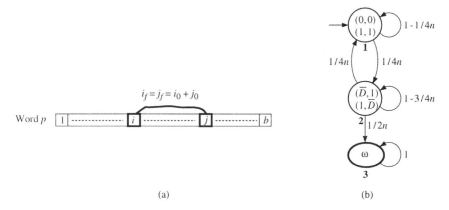

Figure 8.12 (a) 2-bit OR-coupling. (b) Corresponding Markov chain.

---

[7] Notation introduced in Section 2.2.3.1, illustrated in Figure 8.12a.

Property 8.3
a) $H_b(m\text{-bit OR-coupling}) = H_b(m\text{-bit AND-coupling})$.
b) $H_b(2\text{-bit OR-coupling}) = H_1(\text{single cell stuck-at-}\alpha)$.
c) $H_b(m\text{-bit OR-coupling}) < H_b(2\text{-bit OR-coupling})$, for $m > 2$.

Proof
a) Obvious by duality.

b) The fault f = (2-bit OR-coupling) is represented in Figure 8.12a and the corresponding Markov chain in Figure 8.12b. The elementary state can be defined by the pair (α_i, α_j) corresponding to the values in cells i and j. Some states can then be gathered as shown in Figure 8.12b. In the state $\{(0, 0), (1, 1)\}$, there is no error in the memory. One can move to the state $\{(\overline{D}, 1), (1, \overline{D})\}$ if 1) one writes into word p (probability $1/2n$) and 2) the values written into cells i and j are different (probability 0.5). Then the corresponding transition has a probability $1/4n$, and so on. It can be observed that the transition matrix is the same as the transition matrix of a stuck-at fault, i.e., matrix (8.9) in Section 8.2.1.2. The initial probability state depends on initialization of the memory.

c) Let f = (OR coupling of bits i and j in word p) and g = (OR coupling of bits $i, j,$ and k, in word p). Fault g dominates fault f. As a matter of fact there is an error in word p, given f, if $i_0 \neq j_0$, and this implies that there is an error in word p, given g; in both cases there is a detection when reading word p. In addition there are input sequences detecting g which do not detect f (reading when $i_0 = j_0 \neq k_0$). Then, from Property 4.6 (Section 4.4.3.2), the proof of Property 8.3c is obtained.

8.3.2 Multiple faults

Let $F_{AR} = \{F_{sa} \cup F_{to} \cup F_{ic} \cup F_{saPSF} \cup F_{toPSF} \cup F_{icPSF}\}$ denote the set of single faults in the cell array of a n-word by 1-bit RAM. The set of multiple faults whose components are in F_{AR}, is then denoted by F^*_{AR}. Let $f_{1,2,...,r} = [f_1 \& f_2 \& ... \& f_r]$ denote a multiple fault in F^*_{AR} whose components $f_1, f_2, ..., f_r$ are in F_{AR}.

Conjecture 8.1 Any *multiple fault* in F^*_{AR} R-*dominates at least one of its components* for the constant distribution ψ_0.
□

In other words, there is at least one component f_i of $f_{1,2,...,r}$, such that:

$$P(f_{1,2,...,r}, \psi_0, L) \geq P(f_i, \psi_0, L), \quad (8.13)$$

for any test length L. Then, for any confidence level, $H(f_{1,2,...,r}) \leq H(f_i)$.

Even if Conjecture 8.1 has not been shown in all cases, the following properties correspond to proofs in some particular cases.

Property 8.4 Conjecture 8.1 is true for *any double fault* $f_{1,2}$ such that both components are in $\{F_{sa} \cup F_{to} \cup F_{ic}\}$, i.e., are not *PSFs*.

Proof All the cases have been studied thanks to Markov chains [Fu 86a]. Among these faults, there are [(↑i ⇒ ↕j) & (↑k ⇒ ↕j)] and [(↑i ⇒ ↑j) & (↓j ⇒ ↓i)] for example.

Property 8.5 Conjecture 8.1 is true for a multiple fault if *at least one* of its components is a *stuck-at fault*.

Proof Let us assume, without loss of generality, that f_1 corresponds to a stuck-at cell. Fault $f_{1,2,...,r}$ dominates f_1 since any test sequence detecting f_1 also detects $f_{1,2,...,r}$. Then, according to Property 4.6a, $P(f_{1,2,...,r}, \psi, L) \geq P(f_1, \psi, L)$, for any ψ and any L.

□

Before presenting the next property, let us introduce some notation. For a fault f_i, the involved cells may be classed in a set I_i of cells whose values may cause the faults, and a set O_i of cells (singleton for a single fault) whose values may be faulty due to the fault f_i. For example, if $f_i = [\uparrow a \Rightarrow \uparrow b$ when $cd' = 1]$, then $I_i = \{a, c, d\}$ and $O_i = \{b\}$.

Property 8.6 Conjecture 8.1 is true for a multiple fault $f_{1,2,...,r}$ if *at least one* of its components, say f_1, is such that

$$\{I_1 \cup O_1\} \cap \{O_2 \cup ... \cup O_r\} = \emptyset. \tag{8.14}$$

Proof From Equation (8.14) no cell involved in fault f_1 can have its value modified by another fault f_2 or ... or f_r. It follows that any test sequence detecting f_1 also detects the multiple fault, i.e., $f_{1,2,...,r}$ dominates f_1. Then, according to Property 4.6a, $P(f_{1,2,...,r}, \psi, L) \geq P(f_1, \psi, L)$, for any ψ and any L.

□

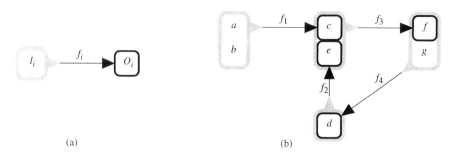

(a) (b)

Figure 8.13 Illustration of Property 8.6. (a) Representation of the sets I_i and O_i for a fault f_i. (b) Example of multiple fault.

Property 8.6 is illustrated in Figure 8.13b. Assume the multiple fault $f_{1,2,3,4}$ such that $f_1 = [\uparrow a \Rightarrow \updownarrow c$ when $b = 1]$, $f_2 = (\downarrow d \Rightarrow \downarrow e)$, $f_3 = [\uparrow e \Rightarrow \uparrow f$

when $c' = 1$], and $f_4 = [\uparrow g \Rightarrow \downarrow d$ when $f = 1$]. Since $I_1 = \{a, b\}$, $O_1 = \{c\}$, $O_2 = \{e\}$, $O_3 = \{f\}$, and $O_4 = \{d\}$, Equation (8.14) is verified. Then $f_{1,2,3,4}$ dominates f_1.

Remark 8.3.
a) For many cases the multiple fault dominates at least one of its components, i.e., R-dominates it, for *any constant distribution* ψ (see proofs of Properties 8.5 and 8.6).

b) For all the multiple faults which have been studied by the author and his team, it has been observed that the multiple fault R-dominates *all its components*, which is stronger than the Conjecture 8.1.

8.4 POWER OF RANDOM TESTING FOR RAMs

Linearity of the test length as a function of the number of cells (and other properties) are presented in Section 8.4.1. In Section 8.4.2, random testing is compared to deterministic testing for RAMS. An example of required test length for a batch of memories which may be affected by various faults is presented in Section 8.4.3.

8.4.1 Linearity of test length as a function of the number of cells

Some results in this section have been formally shown several years ago; they will be presented more briefly in this section. The recent results will be formally shown.

Definition 8.2 A ***k*-bounded** fault is a fault involving k cells, $k \leq n$, such that k is independent of the number n of cells of the memory.

A fault is said to be **bounded** if it is k-bounded for some k.

❏

For example $\uparrow i \Rightarrow \updownarrow j$ is a 2-bounded fault. All the cells of the memory except i and j behave correctly and have no influence on cells i and j. This is apparent in the observer in Figure 8.10a (Section 8.2.3.1) where $a^{\neq i,j}$ (i.e., adressing any cell but i and j) is associated with all the self loops. The reader will easily verify that all the faults listed in Section 8.1.2.2 are bounded. In fact, most of the RAM faults that have been considered in the literature are bounded. An example of an unbounded fault would be the following: [$\uparrow i \Rightarrow \uparrow j$ when *all* the cells of the memory contain the value 0].

214   Chapter 8

Property 8.7 Let f be a non-evasive bounded fault.
a) The transition matrix for probabilistic testing of f with the constant distribution ψ_0 has the form

$$U = I_m + \frac{1}{n} \cdot C \qquad (8.15)$$

where I_m is the $m \times m$ identity matrix (m is the number of states of this Markov chain) and C is a constant matrix.
b) The Markov chain has a single absorbing state which is reachable from every other state.

❏

Property 8.7b is a direct consequence of the non-evasiveness property (see Section 5.3.2). Let us illustrate Property 8.7a with an example. Consider the Markov chain for the toggling fault in Figure 8.10b. The corresponding transition matrix is

$$U = \begin{bmatrix} 1-\frac{1}{4n} & 0 & 0 & \frac{1}{4n} & 0 \\ \frac{1}{4n} & 1-\frac{1}{4n} & 0 & 0 & 0 \\ \frac{1}{2n} & \frac{1}{4n} & 1-\frac{5}{4n} & 0 & \frac{1}{2n} \\ 0 & \frac{1}{2n} & \frac{1}{4n} & 1-\frac{5}{4n} & \frac{1}{2n} \\ 0 & 0 & 0 & 0 & 1 \end{bmatrix} \qquad (8.16)$$

which can be written in the form

$$U = \begin{bmatrix} 1 & 0 & 0 & 0 & 0 \\ 0 & 1 & 0 & 0 & 0 \\ 0 & 0 & 1 & 0 & 0 \\ 0 & 0 & 0 & 1 & 0 \\ 0 & 0 & 0 & 0 & 1 \end{bmatrix} + \frac{1}{n} \cdot \begin{bmatrix} -\frac{1}{4} & 0 & 0 & \frac{1}{4} & 0 \\ \frac{1}{4} & -\frac{1}{4} & 0 & 0 & 0 \\ \frac{1}{2} & \frac{1}{4} & -\frac{5}{4} & 0 & \frac{1}{2} \\ 0 & \frac{1}{2} & \frac{1}{4} & -\frac{5}{4} & \frac{1}{2} \\ 0 & 0 & 0 & 0 & 0 \end{bmatrix}. \qquad (8.17)$$

The reason of this property is that all the transitions between two different states have a probability proportional to $1/n$. As a matter of fact such a transition is the consequence of an operation on a particular involved cell; for every operation, the probability of addressing the corresponding cell is $1/n$, and the probability that the operation is performed, given the cell is addressed, is constant. For example the transition from state 2 to state 1 of the Markov chain is due to the operation w_0^i: the probability of addressing cell i is $1/n$; the probability of writing 0 in cell i, given it is addressed, is $1/4$.

Before proceeding, let us review the notation introduced in Sections 5.4 and 6.1.1.2, respectively.

l_ω: random variable corresponding to the number of steps until detection.

$L(\varepsilon)$: minimum test length ensuring a confidence level at least equal to $1-\varepsilon$.

In [DaBrJü 92 & 97], it was shown that, for a non-evasive bounded fault:

1) the average length to detection is proportional to *the number of cells in the RAM*: $E[l_\omega] = n \cdot K$, where K is a constant value depending on the initial state and the matrix C in Property 8.7;

2) the required test length to obtain the detection of the fault with a probability at least $1-\varepsilon$ is such that: $L(\varepsilon) \leq \dfrac{nK}{\varepsilon} + 1 = O\left(\dfrac{n}{\varepsilon}\right)$. The important point is that the length of the probabilistic test *grows linearly with the memory size n*. However, the bound obtained is not tight: for example, for a cell stuck-at and for $\varepsilon = 10^{-3}$, this bound is $8\,000n$, while $L(\varepsilon)$ is less than $50n$ according to Figure 8.11 in Section 8.2.3.2.

❑

Recent results are presented in the sequel. They are based on the results in Section 7.2.2, and Property 8.7.

Property 8.8 Consider a non-evasive bounded fault in an n-word by one-bit RAM.

a) The jth eigenvalue $\lambda_j(n)$ of the transition matrix U, associated with the Markov chain, depends on n in the following way:

$$\lambda_j(n) = 1 + \frac{\lambda'_j}{n}, \qquad (8.18)$$

where λ'_j is an eigenvalue of matrix C (λ'_j is a constant value less than or equal to 0).

b) If the Markov chain obtained from the observer is such that the submatrix corresponding to transient states is irreducible and regular[8], the coefficient K_2 (defined in Section 7.2.2) is independent of n.

Proof
a) By definition, λ'_j is such that $\det(\lambda'_j I_m - C) = 0$.
It follows that

$$\det\left(\frac{\lambda'_j I_m}{n} - \frac{C}{n}\right) = \det\left(\left(\frac{\lambda'_j}{n}\right)I_m - \frac{1}{n}\cdot C\right) = 0.$$

---
[8] See Appendix C.

Hence,

$$\det\left(I_m\left(\frac{\lambda'_j}{n}+1\right)-I_m-\frac{1}{n}\cdot C\right)=\det\left(I_m\left(\frac{\lambda'_j}{n}+1\right)-U\right)=0, \quad (8.19)$$

which means that $\dfrac{\lambda'_j}{n}+1$ is an eigenvalue of U.

b) Given the initial state i of the Markov chain, K_2 is obtained by (Equation (7.8) in Section 7.2.2):

$$K_2=\left(\frac{z-\lambda_2}{\det(zI_m-U)}\Gamma_{i,m}\right)_{z=\lambda_2}. \quad (8.20)$$

Let α, β, and γ denote constant values, given i and C.
According to (8.18),

$$(z-1)_{z=\lambda_2}=\frac{\lambda'_2}{n}. \quad (8.21)$$

Thus, every element in $(zI_m-U)_{z=\lambda_2}$ has the form $\dfrac{\alpha}{n}$ (α may be equal to 0 for some elements).

It follows that $\left(\dfrac{z-\lambda_2}{\det(zI_m-U)}\right)_{z=\lambda_2}=\beta\cdot n^{m-1}$ and $(\Gamma_{i,m})_{z=\lambda_2}=\dfrac{\gamma}{n^{m-1}}$; hence, from (8.20), $K_2=\beta\cdot\gamma$ is independent of n. The coefficient K_2 depends only on the kind of fault (matrix C) and the initial state of the Markov chain (state i, or more generally states whose components are not 0 in $\pi(0)$).

Property 8.9 Consider a non-evasive bounded fault in an n-word by one-bit RAM. If the Markov chain obtained from the observer is such that the submatrix corresponding to transient states is irreducible and regular:

$$L(\varepsilon)\approx\frac{\ln\varepsilon-\ln(-K_2)}{\lambda'_2}\cdot n, \quad (8.22)$$

and, if ε is very small,

$$L(\varepsilon)\approx\frac{\ln\varepsilon}{\lambda'_2}\cdot n. \quad (8.23)$$

Proof From (8.18), $\lambda_2(n)=1+\lambda'_2/n$. Practically, for the interesting memories, n is large and λ'_2/n is close to 0. It follows that

$$\ln\lambda_2(n)\approx\frac{\lambda'_2}{n}. \quad (8.24)$$

From (7.10) in Section 7.2.2 and (8.24), (8.22) is obtained.

The value of K_2 is not very far from -1 (Property 7.2 in Section 7.2.2). For very small values of ε, $\ln(-K_2)$ is negligible in comparison to $\ln \varepsilon$. Hence, the estimate in (8.23) can be used; this approximate value, which can be derived from (8.22), corresponds also to (7.13) in Section 7.2.2.

❏

From (8.22), it is clear that the test length is proportional to the number n of cells. The length coefficient depends on the fault (λ'_2), the detection uncertainty (ε), and the initial state (K_2).

The approximate value obtained from (8.22) is very accurate. For a stuck-at cell ($\lambda'_2 = -0.1464466$), $\varepsilon = 0.001$, and an average initial state $\pi(0) = (.5, .5, 0)$ (for which $K_2 = -0.8535534$), $L(0.001) \approx 47\,194$ is obtained for a 1 024-cell memory, while the exact value is $L(0.001) = 47\,191$. The approximate value obtained from (8.23) is $L(0.001) \approx 48\,301$.

Remark 8.4 The conditions of Property 8.9 are met for all the faults in Section 8.1.2. According to Remark 7.1 in Section 7.2.2, only a reducible submatrix giving $\lambda_2 = \lambda_3$ could be considered (as a case where all the conditions are not met). Here is an example of an abstract fault with this property. The first reading of cell j, and only this one, changes the value in i; after this time, all the writings in i will write the faulty value (0 instead of 1 and vice-versa). The corresponding transition matrix is, for a 1024-cell RAM,

$$\begin{bmatrix} \dfrac{2047}{2048} & \dfrac{1}{2048} & 0 \\ 0 & \dfrac{2047}{2048} & \dfrac{1}{2048} \\ 0 & 0 & 1 \end{bmatrix}.$$

If the initial state of the Markov process is $\pi(0) = (1, 0, 0)$, $L(0.001) = 18\,906$; if $\pi(0) = (.5, .5, 0)$, $L(0.001) = 17\,554$; if $\pi(0) = (0, 1, 0)$, $L(0.001) = 14\,144$. This fault does not fulfill the conditions in Property 8.9 since the matrix is reducible and $\lambda_2 = \lambda_3 = 2047/2048$ ($\lambda'_2 = \lambda'_3 = -0.5$). However, the approximate value obtained from (8.23) (application to RAMs of the more general result (7.13) in Section 7.2.2) is $L(0.001) \approx 14\,144$, which is consistent with a particular initialization of the Markov chain.

Remark 8.5
a) Properties 8.7 to 8.9 are true even if the probabilities of Read/Write and input data are not 0.5. Only the equal likelihood of all addresses is necessary.

b) Equation (8.23) in Property 8.9 is a proof of the observations in Sections 8.2.1.2 (small influence of initialization), 8.2.1.3 (proportionality to the number of cells), and 8.2.1.4 (proportionality to $\ln \varepsilon$).

c) The above results can be generalized to other kinds of RAMs such as n-word by b-bit memories or multi-port RAMs.

8.4.2 Comparison with deterministic testing

	Fault	Test length		
		Deterministic algorithm	Random testing (calculated with average initialization)	
			$\varepsilon = 10^{-3}$	$\varepsilon = 10^{-6}$
n words by 1 bit	• Stuck-at cell	$4n$	$46n$	$93n$
	• Idempotent coupling	$15n$	$219n$	$445n$
	• Toggling	$10n$	$100n$	$202n$
	• Double toggling	$4n\lceil\log_2 n\rceil + n$	$66n$	$134n$
	• Idempotent coupling plus toggling	?	$81n$	$164n$
	• Active PSF, $V = 3$, with neighborhood	$28n$	$447n$	$905n$
	• Active PSF, $V = 3$, anywhere	$32n\lceil\log_2 n\rceil + n$	$447n$	$905n$
	• Active PSF, $V = 5$, with neighborhood	$195n$	$1805n$	$3625n$
	• Active PSF, $V = 5$, anywhere	?	$1805n$	$3625n$
	• Active PSF, $V = 3$, plus toggling	?	$447n$	$905n$
n words by b bits	• m-bit OR-coupling ($m \geq 2$), in the same word	?	$46n$	$93n$
	• Active PSF, $V = 5$, anywhere	?	$1845n$	$3690n$

Figure 8.14 Random test length versus deterministic test length for some faults in RAMs[9].

In Figure 8.14 deterministic and random test lengths for some faults are presented.

The first general observation is that deterministic testing is much shorter for simple faults (for example stuck-at fault), and may be longer for more complicated faults (for example double toggling). When the faults are very complicated, particularly multiple faults, it becomes practically impossible to obtain a deterministic algorithm, while an upper bound for random testing length

---

[9] This figure is reprinted from [DaFuCo 89] (© 1989 IEEE). Active PSF means toggling or idempotent coupling, for example "active PSF, $V = 3$, anywhere" means any fault in $F_{toPSF,3} \cup F_{icPSF,3}$

can be easily obtained from Conjecture 8.1 (see for example, the fault: active PSF, $V = 3$, plus toggling).

The second observation is that the *test length of random testing grows linearly* with the memory size n. This is clear in Figure 8.14, and has been shown to be true for all the classical fault models (Property 8.9 in Section 8.4.1). For some faults, however, the best algorithm which has been obtained is $O(n \log n)$. Moreover, for several faults, lower bounds of deterministic test lengths have been shown to be $O(n \log n)$. For example, for *double toggling* it has been shown that no deterministic test exists for this fault model with a length shorter than $n \lfloor \log_2 (n-1) \rfloor + 5n$ [CoBr 92]. For small values of n, the random test length, about $66n$ for $\varepsilon = 10^{-3}$, exceeds the deterministic one; however, the random test length drops below the deterministic one for RAMs just a bit larger than 64K. For all the *PSF idempotent couplings*, lower bounds $O(n \log n)$ have been shown in [Co 94]. For example, for PSF idempotent couplings involving $V = 5$ cells anywhere, the lower bound is $8n \log_2 n + 35n$, and the length of the shorter known algorithm is $9.6n (\log_2 n)^{2.322} \pm 70\%$ (note that neither the bound nor this algorithm consider PSF togglings).

8.4.3 Example of application to a batch of circuits

We have obtained results, i.e., test length as a function of the required confidence level (or vice-versa) when some fault is assumed on a RAM. Now, if various faults affect a batch of RAMs, what is the expected defect level, given some test length and assumptions about the faults. This problem is solved by means of results in Chapters 4 and 8. Consider for example the following data.

Data Assume a batch of 1 Mbit RAMs (1Mword by 1 bit), and a random test sequence of length $L = 50 \cdot 10^6$. What is the expected defect level given the following information?

80% of the RAMs are fault-free.
8% of the RAMs are affected by a single stuck-at cell (*).
8%   "   "   by at least two stuck-at cells (*).
1%   "   "   by an idempotent coupling (*).
1%   "   "   by a toggling (*).
1%   "   "   by a double toggling (*).
1%   "   "   by unmodeled faults at least as easy to detect as an active PSF, with $V = 3$.

(*) means that other faults can be present, in addition.

The solution is proposed as an exercise (Exercise 8.5). Note that a similar calculation of defect level, given the fault occurrence probabilities, could be done for a batch of any other kind of circuits.

Naturally, such a calculation cannot easily be derived from a deterministic test since, if a test sequence does not detect all the faults defined by a model, the proportion of undetected faults is usually unknown.

NOTES and REFERENCES

The fault models corresponding to faults in the decoder and read/write logic can be found in [SzFl 71]. An exhaustive study of all the stuck-at faults in the RAM presented in Figure 8.1 [DuVr 75] has found the same kind of faults and observed that the most difficult to detect out of these faults is the stuck-at fault of a cell. This result is partially reported in [ThDa 78a].

The toggling (inversion coupling) and idempotent coupling, single and multiple, have been mainly studied in [NaThAb 78], [Ma 80], and [PaSa 85]. The pattern sensitive faults are considered in [Ha 75a] in which the notion of neighborhood is introduced, and in [SuRe 80]. A synthesis is presented in [DaFuCo 89] which introduces faults in word oriented memories.

The study of random testing of stuck-at cells has been addressed in [DaFe 85], [FuDaCo 86], [BaMcSa 87], and [DaFuCo 89]. Some results are reported in [Go 91]. Most of the results for other faults, extension to word-oriented memories and multiple faults, are drawn from [FuDaCo 86] and [DaFuCo 89].

The linearity of random test length for bounded faults is analyzed in [DaBrJü 93 & 97]. The proof of this property can be found in [DaBrJü 92 & 97]; some of the basic ideas for the proof are due to [Ca 88]. The result in Property 8.9 is given in [Da 97]. Lower bounds for deterministic tests are shown in [CoBr 92] for a double toggling, and in [Co 94] for PSF idempotent couplings.

Relations between single and multiple faults in RAMs are studied in [BrJü 96].

The necessity of random testing for RAMs in order to have very low defect level is analysed in [KrGa 93].

Random testing of dynamic RAMs has been addressed in [Fu 86b].

Testing of RAMs, partially random and partially deterministic, is considered in [SaMcVe 89].

9

Random Test Length for Microprocessors

Although a microprocessor is a sequential circuit, its behavior is too complex to be analysed directly by the tools presented in Chapter 7.

In this chapter, the time unit is associated with an instruction. *A cycle corresponds to a period of a basic clock (several clocks with the same period may exist in a microprocessor). Execution of an instruction requires several cycles, and the number of cycles is not the same for all the instructions in a microprocessor. In this chapter, an instruction I_j then takes the place devoted to an input vector x_j in the preceding ones. The microprocessor is tested by a sequence of random instructions with random data.*

A functional behavior model and functional fault models are considered, which may be used for test analysis in a user environment.

The approach in this chapter can be used for a variety of microprocessor-like circuits, i.e., various circuits whose aim is to perform computations.

Functional models of faults are presented in Section 9.1. An accurate method (Markov chain) and an approximate method (MDTS) can be used for analysis of a fault in a microprocessor: they are compared in Section 9.2. How to estimate of the test lengths for faults in the data processing section and the control section of the microprocessor are presented in Sections 9.3 and 9.4, then the test length for a whole microprocessor is considered in Section 9.5.

9.1 FUNCTIONAL MODELS

A microprocessor is represented as a graph-theoretic model at the register transfer level (Section 9.1.1). This model can be obtained using information about its instruction set and the functions performed. This information is available in the user's manual. The fault model, independent of the implementation details, is also defined on a functional level (Section 9.1.2).

9.1.1 Functional model of a microprocessor

This model is a graph containing two kinds of nodes: registers and operators. The operators are considered at a functional level. The incoming edges of an operator mode correspond to operands and the outgoing edges to results. In this model any edge is between an operator node and a register node (i.e., there is no edge between two registers or between two operators).

This model is illustrated with a hypothetical microprocessor. See Figure 9.1.

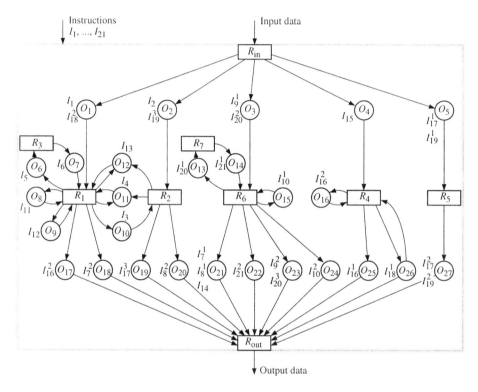

Figure 9.1 Graph model of the example microprocessor[1].

The set of registers includes the registers (denoted by R_i) used in various addressing modes, and the so-called processor status word containing various processor status bits[2].

R_1 : accumulator.

---

[1] This figure is reprinted from [ThDa 81] (© 1981 IEEE). The example microprocessor is taken from [ThAb 80].
[2] This list of registers is reprinted from [ThAb 80] (© 1980 IEEE).

R_2: general purpose register.

R_3 : scratch-pad register.

R_4 : scratch pointer to the top of a last-in first-out (LIFO) stack in main memory.

R_5 : address buffer register to store the address of operands.

R_6 : program counter.

R_7 : subroutine register to save return addresses (allowing a single level of subroutine).

Instructions	Class	# cycles
I_1: Load register R_1 from the main memory using immediate addressing	T	1
I_2: Load register R_2 from the main memory using immediate addressing	T	1
I_3: Transfer the content of register R_1 to register R_2	T	1
I_4: Add the contents of registers R_1 and R_2 and store the result in register R_1	M	1
I_5: Transfer the content of register R_1 to register R_3	T	1
I_6: Transfer the content of register R_3 to register R_1	T	1
I_7: Store register R_1 into the main memory using implied addressing	T	2
I_8: Store register R_2 into the main memory using implied addressing	T	2
I_9: Jump instruction	B	2
I_{10}: Skip if the content of register R_1 is zero	B	2
I_{11}: Left shift register R_1 by one bit	M	1
I_{12}: Complement (bit-wise) the contents of register R_1	M	1
I_{13}: Logical AND the contents of registers R_1 and R_2 and store the result in register R_1	M	1
I_{14}: No operation instruction	B	1
I_{15}: Load the stack pointer (R_4) from the main memory using immediate addressing	T	1
I_{16}: Push register R_1 on the LIFO stack maintained in the main memory	T	2
I_{17}: Store register R_2 into the main memory using direct addressing	T	3
I_{18}: Pop the top of the LIFO stack and store it in R_1	T	2
I_{19}: Load register R_2 from the main memory using direct addressing	T	3
I_{20}: Jump to subroutine (return address is saved in the subroutine register R_7)	B	3
I_{21}: Return from subroutine	B	2

Operators	Number of operand registers	Number of result registers
I_8: Left shift by 1 bit	1	1
I_9: Bit-wise complement	1	1
I_{11}: Bit-wise AND	2	1
I_{12}: Adder	2	1
I_{15}: Increment	1	1
I_{16}: Increment	1	1
I_{26}: Decrement	1	2
Others: Transfer	1	1

Figure 9.2 Instructions and operators of the example microprocessor[3].

---

[3] The first part of this figure is reprinted from [ThAb 80] (© 1980 IEEE).

Two registers named R_{in} and R_{out} are added. They represent respectively the input and output buses where the data[4] come from main memory and input/output devices, and addressed to them.

The operators (denoted by O_k) are activated by instructions (denoted by I_j). The execution of instruction I_j causes data flow, through operators, among the registers. The example microprocessor is worked by 21 instructions activating 27 operators (listed in Figure 9.2). The instructions are classified as Transfer (class T), Manipulation (class M) or Branch (class B) as in [Fl 74], [ThAb 80]. The addressing modes used (immediate, direct, implied [GsMc 75]) are indicated and the number of cycles are given.

An elementary operation occurring during the mth cycle of an instruction is denoted by I_j^m (the top index is omitted if an instruction is executed during a single cycle). A single instruction is enough to exercise several operators. For example, the instruction I_9 (jump) corresponds to I_9^1 (transfer of the jump address from the main memory to R_6: operator O_3) then I_9^2 (transfer of R_6 to the address register of the main memory to read the next instruction: operator O_{23}).

Several operations can drive the same operator. For example, in Figure 9.1, both I_1 and I_{18} drive operator O_1. This means that the operator is physically the same. If I_1 and I_{18} were using two different buses to transfer data from R_{in} to R_1, the model would use two operators O_{1a} and O_{1b}. However, the distinction is only useful for a fault in either O_{1a} or O_{1b}; for any other fault in the microprocessor both cases give the same result.

9.1.2 Fault models

The fault models presented in this section are taken from [ThAb 80]. A microprocessor can be divided into two main parts: the *data processing section* and the *control section*. For the data processing section, we consider faults in the registers and faults in the operators. For the control section, we consider faults in the decoding function on the one hand, and faults affecting instruction decoding and control function on the other hand. The fault models corresponding to these four parts (two for the data processing section and two for the control section) are presented in the sequel.

Let R, O, and I denote respectively the sets of registers, operators, and instructions, i.e., for our example microprocessor:

$R = \{R_1, R_2, ..., R_7, R_{in}, R_{out}\}$,

$O = \{O_1, O_2, ... O_{27}\}$,

$I = \{I_1, I_2, ... I_{21}\}$.

---

[4] Data is used as a generic term referring to the information as well as its address.

A *set of bits* is assigned to a register R_i. A *set of bits* is assigned to a bus. In order to simplify the presentation, we say (by abuse of language) that a bit is stuck-at some value, that two bits are ANDed, and so on.

Let us denote by B_u a **set of bits** and define two fault models related to this set.

$\mathcal{M}_{sa}$: A bit in B_u is stuck-at-0 or stuck-at-1.

$\mathcal{M}_{ao}$: A group of bits in B_u is ANDed (all the bits in the groups have the value 0 if at least one of them has the value 0) or ORed. This is called a **coupling**.

According to our notation (Section 4.2), the following set of faults related to B_u are obtained from these models: $F_{sa}(B_u)$, set a single stuck-at faults; $F_{sa}^*(B_u)$, set of multiple stuck-at faults; $F_{ao}(B_u)$, set of single groups of ANDed or ORed bits; $F_{ao}^*(B_u)$, set of multiple ANDed or ORed groups[5].

Notation 9.1

$$F(B_u) = F_{sa}^*(B_u) \cup F_{ao}^*(B_u).$$

9.1.2.1 Faults in the registers

Let us denote the set of bits in a register R_i by the name of this actual register, i.e., R_i.

We assume that the faults in the registers are in the set

$$F_r = \bigcup_{i: R_i \in R} F(R_i),$$

where $F(R_i)$ corresponds to the set of faults in R_i defined by Notation 9.1.

9.1.2.2 Faults in the operators

We consider two classes of faults: the *external* faults (for all operators) and the *internal* faults (defined only for some operators).

All the faults affecting a bus between the input registers and the output registers of operator O_k are associated with this operator; these are the **external faults** related to O_k. This is illustrated in Figure 9.3 which represents an operator O_k with two input registers (R_1 and R_2) and three output registers (R_1, R_3, and R_4). All the faults $F(B_u)$ such that B_u is the set of bits corresponding to one of the buses β_1 to β_8 are external faults (Figure 9.3b; β_3 is the gathering of β_1 and β_2 at the input of O_k). Note that $F(\beta_3)$ includes both $F(\beta_1)$ and $F(\beta_2)$. On the other hand all the $F(\beta_i)$ are different for $i = 4$ to 8: for example a fault affecting β_4 may lead to faulty values in R_1, R_3, and R_4, while only R_1 can be affected by a fault on β_8.

---

[5] Let us remember that $F_{sa}^*(B_u) \supset F_{sa}(B_\mu)$ and $F_{ao}^*(B_\mu) \supset F_{ao}(B_\mu)$.

226   Chapter 9

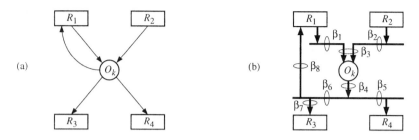

Figure 9.3 Illustration of external faults related to operator O_k. (a) Graph model for O_k. (b) One of the possible implementations (a heavy line represents a bus)[6].

The **internal faults** correspond to the set of faults $F(Y_k)$, where Y_k is the set of bits internal to operator O_k (an internal bit does not appear on a bus external to the operator).

Examples. Arithmetic operators may have internal faults: the carries in an adder are internal bits. A transfer or a Boolean bit-wise operator does not have internal faults.

The set of all the internal and external faults presented in this section is denoted by F_o.

9.1.2.3 *Faults in the register decoding function*

For our purpose, the possible faults in the register decoding function are separated into three models.

$\mathcal{M}_{d1}$: Instead of accessing R_i, no register is accessed. This fault is denoted by $f(R_i / \phi)$.

$\mathcal{M}_{d2}$: Instead of accessing R_i, one or several other registers R_p, R_q ... are accessed. This fault is denoted by $f(R_i / R_p + R_q + ...)$.

$\mathcal{M}_{d3}$: In addition to R_i, one or several registers are also accessed. This fault is denoted by $f(R_i / R_i + R_p + R_q ...)$.

One fault corresponding to every model is illustrated in Figure 9.4. It is clear that all these faults correspond to a modification of the graph. For the fault $f(R_1 / \phi)$, when the value in R_1 is an operand, it is replaced by a string of 0's (or 1's, depending on the technology). For the fault $f(R_1 / R_2)$, I_4 uses a faulty operand if an instruction I_1 has been executed after the last instruction I_2, while I_3 uses a faulty operand if I_2 has been executed after I_1. For the fault $f(R_1 / R_1 + R_2)$, when a result should be written in R_1, it is written into both R_1 and R_2; when the value in R_1 is an operand of I_3, it is replaced by the bit-wise AND function (or OR,

---

[6] This figure is reprinted from [ThDa 81] (© 1981 IEEE).

depending on technology) over values in R_1 and R_2; instruction I_4 uses a faulty operand in R_2 if an instruction I_1 has been executed after the last instruction I_2.

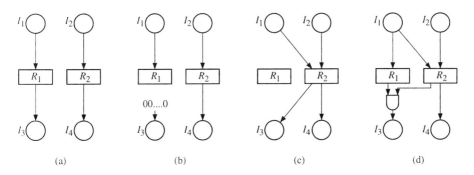

Figure 9.4 Faults in the register decoding function[7]. (a) Fault-free. (b) Fault $f(R_1/\phi)$. (b) Fault $f(R_1/R_2)$. (c) Fault $f(R_1/R_1+R_2)$.

The faults under consideration are the single faults corresponding to the models $\mathcal{M}_{d1}$, $\mathcal{M}_{d2}$, and $\mathcal{M}_{d3}$, for every register R_i in R, and any subset of $R \setminus R_i$ in place of $R_p + R_q + ...$. The set of all these faults is denoted by F_d.

9.1.2.4 Faults in instruction decoding and control function

These faults are separated into three models.

$\mathcal{M}_{i1}$: When instruction I_j has to be executed, no instruction is activated. This fault is denoted by $f(I_j/\phi)$.

$\mathcal{M}_{i2}$: Instead of I_j, some other instruction I_p is executed. This fault is denoted by $f(I_j/I_p)$.

$\mathcal{M}_{i3}$: In addition to I_j, some other instruction I_p is also activated. This fault is denoted by $f(I_j/I_j+I_p)$.

For $f(I_j/I_p)$ and $f(I_j/I_j+I_p)$, instruction I_p is normally executed when it is ordered. According to [ThAb 80], faults such as $f(I_j/I_p+I_q)$ or $f(I_j/I_j+I_p+I_q)$ do not exist.

One fault corresponding to every model is illustrated in Figure 9.5. These faults correspond to a modification of the graph. For faults $f(I_1/\phi)$ and $f(I_1/I_2)$, operation O_1 is never performed (Figure 9.5b and c). For $f(I_1/\phi)$, O_2 is normally performed, while for $f(I_1/I_2)$, operation O_2 is performed both when I_2 is ordered (correct working) and when I_1 should be executed. For $f(I_1/I_1+I_2)$, O_1 is normally performed while O_2 is performed both when I_1 or I_2 is ordered.

---

[7] This figure is reprinted from [ThDa 83] (© 1983 IEEE).

228   Chapter 9

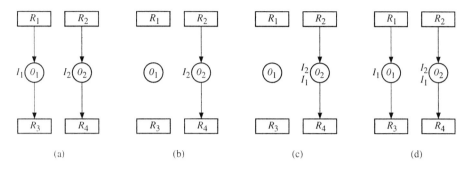

Figure 9.5 Faults in instruction decoding and control function[8]. (a) Fault-free. (b) Fault $f(I_1/\phi)$. (c) Fault $f(I_1/I_2)$. (d) Fault $f(I_1/I_1+I_2)$.

The faults under consideration are the single faults corresponding to the models $\mathcal{M}_{i1}$, $\mathcal{M}_{i2}$, and $\mathcal{M}_{i3}$, for every instruction I_j in I, and every $I_p \neq I_j$. The set of all these faults is denoted by F_i.

9.2 MARKOV CHAINS AND MDTS

Two examples of faults will be analysed in the sequel. It will be shown that a Markov model can be used for an accurate analysis, while the MDTS (Section 7.3.1) can be used for an approximate estimation of the required test length. Since the use of MDTS is much simpler, it is suggested that an approximate study of all faults can be made in this way. A Markov model may then be used for an accurate analysis of the most resistant faults.

Two general hypotheses are made, concerning random values in the registers and propagation of errors.

Hypothesis 9.1 Two random values in registers are different if they are independent from each other.
❑

For b-bit words, the probability that this hypothesis is not correct is 2^{-b}, e.g., $2^{-16} = 1.5 \cdot 10^{-5}$ for 16-bit words.

Hypothesis 9.2 A fault-free operation with exactly one faulty operand gives a faulty result. A fault-free operation with more than one faulty operand has a probability of producing a good result which may be ignored.
❑

---

[8] This figure is reprinted from [ThDa 83] (© 1983 IEEE).

This hypothesis is not always true, particularly for bit-wise AND or OR operations. If operator O_k activated by operation I_j does not fulfil Hypothesis 9.2, this fact is taken into account when the probabilities of propagation are calculated, in the following way: $\Pr[I_j]$ is replaced by $\Pr[I_j] \times \Pr[O_k$ propagates the error]. Then, Hypothesis 9.2 is not a restriction but it simplifies the presentation.

9.2.1 First example

Consider the simple example in Figure 9.6a in which O_1 to O_3 are transfer operators. The fault under consideration is the bit a in R_1 stuck-at-0. The detection process may be modeled by the observer (Section 3.2.2) in Figure 9.6b. A error may occur in register R_1 (when bit a should have the value 1) then be propagated to R_2 and to R_{out} where it is detected. In state b_0 there is no error in the system. In states b_1, b_2, and $b_{1,2}$, there is respectively an error in R_1 only, an error in R_2 only, and an error in both R_1 and R_2 but not in R_{out}. When an error is in R_{out}, the absorbing state ω is reached.

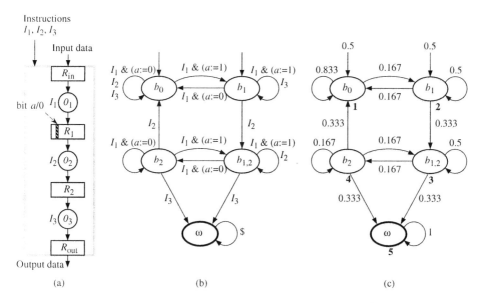

Figure 9.6 (a) Graph model with the fault under consideration[9]. (b) Observer. (c) Markov chain.

---

[9] Figure 9.6a is reprinted from [ThDa 81] (© 1981 IEEE).

Assume the observer is in state b_0. An error may occur in R_1 if two conditions are met: 1) instruction I_1 is executed; 2) the value in R_{in} at this time is such that the bit a in R_1 should take the value 1. The corresponding transition is labelled I_1 & (a:=1). For any other case, i.e., execution of I_2, I_3, or I_1 & (a:=0), the observer remains in state b_0.

From state b_1, the error may either be cancelled (if I_1 & (a:=0)), or be propagated to R_2 in addition to R_1 (if I_2), or remain in R_1 only (if I_3 or I_1 & (a:=1)).

From state $b_{1,2}$, the error in R_1 may be cancelled while an error remains in R_2 (if I_1 & (a:=0)), it may be detected (if I_3), or an error remains in both R_1 and R_2 (if I_2 or I_1 & (a:=1)).

Similarly the states b_0, $b_{1,2}$, b_2 and ω can be reached from b_2 as illustrated in Figure 9.6b.

From this observer, the Markov chain in Figure 9.6c is obtained from the following hypothesis:

1) The three instructions[10] are equally likely, i.e.,

$$\Pr[I_1] = \Pr[I_2] = \Pr[I_3] = 0.333; \tag{9.1}$$

2) The input data are equally likely, i.e.,

$$\Pr[a:=0 \mid I_1] = \Pr[a:=1 \mid I_1] = 0.5; \tag{9.2}$$

3) At initial time, the bit a in R_1 may be either correct or faulty, with the same probability, and there is no error in the other registers, i.e.,

$$\pi(0) = [0.5 \ 0.5 \ 0 \ 0 \ 0]. \tag{9.3}$$

From the Markov chain in Figure 9.6c, according to the calculation explained in Section 7.2, for a detection uncertainty $\varepsilon = 0.001$, the required test length is $L(0.001) = 88$.

Let us now calculate an approximate value using the MDTS (Section 7.3.1). Let Q be the set of possible states of the system in Figure 9.6a, when it is fault-free, and $Q_{a/0}$ the subset of Q such that there is an error in R_1 if the fault $a/0$ is present in the circuit (i.e., $Q_{a/0}$ is the set of states such that a should have the value 1).

Assume that the circuit is in a state $q \in Q_{a/0}$. The error is propagated to R_2 if I_2 is executed, then to R_{out} if I_3 is executed after I_2; thus qI_2I_3 is an MDTS. Furthermore, when the error has been propagated to R_2 (i.e., after I_2), execution of I_1, any number of time, can eventually remove the error from R_1 but not from R_2. It follows that all the transitions sequences in $qI_2I_1^*I_3$ are MDTS. Since this is true for any q in $Q_{a/0}$, the detection set

---

[10] Let us remember that, in this chapter, an instruction I_j takes the place devoted to an input vector in the preceding ones.

$$D_{a/0} = Q_{a/0} \cdot I_2 I_1^* I_3 \qquad (9.4)$$

is obtained, where $Q_{a/0}$ denotes the sum of states in the set. Because of equal likelihood of data, $\Pr[Q_{a/0}] = 0.5$. According to Property 7.5 in Section 7.3.1.2, the following result is obtained:

$$\Pr[D_{a/0}] = \Pr[Q_{a/0}] \frac{\Pr[I_2] \cdot \Pr[I_3]}{1 - \Pr[I_1]} = 0.5 \frac{0.333 \times 0.333}{(1 - 0.333)} = 0.0833$$

From Equation (7.18) in Section 7.3.1.2, given $\Pr[D_{a/0}] = 0.0833$, the test length $L(0.001) \approx 79$ is obtained (to be compared to the exact value 88 obtained from the Markov chain).

9.2.2 Second example

In this example, the propagation through two paths (a path is a string of operators and registers in which each operator and each register appears no more than once) reconverging to R_{out} is considered. In Figure 9.7a, an error in R_1 may be propagated through the path $R_1 O_2 R_2 O_3 R_{out}$ or through the path $R_1 O_4 R_3 O_5 R_{out}$.

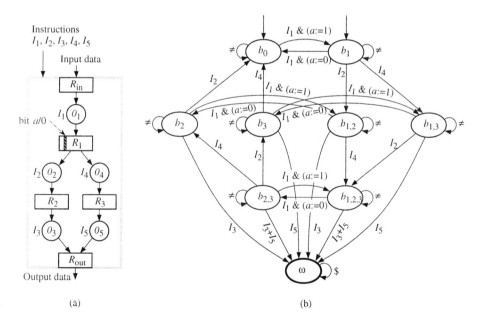

Figure 9.7 (a) Graph model with the fault under consideration[11]. (b) Observer.

---

[11] Figure 9.7a is reprinted from [ThDa 81] (© 1981 IEEE).

232   Chapter 9

The observer corresponding to the fault in Figure 9.7a is presented in Figure 9.7b. Except for b_0 (no error) and ω (fault already detected), the index of a state corresponds to the registers where an error is present. For example $b_{1,3}$ corresponds to an error in R_1 and R_3, and no error in R_2 (and obviously no error in R_{out}, otherwise the state would be ω). The self-loops are labelled "$\neq$" which means "all cases which do not provoke a change of state".

Given the probabilities of the instructions and the input data, a Markov chain can be obtained from the observer in Figure 9.7b (as for the first example). Assuming equal likelihood of input data and instructions, (i.e., $\Pr[I_j] = 0.2$ for $j = 1, ..., 5$), the following is obtained, for example: $\Pr[b_0 \to b_1] = 0.2 \times 0.5 = 0.1$; $\Pr[b_1 \to b_{1,3}] = 0.2$; $\Pr[b_{23} \to \omega] = 0.2 + 0.2 = 0.4$.

From this Markov chain and the initial state $\pi(0) = (0.5, 0.5, 0, 0, 0, 0, 0, 0, 0)$, the test length $L(0.001) = 104$ is obtained.

Let us now calculate an approximate value using the MDTS.

As in the previous example, $Q_{a/0}$ denotes the set of states such that there is an error in R_1.

From a state q such that there is an error in R_1, a string of instructions in $I_2(I_1+I_4+I_5)^*I_3$ detects the error since I_2 propagates the error into R_2, any instruction in $\{I_1, I_4, I_5\}$ does not remove this error, and I_3 propagates the error up to R_{out}. Similarly $I_4(I_1+I_2+I_3)^*I_5$ propagates an error from R_1 to R_{out}. However, all the transition sequences in $D_1 = Q_{a/0} \cdot I_2(I_1+I_4+I_5)^*I_3$ and in $D_2 = Q_{a/0} \cdot I_4(I_1+I_2+I_3)^*I_5$ are not MDTS, according to Definition 7.1 (Section 7.3.1.2). As a matter of fact, for example, $q_1I_2I_4I_5I_3$ is a transition sequence in D_1 if q_1 is in $Q_{a/0}$. Let $q_2 = \delta(q_1, I_2)$. It is clear that q_2 is also in $Q_{a/0}$ since instruction I_2 does not remove the error in R_1. Now $q_2I_4I_5$, which is a subsequence of $q_1I_2I_4I_5I_3 = q_1I_2 \cdot q_2I_4I_5I_3$, is a transition sequence in D_2. Thus, according to Definition 7.1, $q_1I_2I_4I_5I_3$ is not an MDTS.

In the previous example, the problem is that, while we are waiting for the end on a transition sequence propagating the error though a path ($q_1I_2I_4I_5I_3$ propagating through $P_1 = R_1O_2R_2R_{out}$), the propagation has been achieved by the other path ($q_1I_2I_4I_5$ through $P_2 = R_1O_4R_3R_{out}$). To avoid this problem, it is sufficient to restrict the set of strings of instructions in the following way: in the set of strings associated with the path P_1, an instruction of the path P_2 is not allowed, and vice-versa. For our example, let us avoid I_5 (arbitrary chosen between I_4 and I_5) in the expression corresponding to the propagation through P_1, and I_3 in the expression corresponding to the propagation through P_2 (this is easily generalized if there are more than two paths). The following is obtained:

$$D_{a/0} \supset Q_{a/0} \left(I_2(I_1+I_4)^* I_3 + I_4(I_1+I_2)^* I_5 \right). \tag{9.5}$$

All the transition sequences in this expression are MDTS, and their probabilities may then be added [DaTh 80a]. However, not all the MDTS are in this expression, for example $q_1I_2I_5I_3$. From Property 7.5 in Section 7.3.1.2:

$$\Pr[D_{a/0}] \geq \Pr[Q_{a/0}] \cdot \left(\frac{\Pr[I_2] \cdot \Pr[I_3]}{1 - (\Pr[I_1] + \Pr[I_4])} + \frac{\Pr[I_4] \cdot \Pr[I_5]}{1 - (\Pr[I_1] + \Pr[I_2])} \right) \quad (9.6)$$

$$\Pr[D_{a/0}] \geq 0.5 \left(\frac{0.2 \times 0.2}{1 - (0.2 + 0.2)} + \frac{0.2 \times 0.2}{1 - (0.2 + 0.2)} \right) = 0.0667.$$

From Equation (7.18) in Section 7.3.1.2, given $\Pr[D_{a/0}] \approx 0.0667$, the test length $L(0.001) \approx 100$ is obtained (to be compared with the exact value 104 obtained from the Markov chain).

9.3 TEST LENGTH FOR FAULTS IN THE DATA PROCESSING SECTION

The calculation of the test length for a fault in the data processing section is presented with the help of examples: first, faults in the registers (Section 9.3.1), then faults in the operators (Section 9.3.2).

9.3.1 *Faults in the registers*

A set of states Q_f is assigned to a fault f in a register R_i, such that there is an error in R_i, as is also a set of instruction sequences E_i such that all the transition sequences in $Q_f \cdot E_i$ are MDTS. In other words,

$$D_f \supset Q_f \cdot E_i \quad (9.7)$$

since, generally, $Q_f \cdot E_i$ does not contain all the MDTS. The way to obtain an estimate of $\Pr[D_f]$ given $Q_f \cdot E_i$, then the required test length, was illustrated in Sections 9.2.1 and 9.2.2. Our aim is now to find Q_f (this is relatively easy) and E_i.

For example, let us consider E_1 in the example microprocessor of Figure 9.1 in Section 9.1.1, i.e., a set of instruction sequences propagating an error from register R_1 to R_{out}. A way to obtain E_1 is illustrated in Figure 9.8. This figure is similar to Figure 9.1 with emphasis of instructions related to E_1 (the top indices of instructions are not useful for E_1: they are omitted).

As illustrated with heavy lines in Figure 9.8, the propagation from R_1 to R_{out} may be obtained through four paths, namely $P_1 = R_1 O_{17} R_{out}$, $P_2 = R_1 O_{18} R_{out}$, $P_3 = R_1 O_{10} R_2 O_{19} R_{out}$, and $P_4 = R_1 O_{10} R_2 O_{20} R_{out}$. The set E_1 is made of four parts,

$$E_1 = E_{11} + E_{12} + E_{13} + E_{14}, \quad (9.8)$$

where E_{1p} corresponds to a propagation through path P_p.

The following is obtained:

$$E_{11} = I_{16}, \tag{9.9}$$

$$E_{12} = I_{7}, \tag{9.10}$$

$$E_{13} = I_{3} \, G_{13}^{*} \, I_{17}, \tag{9.11}$$

$$E_{14} = I_{3} \, G_{14}^{*} \, I_{8}, \tag{9.12}$$

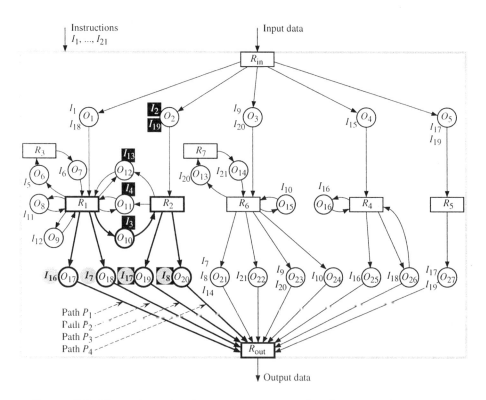

Figure 9.8 Illustration related to an error propagation from R_1 to R_{out}.

where G_{13} and G_{14} are sets of instructions such that the error in R_2 (after execution of I_3) is neither removed nor propagated. Expressions G_{13} and G_{14} have now to be obtained.

The operators in the set $\{O_{17}, O_{18}, O_{19}, O_{20}\}$ are activated by the instructions in the set $\{I_{16}, I_7, I_{17}, I_8\}$.

If no instruction in this set (labelled on grey circles in Figure 9.8) is applied, there is no propagation from R_1 to R_{out}. In each set E_{1p}, only one of them will be allowed in order to avoid propagation through a path other than P_1 (as explained in Section 9.2.2). For example, E_{13} contains I_{17}. It follows that neither I_7 nor I_8 nor I_{16} should be present in E_{13}, thus in G_{13}. Similarly, no instruction in $\{I_7, I_{16}, I_{17}\}$ should appear in G_{14}.

The operators putting a result into R_2 are O_2 and O_{10}, and the operators using the value of R_2 as an operand are O_{11}, O_{12}, O_{19} and O_{20}. If none of these operations is activated, an error in R_2 cannot be removed or propagated. The set of instructions associated with these operators is $C_2 = \{I_2, I_3, I_4, I_8, I_{13}, I_{17}, I_{19}\}$ (labelled on a black square in Figure 9.8). Thus, if no instruction in C_2 is executed, an error in R_2 can neither be removed nor propagated.

The set G_{13} will then contain all the instructions except those which are in $\{I_7, I_8, I_{16}\}$ (to avoid propagation through a path other than P_3) or in C_2, i.e,

$$E_{13} = I_3(I_1 + I_5 + I_6 + I_9 + I_{10} + I_{11} + I_{12} + I_{14} + I_{15} + I_{18} + I_{20} + I_{21})*I_{17}. \quad (9.13)$$

In this example $G_{14} = G_{13}$. Hence, from (9.8), (9.9), (9.10), (9.13), and (9.12) given $G_{14} = G_{13}$, (9.7) may be written as:

$$D_f \supset Q_f \cdot (I_{16} + I_7 + I_3(I_1 + I_5 + I_6 + I_9 + I_{10} + I_{11} + I_{12} + I_{14} + I_{15} + I_{18} + I_{20} + I_{21})*(I_{17} + I_8)) \quad (9.14)$$

Assuming $\Pr[Q_f] = 0.5$ (stuck-at of a bit in R_1 and equal likelihood of data in R_1) and $\Pr[I_j] = \frac{1}{21}$ for all j,

$$\Pr[D_f] \geq 0.5 \left(\frac{1}{21} + \frac{1}{21} + \frac{\frac{1}{21} \times \frac{2}{21}}{1 - \frac{12}{21}} \right) = 0.053$$

is obtained from Properties 7.4 and 7.5 (Section 7.3.1.2). Given $\Pr[D_f] \approx 0.053$, the test length $L(0.001) \approx 127$ is obtained.

Remark 9.1 A fault g in register R_5 is a particular case (see Figure 9.1 in Section 9.1.1): an instruction propagating an error from R_5 (i.e., I_{17} or I_{19}) is also an instruction which modifies R_5 before the propagation. There is a transfer from R_{in} to R_5 during the first cycle, then from R_5 to R_{out} during the second cycle. The set of states Q_g allowing the detection is related to register R_{in}. We obtain $D_g = Q_g \cdot (I_{17} + I_{19})$. This is equivalent to considering the set $\{O_5, R_5, O_{27}\}$ as a single operator and that the fault g is in this operator.

A fault in register R_6 is also a special case; $\{O_5, R_5, O_{27}\}$ may be condidered as a single operator (I_9 and I_{20} modify R_6 before the propagation).

❑

In section 9.1.2, it is assumed that a fault in a register is either a single or multiple stuck-at of bits of the register, or a single or multiple coupling of bits of the register (for a single coupling, several bits are ANDed or ORed). The two properties in the sequel show that the most resistant fault in a register is a single stuck-at. These properties are based on the fact that, under Hypothesis 9.2, all the errors in a register R_i caused by a fault in this register are propagated to R_{out} by the same set of instruction sequences. Then, only the probabilities of errors in the register R_i have to be compared[12].

Property 9.1 For any multiple stuck-at fault f in R_i, there is a single stuck-at fault in R_i which is *dominated* by f.

Proof Assume $f = [g_1 \& g_2 \& ...]$ where each g_i is a single stuck-at fault in R_i. Fault f *dominates* g_1 since, under Hypothesis 9.2, any test sequence detecting g_1 also detects f.

Property 9.2 For any single or multiple coupling f in R_i, there is a single stuck-at fault in R_i which is *R-dominated* by f.

Proof For any multiple coupling f, there is a single coupling which is dominated by f: the proof is similar to the proof of Property 9.1. The case of a single coupling now has to be considered.

Let f be the AND coupling of two any bits a_1 and a_2 in R_i. Let $p_1 = \Pr[a_1 = 1]$ and $p_2 = \Pr[a_2 = 1]$. A state in Q_f is such that $a_1 a_2 = 01$ or 10 in the fault-free machine. Then $\Pr[Q_f] = (1-p_1) p_2 + p_1 (1-p_2) = p_1 + p_2 - 2 p_1 p_2$.

Let $g_1 = a_1/0$ and $g_2 = a_1/1$. Then $\Pr[Q_{g_1}] = p_1$ and $\Pr[Q_{g_2}] = (1 - p_1)$. If $p_1 \leq 0.5$, $\Pr[Q_{g_1}] \leq \Pr[Q_f]$, and if $p_1 \geq 0.5$, $\Pr[Q_{g_2}] \leq \Pr[Q_f]$.

The proof is similar for the coupling of several bits, and for an OR coupling.

9.3.2 Faults in the operators

The analysis of faults in the operators and in the registers present similar features.

Let us introduce a notation which is helpful for both.

Notation 9.2 The set of instructions which read or write into register R_i is denoted by C_i. The set of all instructions which are not in C_i is denoted by K_i (set of **keeping instructions**).

☐

Let us consider a fault f in operator O_4 in Figure 9.1: $f = a/0$ where a is a bit of the transfer bus. This fault produces an error in register R_4 if two conditions are met: 1) instruction I_{15} is executed; 2) the value in R_{in} at this time is such that the bit a transfered to R_4 should be 1 (state among a set denoted by Q_f).

Then, after a transition sequence in $Q_f \cdot I_{15}$, there is an error in R_4. The set of instructions which write or read a value in R_4 is $C_4 = \{I_{15}, I_{16}, I_{18}\}$. If none of these instructions is executed (i.e., if only instructions in K_4 are executed), the

---

[12] The concepts of dominance and R-dominance are defined in Sections 3.2.4.1 and 4.4.3.2.

error in R_4 can neither be removed nor propagated. An error in R_4 can be propagated to R_{out} by execution of I_{16} or I_{18}. Then, according to Notation 9.2, the following is obtained:

$$D_f \supset Q_f \cdot I_{15} K_4^*(I_{16}+I_{18}),$$

i.e.,

$$D_f \supset Q_f \cdot I_{15}(I_1+I_2+I_3+I_4+I_5+I_6+I_7+I_8+I_9+I_{10}+I_{11}+I_{12} \\ +I_{13}+I_{14}+I_{17}+I_{19}+I_{20}+I_{21})*(I_{16}+I_{18}). \tag{9.15}$$

Assuming equal likelihood of instructions and data,

$$\Pr[D_f] \geq 0.5 \ \frac{\frac{1}{21} \times \frac{2}{21}}{1-\frac{18}{21}} = 0.0159,$$

then $L(0.001) \approx 432$, are obtained for this fault.

Let us now consider a fault g in operator O_{12}: $g = a/0$ where a is a bit of the output bus of O_{12}. This operator produces the bit-wise AND of the contents in R_1 and R_2. An error appears in R_1 after a transition sequence in $Q_g \cdot I_{13}$, where Q_g is such that both bits in R_1 and R_2 ANDed to produce a have the value 1 in the fault-free circuit. Thus,

$$D_g \supset Q_g \cdot I_{13} K_1^* E_1, \tag{9.16}$$

where E_1 is a set of instruction sequences propagating an error from R_1 to R_{out} (obtained in Section 3.2.1).

Remark 9.2 The detection of a fault in operator O_j makes it necessary to 1) produce an error in a register R_i by executing an instruction activating O_j (when the state of the system allows this error to be obtained), and 2) propagate the error from R_i to R_{out}. On the other hand, the detection of a fault in R_i makes it necessary only (when the state is such that there is an error in R_i) to propagate the error from R_i to R_{out}.

This is clear for the two faults f and g studied above. For example $Q_g \cdot I_{13} K_1^* E_1$ (for the previous fault g in operator O_{12}) may be compared to $Q_h \cdot E_1$ (for a fault h in register R_1). The probabilities $\Pr[Q_g]$ and $\Pr[Q_h]$ usually have the same order of magnitude for single faults and $\Pr[I_3 K_1^*] \ll 1$. It follows that fault g requires a longer test sequence than h.

Although this example is not a general proof, one may expect that the faults in operators are much more difficult to detect than the faults in the registers. This observation will be confirmed by numerical results (Sections 9.5.1.4 and 9.5.2.2).

Remark 9.3 Properties 9.1 and 9.2 (presented for faults in the registers) are also related to all *external faults* of operators (faults on the input and output buses).

☐

Let us now give some general results concerning the faults in usual operators.

Operator O_{16} in Figure 9.1 is an increment operator (the number N in R_4 is converted into the number $N+1$). Let us assume that register R_4 is made of b bits, $a_{b-1} \ldots a_1 a_0$ such that $N = \sum_{i=0}^{b-1} a_i \cdot 2^i$. The carry between a_j and a_{j+1} is 1, when instruction is executed, if and only if $a_j = a_{j-1} = \ldots = a_0 = 1$.

If the carry between a_j and a_{j+1} is stuck-at-0, the probability that an error is put into R_4 is $\Pr[a_j = \ldots = a_0 = 1]$. This probability is minimum for the carry between a_{b-2} and a_{b-1}.

If the carry between a_j and a_{j+1} is stuck-at-1, the probability that an error is put into R_4 is $1 - \Pr[a_j = \ldots = a_0 = 1]$, which is minimum for the carry between a_0 and a_1. This case occurs with the probability $1 - \Pr[a_0 = 1] = \Pr[a_0 = 0]$ which is the probability of an error due to the stuck-at-1 of the bus line feeding cell a_0 in O_{16} (external fault). These observations are summarized in Property 9.3d (i.e., case d in Figure 9.9).

Operator	Set of faults containing the most resistant for any probabilities of the data (b-bit words: $a_{b-1}\, a_{b-2} \ldots a_1\, a_0$; α in $\{0, 1\}$)
(a) Transfer (b) Bit-wise complement	Single stuck-at-α of a line of the output (or input) bus
(c) Left shift by 1 bit	Single stuck-at-α of an input line in $\{a_{b-2}, \ldots, a_1, a_0\}$ + AND or OR coupling of lines a_{b-1} and a_{b-2} in the input bus *Note: The stuck-at-α of the left-most input line and the stuck-at-0 of the right-most output line are not detectable*
(d) Increment (e) Decrement	Single stuck-at-α of a line of the input bus + Single stuck-at-α of a line of the output bus + Stuck-at-0 of the left-most carry *If equal likelihood of data: stuck-at-0 of the left-most carry*
(f) Adder	Single stuck-at-α of a line of one of the two input buses + Single stuck-at-α of one of the carries *If equal likelihood of data: stuck-at-0 of the right-most carry*
(g) Bit-wise AND	Single stuck-at-α of a line of one of the two input buses

Figure 9.9 Subsets of faults containing the most resistant fault for some usual operators (Property 9.3).

Property 9.3 The sets of faults containing the most resistant (among the faults presented in Section 9.1.2.2) are given in Figure 9.9 for some usual operators.

❑

This property is taken from [ThDaJo 81]. It is given without complete proof. The Property 9.3d has been previously explained. Let us simply add the basis of the proof of Property 9.3f (adder):
First case. Any probabilities: similar to the proofs of Properties 9.1 and 9.2 in Section 9.3.1.
Second case. The probability of each line of the input buses is 0.5 (and they are independent of each other):

> The stuck-at-0 of the right-most carry is detected if both right-most bits of the operands are 1 (Probability = 0.25).
> For the stuck-at-1 of the same carry: probability = 0.75.
> For the stuck-at-0 or 1 of other carries: probability > 0.25.

9.4 TEST LENGTH FOR FAULTS IN THE CONTROL SECTION

Just like a fault in the data processing section, the detection of a fault in the control section requires two steps: 1) *provoking* an error into one (or several) register(s) (this corresponds to the usual *controllability* concept); 2) *propagating* the error to the register R_{out} (*observability* concept).

The propagation (presented in Section 9.3) is similar for all the faults. The conditions for provoking an error are different. These conditions are more complicated to study for the faults in the control section (because the graph model is modified if such a fault is present), but the faults are *easier to detect* (the required test lengths are shorter because of the strong perturbation of the behavior produced by this fault). Numerical examples are given in Section 9.5.

According to Section 6.1.2.1, the faults relatively easy to detect have practically no influence on the no-coverage probability. Since the faults in the control section of a microprocessor have this feature, only a single example of a fault is presented in this section. For more information, the reader is referred to [ThDa 83].

Consider the fault $g = f(R_1 / R_2)$ illustrated in Figure 9.10. Both registers R_1 and R_2 are disturbed by this fault. Accordingly the fault may be detected by reading out (i.e., propagating up to R_{out}) either R_1 or R_2. Let V_i and $V_{i,g}$ denote respectively the values in R_i for the fault-free circuit and for the circuit affected by fault g (see Figure 9.10).

Execution of I_3 propagates V_1 to R_{out} in the fault-free circuit and $V_{2,g}$ to R_{out} in the faulty one. Execution of I_4 propagates V_2 to R_{out} in the fault-free circuit and

$V_{2,g}$ to R_{out} in the faulty one. Let $Q_{1\neq 2,g}$ denote the set of states such that $V_1 \neq V_{2,g}$, and $Q_{2\neq 2,g}$ the set of states such that $V_2 \neq V_{2,g}$. The detection set D_g associated with the fault g is

$$D_g = Q_{1\neq 2,g} \cdot I_3 + Q_{2\neq 2,g} \cdot I_4. \tag{9.17}$$

It follows that:

$$\Pr[D_g] = \Pr[V_1 \neq V_{2,g}] \cdot \Pr[I_3] + \Pr[V_2 \neq V_{2,g}] \cdot \Pr[I_4]. \tag{9.18}$$

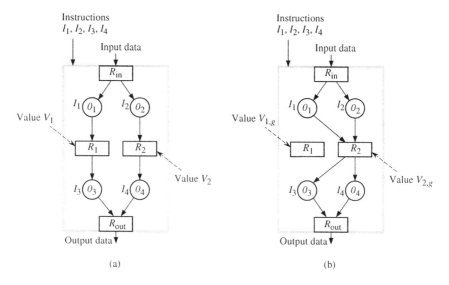

Figure 9.10 Example of a fault in the control section[13]. (a) Fault-free circuit. (b) Circuit affected by the fault $g = f(R_1/R_2)$.

If we assume the equal likelihood of data and instructions, and according to Hypothesis 9.1:

$\Pr[V_1 \neq V_{2,g}] = \Pr[\text{last execution of } I_2 \text{ after last execution of } I_1] = 0.5;$ (9.19)

$\Pr[V_2 \neq V_{2,g}] = \Pr[\text{last execution of } I_1 \text{ after last execution of } I_2] = 0.5;$ (9.20)

$\Pr[I_3] = \Pr[I_4] = 0.25.$ (9.21)

From (9.19) to (9.21), Equation (9.18) becomes

$\Pr[D_f] = 0.5 \times 0.25 + 0.5 \times 0.25 = 0.25.$

---

[13] This figure is reprinted from [ThDa 83] (© 1983 IEEE).

From this value, the required test length $L(0.001) \approx 24$ is obtained by Equation (7.18) in Section 7.3.1.2.

Let us illustrate the fact that the faults in the operators are more difficult to detect than the faults in the control section with the previous example. For a single stuck-at fault in operator O_1 of Figure 9.10a, say h, $D_h = Q_h \cdot I_1(I_2 + I_4)^* I_3$. Under the equal likelihood assumption, $L(0.001) \approx 107$ is obtained for this fault. This length is significantly greater than $L = 24$ obtained for the fault $f(R_1/R_2)$.

9.5 TEST LENGTH FOR A MICROPROCESSOR

The test length for a whole microprocessor is considered in this section. Various hypotheses are made, for the example microprocessor in Section 9.5.1, and for the Motorola 6800 in Section 9.5.2.

The set of faults under consideration is

$$F = F_r \cup F_o \cup F_d \cup F_i, \qquad (9.22)$$

where, according to Sections 9.1.2.1 to 9.1.2.4, F_r, F_o, F_d, and F_i, are respectively the sets of faults in the registers, in the operators, in the register decoding function, and in the instruction decoding and control function.

The notation $L_m(\varepsilon)$, the test length of a random sequence that ensures a level of confidence $1 - \varepsilon$ for the most resistant fault in the set F of faults under consideration was introduced in Section 6.1.2.2. Let us introduce the notation $L_m(F_k, \varepsilon)$: same definition for the subset F_k.

9.5.1 Example microprocessor

As shown in Sections 9.3 and 9.4, all the probabilities of fault detections are estimated from the probabilities of error propagations from registers R_i to R_{out}. For each register R_i, the propagation expression may be calculated as shown for E_1 in Section 9.3.1; thus, given the probabilities of the various instructions the values $\Pr[E_i]$ can be calculated. For example, the values of the various $\Pr[E_i]$, given $\Pr[I_j] = 1/21$ for each instruction, are given in Figure 9.11.

R_i	R_1	R_2	R_3	R_4	R_5	R_6	R_7	R_{in}	R_{out}
$\Pr[E_i]$	0.106	0.110	0.011	0.095	0.095	0.190	0.048	> 0.22	1

Figure 9.11 Probabilities of the propagation expressions of the registers in the example microprocessor, given equal likelihood of the instructions.

9.5.1.1 Faults in the registers

It is clear from the results in Figure 9.11 that the lower probability of error propagation is obtained for the register R_3: $\Pr[E_3] = 0.011$. Assuming an equal likelihood of data, for any single stuck-at fault f in R_3, we obtain

$$\Pr[D_f] \geq \Pr[Q_f] \cdot \Pr[E_3] = 0.5 \times 0.011 = 0.0055, \tag{9.23}$$

then $L(0.001) \approx 1300$ by (7.18). Since this fault is the most difficult to detect in F_r, the following is obtained (according to the notation at the beginning of Section 9.5):

$$L_m(F_r, 0.001) \approx 1300. \tag{9.24}$$

9.5.1.2 Faults in the operators

Let us assume that the data are equally likely. Then, the most resistant fault in F_o is the stuck-at-0 of the most left carry of increment operators O_{15} or O_{16} (see Property 9.3). Operator O_{16} is activated by instruction I_{16}, while O_{15} is activated by I_{10} *when the content of R_1* (denoted by V_1) *is zero*. This condition has a low probability: $\Pr[V_1 = 0] = 1/2^8 \approx 0.004$ if the microprocessor is assumed to contains 8-bit registers.

Assumption 1. Operators O_{15} and O_{16} are physically different. In this case both operators must be tested and the worst case is obtained for O_{15}. Let g be the stuck-at-0 of the most left carry in O_{15}. This fault produces an error in R_6 if 1) the content in R_6 is in $\{01111111, 11111111\}$, 2) I_{10} is executed, and 3) $V_1 = 0$. Let us observe that (see Figure 9.1 in Section 9.1.1), if I_{10} is executed, the value which is put into R_6 at the first cycle (I_{10}^1, operator O_{15}) is immediately propagated to R_{out} at the second cycle (I_{10}^2 operator O_{24}). It follows that, for the most resistant fault[14] g:

$$D_g = Q_g \cdot (I_{10} \& (V_1 = 0)) \tag{9.25}$$

Given $\Pr[Q_g] = 2^{-7}$ (corresponding to two states among 2^8 possible in R_6), $\Pr[I_{10}] = 1/21$, and $\Pr[V_1 = 0] = 2^{-8}$, the value $\Pr[D_g] \approx 1.5 \times 10^{-6}$ is obtained and leads to $L(0.001) \approx 4\,800\,000$.

Let us remark that if the probabilities of instructions are modified in such a way that I_{10} has a higher probability, the test length can be reduced. For $\Pr[I_{10}] = 0.9$ and $\Pr[I_j] = 0.005$ for $j \neq 10$, the length $L(0.001) \approx 250\,000$ is obtained for the fault under consideration (which remains one of the most difficult to detect).

---

[14] The detection of a fault in O_{16} would require two instructions. However, this detection remains easier because there is not the condition $V_1 = 0$.

Assumption 2. Operators O_{15} and O_{16} correspond to the same physical increment operator (realistic assumption). In this case it is sufficient to test operator O_{16}. The most resistant fault is then the stuck-at-0 of the left-most carry of this operator, say h. After an error has been put into R_4, any string of instruction in K_4 (Notation 9.2 in Section 9.3.2) may be executed before propagation by a string in E_4. Then,

$$D_h = Q_h \cdot I_{16} K_4^* E_4. \tag{9.26}$$

Given $\Pr[Q_h] = 2^{-7}$, $\Pr[I_{16}] = 1/21$, $\Pr[K_4] = 18/21$, and $\Pr[E_4] = 0.095$ (Figure 9.11), the value $\Pr[D_h] \approx 2.5 \times 10^{-4}$ is obtained and leads to $L(0.001) \approx 28\,000$. Thus

$$L_m(F_o, 0.001) \approx 28\,000. \tag{9.27}$$

9.5.1.3 Faults in the control section

From the method briefly presented in Section 9.4, the following results have been obtained [Th 83]:

$$L_m(F_d, 0.001) \approx 1\,500. \tag{9.28}$$

$$L_m(F_i, 0.001) \approx 2\,800. \tag{9.29}$$

9.5.1.4 Faults in the whole microprocessor

The test length for the most resistant fault in the microprocessor is obtained by:

$$L_m(\varepsilon) = \max(L_m(F_r, \varepsilon), L_m(F_o, \varepsilon), L_m(F_d, \varepsilon), L_m(F_i, \varepsilon)). \tag{9.30}$$

From (9.24), (9.27), (9.28), and (9.29), Equation (9.30) may be written as

$$L_m(0.001) \approx \max(1\,300, 28\,000, 1\,500, 2\,800) = 28\,000. \tag{9.31}$$

It is clear that the most resistant faults are in the operators (Remark 9.2 in Section 9.3.2).

According to Property 4.3 in Section 4.4.3.1, given a test length, P_m is a lower bound of the fault coverage, then $L_m(\varepsilon)$ is an upper bound of the test length required for $P = 1 - \varepsilon$. Since many faults are easy to detect (all the faults in F_r, F_d, and F_i) a length shorter than $L_m = 28\,000$ could be obtained from some hypothesis of the occurrence probabilities of faults. An example of calculation will be presented in Section 9.5.2.2.

9.5.2 Microprocessor Motorola *6800*

This microprocessor is a 40-pin package. According to Appendix H, its inputs consists of an 8-bit bidirectional data bus and five control lines: *RST*, *HLT*, *NMI*, *IRQ*, and *TSC*. The three types of outputs are as follows: 8-bit bidirectional data bus when the microprocessor is in write state; 16-bit address bus; three status outputs (*VMA*, *BA*, and *R/W*).

Basic results assuming equal likelihood of instructions and equal likelihood of data are presented first. Then, further results are given, particularly the test length of the most resistant fault obtained from a Markov chain, and two approaches to shorten the required test length.

9.5.2.1 Basic results

The main features of this microprocessor, designed by the Society Motorola and manufactured by several comparies [Mo 79] [Th 78], are presented in Appendix H.

From the knowledge available to a user [Mo 79], i.e., information presented in Figures H.1 to H.3 and the execution abstract of each instruction (two examples of which are presented in Figure H.4), the graph model of the microprocessor can be obtained [Th 83]. This graph contains 10 registers; after gathering all the operators having the same set of operand registers and the same set of result registers into a single "operator", there are 38 "operators" (for example, if this simplification were applied to the graph in Figure 9.1, O_{11} and O_{12} would be represented by a single "operator", and the set $\{O_{21}, O_{22}, O_{23}, O_{24}\}$ too; this simplification is consistent with Hypothesis 9.2 in Section 9.2). The following features may be pointed out:

1) There are two output registers, namely *TE.W* and *OB* (Figure H.2). For calculation of the test length, faulty values on the state outputs are not taken into account.

2) The content of register *RI* cannot be observed by propagation through the data flow (no path from *RI* to *TE.W* or *OB*). Its faulty behavior corresponds to faults in the control section.

3) The control line *DBE* is connected to CK_2, in normal operation and during a test experiment. Five control inputs remain which are considered as 5 hardware instructions; since there are 197 software instructions (Appendix H), when the instructions are equally likely:

$$\Pr[RST] = \Pr[HLT] = \Pr[NMI] = \Pr[IRQ] = \Pr[TSC] = \frac{1}{202} \qquad (9.32)$$

$$\Pr[I_j] = \frac{1}{202}, j = 1, ..., 197. \qquad (9.33)$$

4) For each bit c_s in the register CCR, $\Pr[c_s = 1]$ is based on

$$\Pr[c_s = 1] = \sum_{j:I_j \text{ affects } c_s} \Pr[I_j] \cdot \Pr[c_s = 1 \mid I_j]. \tag{9.34}$$

Let us assume that the data are equally likely (i.e., every 8-bit value has a probability $1/2^8$ of being in $TE.R$) and that the instructions are equally likely too. The corresponding probability of the bits in the condition code register CCR and of the propagation expressions of the various registers are given in Figure 9.12a and b.

(a)

c_s	c_7	c_6	$c_5 = H$	$c_4 = I$	$c_3 = N$	$c_2 = Z$	$c_1 = V$	$c_0 = C$
$\Pr[c_s=1]$	1	1	0.47	0.76	0.46	0.045	0.15	0.45

(b)

R_i	TE.R	ACCUA	ACCUB	IX	SP	PC	CCR	RI	OB	TE.W
$\Pr[E_i]$	0.55	0.040	0.041	0.23	0.082	0.97	0.020	0	1	1

Figure 9.12 Microprocessor *6800*, equal likelihood of data and instructions. (a) Probability of bits in *CCR*. (b) Probability of the propagation expressions related to the registers.

The following test lengths have been obtained [ThDaJo 81], [Th 83].

$$L_m(F_r, 0.001) \approx 7600, \tag{9.35}$$

$$L_m(F_o, 0.001) \approx 5\,500\,000, \tag{9.36}$$

$$L_m(F_d, 0.001) \approx 380, \tag{9.37}$$

$$L_m(F_i, 0.001) \approx 460\,000. \tag{9.38}$$

Thus, for the most resistant fault in the microprocessor *6800*,

$$L_m(0.001) \approx 5\,500\,000, \tag{9.39}$$

which corresponds to a probability $p_{\min} \approx 1.3 \times 10^{-6}$ (Equation (6.33) in Section 6.1.2.2).

It is clear, in this example too, that the most resistant faults affect the operators (Remark 9.2 in Section 9.3.2).

9.5.2.2 *Further results*

The approach used to obtain the results in Section 9.5.2.1, explained in Section 9.5.1, is relatively easy to use but is not accurate. The Markov-chain approach

(Section 9.2) is accurate but too complicated to be applied to many faults. However, it can be used for some specific faults.

The most resistant fault which has been found in the operators, denoted by f_{IZ}, is the following. There is a functional operator putting the value 0 in the bit I in CCR (correct behavior), and this operator *also* puts the value 0 in the bit Z in CCR (faulty behavior). This fault was analysed with a Markov chain and the test length $L(0.001) \approx 4\,950\,000$ was obtained [Fé 84]. This length is to be compared with the value $5\,500\,000$. This approximate value was obtained using several simplications; some of them tends to increase the length, and others tends to decrease it.

Remark 9.4 From confidential data (on the internal architecture and on the micro-operations executing the instructions), which are not known in a user's environment, a more complex graph model was obtained. From this model, the required worst case test length $L_m(0.001) \approx 6\,300\,000$ was obtained [ThDaJo 81][Th 83] (for the same fault f_{IZ}). It is not worth making an analysis of this more complex model.

□

There are several ways to reduce the test length, and they can be combined.

A first way consists of changing the probabilities of instructions and/or data (i.e., using weigthed test patterns: Sections 10.3.2.2, 11.2, 11.3.1, and L.1 in Appendix L). For example, consider the fault f_{IZ}. This fault can produce an error only if the instruction CLI is executed (only instruction clearing the value in I). If the probability of this instruction is increased (and the others decreased) the worst case test length may be shortened. The test length $L_m(0.001)$ is divided by about 1.8 if $\Pr[CLI]$ is multiplied by about 5 while all the other instructions have the same probability [ThDaJo 81].

Another way takes into account that many faults are easy to detect. Some information or assumptions allow a test length to be estimated from the weighted minimum testing probability $P_{wm}(\rho)$ presented in Section 4.4.2. Let

$$P_{wm}(\rho) = P_m[F_1] \cdot \Pr[F_1 \mid F] + P_m[F_2] \cdot \Pr[F_2 \mid F], \tag{9.40}$$

where $\rho = \{F_1, F_2\}$ and $F_1 = F_r \cup F_d \cup F_t$, i.e., F_1 is a set of faults easy to detect in the Motorola 6800 (according to (9.35), (9.37), and (9.38)), and $F_2 = F_s$. It follows that

$$P_m[F_1] \approx 1. \tag{9.41}$$

If we assume that the faults in different parts (registers, operators...) are proportional to the areas occupied on the chip by these parts, the following is obtained (from some information on the implementation):

$$\Pr[F_1 \mid F] \approx 0.8 \text{ and } \Pr[F_s \mid F] \approx 0.2. \tag{9.42}$$

From (9.41) and (9.42), (9.40) can be rewritten

$$P_{wm}[\rho] \approx 0.8 + P_m[F_s] \times 0.2. \tag{9.43}$$

Since the most resistant fault is in F_s: $P_m[F_s] = 1 - (1 - p_{min})^L$. It was found that $P_{min} \approx 1.3 \times 10^{-6}$. If $P_{wm}(\rho) = 0.999$ is required, the test length $L \approx 4\,100\,000$ is then obtained from (9.43).

Now, there are certainly many easy faults in F_s. If some other assumptions are made about their occurrence probabilities, the length can still be reduced.

Hence, this last approach consists in several steps. In [Ja 89], it was applied to a microprocessor whose random test length was initially studied in [Th 87], and the required test length was roughly divided by fifteen.

NOTES and REFERENCES

The functional fault models which are considered in this chapter are taken from [ThAb 80]. The graph model is different: in [ThAb 80] the data flow among registers is represented; we have added nodes corresponding to operators [ThDa 81]. All the information required to obtain this model is avalaible in the user's manual.

More details on the ways to calculate the test lengths for faults in the data processing section and the control section of a microprocessor can be found in [ThDa81] and [ThDa 83]. This work was used in [Be 87].

In [Kl 88], a simplified way to estimate a test length is applied to the microprocessor Intel *8086*: the average rate of accesses to the various registers is estimated from a simulation; the test length is then calculated to have an access to every register with a probability 0.999.

In [AbTh 88], the analysis of a board containing a microprocessor and a ROM (read only memory) is presented.

Given a register transfer level model of a microprocessor, a structural analysis of fault propagation probabilities could be considered. Consider for example a 2-input multiplexer (each input corresponding to a bus). The probability that a combination (or an error) is present on the output depends on the probabilities that the combination (or the error) is present on each input and on the probability of the control line. This approach is interesting and difficult [Ma 95], [MaPu 96].

10

Generation of Random Test Sequences

An ideal random test (Section 5.2) requires a random test sequence allowing testing of all the specified cases and only the specified cases corresponding to the circuit under test. In practice, the generation is based on some algorithm generating a pseudorandom test sequence. Such a sequence has two features: 1) it is not truly random since the generation is algorithmic (from some initial state of the generator, the pseudorandom sequence is always the same); 2) it presents some randomness properties such that it is a good approximation of a truly random one. We shall mainly use the word random to qualify the test sequences. The word pseudorandom will be used in some specific cases.

This chapter presents the needs in Section 10.1 (i.e., the conditions to be satisfied by the random test sequences). Software and hardware generations taking these needs into account are presented in Sections 10.2 and 10.3 respectively.

10.1 NEEDS

The needs depend on the problem to be tackled. If only combinational faults are concerned, a set of vectors is required (Section 10.1). For sequential faults (Section 10.2), some subsequences should be present in the test sequence, while some sequences are not allowed because the corresponding behavior is not specified.

10.1.1 Set of vectors for combinational faults

If it is assumed that only combinational faults can be present in the circuit under test, then only the set of vectors which are applied is significant. 1) The order of their application is not significant. 2) If a vector is repeated in a test sequence, the set of detected faults remains unchanged. Let $T(S_i)$ denote the set of vectors present in the test sequence S_i. Assume a 2-input combinational circuit C. The test sequences $S_1 = x_0 x_1 x_3$, $S_2 = x_3 x_0 x_1$ and $S_3 = x_0 x_1 x_0 x_3 x_0 x_3$ detect exactly the same combinational faults[1] in the circuit since $T(S_1) = T(S_2) = T(S_3) = \{x_0, x_1, x_3\}$.

10.1.1.1 Equal likelihood of all the input vectors

This is the most usual generation, which may be used if the behavior of the CUT is specified for all the input vectors and if there is no obvious reason to weight the drawing probability. For an n-input CUT, there are 2^n possible input vectors. At any time, each input vector has the same probability to be drawn:

$$\Pr[x(t) = x_i] = \frac{1}{2^n}, \text{ for any time } t \text{ and any vector } x_i. \tag{10.1}$$

This generation corresponds to the constant distribution

$$\psi_0 = \left(\frac{1}{2^n}, \frac{1}{2^n}, \ldots, \frac{1}{2^n}\right). \tag{10.2}$$

10.1.1.2 Constant but not equally likely distribution

There are two main cases where a constant but not equally likely distribution may be useful.

First case: the behavior of the CUT is not specified for some input vectors, then these input vectors must have a zero probability. Consider for example the circuit defined in Figure 5.2a (Section 5.2.1). The behavior is not specified for the input vectors x_{10} to x_{15}. In order to test *only the specified cases*, these vectors must have a zero probability. The constant distribution may be for example:

$$\psi = (.1, .1, .1, .1, .1, .1, .1, .1, .1, .1, 0, 0, 0, 0, 0, 0) \tag{10.3}$$

Second case: it would be useful to have a high probability for some input vectors and a low probability for others, in order to shorten the test length. Consider for example a circuit made up of a single 4-input AND gate and the set of stuck-at faults. According to Remark 3.3b in Section 3.1.4.1, after fault dominance collapsing, 5 faults have still to be considered: stuck-at-1 of every input which are detected by $x_7 = 0111$, $x_{11} = 1011$, $x_{13} = 1101$, and $x_{14} = 1110$, and stuck-at-0 of the output z, which is detected by the input vector $x_{15} = 1111$. Let us denote by f_i the fault detected by x_i.

---

[1] For combinational faults, random drawing without replacement may be used (Appendix I).

Generation of Random Test Sequences 251

If an equally likely constant distribution were used, then the probability of detecting every considered fault by a random test vector would be $p_7 = p_{11} = p_{13} = p_{14} = p_{15} = 1/16 = 0.0625$. One can observe that the "useful" test vectors contain many inputs at value 1, particularly x_{15}. It would then be possible to have an input pattern generation such that for each input x_j, $\Pr[x_j = 1] > 0.5$. For example, if $\Pr[x_j = 1] = 0.75$ for $i = 1, 2, 3, 4$, one obtains $p_7 = p_{11} = p_{13} = p_{14} \approx 0.105$ and $p_{15} \approx 0.316$. The reader may verify that, for a minimum testing probability $P_m = 0.999$, the required test length is $L = 108$ if the input vectors are equally likely and $L = 62$ otherwise.

An important point which must now be brought up is: is it more pertinent to affect probabilities to individual inputs x_j or to input vectors x_i? The answer is in Property 10.1.

Assume an n-input circuit and a random generator of inputs to it. Let us define two kinds of generators.

The **input-vector generator** produces each input vector x_i with a given probability $\Pr[x_i] = \psi(x_i)$, $i = 1, ..., 2^n$, with $\sum_i \psi(x_i) = 1$.

The **input-bit generator** produces each bit with a given probability, $\Pr[x_j = 1] = p_{(j)}$, $j = 1, ..., n$, with $\Pr[x_j = 0] = 1 - \Pr[x_j = 1]$.

Property 10.1

a) Given a set of probabilities of the n input bits, *there is an input-vector generator* which can generate it.

b) Given a set of input-vector probabilities (i.e., a constant distribution), *there is not always an input-bit generator* which can generate it.

Proof

a) Consider any input vector, for example $x_i = x_1 x'_2 ... x_n$.
Given the probabilities of every bit $\Pr[x_j = 1] = p_{(j)}$, the following can be obtained:

$$\psi(x_i) = p_{(1)} \cdot (1 - p_{(2)}) \cdot ... \cdot p_{(n)},$$

and then defines the corresponding input-vector generator.

b) Consider for example the distribution $\psi = (\psi(x_0), \psi(x_1), \psi(x_2), \psi(x_3))$. We have three independent equations:

$\Pr[x_1 = 0 \text{ AND } x_2 = 0] = \psi(x_0),$

$\Pr[x_1 = 0 \text{ AND } x_2 = 1] = \psi(x_1),$

$\Pr[x_1 = 1 \text{ AND } x_2 = 0] = \psi(x_2)$

($\Pr[x_1 = 1 \text{ AND } x_2 = 1] = \psi(x_3)$ is obtained from the three other equations), and only two unknowns: $\Pr[x_1 = 1]$ and $\Pr[x_2 = 1]$. There is generally no solution. ❑

Let us illustrate Property 10.1.b with an example. Assume the distribution $\psi = (\psi(x_0), \psi(x_1), \psi(x_2), \psi(x_3)) = (.2, .5, .2, .1)$ is wanted. Only two equations are required to obtains values for $\Pr[x_1 = 1]$ and $\Pr[x_2 = 1]$; for example, $\Pr[x_1 = 1] = \psi(x_2) + \psi(x_3) = 0.3$ and $\Pr[x_2 = 1] = \psi(x_1) + \psi(x_3) = 0.6$. Unfortunately, from these values, $\psi(x_0) = 0.28$, $\psi(x_1) = 0.42$, $\psi(x_2) = 0.12$, and $\psi(x_3) = 0.18$ are obtained (for example $\psi(x_1) = \Pr[x_1 = 0] \cdot \Pr[x_2 = 1] = 0.7 \times 0.6 = 0.42$).

The conclusion is that, when some primary inputs should have low or high probabilities, it is always possible to generate them from a constant distribution of the input vectors. On the other hand, if some constant distribution is required, it is not always possible to obtain it from probabilities of every input line. For example the distribution (10.3), where some input vectors must have a null probability, cannot be obtained from probabilities associated with every input line.

10.1.2 Set of subsequences for sequential faults

For sequential faults (Definition 3.10 in Section 3.2.4.1), it is not sufficient to apply a set of input vectors. The order of their application may be important: the input sequence $S_1 = x_0 x_1 x_3 x_2$ detects the fault in Figure 2.15 (Section 2.3.3) because x_2 is applied after x_3, while $S_2 = x_0 x_2 x_3 x_1$ does not. Other examples are presented and commented in Appendix K.

According to Definition 5.1 in Section 5.2, an ideal random test requires that, informally, 1) all the input sequences which could occur in normal operation have a non-zero probability, and 2) any input sequence which cannot occur in normal operation has a zero probability. The first requirement (Section 10.1.2.1) may be difficult to satisfy in practice since the generators are not purely random. Theoretically, the second requirement (Sections 10.1.2.2 and 3) can always be satisfied thanks to a Markov source (Appendix A); in practice, a special purpose solution is adapted to the problem to be solved (Section 10.3.4)

10.1.2.1 Synchronous test

Many faults require a subsequence of length 2 for being detected, even in combinational circuits. This is true for stuck-open faults (for the fault in Figure 2.15 in Section 2.3.3, the subsequence $x_3 x_2$ is required) or for delay faults (see Appendix J). In some cases, a subsequence of length greater than 2 may be required (see Appendix K).

For a synchronous circuit whose set of input states is X, if there is no restriction on the input language, every input $x(t)$ may be any input vector x in X. This is also true for a combinational circuit. Then, if the length is not bounded, any input sequence in X^* can be applied.

In practice, since the generators are not purely random, any sequence in X^* cannot be generated, and we can be satisfied if all the *subsequences of length v* have a non-zero probability of being present in the test sequence. A **subsequence** is a string of input vectors which is not necessarily a prefix of the considered input sequence (see Section 2.1.3.1). The input sequence $S = S_1 x_1 x_3 S_2$ contains the subsequence $x_1 x_3$. For example $S = x_2 x_1 x_1 x_3$ contains the subsequence $x_1 x_3$ (with $S_1 = x_2 x_1$ and $S_2 = \lambda$). Let $Sub(S)$ denote the set of subsequences in S. For example $Sub(x_2 x_1 x_1) = \lambda + x_2 + x_1 + x_2 x_1 + x_1 x_1 + x_2 x_1 x_1$. With this notation: if $Sub(S) \supset X$, then all the input vectors are in S; if $Sub(S) \supset X^v$, then all the subsequences of v vectors in X, i.e., of length v, are contained in S.

10.1.2.2 Asynchronous test (adjacent vectors)

A test sequence made up of adjacent vectors is required for an asynchronous test, i.e., from $x(t)$ to $x(t+1)$ only one input x_i changes (Sections 5.2.2 and 7.1). A sequence of adjacent vectors may also be useful, although not essential, to shorten a test sequence for delay faults (Remark J.1 in Appendix J).

Let X be the set of input states. In an asynchronous test, if $x(t) = x_i$, then $x(t+1)$ must be an input vector x_j adjacent to x_i; this means that exactly one input variable is different between x_i and x_j. Consider for example a 4-input circuit. If $x(t) = 0001$, then $x(t+1)$ will be randomly drawn in the set {1001, 0101, 0011, 0000}.

The concepts of Markov source (generalized distribution, Appendix A), of subsequences, and of language restriction (for example the subsequence $x_i x_j$ has a zero probability although x_i and x_j are adjacent), are easily adapted to this case. Consider for example a 2-input circuit; the sequence $S = x_0 x_1 x_3 x_2 x_0 x_2 x_3 x_1 x_0$ contains all the subsequences of length 2 since a subsequence such as $x_0 x_3$ cannot occur in an asynchronous input sequence ($x_0 = 00$ and $x_3 = 11$ are not adjacent).

A sequence of adjacent test vectors corresponds to a particular case of generalized distribution. The general case is presented in the sequel.

10.1.2.3 Generalized distribution

In some cases, a generalized distribution may be used to shorten the test sequence (see comment on [GrSt 93] in 'Notes and references' of Chapter 4).

Such a distribution may also be *required* in order to test only specified cases: a circuit may be such that, when it is in some state q_i, only a subset $X(q_i) \subset X$ of the input vectors can be applied to it. The microprocessor presented in Appendix H is an example: when an instruction has been executed, the microprocessor is waiting for an operation code on its data bus. Only 197 out of the 256 eight-bit vectors correspond to an operation code. A solution for this example will be presented in Section 10.3.4.

10.2 SOFTWARE GENERATION

This section presents the generation of a constant distribution, the generation of a generalized distribution and then provides some comments on the software generation.

A software generation can be performed thanks to an instruction "random", drawing a random number uniformly distributed in the range [0, 1), which can be found in any programming language. Let us call this instruction **rand**.

10.2.1 Constant distribution

Assume the constant distribution $\psi_0 = (.25, .25, .25, .25)$ corresponding to a 2-input circuit. A random test sequence of length L with this distribution may be obtained by Algorithm 10.1.

Algorithm 10.1
   for $t = 1$ to L
       $Y = $ **rand**
       if $Y < 0.25$ **then** $x(t) = x_0$
       else if $Y < 0.50$ **then** $x(t) = x_1$
           else if $Y < 0.75$ **then** $x(t) = x_2$
               else $x(t) = x_3$
               end
           end
       end
   end

It is clear that any constant distribution with different weights can be obtained. For example $\psi = (.5, .3, 0, .2)$ can be obtained with $x(t) = x_0$ for $Y < 0.5$, $x(t) = x_1$ for $0.5 \le Y < 0.8$, and $x(t) = x_3$ otherwise.

Remark 10.1
a) Simplest variants may be obtained for some specific cases. For example if h different input vectors $x_0, ..., x_{h-1}$ must be obtained with the same probability, then $x(t) = x_i$ where $i = \lfloor Y \cdot h \rfloor$.

b) Generally, if a repartition function $F(x)$ is required for a random variable X, i.e., $\Pr[X < x] = F(x)$ (for our purpose, X may be drawn in an ordered set of possible input vectors), it can be obtained from a variable Y uniformally distributed on the range [0, 1] by the inverse function $X = F^-(Y)$.

10.2.2 Generalized distribution

A generalized distribution is obtained from a Markov source modeled by a Moore machine as explained in Appendix A. In the general case a part of the program is associated with every state of the Moore model: in this part a random drawing determines both the new input vector and the next state.

Let us illustrate this process with the 2-internal states example in Figure A.1d. There are 4 possible input vectors. In state r_0, all the input vectors are drawn with the same probability 0.25, and r_1 is reached if x_0 is drawn; in state r_1, x_1 cannot be drawn, the other input vectors are drawn with the same probability 0.33, and r_0 is reached again. Given the variable r has the value k when the internal state is r_k, a random sequence of length L with the generalized distribution of Figure A.1d may be obtained as shown in Algorithm 10.2 (taking into account Remark 10.1a).

Algorithm 10.2
```
r = 0
for t = 1 to L
    Y = rand
    if r = 0 then    %² part corresponding to state r₀
        i = ⌊4 · Y⌋
        x(t) = xᵢ
        if i = 0 then r = 1    % otherwise r remains 0
        end
    else % part corresponding to state r₁
        i = ⌊3 · Y⌋
        if i = 0 then x(t) = xᵢ
        else j = i + 1    % because x₁ has a zero probability
            x(t) = xⱼ
        end
        r = 0
    end
end
```

In some specific cases the software generation may be simplified, for example if there are several parts in the random sequence where each part corresponds to a constant distribution. This is the case of Figure A.1b where the constant distribution ψ_3 is considered for the first three drawings and the distribution ψ_4 for the next ones. For any $L \geq 4$ the generation can be performed in the following way.

Algorithm 10.3
```
for t = 1 to 3
    random drawing of x(t) with constant distribution ψ₃
end
for t = 4 to L
    random drawing of x(t) with constant distribution ψ₄
end
```

---

[2] Let us remind that % precedes a comment which is not executed in the program.

10.2.3 Comments on software generation

A software random generation is easy to perform since it is very flexible. An asynchronous random test (successive adjacent vectors) is easy to obtain. The randomness of the obtained test sequence is usually good. However, there are some limitations described in the following two remarks.

Remark 10.2 If the numbers are coded by 32 bits, a random number Y in the range [0, 1) actually corresponds to a string of 31 bits y_j such that[3]

$$y_{30}y_{29}...y_1y_0 = Y \cdot 2^{31} = \sum_{i=0}^{i=30} 2^i \cdot y_i;$$

the 32nd bit is a sign, and the range is, more specifically, $\left[0, \dfrac{2^{31}-1}{2^{31}}\right]$.

Consider the example in Algorithm 10.1: $Y < 0.25$ if $y_{30}y_{29} = 00$, $0.25 \le Y < 0.5$ if $y_{30}y_{29} = 01$, and so on. It follows that the two random values applied to the input lines x_1 and x_2 of the circuit under test correspond respectively to y_{30} and y_{29}. More generally, if all the input vectors applied to an n-input circuit are equally likely, the random values y_{30}, ..., y_{30-n+1} are applied to the input lines x_1, ..., x_n. Thus, this kind of generation is limited to $n \le 31$ in our example (it can be generalized to numbers coded by 64, 128 bits ...).

Remark 10.3 The numbers Y are not purely random. The method which forms the basis of the **rand** instruction of a large proportion of present-day programming languages uses a recurrent relation producing a sequence called **linear congruential** [Kn 69], [YaDe 88]. According to the last reference, a sequence of bits $y_j(1)y_j(2)...$ is periodic and the period may be small for small values of j. For the **rand** instruction of the C programming language, the following periods have been observed [Pr 97]: 2 for y_0 (repetition of 01), 4 for y_1 (repetition of 1100), 8 for y_2 (repetition of 11110000), 16 for y_3 (repetition of 0001000111101110).

Thus, if sequential faults are considered (Section 10.1.2), it would be recommended not to use the generation explained in Remark 10.2 for $28 < n \le 31$ (for example, $y_1y_0 = 00$ is always followed by $y_1y_0 = 01$). For $n > 28$, several random numbers Y can be used to obtain a pseudorandom n-bit word.

❏

A random test sequence may be software generated off-line, then stored and used like a deterministic test sequence, after having or not verified the fault coverage of the test sequence obtained: the software generation presented in this Section 10.2 may be used for steps 4 in Figures 3.25 in Section 3.2.4.3 (if *only*

---

[3] This notation "little endian" corresponds to decreasing weights from left to right. The inverse "big endian" notation is also encountered: the weights decrease from right to left.

the detection of *combinational faults* is concerned, the *random generation* of a test sequence may be biased by cancelling some redundant vectors, as shown in Figure 3.25a).

Remark 10.4 The *hardware generation* presented in the next section *can obviously be programmed*. This provides another kind of software generation. It may be useful for circuits with a large number of inputs.

10.3 HARDWARE GENERATION

First, the properties of the linear feedback shift registers generating pseudorandom sequences are presented in Section 10.3.1. Based on these properties, Sections 10.3.2 and 10.3.3 are then devoted to generation of constant and generalized distributions respectively.

10.3.1 Basic properties of LFSRs and M-sequences

Pseudorandom bit sequence generators are usually based on linear feedback shift registers (LFSRs) generating **maximum length sequences**, also called **M-sequences**. A **shift register** is made up of a string of flip-flops as illustrated in Figure 10.1a. The feedback is **linear** if every input of flip-flop is the sum modulo 2 of some outputs of flip-flops. This function is performed by an EXOR gate whose output is 1 when the number of inputs with the value 1 is odd: if there are 2 inputs, a and b, the output is $z = a \oplus b = ab' + a'b$. Let the time t be defined by the clock CK. The LFSR in Figure 10.1a is such that

$Q_3(t+1) = Q_2(t)$, i.e., shifting by one bit;

$Q_2(t+1) = Q_1(t)$, i.e., shifting by one bit;

$Q_1(t+1) = Q_2(t) \oplus Q_3(t)$, i.e., linear function.

In order to simplify the figures, the circuit in Figure 10.1a will be represented as shown in Figure 10.1b. Let us consider the behavior of the 3-stage LFSR when the initial state is $Q_1(1)Q_2(1)Q_3(1) = 111$ as illustrated in Figure 10.1c. At time 2, $Q_1(2) = Q_2(1) \oplus Q_3(1) = 0$, $Q_2(2) = Q_1(1) = 1$, and $Q_3(2) = Q_2(1) = 1$; then the LFSR contains the pattern 011. The next input to the LFSR will be $Q_1(3) = Q_2(2) \oplus Q_3(2)$ and so on. This process defines a periodic input sequence whose length is 7. Let $b(1)b(2)... b(7)$ denote this sequence represented horizontally in Figures 10.1c and d. As illustrated in Figure 10.1c, this sequence corresponds to the output string which can be observed at Q_3 (or at Q_2 or Q_1 with some shifting).

This 7-bit sequence is called an *M*-sequence as we shall specify. The *M*-sequences have very interesting properties for generating almost random sequences.

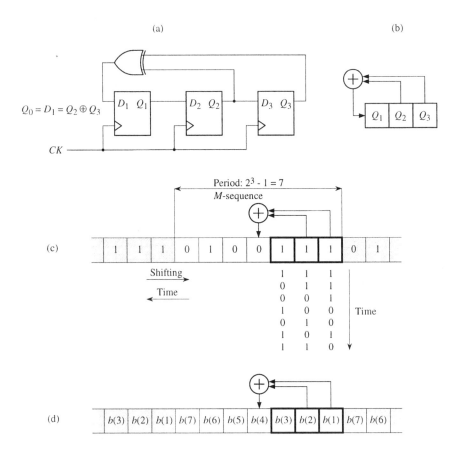

Figure 10.1 (a) Three-stage LFSR. (b) Simplified representation. (c) *M*-sequence. (d) Bit notation in the *M*-sequence.

Although LFSRs are very easy to implement, they are based on a rather complex mathematical theory. References are given at the end of this chapter, where the reader can find all the properties of the *M*-sequences and their proofs. In this section we shall present informally the basic ideas and recall the main properties.

A binary vector $a_m a_{m-1} \ldots a_0$ can be expressed as a **polynomial** $a_m x^m \oplus a_{m-1} x^{m-1} \oplus \ldots \oplus a_0$, where the product is the usual product and the sum is modulo 2 (represented by $\oplus$). For example, a vector 111001 can be expressed by $x^5 \oplus x^4 \oplus x^3 \oplus 1$. Similarly an LFSR can also be expressed as a polynomial qualified characteristic. The *characteristic polynomial* of the LFSR in Figure 10.1 is $1 \oplus x^2 \oplus x^3$, where a term x^i is present if the corresponding Q_i is used in the feedback. The **degree** of polynomial is the superscript of the highest non-zero term. We can add or multiply polynomials. For example consider the polynomials $1 \oplus x$ and $1 \oplus x \oplus x^3$:

$$(1 \oplus x) \oplus (1 \oplus x \oplus x^3) = 1 \oplus x \oplus 1 \oplus x \oplus x^3 = x^3; \qquad (10.4)$$

$$(1 \oplus x) \cdot (1 \oplus x \oplus x^3) = 1 \oplus x \oplus x^3 \oplus x \oplus x^2 \oplus x^4 = 1 \oplus x^2 \oplus x^3 \oplus x^4. \qquad (10.5)$$

Since multiplication exists, division exists too. According to (10.5) the division of $1 \oplus x^2 \oplus x^3 \oplus x^4$ by $1 \oplus x$ gives $1 \oplus x \oplus x^3$. Division of a polynomial, $\alpha(x)$, by another, $\beta(x)$, produces a quotient polynomial, $\theta(x)$, and, if the division is not exact, a remainder polynomial $\rho(x)$ whose degree is less than the degree of $\beta(x)$:

$$\alpha(x) = \beta(x) \cdot \theta(x) \oplus \rho(x), \qquad (10.6)$$

and this couple $(\theta(x), \rho(x))$ is unique.

For example, let $\alpha(x) = x^3 \oplus x^2 \oplus 1$ and $\beta(x) = x^2 \oplus 1$. The results $\theta(x) = x \oplus 1$ and $\rho(x) = x$ are obtained.

A polynomial which cannot be factored, i.e., which cannot be exactly divided by another one (except the polynomial 1) is said to be **irreducible**. In order to produce an *M*-sequence, the polynomial of the LFSR must be irreducible. However, this condition is not sufficient: in addition it must be *primitive*. A polynomial $\varphi(x)$ of degree m is **primitive** if the lower value of d such that $x^d \oplus 1$ is divisible by $\varphi(x)$ is $d = 2^m - 1$.

An *M-sequence* is generated by an LFSR whose degree of the characteristic polynomial is m if and only if this *polynomial is primitive*; the length of the *M*-sequence is then $M = 2^m - 1$. In other words, the string of bits in a stage, for example Q_3, is periodic with a period M. This *M*-sequence is obtained from any initial state of the LFSR (with some shifting), except the all-zero state. For example, the characteristic polynomial of the LFSR in Figure 10.1 is $1 \oplus x^2 \oplus x^3$, which is primitive: the *M*-sequence generated has a length $M = 2^3 - 1 = 7$. For the initial state $Q_1Q_2Q_3 = 111$, or 101, or 000 the output strings are, respectively (a period is presented in bold characters):

$$Q_3(1)Q_3(2)\ldots = \mathbf{1110010}1110010\ldots \qquad (10.7)$$

$$Q_3(1)Q_3(2)\ldots = \mathbf{1011100}1011100\ldots \qquad (10.8)$$

$$Q_3(1)Q_3(2)\ldots = \mathbf{0000000}0000000\ldots \qquad (10.9)$$

In the first two cases (10.7) and (10.8) the *M*-sequence is the same (except a shifting, or cyclic permutation). From the all-zero state, all the inputs generated are 0.

The following properties can be observed in Figure 10.1. In $Q_1Q_2Q_3$, every 3-tuple is seen exactly once in a period except 000 (this corresponds to sliding a

260   Chapter 10

window of width 3 along the M-sequence). In Q_1Q_2, every 2-tuple is seen exactly twice in a period except 00 which is seen once (this corresponds to sliding a window of width 2). In Q_1, the value 1 is seen 4 times and the value 0 three times in a period (window of width 1). This property, **Window Property**, is generalized as follows.

Property 10.2 Assume a window of width w is slid along an M-sequence of length $2^m - 1$.

a) If $w = m$, every m-tuple is seen once, except the all-zero m-tuple which is not seen.

More generally:

b) If $w = m - i$, $0 \le i \le m - 1$, every w-tuple is seen 2^i times, except the all-zero w-tuple which is seen $2^i - 1$ times.

❏

Property 10.2a is a particular case of Property 10.2b. We have presented it separately because it is more often used.

Remark 10.5   An m-stage LFSR is a synchronous Moore automaton (without input) whose state is coded with m binary variables. Thus, its number of states cannot be greater than 2^m. This observation is consistent with the following ones.

a) If the polynomial is primitive, all the states are reached except the all-zero state.

b) If a window of width $w > m$ were used, only $2^m - 1$ out of the 2^w w-tuples would be seen once.

c) If the polynomial is not primitive, only a relatively small subset of states is reached (the subset depends on the initial state).

❏

Given m, several M-sequences of length $2^m - 1$ exist for $m > 2$. If $\varphi(x)$ is a primitive polynomial, then $\varphi^-(x)$ is the **reciprocal polynomial**. The reciprocal polynomial is obtained by replacing each term x^j in $\varphi(x)$ by x^{m-j}. For example $x^3 \oplus x \oplus 1$ is the reciprocal polynomial of $1 \oplus x^2 \oplus x^3$;

$1 \oplus x^2 \oplus x^3$ produces the M-sequence 1110010;   (10.10)

$1 \oplus x \oplus x^3$ produces the M-sequence 0100111.   (10.11)

The M-sequence (10.11) is reciprocal to (10.10) but is not a shift of it. In addition, other primitive polynomials can exist for some m, and then other M-sequences. For $m = 3, 4, 5, 6, 7, 8, 9$ there are respectively 2, 2, 6, 6, 18, 16, 48 M-sequences.

Here are two other interesting properties of M-sequences.

Generation of Random Test Sequences 261

Let B be any M-sequence of length $M = 2^m-1$, and B_a a a-bit shift of B. If a is neither 0 nor a multiple of M, then the bitwise sum $B \oplus B_a$ is another shift B_b of B. This is called the **Shift-and-Add Property**[4].

Let $B = b(1)b(2).....b(M)$ be an M-sequence. The sequence $B_{(a)} = b(a)b(2a)b(3a).....$ is called a decimation of B. The period of $B_{(a)}$ is the greatest common factor of M and a. If a is relatively prime to the period M, the decimation is said to be *proper* or *normal*. Any *proper decimation* of an M-sequence is itself an M-sequence; this is the **Decimation Property**.

Remark 10.6 From an LFSR producing an M-sequence of length $M = 2^m-1$, a simple modification (non-linear) allows a periodic sequence of length 2^m to be obtained, where every m-tuple appears exactly once (i.e., even the all-zero m-tuple). The modification is as follows: a NOR gate is added whose inputs are $Q_1, ..., Q_{m-1}$, and whose output is an additional input of the feedback EXOR gate. After the pattern 0...01, the pattern 10...0 should be obtained in the register (without modification). When the pattern is 0...01, the output of the NOR gate is 1; two inputs of the EXOR gate are 1 (output of the NOR gate and Q_m), then the next pattern is 0...0. Now, the EXOR gate has a single 1 input (output of the NOR gate) and the following pattern will be 10...0.

This modified FSR has the advantage that all the m-tuples appear: this may be used for testing combinational faults in an m-input circuit. On the other hand some of the "magic" linear properties of LFSRs are not preserved, particularly the Add-and-Shift and the Decimation properties.

10.3.2 Constant Distribution

Equally likely distribution is firstly detailed. Then generation of weighted test vectors is presented.

10.3.2.1 Equally likely distribution

There are two main cases to be considered: a set of vectors is required for combinational faults (Section 10.1.1) or a set of subsequences is required for sequential faults (Section 10.1.2).

Set of vectors for combinational faults

According to Figure 5.1 in Section 5.1, the test procedure can be represented as shown in Figure 10.2. Given the (pseudo) random pattern source has m stages and the CUT has n inputs, $u = m - n$ bits of the generator are unused. The first

---
[4] The sum of two M-sequences is a M-sequence: M-sequences are a linear space.

Chapter 10

vector is a string of n successive bits in the register, whose position is arbitrary (since $m > n$). In order to simplify the notation, we chose to use the first n bits of the generator, i.e., $x_i = Q_i$, as illustrated in Figure 10.2. Thus,

$$x = Q_1 \ldots Q_n. \tag{10.12}$$

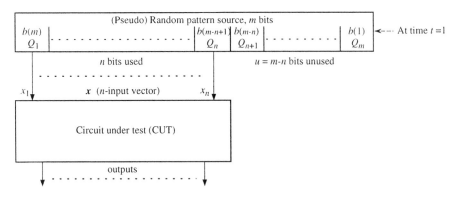

Figure 10.2 Test procedure

Usually the test sequence is obtained by a simple shifting at each time unit. Let us call this the **ordinary test sequence**. From (10.12), according to Figure 10.2, the ordinary test sequence is:

$x(1) = b(m)b(m-1) \ldots b(m-n+1)$,

$x(2) = b(m+1)b(m) \ldots b(m-n+2)$,

$\ldots\ldots\ldots\ldots\ldots\ldots\ldots\ldots\ldots\ldots\ldots\ldots\ldots$

$x(t) = b(m+t-1)b(m+t-2) \ldots b(m-n+t)$.

Because of the periodicity, $b(i)$ must be understood as $b(i \bmod 2^m-1)$. This is illustrated in Figure 10.3 for $m = 5$ (polynomial $1 \oplus x^2 \oplus x^5$) and $n = 2$. The *ordinary test sequence* is presented in Figure 10.3b.

According to Property 10.2, the ordinary test sequence contains all the vectors x_i in a period, and each vector appears 2^u times (where $u = m - n$, is the number of bits unused as illustrated in Figure 10.2), except the all-zero vector which appeared $2^u - 1$ times. If only combinational faults are considered, there is a redundancy of test vectors. In the case $m = n$ (i.e., $u = 0$), every vector appears once, except the all-zero vector which can be obtained according to Remark 10.6. This is called a **pseudorandom test**. It corresponds to an exhaustive testing for combinational faults (Section 2.3.3) if all the vectors in a period are applied. The analysis of the required test length as a function of u is presented in Appendix I.

Set of subsequences for sequential faults

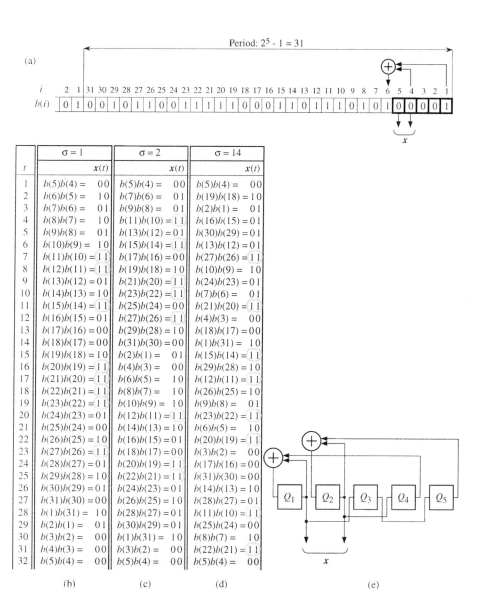

Figure 10.3 Illustration of the generation of a pseudorandom test sequence where $m = 5$ and $n = 2$. (a) The M-sequence. (b) Ordinary test sequence. (c) Test sequence with a shifting $\sigma = 2$ bits. (d) Test sequence with a shifting $\sigma = 14$ bits. (e) Circuit producing the sequence for $\sigma = 2$.

Because of the Window Property (Section 10.3.1), for combinational faults, the ordinary test sequence (and even the pseudorandom test), is sufficient to obtain an ideal random test (Definition 5.1 in Section 5.2). However, according to Section 10.1.2 and Appendix K, this kind of generation does not allow an ideal random test to be obtained for sequential faults: for some faults a subsequence $x_j x_k$ is required.

Consider the string of vectors generated by the ordinary test sequence in Figure 10.3b. One can observe that if $x(t) = 11$, $x(t+1) = 01$ or 11. In other words, the subsequences $x_3 x_1$ and $x_3 x_3$ can be obtained, but the subsequences $x_3 x_0$ and $x_3 x_2$ cannot. The reason is obvious: after one shifting, the vector $b(i+2)b(i+1)$ takes the place of $b(i+1)b(i)$, and the bit $b(i+1)$ appears in both vectors.

Let us now assume that there are two shiftings before using a new test vector. The vector $b(i+3)b(i+2)$ then takes the place of $b(i+1)b(i)$: there is no bit $b(j)$ appearing in both vectors. Let us denote by σ the number of shiftings before using a new test vector. The case $\sigma = 2$ is shown in Figure 10.3c. One can observe that if $x(t) = 11$, $x(t+1)$ may be either 00, or 01, or 10, or 11, i.e., the subsequences $x_3 x_0$, $x_3 x_1$, $x_3 x_2$, and $x_3 x_3$ (and all the other subsequences of length 2) can be obtained.

If a subsequence is present in the period, for example $x_3 x_2$, this subsequence has a non-zero probability of appearing in a random test sequence whose length is shorter than the period (see Definition 5.1 in Section 5.2).

Before discussing the values of σ, let us show how a modification of the LFSR allows us to obtain the result corresponding to σ shiftings in a single clock period. Let

$$Q(t) = [Q_1 \ Q_2 \ ... \ Q_m]^T \qquad (10.13)$$

denote the column vector corresponding to the state of the LFSR at time t. If $\sigma = 1$, the state at time $t+1$ is obtained by the matrix multiplication

$$Q(t+1) = U \cdot Q(t), \qquad (10.14)$$

where the matrix U (called the **companion matrix** of the polynomial) is made up of Boolean 0's and 1's corresponding to the connections among cells of the LFSR, and the product and sum are respectively the Boolean product and the sum modulo 2. It follows that σ shiftings in a single clock period are obtained by

$$Q(t+1) = U^\sigma \cdot Q(t). \qquad (10.15)$$

For example the LFSR in Figure 10.3a corresponds to

$$Q(t+1) = \begin{bmatrix} 0 & 1 & 0 & 0 & 1 \\ 1 & 0 & 0 & 0 & 0 \\ 0 & 1 & 0 & 0 & 0 \\ 0 & 0 & 1 & 0 & 0 \\ 0 & 0 & 0 & 1 & 0 \end{bmatrix} \cdot Q(t). \tag{10.16}$$

For $\sigma = 2$,

$$Q(t+1) = \begin{bmatrix} 0 & 1 & 0 & 0 & 1 \\ 1 & 0 & 0 & 0 & 0 \\ 0 & 1 & 0 & 0 & 0 \\ 0 & 0 & 1 & 0 & 0 \\ 0 & 0 & 0 & 1 & 0 \end{bmatrix}^2 \cdot Q(t) = \begin{bmatrix} 1 & 0 & 0 & 1 & 0 \\ 0 & 1 & 0 & 0 & 1 \\ 1 & 0 & 0 & 0 & 0 \\ 0 & 1 & 0 & 0 & 0 \\ 0 & 0 & 1 & 0 & 0 \end{bmatrix} \cdot Q(t), \tag{10.17}$$

which corresponds to the generator in Figure 10.3e.

The question now arises: what would be a "good" number σ of shiftings?

The first answer is that the *number σ of shiftings and $2^m - 1$ must not share a common factor*, in order that the period of vectors generated remains $2^m - 1$. For the example in Figure 10.3b, $\sigma = 2$ does not share a common factor with $2^5 - 1 = 31$ and the period of vector generated is 31 (the vector $b(5)b(4)$ on the grey line 32 is the beginning of the second period). The reader may verify that, for $m = 4$ and $\sigma = 3$, the period of vector generated is only 5 because $2^4 - 1 = 15$ and $\sigma = 3$ share the common factor 3.

The second answer is that $n \leq \sigma \leq (2^m - 1 - n)$ is required[5] in order to prevent a bit $b(j)$ from appearing in two successive vectors.

However, not all the values σ fulfilling these constraints allow a "good" generation. For example $\sigma = 14$ (illustrated in Figure 10.3d) is such that 14 does not share a common factor with 31, and $2 \leq 14 \leq 29$; however, in the sequence obtained, $x(t) = 11$ is always followed by $x(t+1) = 00$ or 10; thus the subsequences $x_3 x_1$ and $x_3 x_3$ never appear.

Property 10.3 If n and $2^m - 1$ do not share a common factor and $\dfrac{m}{n} > a$ (where a is an integer), the test sequence obtained from a shifting $\sigma = n$ contains all the subsequences of length a in a period.

❑

Before the proof, let us illustrate this property with the example in Figure 10.3, where $n = 2$ and $m = 5$. The values 2 and 31 do not share a common factor and $\dfrac{m}{n} \geq 2$. According to Property 10.3, all the subsequences of length 2 are generated if $\sigma = n = 2$. This is illustrated in Figure 10.3c.

---

[5] Modulo $2^m - 1$ of course.

Let us say that a w-tuple is *generated* if there is some time τ when $Q_1(\tau)...Q_w(\tau)$ corresponds to this w-tuple (generated by Equation (10.15)).

Proof of Property 10.3 Since $m > a \cdot n$, according to Property 10.2, all the $(a \cdot n)$-tuples are seen at least once in the M-sequence.

Since $\sigma (= n)$ does not share a common factor with $2^m - 1$, the period of the $(a \cdot n)$-tuples generated is $2^m - 1$. Thus, all the $(a \cdot n)$-tuples in the M-sequence are generated.

Let $x_{(i)}$ denote an n-tuple. An $(a \cdot n)$-tuple can be split into a successive n-tuples $x_{(1)}x_{(2)}...x_{(a)}$. If this $(a \cdot n)$-tuple is present in the register at time $t \geq a$, then the successive test vectors were $x(t) = x_{(1)}$, $x(t-1) = x_{(2)}$, ... $x(t-a+1) = x_{(a)}$. Thus the subsequence $x_{(a)}...x_{(2)}x_{(1)}$ is in the test sequence.

Property 10.3' If σ and $2^m - 1$ do not share a common factor, $\sigma \geq n$, and $\dfrac{m}{\sigma} > a$, the test sequence contains all the subsequences of length a in a period.

☐

This is a generalization of Property 10.3. The proof is similar.

The shifting $\sigma = n$ presented in Property 10.3 allows all the subsequences of $a = \left\lfloor \dfrac{m}{n} \right\rfloor$ vectors of n bits to be generated. This is a good generation since it is not possible with an automaton coded with m binary variables to produce all the subsequences of $(a + 1)$ vectors of n bits (an automaton able to produce that required at least $2^{(a+1)n}$ states). There are certainly other shiftings than $\sigma = n$ which produce the same result, but they are not so easy to obtain! We then suggest using the shifting $\sigma = n$.

$x^2 \oplus x^1 \oplus 1$	$x^{31} \oplus x^3 \oplus 1$	$x^{103} \oplus x^9 \oplus 1$	$x^{223} \oplus x^{33} \oplus 1$
$x^3 \oplus x^1 \oplus 1$	$x^{41} \oplus x^3 \oplus 1$	$x^{113} \oplus x^9 \oplus 1$	$x^{233} \oplus x^{74} \oplus 1$
$x^5 \oplus x^2 \oplus 1$	$x^{47} \oplus x^5 \oplus 1$	$x^{119} \oplus x^8 \oplus 1$	$x^{239} \oplus x^{36} \oplus 1$
$x^7 \oplus x^1 \oplus 1$	$x^{49} \oplus x^9 \oplus 1$	$x^{127} \oplus x^1 \oplus 1$	$x^{241} \oplus x^{70} \oplus 1$
$x^{11} \oplus x^2 \oplus 1$	$x^{71} \oplus x^6 \oplus 1$	$x^{137} \oplus x^{21} \oplus 1$	$x^{247} \oplus x^{82} \oplus 1$
$x^{17} \oplus x^3 \oplus 1$	$x^{73} \oplus x^{25} \oplus 1$	$x^{151} \oplus x^3 \oplus 1$	$x^{257} \oplus x^{12} \oplus 1$
$x^{23} \oplus x^5 \oplus 1$	$x^{79} \oplus x^9 \oplus 1$	$x^{167} \oplus x^6 \oplus 1$	$x^{263} \oplus x^{93} \oplus 1$
$x^{25} \oplus x^3 \oplus 1$	$x^{89} \oplus x^{38} \oplus 1$	$x^{169} \oplus x^{34} \oplus 1$	$x^{271} \oplus x^{58} \oplus 1$
$x^{29} \oplus x^2 \oplus 1$	$x^{97} \oplus x^6 \oplus 1$	$x^{191} \oplus x^9 \oplus 1$	$x^{281} \oplus x^{93} \oplus 1$
		$x^{193} \oplus x^{15} \oplus 1$	$x^{289} \oplus x^{21} \oplus 1$
		$x^{199} \oplus x^{34} \oplus 1$	

Figure 10.4 Primitive 3-term polynomials such that the smallest factor of $2^m - 1$ is greater than m (for $m \leq 300$).

In Figure 10.4, a list of primitive polynomials with the two following features is given.

1) The polynomial contains only three terms[6]. For a polynomial $x^m \oplus x^j \oplus 1$, performing σ shiftings in a single clock pulse requires only σ 2-input EXORs for $1 \leq \sigma \leq m - j$.

2) The smallest factor[7] of $2^m - 1$ is greater than m. Thus, for any $\sigma \leq m$, σ does not share a common factor with $2^m - 1$. For example, $2^{11} - 1 = 23 \times 89$; not any $\sigma \leq 11$ can share a common factor with $2^{11} - 1$ since $11 < 23$.

10.3.2.2 Weighted test vectors

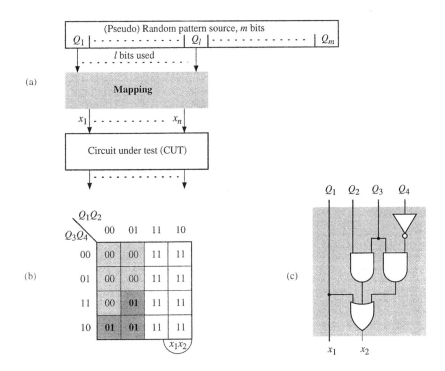

Figure 10.5 Constant distribution of weighted test vectors. (a) General scheme. (b) Example of 4-input, 2-output mapping. (c) A possible building of mapping in (b).

As explained in Section 10.1.1.2, it may be useful to apply weighted test vectors to a circuit under test. A general scheme is presented in Figure 10.5a.

---

[6] Lists of polynomials for every m up to 168 and up to 300 can be found in [St 73] and [BaMeSa 87] respectively. The polynomials in Figure 10.4 are taken from these lists.

[7] This information is drawn from [Br et al. 88].

The n-bit test vectors are generated from l-bit equally likely random vectors through a mapping.

An example of 4-input 2-output mapping is presented in Figure 10.5b. When a random vector $Q_1Q_2Q_3Q_4$ is drawn, the corresponding entry specifies the value of the vector x_1x_2. The probability of a vector x_i is then proportional to the corresponding number of entries. The positions of these entires in the mapping have no importance since the successive vectors $Q_1Q_2Q_3Q_4$ are random and independent.

According to Figure 10.5b, $\psi(x_0) = 5/16 = 0.3125$, $\psi(x_1) = 3/16 = 0.1875$, $\psi(x_2) = 0$, and $\psi(x_3) = 8/16 = 0.5$. This mapping may be obtained from a combinational circuit (Figure 10.5c) or from a 2^4-word by 2-bit memory (ROM, i.e., read only memory, or RAM).

If a constant distribution is required, it is not always possible to have the exact distribution, but the accuracy may be improved by increasing the number l of used bits. For example, assume that the constant distribution $\psi = (.3, .2, 0, .5)$ is required. The mapping in Figure 10.5b provides an approximate distribution. If 6 bits Q_1 to Q_6 were used instead of 4, the more accurate distribution $\psi = (.297, .203, 0, .5)$ could be obtained by allocating 19 entries to x_0, 13 to x_1, and 32 to x_3.

Remark 10.7 The scheme explained in this section corresponds to *an input-vector generator*.

It is more usual to build *input-bit generators* (in which the probabilities of the input lines are generated independently of one another) [ScLiCa 75], [BaMeSa 87]: a bit with a low probability of having the value 1 can be obtained from an AND gate (for example $x_1 = Q_1Q_2Q_3$ implies $\Pr[x_1 = 1] = 0.125$); a bit with a high probability of having the value 1 can be obtained from an OR gate ($x_2 = Q_4 + Q_5$ implies $\Pr[x_2 = 1] = 0.75$). According to Property 10.1, the input-bit generator is less convenient than the input-vector generator.

10.3.3 Sequence of adjacent vectors

As explained above (Section 10.1.2.2), a sequence of adjacent vectors is a *particular case of generalized distribution*. It is presented in the sequel. Afterwards, a more general case is illustrated with an example: a random tester for the Motorola *6800* microprocessor.

A sequence of adjacent vectors may be useful for testing an asynchronous sequential circuit (Section 5.2.2) or delay faults (Remark J.1).

The basic idea is that the vector $x(t+1)$ is obtained from $x(t)$ by changing one input variable x_i randomly chosen among the n input variables. The principle is illustrated in Figure 10.6a where the vector $x(t)$ is stored in a register made of n T flip-flops. A random l-bit vector is transcoded into a n-bit vector $(T_1, ..., T_n)$, in which exactly one component has the value 1. For example, if $T_i = 1$, $T_1 = ... = T_{i-1} = T_{i+1} = ... T_n = 0$. The output of the corresponding trigger changes on the rising edge of the clock. Thus, the next test vector $x(t+1)$ differs from $x(t)$ only by the input x_i.

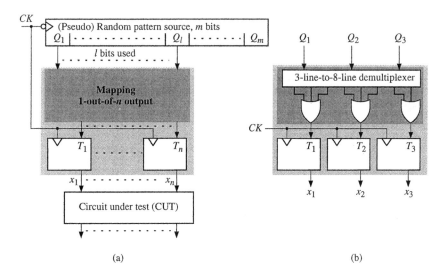

Figure 10.6 Generation of a random sequence of adjacent vectors. (a) General scheme. (b) Example using a demultiplexer.

The mapping in Figure 10.6a may be carried out by general techniques as in Section 10.3.2.2. It can also be built from a demultiplexer as illustrated in Figure 10.6b. From the random 3-bit vector $Q_1Q_2Q_3$, the 3-line-to-8-line demultiplexer produces $2^3 = 8$ outputs exactly one of which has the Boolean value 1. From these 8 lines, three outputs T_1, T_2 and T_3 are obtained by OR gates. For the example in Figure 10.6b, $\Pr[T_1 = 1] = \Pr[T_2 = 1] = 3/8 = 0.375$ and $\Pr[T_3 = 1] = 2/8 = 0.25$. The three variables x_1, x_2, and x_3, do not exactly have the same probability of change. These probabilities could become closer to each other by using more bits Q_j.

Note that this kind of generation can also be weighted. For example, $\Pr[T_1 = 1] = 5/8$, $\Pr[T_2 = 1] = 1/8$, $\Pr[T_3 = 1] = 2/8$ can be obtained. In this case, x_1 changes more often than the other input variables.

10.3.4 Generalized distribution. Example for the Motorola *6800* microprocessor.

The generation of a generalized distribution requires some memorization for the state of the generator (in addition to the equally likely random source such as an LFSR). See Appendix A.

270   *Chapter 10*

There are many ways to build this sequential part of the generator. The solution chosen greatly depends on the features of the considered generation.

For example, if the test sequence is made up of several parts, each one with a constant distribution (e.g., the inputs having a high probability of being 0 in the first part and of being 1 in the second part [GrSt 93]), the memory may be simply a counter measuring the number of tests vectors in each part.

A more complex example for the Motorola *6800* microprocessor is presented in the sequel In this example the random vector on the data bus is sometimes any 8-bit vector (data) or a valid operation code (197 out of the 256 possible input vectors according to Appendix H).

10.3.4.1 *General description of the random test machine*

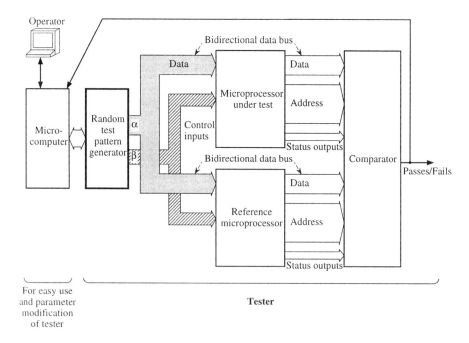

Figure 10.7 General description of the test machine[8].

This machine is an example in which the generator is based on *weighted patterns* fulfilling the requirements of *generalized distribution*.

---

[8] This machine was presented in [FéDa 86] and Figure 10.7 is reprinted from this paper (© 1986 IEEE).

According to Appendix H, the inputs of the Motorola *6800* microprocessor consists of an 8-bit bidirectional data bus (when the microprocessor is in read state) and six control lines. The control line *DBE* is connected to the clock CK_2, in normal operation and during a test experiment; five control inputs remain: *RST* (Reset), *HLT*, *NMI*, *IRQ*, and *TSC*. The three types of outputs are as follows: 8-bit bidirectional data bus when the microprocessor is in write state; 16-bit address bus; three status outputs (*VMA, BA, R/W*).

The general scheme of the test machine is shown in Figure 10.7. The test pattern generator is a hardware generator which produces the random patterns α and β. These patterns are applied on the reference microprocessor and the microprocessor under test. All the outputs are compared at each cycle. A fault is detected when at least one output line of the microprocessor under test does not have the same value (0, 1, or tri-state[9]) as the reference. The structure of the generator allows the operator to modify the probabilities of the patterns α and β between two test experiments (using a microcomputer). Note that a signature analysis can be performed instead of the comparison with a reference circuit (Section 11.4.2 and Chapter 12).

10.3.4.2 Principle of the input sequence generation

The input consists of two input patterns, the 8-bit data pattern, and a 5-bit control pattern, denoted by α and β respectively in Figure 10.7. The basic principle is to execute random instructions on random data. We will consider first the generation on the data bus, then the generation on the control lines.

According to Definition 5.1 in Section 5.2, an *ideal random test* is such that 1) any sequence which is possible according to the specifications provided by the manufacturer should have a non-zero probability and 2) any sequence for which the behavior is not specified should have a zero probability (i.e., the test sequence is syntaxically correct). Accordingly, when the device is waiting for an operation code, it is necessary to send one of the 197 valid operation codes on the data bus (see Appendix H, the behavior is not specified for 59 "invalid" operation codes). Any one of the $2^8 = 256$ possible data patterns may be sent when the circuit is waiting for data. Each instruction is characterized by an operation code I_i and a number of cycles n_i necessary to execute the instruction. Let us consider the instruction *ADD A DIR* (addition to Accumulator A, direct addressing) presented in Figure H.4. This instruction needs three cycles: 1st cycle, operation code *9B* is present on the data bus; 2nd cycle, address of operand on the data bus; and 3rd cycle, value of operand on the data bus. At the following cycle, the microprocessor is waiting for another operation code. Then for $I_i = 9B$, $n_i = 3$.

---

[9] The three states are coded with two Boolean value (00 for 0, 11 for 1, and 01 for high impedance state): the output is connected to V_{SS} and V_{DD} through two resistances and the voltage is compared with two thresholds.

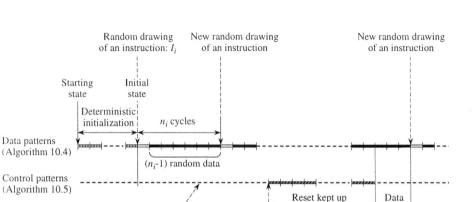

Figure 10.8 Illustration of Algorithms 10.4 and 10.5: data pattern and control pattern generations[10].

The string of data patterns is built in the following way (see Figure 10.8).

Algorithm 10.4 *Data pattern generation*[11].
Step 1. Deterministic initialization. This refers to the string of instructions which loads known values in all the registers of both microprocessors.
Step 2. Draw an instruction I_i randomly from the set $\{I_1, ..., I_{197}\}$. Let $N = n_i$.
Step 3. Decrement N. If $N = 0$, go to step 2.
Step 4. Draw data randomly in the set of 256 possible data. Go to step 3.
❏

It is conceivable that the microprocessor does not require data at each cycle (for example, see instruction *JSR IND*, code *AD*, in Figure H.4), but it does not matter. When the microprocessor needs data, it reads a random pattern.

Let us now consider the generation of control inputs. The way in which the control patterns are generated is explained in Algorithm 10.5 and illustrated in Figure 10.8. At each cycle a random pattern is randomly drawn from the set $\{Nothing, RST, HLT, NMI, IRQ, TSC\}$. When a control input has been drawn, a treatment depending on this control variable is performed (step 2 of Algorithm 10.5). More details can be found in [Mo 79] as to how these variables work. Let us only recall what is necessary for understanding Algorithm 10.5. A Reset function (*RST*) is executed correctly if the Reset variable is active for at least 8

---

[10] This figure is reprinted from [FéDa 84] (© 1984 IEEE).
[11] This algorithm is reprinted from [FéDa 86] (© 1986 IEEE).

cycles. The Reset function is achieved in 3 cycles after deactivation of the Reset variable. Then the microprocessor waits for a new instruction. The Halt function (*HLT*) works in the same way with 7 and 2 cycles respectively. The control *TSC* must be maintained at most 4 cycles during which the current instruction is in abeyance. The control *NMI* is an interruption, the end of which is characterized by the string of hexadecimal values *FFFA · FFFB* on the address bus (this is the corresponding interrupt address). Then, the microprocessor waits for a new instruction. The control *IRQ* behaves like *NMI* with the string *FFF8 · FFF9*. These strings may be observed on the reference microprocessor whenever necessary.

After initialization of the microprocessor, the drawing of control inputs at each cycle corresponds to the following algorithm.

Algorithm 10.5   *Control pattern generation*[12].

Step 1. Draw randomly a control pattern from the set {*Nothing, RST, HLT, NMI, IRQ, TSC*}. If *Nothing* is drawn go to step 1. If *RST* is drawn go to step 2.a; if *HLT* step 2.b; if *NMI* step 2.c; if *IRQ* step 2.d; if *TSC* step 2.e.

Step 2.
   Step 2.a. Maintain *RST* for at least 8 cycles[13]. Then put Algorithm 10.4 in its step 4 with $N = 3$. Wait for 3 cycles and go to step 1.
   Step 2.b. Maintain *HLT* for at least 7 cycles[13]. Then put Algorithm 10.4 in its step 4 with $N = 2$. Wait for 2 cycles and go to step 1.
   Step 2.c. Maintain *NMI* until the string *FFFA · FFFB* occurs on the address bus. Then put Algorithm 10.4 in its step 2. Go to step 1.
   Step 2.d. Maintain *IRQ* until the string *FFF8 · FFF9* occurs on the address bus. Then put Algorithm 10.4 in its step 2. Go to step 1.
   Step 2.e. Add 1, 2, 3, or 4 to N in Algorithm 10.4. The control *TSC* is respectively[13] maintained for 1, 2, 3, or 4 cycles. Then go to step 1.

❑

This approach takes benefit from the *independence between data and control generations*. Algorithm 10.5 modifies the autonomous behavior of Algorithm 10.4, only to ensure that the microprocessor is tested for a specified function.

It is easy to implement a software generator based on the principle indicated above. However, an on-line generation would be slower than a hardware generation (using similar technologies), and storing an off-line generated test sequence may require a large capacity of memory. A hardware generator is presented in the next section.

---

[12] This algorithm is reprinted from [FéDa 86] (© 1986 IEEE).
[13] When some values are not completely defined, several solutions can be chosen. For example, the Reset is maintained during N_R cycles. This number may be a random value such that $N_R \geq 8$. From practical considerations of simplicity, $N_R = 12$ has been chosen in the generator presented in Section 10.3.4.3.

10.3.4.3 Hardware test pattern generator

This section presents a hardware implementation corresponding to the principle of the input sequence generation presented in the above section.

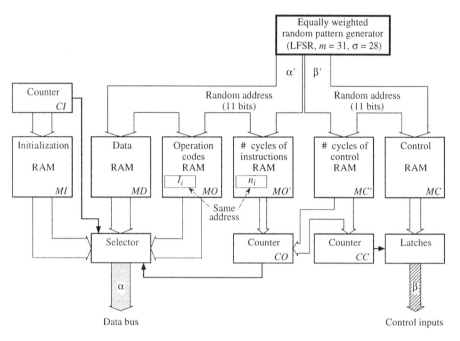

Figure 10.9 Random test pattern generator[14].

The generator presented in Figure 10.9 is mainly based on six memories, all of which are RAM, denoted by *MI*, *MD*, *MO*, *MO'*, *MC*, and *MC'*. The memory *MI* contains the initialization sequence. After initialization, the selector sends α which can be either a word coming from *MO* or a word coming from *MD*. An equally weighted word α' addresses simultaneously *MD*, *MO*, and *MO'*.

Memory *MO* is a 2 048-word by 8-bit RAM (addressed by a 11-bit random address since $2^{11} = 2048$). In this memory, the operation codes are stored according to the chosen probabilities (see Section 10.3.2.2, Figure 10.5). Let α_1 to α_{256} denote the possible 8-bit words on the data bus, and let α_i be the code of instruction I_i. If one requires $\Pr[I_i] = 1/197 \approx 0.0051$ for $i = 1, ..., 197$ (step 2 in Algorithm 10.4), the memory is loaded such that it is nearly equally distributed over α_1 to α_{197}. The following probabilities are obtained:

---

[14] This figure is reprinted from [FéDa 86] (© 1986 IEEE).

$$\Pr[\alpha_1] = \ldots = \Pr[\alpha_{119}] = \frac{10}{2048} \approx 0.0049,$$

$$\Pr[\alpha_{120}] = \ldots = \Pr[\alpha_{197}] = \frac{11}{2048} \approx 0.0054,$$

$$\Pr[\alpha_{198}] = \ldots = \Pr[\alpha_{256}] = 0.$$

The corresponding number of cycles is stored in MO'. According to Algorithm 10.4, when an instruction I_i is drawn from MO, the corresponding number of cycles n_i is read in MO' and stored in the counter CO. This number is decremented at each cycle. As long as the value in the counter CO is not zero, the selector sends data coming from MD to the bus α (these data may or may not be equally likely; they have the same probability if each one is written into 8 addresses since $2048 / 256 = 8$).

The part concerning the control bus β works in a similar fashion, as defined in Algorithm 10.5.

The equally weighted random patterns used as random addresses to the memories are generated by an LFSR (Section 10.3.2.1) of length $m = 31$ with $\sigma = 28$ shiftings at each cycle. The random address α' corresponds to bits Q_1 to Q_{11}, and β' corresponds to bits Q_{12} to Q_{22}.

The console and microcomputer in Figure 10.7 allow the following choices by modifying the random test pattern generator in Figure 10.9:

1) the *initialization of the two microprocessors* (the desired values in the registers of the microprocessors are stored into the appropriate addresses in RAM *MI*, from the console);

2) the *initialization of the LFSR* in the generator, which implies a pseudorandom test sequence (according to the notation in Section 10.3.1, bits $b(1)$ to $b(31)$ may be chosen. Hence $2^{31} - 1$ different pseudorandom sequences are possible);

3) the set of *instructions* and *data* and their *probabilities*;

4) the *probabilities* of *control inputs*;

5) the *length of the test sequence*, i.e., the number of random instructions;

6) *stopping the test* either at the first detection, or recording the number of detections up to the end of the input sequence.

All the possibilities provided by the test machine that we have presented are not necessary to perform a test (for example, equally likely data could be obtained by removing the RAM *MD* whose utility is to produce weigthed data patterns; 8 bits comming directly from the LFSR would replace the output of this RAM to the selector). This machine has been built to perform a wide range of experiments. Some experimental results are discussed in Section 11.3. A *simplified version* of this test machine is presented in Section 11.4.2.

NOTES and REFERENCES

Theoretical properties of linear congruential sequences used for software generation of random numbers are introduced in [Le 51], widely developped in [Kn 69] and can be found in [YaDe 88].

The references [Go 82], [Mc 87], and [YaDe 88] present the theoretical properties of hardware generation by LFSRs. Variants of pseudorandom generators based on LFSRs are presented in [BaMcSa 87]. The modification of an LFSR for introducing the all-0 pattern in the period is presented in [Vi 71].

Shifting of σ positions ($\sigma \geq n$) in a single clock pulse was used in [Te 74] [FéDa 86]. Random test sequences of adjacent vectors were performed by the random test machine presented in [Te 74]. The random tester for Motorola *6800* microprocessor was presented in [FéDa 86].

Hardwired generation of random sequences is also performed by cellular automata [HoMcCa 89], [NaCh 96].

Software generation of (pseudo)random test vectors using functional blocks already available on the chip have been proposed [SoKu 96], [RaTy 96].

Various works are devoted to generation of subsequences of two patterns for testing stuck-open and delay faults [FuMc 91], [ChGu 96], [DuZo 97].

11

Experimental Results

This chapter presents experimental results obtained from true *circuits. In most cases the results obtained from deterministic and random testing are compared.*

We insist on the fact that these results are not obtained from simulation but from circuits. Simulation is an extremely powerful tool which is very efficient for example for calculating a fault coverage or estimating detection probabilities. However, it is not adapted to comparing random and deterministic tests. As a matter of fact, a strong point of random testing is its ability to detect faults even if they are not targeted (Property 5.2 in Section 5.3.2). On the other hand, any comparison between random and deterministic testing based on simulations is biased since a simulation can only take into account faults devised by somebody.

We first present results obtained by the author's team, in cooperation with various companies, in Sections 11.1 to 11.3. Other results presented in the literature are then given in Section 11.4.

11.1 TTL CIRCUITS

Several hundreds of combinational and sequential circuits (SSI and MSI in TTL technology) were tested by both a deterministic test (in a French Company) and a random test (author's team). After presentation of the context of these experiments, the detailed test results on a batch of combinational circuits are compared. Then, experimental observation of the memory effect is pointed out on some faults in sequential circuits.

The random test experiments presented in this section were obtained using the random test machine presented in [Te 74]. This machine presents the following features.

1) Maximum of 8 inputs and 8 outputs.
2) LFSR where $m = 28$ (primitive polynomial $x^{28} \oplus x^3 \oplus 1$), $\sigma = 8$ (8 does not share a common factor with $2^{28} - 1$, see Section 10.3.2.1).

3) Choice between an input language $I = X^*$ or $I \subset X^*$ (in this case, the language is defined by an hardwired state graph).

4) Choice between synchronous test or asynchronous test (i.e, string of adjacent vectors). The second one can be used for any combinational or sequential circuit but the first one does not apply to asynchronous sequential circuits.

5) Various countings and criteria to stop the test experiments.

Experiments related to four types of combinational circuits and two types of sequential circuits are reported in [DaFoTe 75]. The combinational circuits are 57 BCD (binary-coded decimal) to decimal decoders (reference *442 E*), 61 four-bit binary full adders (*483 E*), 52 four-bit magnitude comparators (*485 H*), and 65 quadruple 2-input exclusive OR gates (*486 E*). The sequential circuits are 81 dual JK master-slave flip-flops (*473 E*), and 50 synchronous 4-bit up/down BCD counters (*4192 E*). The author's team was provided with these circuits by the Sescosem Division of Company Thomson-CSF[1]. All these circuits, previously tested in this company, presented either a parametric or a logic faulty behavior. In the sequel, random and deterministic test results are compared on the set of combinational circuits *483 E*, then the memory effect (Section 5.4) is illustrated with test experiments related to two sequential circuits.

11.1.1 Batch of 61 circuits reference *483 E*

In this section, the deterministic and random test results related to a batch of circuits are presented and commented. Similar results have been obtained for other batches.

The results are presented in Figure 11.1. The *deterministic test* performed by the Company Thomson-CSF consisted of 19 test vectors. It is able to test and diagnose every single stuck-at fault of a connection. The test program furnishes a list of single stuck-at faults associated with the corresponding faulty behaviors. The test sequence may detect some multiple faults, but cannot diagnose them (denoted by ? in the last column of Figure 11.1a).

Among the 61 circuits, 40 have been found to have a logic fault (for deterministic testing, a class corresponds to a set of circuits which produce the same behavior for every 19 input vectors), and 21 have been found to have only parametric failures (obtained by *means other than the logic test* sequence). All the circuits have been tested by both deterministic and random testing, but we are interested in comparing the tests for logic faults only.

The *random test* performed is as follows. The test is synchronous (i.e., the vector $x(t)$ is completely independent from $x(t-1)$). If no fault is detected, the test length is $L = 10^4$, corresponding to $\varepsilon = 3 \times 10^{-9}$ for the worst case fault. If a

---

[1] Thanks to a fruitful cooperation with J.-C. Jullien, J.-P. Lusincki, and J. Zirphile, working in this Company. The circuits are described in [Se 75].

fault is detected, the test continues and up to 30 detections are obtained (i.e., 30 vectors in the test sequence have produced at least one faulty output). A faulty circuit i is characterized by a set of pairs

$$C(i) = \{(l(j), Z(j))\}, \; j = 1, \ldots, 30,$$

where $l(j)$ is the length, and $Z(j)$ the set of faulty outputs, corresponding to the jth detection. For random testing, all the circuits with the same $C(i)$ are in the same class.

Class	Circuits	Length to detection	Faulty outputs $z_1\,z_2\,z_3\,z_4\,z_5$	Number of possible faults
1	1,37,46,53	15	————×	5
2	3,15	1	×××××	?
3	6	1	××	?
4	7	16	—×—	1
5	8,11	5	××××	1
6	13	6	××××	1
7	17	1	———××	?
8	19	1	×	?
9	21	5	—×××—	1
10	22	10	————×	6
11	23	5	———×—	2
12	25	1	××××	3
13	26,43	9	—×××	4
14	28	1	×××××	?
15	30	1	———××	?
16	31	1	×××××	?
17	32	1	×××××	?
18	35	1	—×—××	?
19	36	5	—×××—	?
20	38,60	1	———××	1
21	40	1	×××××	?
22	44	7	—×—	?
23	45	1	×××××	?
24	47,57	10	————×	1
25	48,51,59,61	8	————×	3
26	49	13	————×	2
27	52	6	—×—	2
28	54	1	———×—	6
29	55	5	—×—	1
30	21 others	Parametric faults		

(a)

Class	Circuits	Length to detection l_ω	Length to 30 detect.	Faulty outputs $z_1\,z_2\,z_3\,z_4\,z_5$
1	1,37,46,53	142	2378	————×
2	3	1	30	×××××
2'	15	3	45	×××××
4	7	1	32	×××××
5	8,11	1	47	×××××
6	13	4	55	—××××
7	17	1	30	———××
8	19	23	?	
9	21	1	88	—×××—
10	22	169	965	————×
11	23	2	83	———×—
12	25	4	62	—××××
13	26,43	3	68	——×××
14	28	1	33	×××××
15	30	2	41	———××
16	31	1	30	×××××
18	35	1	30	———××
19	36	3	68	—×××—
20	38,60	1	65	———××
21	40	1	54	×××××
22	44	1	496	—×—
23	45	1	30	×××××
24	47,57	142	799	————×
25	51,59,61	94	589	————×
26	49	44	?	
27	52	4	?	—×—
28	54	4	71	————×
29	55	1	77	—×—
30	24 others	Parametric faults ?		

(b)

Figure 11.1 Test of 61 circuits (reference *483 E*). (a) Deterministic test. (b) Random test.

In order to state if a fault is permanent or intermittent, the same random test sequence was applied at least two times to each circuit. Among the 61 circuits, 34 have been found to have a permanent logic fault, and 3 an intermittent logic fault (such a fault is identified if the set $C(i)$ is not always the same when the same input sequence is applied).

Comparison of the results.

1) Some faults are parametric. The detection of logic faults may depend on some parametric conditions, for example the supply voltage. This is an accounts for the non-detection of faults in circuits number 6, 32, and 48 (detected in the deterministic but not in the random test). One can imagine that if the conditions were slightly different (supply voltage for example), the result would be different.

More outputs were found faulty by random test for classes 4 and 5. This could be for a similar reason or because the random test is longer.

2) The previous comment is supported by the fact that 3 circuits (numbers 19, 49, and 52) have been found to have intermittent faults by random testing. They could correspond to parametric faults at the limit: sometimes a logic fault appears, sometimes it does not.

3) All the circuits found with a permanent fault correspond to the same classes for both deterministic and random testing, except for the faults in circuits 3 and 15 which are distinguished by the random test.

4) For the random test, the length to detection is less than 19 (length of the deterministic test sequence) for 22 out of 28 classes.

5) When these experiment were performed, computation of the deterministic test sequence required 3 minutes on a computer *IRIS 80*, the random test length was obtained in a few seconds without computer: the circuit under consideration is a 9-input combinational circuit then $p_{min} \geq 1/2^9$, which leads to $L_{ni}(0.001) \approx 3\,300$ (Equation (6.33) in Section 6.1.2.2).

6) From the previous points 1 to 5, we can conclude that the results are almost similar for both deterministic and random testing. The main differences are that: i) deterministic testing is shorter and provides a partial diagnosis of faults for about half the circuits; ii) random testing is easier and less costly to implement.

11.1.2 Memory effect in sequential circuits

An asynchronous random test was performed on the sequential circuits references *4192 E* and *473 E*. According to Section 7.1 and Appendix F, the test sequence is a string of randomly drawn adjacent vectors (Section 10.3.2.3). For the faulty circuits, the test experiments last up to the 30th detection. The results are presented in Figure 11.2a and b for a circuit *4192 E* and a circuit *473 E* respectively. A vertical stroke corresponds to a detection (30 detections for each case).

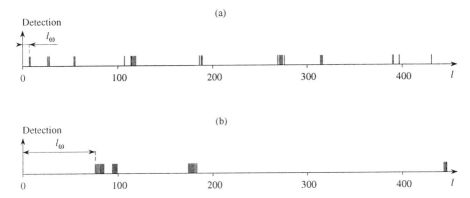

Figure 11.2 Asynchronous test of two sequential faulty circuits. (a) A circuit *4192 E*. (b) A circuit *473 E*.

For the fault in the circuit *473 E*, the memory effect (Section 5.4.2) is particularly obvious. This effect depends on both the circuit and the fault; it is often more important for asynchronous testing than for synchronous testing [DaTh 81] because only one input changes from time t to $t + 1$ in the case of asynchronous testing.

11.2 LSI CMOS CIRCUITS

This section presents results obtained for the testing (deterministic and random) of two kinds of circuits manufactured by the CNET[2] (Centre National d'Etudes des Télécommunications).

Both tests (deterministic and random) were performed on the same general purpose Tektronix *S3270* tester (number of inputs plus number of outputs = 128. Maximum frequency: 20 MHz). The random test sequences were software generated off-line (Section 10.2), then stored.

The random test sequences are syntaxically correct (i.e., belong to the input languages in normal working). Their length was calculated using the methods for microprocessors (Section 9.3 and 9.4), even if the circuits are not microprocessors; for this purpose "functional" instructions (examples will be given in the sequel for the second kind of circuits) and fictitious registers were introduced (a similar approach is proposed in Exercise 9.4).

More details on this section can be found in [DeThBe 84].

---

[2] This work was achieved thanks to the Convention n° 83 3B 013 between the CNET and the Institut National Polytechnique de Grenoble: cooperation of the author's team with L. Beghin, J.-L. Rainard, and J. Gobbi from the CNET Grenoble.

(a)

	Tested	Number of circuits			Test length	
		Intermittent faults	Fault-free	Faulty (permanent)	Fault-free circuit	Max up to detection
Deterministic test	699	0	521	178	23	≤ 23
Random test			520	179	78	22

(b)

	Tested	Number of circuits			Test length	
		Intermittent faults	Fault-free	Faulty (permanent)	Fault-free circuit	Max up to detection
Deterministic test	399	7	49	343	2 415	≤ 2 415
Random test			28	364	4 400	2 243

Figure 11.3 Test results about circuits manufactured by the CNET. (a) Circuits *AUT*. (b) Circuits *MTA*.

First kind of circuits. The circuit called *AUT* is a 6-input, 2-output self-checking bus controller containing more than one hundred gates.

The deterministic test contains 23 input vectors. For $P_m = 0.999$ (worst case fault, according to the notation introduced in Section 4.4.2), the random test length is 78 input vectors (generated as a string of 52 instructions of one or two cycles each; the data are equally likely and the instruction probabilities are weighted in order to minimize the test length [DeThBe 84]).

The test results are presented in Figure 11.3a. The set of 179 circuits detected as faulty by the random test includes the set of 178 circuits detected as faulty by the deterministic test. When the circuit is faulty the random test length to detection is short (often 1 or 2 vectors, maximum 22 vectors).

For a wafer containing 253 chips, out of which roughly 20% are faulty, the total test time is 77 seconds for random testing and 75 seconds for deterministic testing (this time includes mechanical moving which is similar for both).

Second kind of circuits. The circuit called *MTA* is a 21-input, 19-output self-checking multiplier containing more than one thousand gates. It can perform either serial or parallel multiplication. Here are two "functional" instructions out of the five which were defined for the analysis of the test length: I_3, read the accumulator in parallel mode (1 cycle); I_5, multiplication in serial mode (16 cycles).

The deterministic test contains 2415 input vectors. For $P_m = 0.999$, the random test length contains 4400 vectors corresponding to 527 random instructions with random data.

The test results are presented in Figure 11.3b. The comparison between deterministic and random testing is made only for circuits without intermittent faults (such a circuit does not always produce the same result for the same input sequence). The set of 364 circuits detected as faulty by the random test includes the set of 343 circuits detected as faulty by the deterministic test.

Experimental Results **283**

After these results, an analysis was made to determine why the difference between deterministic and random testing was so great (21 circuits). It was then found that at least two features were not taken into account in the deterministic test sequence: 1) multiplication was not tested in serial mode; 2) all the input data had a fault-free parity bit in the deterministic test sequence (this bit had a 0.5 probability of being faulty in the random test sequence). This is an example showing that any test designer can forget to test some features.

Implementation time. The implementation time was shorter for the random test (about one week for each circuit, without computer aid, to calculate the test length starting from a detailed functional description of the circuits). Initially, a comparison of random and deterministic tests was scheduled for a third kind of circuits, a self-checking Hamming coder containing more than five thousand gates; the random test experiment was ready to start (length 6 400 vectors) but the deterministic test sequence was never finished.

11.3 MOTOROLA *6800* MICROPROCESSOR

Experiments have been conducted with different sets of *6800* microprocessors (the main features of these microprocessors are presented in Appendix H) supplied by different French Companies (several hundreds of circuits). After the detection results (passed or failed) for a set of 24 new microprocessors, detailed results concerning a set of 60 circuits previously tested by a deterministic sequence are presented in Section 11.3.1. A comparison between theory and experiments is then made in Section 11.3.2.

	Tested	Good	Faulty
Manufacturer *A*	8	8	0
Manufacturer *B* (batch *a*)	7	7	0
Manufacturer *B* (batch *b*)	1	1	0
Manufacturer *B* (batch *c*)	1	1	0
Manufacturer *B* (batch *d*)	5	2	3
Manufacturer *B* (batch *e*)	2	0	2
Total	24	19	5

Figure 11.4 Random testing results on a set of 24 new microprocessors[3].

Section 11.3.1 is related to a set in which all the circuits are found faulty. Fortunately, it is not the case for other sets of circuits. The results in Figure 11.4

---

[3] This figure is reprinted from [FéDa 86] (© 1986 IEEE).

relate to 24 new microprocessors (just bought, never tested to our knowledge). The proportion of good and faulty circuits in this set is not significant, because this set is too small.

11.3.1 Experimental results for a set of 60 microprocessors

We should consider that: 1) the number of faulty circuits in this set has no meaning because the microprocessors are drawn from a large batch which is known to be particularly deficient; 2) the quality of the deterministic tester is not questionable; 3) the proficiency of the persons concerned by design and use of the deterministic sequence is not doubtful. Hence, we only want to compare the results which can be obtained from random testing, with results which have been obtained from deterministic testing.

Out of a batch of 3 000 microprocessors, which were already tested by a specialized French Company, 60 were selected for further testing by random testing. The original testing was performed on a Fairchild tester using a deterministic sequence of about 1 000 instructions. Out of the 60 microprocessors 30 were found to be faulty, and the rest passed the test without detection of errors.

The random test experiments were performed with the tester presented in Section 10.3.4. When nothing is specified, i.e., under *normal conditions*, the results presented correspond to equally likely instructions (probability about 1/197). Each control input (i.e., *RST*, *HLT*, *NMI*, *IRQ*, and *TSC*) has approximately the same probability of being executed as an instruction, i.e., roughly 1/800 at each cycle because the average length of an instruction corresponds to 4 cycles (Appendix H). Then the drawing of *Nothing* in Algorithm 10.5 (Section 10.3.4.2), i.e., no control input is activated, has approximately the probability $1 - 5/800 \approx 0.994$ at each cycle.

The following results are found:

1) for the original 30 failing circuits, the test confirmed all to be faulty using a random test sequence of less than 25 000 instructions;

2) for the original 30 passed circuits, the test found, on the contrary, that all were faulty according to random testing of less than 80 000 instructions (equivalent to approximately 0.3 second, with the tester working at 1 MHz).

NB: the proportion of faulty circuits in this set is not significant.

Remark 11.1 These results do not mean that the deterministic tester is wrong, but that actual faults are not tested by the deterministic test sequence because they are not included in the set of *prescribed faults*: some actual faults are *non-target faults* (Section 4.2). According to Section 5.3 (specifically Properties 5.2 and 5.3), a random test has a chance of detecting a fault which is not detected by a deterministic test (note that there is a contradiction between the notion of *unknown* non-target faults and *deterministic* test sequence).

❏

---

f_1: Reset during an operation on stack pointer (SP) $\Rightarrow$ the hexadecimal value $FFFF$ is loaded into SP.
f_2: Reset when $WAI \Rightarrow$ write instead of read at the end of Reset.
f_3: No incrementation of program counter during instruction BCS.
f_4: For instruction SWI, the values $FFF8$ and $FFF9$ are read instead of $FFFA$ and $FFFB$.
f_5: Bridging between a_{10} and a_{11} in the address bus.
f_6: Some improper functioning of control inputs.
f_7: No incrementation of the program counter (PC) when new instruction.
f_8: Decrementation of PC instead of incrementation.
f_9: Always in a write state.
f_{10}: Instruction $PSH\ B \Rightarrow$ value FF is read instead of the content of accumulator B.
f_{11}: Dynamic faulty behavior. Apparently an intermittent fault (not always detected at the same length).
f_{12}: Faulty value in PC after instruction CPX in indexed addressing mode.
f_{13}: Wrong addition of offset on index register during instruction JSR in indexed addressing mode.
f_{14}: Three state problem.
f_{15}: Improper Halt timing.

Figure 11.5 The faults which have been observed[4].

Figure 11.5 lists all the faults found. When a fault is detected by a short input sequence, for example f_5, one can easily observe on a logic analyser what the faulty behavior is. For other faults, the diagnosis required several experiments, as explained in Appendix L.

A single circuit may be affected by several faults, as shown in Figure 11.6 (when the experimental conditions are modified, the first fault found may not be the same). In this figure the 60 microprocessors have been sorted out into 17 classes corresponding to observed faults. The lengths to first detections corresponding to *normal conditions* (behavior explained in Section 10.3.4, with equal likelihood of instructions, and equal likelihood of data) are given in Figure 11.6.a.

According to Section 10.3.4.3, the probabilities of instructions and data can be modified by changing the values in the RAMs. Experiments in specific conditions have been performed and are presented in the sequel. In *normal operation*, the data inputs are randomly drawn either from the 197 operation codes or from the 256 possible patterns. The tester should be very simple if the data are always to be drawn from the same set. Some experiments were carried out to observe the consequence of such a simplification.

---

[4] This figure is reprinted from [FéDa 86] (© 1986 IEEE).

1) *Application of invalid operation codes*

When any of the 256 patterns are sent at any time, the microprocessor may read an invalid operation code. The first question that arises is: does the microprocessor behave in a deterministic way for an invalid operation code? If the response is no, some good circuits could be erroneously found faulty (Section 5.2). For the *6800* microprocessor, the answer is always affirmative; this is not known from the user's manual but was shown in [Ne 79]. Therefore, any pattern can be sent while waiting for an operation code. The second question concerns the test length for the prescribed confidence level. In the set of 59 invalid operation codes, 55 behave like valid operation codes, and four lead the microprocessor into a loop [Ne 79]. When the loop has been reached, the test is virtually useless unless a *Reset* input is randomly drawn. Hence, one can expect that the test lengths required are greater than for the normal operation. The results are shown in Figure 11.6b. Most of the faulty microprocessors require larger lengths (up to 5.3 times longer for class 14). It must be noted that the fault f_1, which is the hardest to detect, requires a larger length.

Class	Faults in the circuits	Number of circuits in the class	Normal conditions			Any operation code (valid or invalid) K_1 = Av. length / L_a	Data restricted to operation codes K_2 = Av. length / L_a
			Min(l_ω)	Av(l_ω) = L_a	Max(l_ω)		
1	f_1	6	103	10 501	28 885	3.8	1.0
2	f_1, f_2	28	103	8 188	24 102	0.3	1.0
3	f_1, f_3, f_4, f_{14}	10	103	2 111	10 419	0.2	1.7
4	$f_1, f_3, f_4, f_{14}, f_{11}$	1	103	1 075	3 319	0.4	0.8
5	$f_1, f_3, f_4, f_{14}, ?$	1	60	911	1 507	0.5	1.0
6	$f_1, f_3, f_4, f_{14}, f_{15}$	2	103	657	2 519	0.6	1.5
7	$f_6, ?$	1	7	215	861	1.5	3.2
8	f_1, f_{10}	1	6	207	394	4.3	1.0
9	$f_1, f_{14}, ?$	1	7	199	861	1.4	1.3
10	f_1, f_6, f_{12}	1	10	188	294	2.3	1.0
11	$f_{13}, ?$	1	17	145	360	1.3	0.9
12	f_1, f_2, f_{14}	1	8	91	226	4.3	1.0
13	$f_7, ?$	1	3	28	84	5.2	2.3
14	$f_5, ?$	1	4	13	24	5.3	1.3
15	$f_{11}, ?$	2	2	3	6	1.0	1.0
16	$f_9, ?$	1	1	1	1	1.0	1.0
17	$f_8, ?$	1	1	1	1	1.0	1.0
	(a)		(b)				(c)

Figure 11.6 The classes of observed microprocessors. In every case (a), (b), and (c), the results are based on 15 different experiments for each class[5].

---

[5] This figure is reprinted from [FéDa 86] (© 1986 IEEE).

However, some of them require shorter lengths. The reason is that some of the invalid operation codes behave like valid operation codes. For example, one of them performs the same operations as the operation code allowing detection of f_3 (instruction *BCS*, see Figure 11.5). Thus, the probability of this operation code is multiplied by two, and the test lengths are shorter for the corresponding classes i.e., classes 3 to 6.

2) **Restriction of possible data**

In this case, the data inputs are always drawn from the field of 197 valid operation codes. This means that when the instruction codes correspond to α_1 to α_{197}, the data are randomly drawn from this set of patterns. Then patterns α_{198}, ..., α_{256} are never found. However, if each bit has approximately the same probability of occurring (0.5), the result may not be strongly affected. This observation, confirmed by the results in Figure 11.6c, suggests that it is possible to make a simpler test machine without affecting performance. The machine is simpler because RAMs *MD* and *MO'* and counter *CO* in Figure 10.9 (Section 10.3.4) are no longer useful.

11.3.2 Experiments versus theory

This section is based on additional experiments performed on the set of microprocessors presented in Section 11.3.1. The first part is based on a detailed analysis of the fault f_1 (circuit of class 1). The second part is a comparison between the length to first detection and the length between two detections on all the classes.

According to the definition of f_1 in Figure 11.5, the transition matrix of the Markov chain associated with the detection process of this fault is

$$U = \begin{bmatrix} 1-a & a & 0 \\ b & 1-b-c & c \\ 0 & 0 & 1 \end{bmatrix} \qquad (11.1)$$

where a is the probability of applying the *Reset* input during an operation on the stack pointer (*SP*), b is the probability of loading a new value in the stack pointer, and c is the probability of observing the value stored in the stack pointer. Let us show how these values can be evaluated:

$$a = \Pr[RST] \times \Pr[\text{Operation on } SP]; \qquad (11.2)$$

$$\Pr[\text{Operation on SP}] = \sum_i \left(\Pr[I_i] \times \frac{\text{\# operations on } SP \text{ during } I_i}{\text{\# cycles in } I_i} \right). \qquad (11.3)$$

For example, according to Figure H.4, there are three operations on the *SP* during the eight cycles of instruction *JSR IND*. From the experimental conditions (probabilities of *RST* and the various instructions), the value $a = 9.2 \times 10^{-5}$ is found.

As there are five instructions loading *SP*, the value $b = 5/197 \approx 0.025$ is obtained. Seventeen instructions and two control inputs can lead to observation of *SP*; from the various probabilities, $c = 0.093$ is found.

Hence,

$$U = \begin{bmatrix} 0.999908 & 0.000092 & 0 \\ 0.025 & 0.882 & 0.093 \\ 0 & 0 & 1 \end{bmatrix}. \qquad (11.4)$$

The theoretical results obtained from this Markovian model are compared with the experimental results in Figure 11.7a.

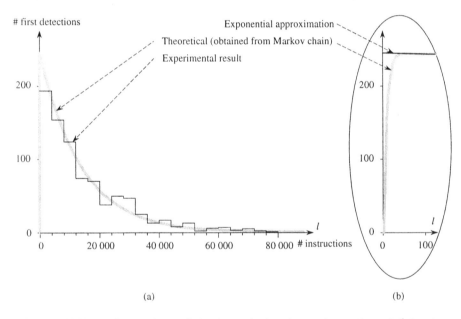

Figure 11.7 (a) Comparison of the theoretical and experimental results[6] for the fault f_1. (b) Exponential approximation versus exact theoretical result (expansion of the left part of (a)).

A set of 853 different pseudorandom test sequences is performed. The experimental results are presented in the following way: the number of test

---

[6] Part (a) of the figure is reprinted from [FéDa 86] (© 1986 IEEE).

sequences out of the 853 trials such that $l_\omega \leq 4\,000$ is 194; the number of test sequences such that $4\,000 < l_\omega \leq 8\,000$ is 154, and so on.

Given that the initial state of the Markov chain is $\pi(0) = (1, 0, 0)$, the probability $\Pr[l_\omega = l]$ that the first detection occurs at time l is obtained from Equation (7.7) in Section 7.2.2. This result is presented in Figure 11.7a, where the scale is adapted taking into account the number 853 of experiments and the number 4 000 corresponding to the width of ranges of experimental results. From the Markov chain, the detection probability $P(f, L) = 0.999$ is obtained for $L \approx 96\,000$. The longest test sequence for detection is roughly 80 000 for the set of 853 experiments.

In Section 7.2.2, it is shown that the approximate value $\Pr[l_\omega = l] \approx \lambda_2^{l-1} \cdot (1 - \lambda_2)$ can be used (where λ_2 is the greatest eigenvalue of the transition matrix except for the value 1). For the matrix U in (11.4), $\lambda_2 = 0.9999275$. The corresponding approximation is illustrated in Figure 11.7b. The approximate value $\Pr[l_\omega = l] \approx 0.0000725 \times 0.9999275^{l-1}$ can be distinguished from the exact value only for small values of l ($l < 30$).

Now the comparison of length to first detection (l_ω) and length between detections (l_τ) is presented in Figure 11.8. The value $Av(l_\omega)$ is the average value of l_ω for 15 experiments already presented in Figure 11.6a. The value $Av(l_\tau)$ is obtained from the following experiment: a test sequence of 10 million instructions is performed and the number of detections is counted (for every class of faults). The memory effect (Section 5.4.2) is quite clear. For most classes of circuits (but not all), $Av(l_\omega) > Av(l_\tau)$, and for some cases $Av(l_\omega)/Av(l_\tau) > 10$.

Class	Length to first detection $Av(l_\omega)$	Length between detections $Av(l_\tau)$
1	10 501	965
2	8 188	615
3	2 111	161
4	1 075	121
5	911	160
6	657	159
7	215	160
8	207	349
9	199	157
10	188	8
11	145	144
12	91	10
13	28	5
14	13	2
15	3	4
16	1	1
17	1	1

Figure 11.8 Observation of the memory effect for the classes of faulty microprocessors in Figure 11.6.

Chapter 11

11.4 OTHER EXPERIMENTAL RESULTS

In this section, several random testing experiments found in the literature are reported. Each of them emphasizes either the *low cost* mainly due to a shorter development time, or (inclusive) the *high level of defect coverage*.

For all these experiments, the defects are *unknown* (simulation with fault injection is not considered, according to the presentation at the beginning of this chapter).

11.4.1 Experiments by W. Luciw: Intel *8080* microprocessor

The paper [Lu 76] presents an experimental random tester built by the Company Sperry Univac for low-cost incoming inspection of Intel *8080* microprocessors. The main motivation is a *low cost* tester, making it unnecessary to rely entirely on vendors test credibility.

The test contains an initialization program, followed by a random test sequence. The behavior of the circuit under test is compared dynamically for parametric and functional coincidence with reference circuit.

The initialization program is provided by a ROM of 256 × 8 bits (the input of the *8080* microprocessor is a 8-bit bus). The main goal of this program is to preset both *8080* (reference circuit and CUT) to a known state. This initialization program also allows measurement of absolute output delays of the circuit under test, and includes some deterministic tests (particularly for increment and decrement of registers, and some counters).

The basic philosophy of the random test is to execute random instructions with random data. The string of vectors is generated by a 32-bit LFSR. The 8 bits of the input vector are obtained by tapping off appropriate cells of the LFSR, and the adjacent bits of a byte are separated by EXOR gates for the purpose of removing the correlation between successive bytes.

In the paper, the author does not give numerical results. However, he is clearly satisfied by the results which were obtained when he writes: "The performance of this tester has exceeded all expectations. Available *8080* CPU samples, with known defects, have been readily segregated by the tester within one to four seconds of PRSG time" (PRSG is used for Pseudo-Random Sequence Generator).

11.4.2 Experiments by A. Laviron *et al.*: Motorola *6800* microprocessor

Research is being conducted in France at the CEA (Commissariat à l'Energie Atomique), to measure the effect of low irradiation on microprocessors

[GéRaLa 83], [LaGéHe 89]. The aim is to identify the behavior of circuits which are installed in a nuclear plant. There are two main kinds of experiments. First kind: some dozen of microprocessors (identical and functioning correctly) are placed in an enclosure with a low level irradiation rate. After a certain time, these microprocessors are taken out of the enclosure and tested. In this way, the proportion of circuits becoming faulty in some defined experimental conditions is estimated. Second kind of experiment: only one microprocessor is placed in an enclosure as mentioned above. A tester, outside the enclosure and connected to the circuit, tests it periodically. We thus estimate the moment when the circuit becomes faulty. In these experiments, *the tester is an essential measurement tool* [GéRaLa 83]. It "decides" whether or not a circuit is correct in the first kind of experiments, and when a microprocessor becomes faulty in the second kind.

In 1989, the CEA engineers published comparative experimental results on fault detection of some circuits. For a batch of fifty *6800* microprocessors, three deterministic test sequences and one random test sequence were used.

Figure 11.9 presents the summary of the results. The test sequences called LIF, FAC, and UGM are deterministic. The test sequence called LAG is a random one.

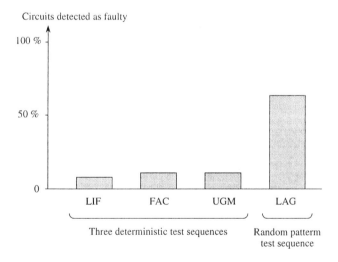

Figure 11.9 Results obtained by A. Laviron *et al.* on a batch of 50 Motorola *6800* microprocessors[7].

LIF contains about 50 bytes (about 13 instructions).
FAC contains 2 K bytes (about 500 instructions).

---

[7] This figure is reprinted from [LaGéHe 89] with permission of the authors.

UGM contains about 60 K bytes (about 15 000 instructions). It is based on functional fault hypotheses. However, the data are random. For example: when addition of the numbers in registers R_1 and R_2 is tested, the operands put into these registers are random [BeVe 84].

LAG is a random tester adapted to provide the experimental needs of CEA explained at the beginning of this section: long input and output buses to keep a circuit under test in the enclosure and the tester outside; periodic testing automatically performed; printing of results etc.

CEA random tester. This random tester [Fu 84] is a simplified version of the tester presented in Section 10.3.4. The following simplifications have been made.

1) The initial state of the LFSR is always the same. The test length is roughly 4 millions instructions[8] (16 seconds). Hence the test sequence is *always the same*.

2) The data on the data bus are always drawn from the 197 valid operation codes. According to Section 11.3.1 (part *Restriction of possible data*), the machine is simplest without loss of performance.

3) Since the sequence of vectors produced by the LFSR is always the same, it is possible to find (by simulation) the addresses in the memory *MO* (Figure 10.9 in Section 10.3.4.3) which are addressed at times 1, 2, 3, ... The values written in these addresses are chosen in such a way that the beginning of the random sequence provides an initialization of the microprocessor (the other addresses are then written with values such that the 197 values are nearly equally likely, according to Section 10.3.4.3).

In such a way, RAM *MI* and Counter *CI* (Figure 10.9) are no longer needed.

4) The tristate outputs are transformed into outputs with the value 1.

5) A 23-bit signature analyser (Figure 5.1 in Section 5.1, and Chapter 12) is used instead of a reference microprocessor.

Remark 11.2. A slight modification of *6800* microprocessors was made in 1977. The microprocessors manufactured after November 1977 do not have the same signature as the microprocessors manufactured earlier. Hence, two signatures are stored in the test machine. The circuit is considered as faulty if and only if it does not have one of the two signatures[9].

Remark 11.3. The number of circuits detected as faulty by the three deterministic test sequences does not increase significantly when the length of the sequence is greater (Figure 11.9: the length of the test sequence UGM is about 30 times the length of FAC and about 1 200 times the length of LIF). The random test sequence has detected more faulty circuits than the other sequences. After the

---

[8] Another test length about 250 000 instructions (1 second) may be chosen by the user.
[9] Signatures for intermediary test length are also available (16, 250, 8 000, 250 000, or 10^6 instructions) if the user wants to know approximately at what length the fault was detected.

experiments by A. Laviron *et al.*, all the defects detected by the random tester have been diagnosed[10] [HuDa 92]; one of the faults was intermittent.

11.4.3 Experiments by R. Velazco *et al.*: Motorola 6800 microprocessor

The experimental results in this section are taken from [VeBeMa 90].

A single defect in a metal or a polysilicon wire (Section 2.2.2) was randomly introduced in 75 "good" *6800* circuits (i.e., having passed all the manufacturer's tests). Defects were created by means of a laser, allowing cutting either metal or polysilicon tracks. Beam position accuracy is 0.1 µm for X and Y coordinates. Random pairs of X and Y coordinates were first generated. For a given (X, Y), the laser beam was positioned at the exact position; then, the nearest metal or polysilicon track (except for power supply tracks) was looked for and cut. This technique thus provides the experimenter with a sample of defective circuits that is both realistic and non-biased with respect to test methods.

According to the available hardware and software, it has been possible to compare various test approaches using various testers. The test methods called minimal, tailored, systematic, manufacturer, and random [VeBeMa 90] are presented in the sequel[11].

Test methods

Minimal test. This is a program consisting of a few instructions and allowing detection of defects that are trivialy detected.

Tailored test. A specific functional test program for the *6800* was available [MiJa 87]. This program is made up of 9 modules, each one verifying a given subset of the *6800* instructions.

Systematic test. This functional test program is generated from a high level instruction set [BeVe 84]. The data are random.

Manufacturer's test. This test sequence is the standard Motorola test sequence improved with specific test patterns (added to detect defects which were not initially found by the test sequence).

Random. This test is performed by the CEA random tester (presented in Section 11.4.2).

Testers

Minimal. This is a simple tester allowing execution of a program stored in the tester's RAM along with operands, the test results being the values contained in

---
[10] With the collaboration of X. Fédi, R. Velazco, J.-P. Acquadio, and the CIMF of Grenoble (*Centre Interuniversitaire de Micro-Electronique*).
[11] The names of the methods are those used in the papers referred to. Some tests were used with a different name in [LaGéHe 89] (Section 11.4.2): Minimal test = LIF, Tailored test = FAC, Random = LAG, Systematic test ≈ UGM (several variants are available for this test, but the others are exactly the same).

the same memory after test program execution. Such a tester can perform only a restricted logic testing.

TEMAC. This tester is an improvement on the *Minimal* tester [BeVe 84].

GR 125 (Genrad) and *Sentry 20* are two classical component testers able to perform logic, parametric, and dynamic testings. However, only logic testing was performed with the *GR 125*.

Random refers here to the CEA random tester presented in the above section. ❑

Test method	Tester	Number of passed circuits	Defect coverage	Passed circuits
Minimal	Minimal	30	60 %	List not specified
Tailored	Minimal	12	84 %	10,14,15,19,20,32,33,41,50,64,67,75
	TEMAC	11	85.3 %	10,14,15,19,20,32,41,50,64,67,75
	GR 125	8	89.3 %	10,14,20,32,41,50,64,67
Systematic	Minimal	12	84 %	10,14,15,19,20,32,33,41,50,64,67,75
	TEMAC	10	86.7 %	10,14,15,19,20,32,41,64,67,75
	GR 125	7	90.7 %	10,14,15,20,41,64,67
Manufacturer	Sentry 20	4	94.7 %	10,19,64,75
Random	Random	5	93.3 %	14,15,19,64,75

Figure 11.10 Results for the 75 circuits with random defects (only the experiments with the *Sentry 20* included parametric and dynamic testings)[12].

Test method	Number of passed circuits	Logic fault coverage	Passed circuits
Minimal	24	65.2 %	Not specified
Tailored	5	92.7 %	10,20,41,50,67
Systematic	4	94.2 %	10,20,41,67
Manufacturer	1	98.5 %	10
Random	0	100 %	

(a)

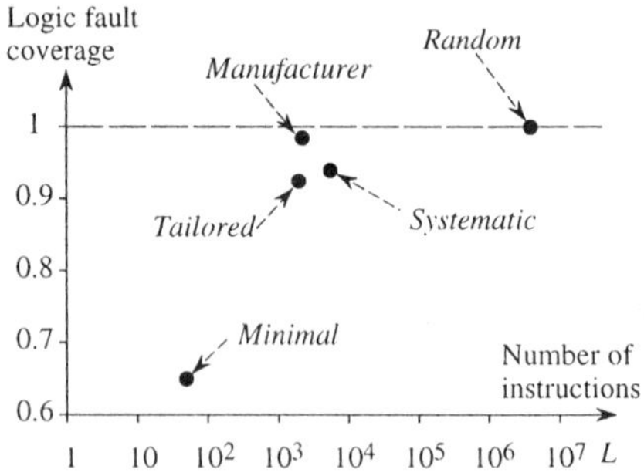

(b)

Figure 11.11 Test experiments for the 69 circuits with a logic fault[13]. (a) Results. (b) Fault coverage versus test lengths.

The *Tailored* and *Systematic* tests were implemented on several testers (implying some variants). The results concerning the 75 circuits with random

---

[12] This figure is reprinted from [VeBeMa 90] (© 1990 IEEE).
[13] This figure is reprinted from [VeBeMa 90] (© 1990 IEEE).

defects (as explained at the beginning of this section) are presented in Figure 11.10. In this set of experiments, the *Manufacturer* test, by the *Sentry 20* tester, performed logic, parametric, and dynamic testings. The others all performed only logic testing. It was found that 69 circuits were affected by a logic fault, five circuits were affected by a dynamic or parametric fault (numbers 14, 15, 19, 32, 75). No defect was detected, by any test, in the circuit number 64. Is the defect redundant, i.e., undetectable?

The results for the 69 circuits affected by logic defects are presented in Figure 11.11. A 100% logic defect coverage was obtained only by the random test. This test is longer as is illustrated in Figure 11.11b.

11.4.4 Experiments by D.A. Wood *et al.*: multiprocessor cache controller

The experiments in this section are reported in [WoGiKa 90]. They relate to the cache controller for SPUR, short for symbolic processing using RISC (reduced-instruction-set computer), a multiprocessor workstation developed at the University of California Berkeley. The strategy was to develop a random tester that would generate and verify the complex interactions between multiple processors in functional simulation. Replacing the CPU (central processing unit) model, the tester generated memory references by randomly selecting from a script of actions and checks.

Coherency protocols for write-back caches is notoriously difficult to design and debug. Testing these protocols by an extensive simulation can significantly reduce the time to market because it is far easier to discover and repair errors during simulation than in hardware (note that this simulation to *uncover errors in the design* is different from a simulation to test ability of a sequence to test injected faults). According to the authors: "Specifying the input to a simulation, which is usually test vectors of some high-level format, ranges from time-consuming to impossible. The complexity of cache controller designs makes it difficult to develop a comprehensive set of design verification tests [...] Coherency protocols are very sensitive to specific sequencing activity [...] Among the myriad sequence-specific events, we can easily miss important test cases."[14]

Before designing the random tester, the authors spent several person-months writing design-verification tests and diagnoses. However, the set of tests covered none of the multiprocessor cases (only a part of the uniprocessor case). On the other hand, *the total design effort time for the random tester was about two person-months* (an initial version requiring only one week to develop from scratch uncovered numerous errors).

The generation is such that all the multiprocessor interaction cases have a non-zero probability of occuring during the simulation (coherent with Section 5.2).

---

[14] Reprinted from [WoGiKa 90] (© 1990 IEEE).

The detailed functional simulation of the SPUR cache controller was based on the use of Endot's *N.2* system [RoOrPa 83]. Between 50 million and 100 million cycles were simulated. From the detailed records on the bugs discovered during simulation, the bugs were classified by the following types:

1) *Functional*: produced incorrect results.
2) *Performance*: produced correct results but slowly.
3) *N.2 artifact*: simulation artifact.
4) *Test*: errors in the test structure.

	Types of bugs				
	Functional	Performance	N.2 Artifact	Test	Total (%)
Random tester	25	2	2	9	38 (55%)
Inspection	13	2	2	0	17 (25%)
Other	9	1	4	0	14 (20%)
Total (%)	47 (68%)	5 (7%)	8 (12%)	9 (13%)	69 (100%)

Figure 11.12 Breakdown of functional simulation bugs[15].

Figure 11.12 summarizes the bug reports. According to the authors: "The random tester uncovered over half the functional bugs in the design. More important, it detected a number of complex sequencing and interaction problems that we believe would not otherwise have been found until after fabrication."[16]

After simulation, the multiprocessor cache controller was manufactured (enhancements to the *N.2* system allow automatic generation of much of the logic control [Ko *et al.* 87]), and tested. An assembly language version of the random tester was used. A few minor design errors were discovered during testing of hardware. They were related to electrical and timing problems; these errors were not picked up by random testing of the simulation since electrical information is not represented in functional simulation.

NOTES and REFERENCES

The random tester presented in [Te 74] was able to perform synchronous and asynchronous tests, and to apply test sequences in an input language I which is a proper subset of X^*. To the best of our knowledge, this machine built in 1970-73

---

[15] This figure is reprinted from [WoGiKa 90] (© 1990 IEEE).
[16] Reprinted from [WoGiKa 90] (© 1990 IEEE).

was the first random tester. The results in Section 11.1 have been obtained with it.

The results in Sections 11.1 [DaFoTe 75], 11.2 [DeThBe 84], and 11.3 [FéDa 86] were obtained by the author's team for the random testing, and by various French Companies for the deterministic testing.

The experiments in Sections 11.4.1 [Lu 76], 11.4.2 [LaGéHe 89], 11.4.3 [VeBeMa 90], and 11.4.4 [WoGiKa 90] were obtained by various other teams.

All these results show experimentally that the random test is longer than the deterministic test length but that random testing requires a shorter development time and provides us with a greater logic defect coverage.

12

Signature Analysis

Signature analysis consists of mapping the response of a circuit under test into a compact pattern called signature. There is some loss of information and a faulty response may correspond to a fault-free signature; this phenomenon is called aliasing.

General features of signature analysis are presented in Section 12.1. For a single output circuit, a single input signature analyser is used; the properties of this kind of signature analysers are studied in Section 12.2. Section 12.3 is devoted to multiple input signature analysers, which are used for multiple output circuits.

12.1 GENERAL FEATURES

Let us assume a single output combinational circuit, whose most resistant fault has a detection probability $p_{min} = 0.0002$, and that the test length is based on this fault (Section 6.1.2.2). For a level of confidence 0.999, according to Equation (6.16) (Section 6.1.1.2), the test length is $L(0.001) \approx 7/p_{min} = 7/0.0002 = 35\,000$. If this fault is present in the circuit under test, the expected number of detections is 7; in other words, in the 35 000-bit response, it is expected that 7 bits are faulty and 34 993 are fault-free. Then, the average "information" by bit, concerning the presence of the fault, is very small. Note that the proportion of faulty bits in the response may be still smaller for a multiple output circuit. For example, assume that the preceding circuit is a 10-output circuit instead of a single output one, and that only one output is faulty when there is a detection. Then, the response contains $10 \times 35\,000 = 350\,000$ bits out of which 7 are expected to be faulty! It is clearly interesting to obtain a compact function of the response, containing far fewer bits and virtually without loss of information, i.e., without *aliasing*.

299

In the sequel, first the notions of aliasing and non-revelation are defined in Section 12.1.1. A general property of signature analysis is given in Section 12.1.2, and the signature analyser needs are presented in Section 12.1.3.

12.1.1 Aliasing and non-revelation

The principle of signature analysis is illustrated in Figure 12.1a. When a test sequence S is applied to the circuit under test, a response R is obtained. This response depends both on S and on the fault f_i in the circuit. Note that this is true both for deterministic testing and random testing. For a m-output circuit and a test sequence of length L, the response R contains[1] $m \times L$ bits; signature analysis consists of mapping R into a k-bit signature denoted by A, which may be specified as $A(R)$.

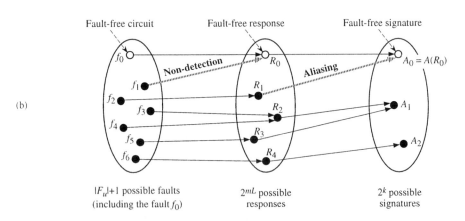

Figure 12.1 Signature analysis. (a) General principle. (b) Illustration of aliasing.

The concept of **aliasing** is illustrated in Figure 12.1b. A fault-free circuit, corresponding to the fault f_0 (Section 4.4), produces a fault-free response R_0, and

---

[1] To simplify presentation, it is assumed that each output takes its value from the set $\{0, 1\}$. An output which may have a high impedance state can be coded with two Boolean values (Section 10.3.4.1).

the signature A_0. Assume a circuit is faulty, i.e., contains a fault in F_u (the universal fault set, Section 4.2). This circuit produces a response which may be either fault-free ($R = R_0$ because the test sequence does not detect the fault under consideration) or faulty ($R \neq R_0$). If the response is fault-free, the signature is obviously fault-free. If the response is faulty, either the signature is faulty (for example $A(R_2) = A_1$ in Figure 12.1b) or the signature is fault-free (for example $A(R_1) = A_0$ in Figure 12.1b).

This last situation is called aliasing (i.e., *the test sequence is able to detect the fault* but the *signature is fault-free*).

Let us denote by AL the probability of aliasing. The explanation above leads to the following definition.

Definition 12.1 The **probability of aliasing** is

$$AL = \Pr[\text{aliasing}] = \Pr[A = A_0 \mid R \neq R_0]. \tag{12.1}$$

❏

From this definition, the following property is obtained.

Property 12.1 The *probability of aliasing* is equal to

$$AL = \frac{\Pr[A = A_0] - \Pr[R = R_0]}{\Pr[R \neq R_0]}. \tag{12.2}$$

Proof By definition,

$$AL = \Pr[A = A_0 \mid R \neq R_0] = \frac{\Pr[A = A_0 \text{ AND } R \neq R_0]}{\Pr[R \neq R_0]}. \tag{12.3}$$

$\Pr[A = A_0]$ may be developped as:

$$\Pr[A = A_0] = \Pr[A = A_0 \text{ AND } R = R_0] + \Pr[A = A_0 \text{ AND } R \neq R_0]. \tag{12.4}$$

Since $R = R_0$ implies $A = A_0$, (12.4) may be rewritten as:

$$\Pr[A = A_0] = \Pr[R = R_0] + \Pr[A = A_0 \text{ AND } R \neq R_0]. \tag{12.5}$$

From (12.5),

$$\Pr[A = A_0 \text{ AND } R \neq R_0] = \Pr[A = A_0] - \Pr[R = R_0] \tag{12.6}$$

is obtained. Hence, (12.2) is obtained from (12.3) and (12.6).

❏

Remark 12.1 The word *aliasing* was introduced in [Wi et al. 86 & 88]. In these papers, the probability of aliasing is calculated by (using our notation and a * to distinguish this value from the one previously defined):

$$AL^* = \Pr[A = A_0] - \Pr[R = R_0]. \tag{12.7}$$

Although this expression (which has been used by many people) is not exact, the value obtained from (12.7) is close to the result obtained from (12.2) when $\Pr[R \neq R_0]$ tends to 1 (i.e., for long test sequences).

❏

In order to *avoid confusion*, let us use two different words for saying that the response is faulty and that the signature is faulty.

Definition 12.2 A fault is **detected** (or *tested*) if the *response is faulty*, i.e., $R \neq R_0$ (this is consistent with the previous chapters).

A fault is **revealed** if the *signature is faulty*, i.e., $A \neq A_0$ (there is **revelation**).

Accordingly, there is aliasing when a fault is detected but not revealed.

❏

According to (12.2), the probability of non-revelation is obtained from the probability of testing and the probability of aliasing as expressed in the following property.

Property 12.2 The **probability of non-revelation** is equal to

$$\Pr[A = A_0] = \Pr[R = R_0] + \Pr[R \neq R_0] \cdot AL. \qquad (12.8)$$

12.1.2 General property

This section shows that if r different signatures are possible, the minimum probability of non-revelation is obtained if and only if all the signatures are equally likely.

Definition 12.3 Let C be a logic circuit, and f and g be two faults in $F_u \cup \{f_0\}$. If an input sequence S is applied to the circuit affected by the fault f, denoted by C_f, the signature $A(f)$ is observed. The **distinction potential** is the probability that $A(f) \neq A(g)$.

❏

Clearly, the distinction potential is a function of the faults f and g, of the input sequence S, and of the signature analyser. If each of these parameters is known, then the distinction potential is either 0 or 1. If at least one of these parameters is unknown, then the distinction potential gives an *a priori measure of the performance of the test device*. Let us illustrate this idea by Figure 12.2. It is assumed that *the signature analyzer is fixed* while the test sequence and the faults are not fixed.

The scheme in Figure 12.2a shows that, when the same input sequence S is applied to both circuits C_f and C_g, the signatures $A(f)$ and $A(g)$ are obtained. Figure 12.2b illustrates that a sequence S_a may lead to two different signatures: $A(f) = A_1 \neq A(g) = A_3$, while S_b may lead to the same signature:

$A(f) = A(g) = A_2$. Hence, the distinction potential is the probability that the sequence S behaves like S_a.

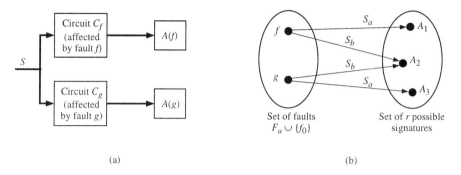

Figure 12.2 Illustration of distinction potential. (a) A random input sequence leads to signatures depending on the faults. (b) Two different faults may have either different signatures (for the test sequence S_a) or the same signature (for the test sequence S_b).

Let $\mathcal{A} = \{A_0, A_1, \ldots A_{r-1}\}$ denote the set of r possible signatures. The indices i of A_i are presently arbitrary (when the sequence S is fixed, it is convenient to assign index 0 to the fault-free signature, i.e., $A_0 = A(f_0)$; but presently S is not fixed and the index 0 is like any other one).

Notation 12.1 Assume that the sequence S is in a set $\mathcal{S}$ of input sequences, and that there is a probability $\Pr[S = S_i]$ for each S_i in $\mathcal{S}$ ($\sum_i \Pr[S = S_i] = 1$). This implies that, given a fault f, each signature has a conditional probability to be obtained: $\Pr[A_j | f]$, $j = 0, \ldots, r-1$, and $f \in F_u \cup \{f_0\}$ (since, given S_i and f, the signature is unique).

Let $p_j^f = \Pr[A_j | f] = \Pr[A(f) = A_j]$.

Let $\mathcal{A}_v$ denote a subset of v signatures in $\mathcal{A}$ (i.e., $\mathcal{A}_r = \mathcal{A}$).

Let $D(\mathcal{A}_v, f, g) = \Pr[A(f) \neq A(g) | A(f), A(g) \in \mathcal{A}_v]$.

It appears that, if $\mathcal{A}_v = \mathcal{A}$, $D(\mathcal{A}_v, f, g) = D(\mathcal{A}, f, g)$ and this is the distinction potential.

Property 12.3 Given

$$\sum_{j:A_j \in \mathcal{A}_v} p_j^f = \alpha \text{ and } \sum_{j:A_j \in \mathcal{A}_v} p_j^g = \beta, \alpha \text{ and } \beta \leq 1,$$

$D(\mathcal{A}_v, f, g)$ is maximum if and only if

$$p_j^f = \frac{\alpha}{v} \text{ and } p_j^g = \frac{\beta}{v} \text{ for all } A_j \text{ in } \mathcal{A}_v.$$

Note that $\alpha = \beta = 1$ when $\mathcal{A}_v = \mathcal{A}$.

Proof See Section G.3 in Appendix G.

Property 12.4 The *probability of non-revelation is minimal* if and only if $p_j^f = \frac{1}{r}$ for all signatures A_j in $\mathcal{A}$, and all faults f in $F_u \cup \{f_0\}$, i.e., *if and only if all the signatures are equally likely for all the faults*.

Proof This is a direct consequence of Property 12.3: $D(\mathcal{A}, f, g)$ is maximal if $p_j^f = p_j^g = 1/r$. The revelation corresponds to the case where $f = f_0$ (fault-free circuit) and g is any fault in F_u.
❏

In the sequel, several useful consequences will be drawn from Property 12.4.

12.1.3 Choice of k

The goal of signature analysis is to compare a few bits (the k bits of the signature) instead of the $m \times L$ bits of the whole response (Figure 12.1a). For an efficient signature analysis, the probability of aliasing should be low. But, what does "low" mean?

If the test sequence is convenient, it is expected that $\Pr[R \neq R_0] \approx 1$ for a faulty circuit. Hence, from (12.8)

$$\Pr[A = A_0] \approx \Pr[R = R_0] + AL \tag{12.9}$$

is obtained, which means that the probability that a defective circuit passes is the sum of 1) the probability that the fault is not detected by the test sequence, and 2) the probability of aliasing.

Remark 12.2 If $\Pr[R \neq R_0] \approx 1$, then $\Pr[R = R_0] \approx 0$; however, (12.9) cannot be reduced to $\Pr[A = A_0] \approx AL$ because $AL \approx 0$ too. In fact, as will be shown in the sequel, AL should be smaller than $\Pr[R = R_0]$ for an efficient signature analysis.
❏

The sum (12.9) applies to a fault. It may also be applied to a set of faults. If it is applied to the *set of faults* considered for the circuit, then $\Pr[R = R_0]$ is the probability of *fault uncoverage* (complement of the *fault coverage* which corresponds to $\Pr[R \neq R_0]$). Hence, from (12.9) we may conclude that, in order

Signature Analysis 305

not to penalize the test performance, the probability of aliasing should be small in comparison with the fault uncoverage[2].

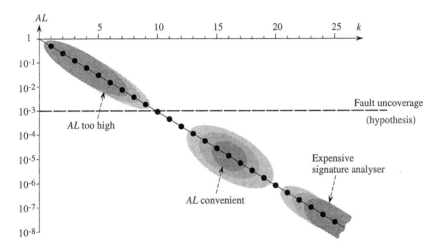

Figure 12.3 Dimensioning of the signature analyser.

As will be seen in the following sections, the probability of aliasing is roughly $AL \approx 1/2^k$ for a "good" signature analyser made up of k bits (for all faults). The choice of the value k is illustrated in Figure 12.3. Assume a test sequence such that the fault coverage is 0.999, i.e., $\Pr[R = R_0] = 0.001$ for a faulty circuit under test (deterministic or random test sequence). For this example, (12.9) may be rewritten as

$$\Pr[A = A_0] \approx 0.001 + 1/2^k.$$

Let us consider $\Pr[A = A_0]$ for several values of k.

If k is too small, the probability of aliasing may be greater than the fault uncoverage. For example, for $k = 6$,

$$\Pr[A = A_0] \approx 0.001 + 0.016 = 0.017.$$

For a convenient value of k, for example $k = 16$, the probability of aliasing does not penalize the test performance:

$$\Pr[A = A_0] \approx 0.001 + 0.000015 = 0.001015.$$

For a greater value of k, for example $k = 23$, the test performance is obviously not penalized, since

---

[2] A similar reasoning may be applied for another measurement such as the *faulty circuit coverage* (Section 4.4.1).

$\Pr[A = A_0] \approx 0.001 + 0.00000012 = 0.00100012$,

but the signature analyser would be more expensive than necessary (the price, roughly proportional to k will be discussed in Section 12.2.4).

The values of k which are discussed above are based on an hypothesis of a 0.999 fault coverage.

If the fault coverage were different, for example $1 - 10^{-6}$, the convenient values of k would be greater (see Exercise 12.1).

12.2 SINGLE INPUT SIGNATURE ANALYSERS

There are two main ways to obtain a signature: counting methods and use of linear feedback shift registers. The counting methods, which were basically developed for deterministic testing are presented in Section 12.2.1. The principle of signature analysis based on an LFSR is presented in Section 12.2.2, and the main properties are developed in Section 12.2.3.

12.2.1 Counting methods

In this section, the counting methods are briefly explained, and bibliographic notes are given in "Notes and References" at the end of the chapter. These methods were basically developed for deterministic testing (mainly for combinational circuits): the test vectors are ordered in such a way that aliasing is avoided for many faults. An important feature of these deterministic methods is that *the test sequence* AND the *signature* are studied at the same time. It is practically impossible to use them for large circuits.

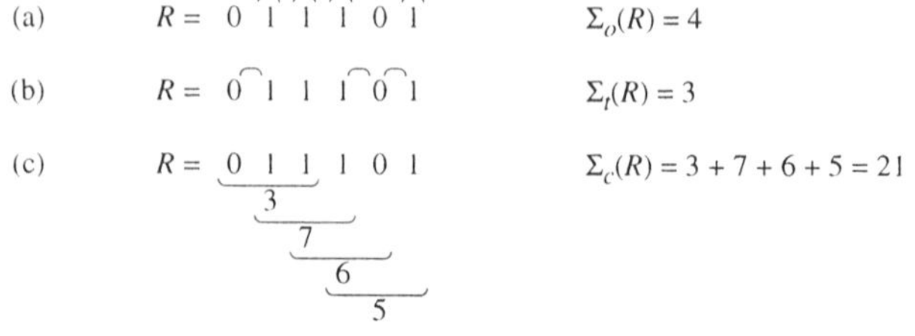

(a) $R = 0\ \widehat{1}\ \widehat{1}\ \widehat{1}\ 0\ \widehat{1}$ $\Sigma_o(R) = 4$

(b) $R = 0\ \widehat{1\ 1}\ 1\ \widehat{1\ 0}\ 1$ $\Sigma_t(R) = 3$

(c) $R = 0\ \underbrace{1\ 1}_{3}\ 1\ 0\ 1$ $\Sigma_c(R) = 3 + 7 + 6 + 5 = 21$
$\underbrace{}_{7}$
$\underbrace{}_{6}$
$\underbrace{}_{5}$

Figure 12.4 Countings. (a) Counting of ones. (b) Countings of transitions. (c) An example of check-sum.

Counting methods are illustrated in Figure 12.4: the test length is $L = 6$ and the response of the circuit under test is $R = 011101$.

Counting of ones is illustrated in Figure 12.4a. There are four 1 in R; this is denoted by $\Sigma_o(R) = 4$.

In Figure 12.4b, it is shown that a transition is counted each time that $R(l + 1) \neq R(l)$, $1 \leq l < L$. The sum $\Sigma_t(R) = 3$ is obtained.

A set of testing methods called check-sum tests is presented in [Ha 76a]. The principle is as follows. A string of u bits is assigned a number (non-negative and integer). Numbers are associated with the strings $R(1)...R(u)$, $R(2)...R(u + 1)$ etc, and the check-sum is the sum of these numbers. This is illustrated in Figure 12.4c where $u = 3$ and the number associated with a string is the decimal number corresponding to the binary coding. For example $R(1)R(2)R(3) = 011$ corresponds to the number 3, $R(2)R(3)R(4) = 111$ corresponds to 7 and so on. The check-sum is $\Sigma_c(R) = 3 + 7 + 6 + 5 = 21$. One can use different values of u and different methods to associate a number with the string. Counting of 1 and counting of transitions appear to be particular cases of check-sum tests.

A conclusion in [Ha 76a] is that counting of 1 is better than counting of transitions, particularly because it does not depend on the order of test vectors in the test sequence S (only combinational circuits were concerned).

If the test sequence S is pseudorandom, the counting methods can be used. However, *these methods do not have good performances because they do not provide equally likely signatures* (Property 12.4 in Section 12.1.2) as explained in the sequel.

Assume the signature is the counting of 1 and a random test whose length is $L = 6$ is applied to the circuit under test. The response $R = 011101$ has $\Sigma_o(R) = 4$ as signature. However, there are many other responses with the same signature, for example $R = 111100$. In fact there are

$$\binom{6}{4} = \frac{6!}{2!\,4!} = 15$$

responses with this signature (by symmetry, there are $\binom{6}{2} = \frac{6!}{4!\,2!} = 15$ responses with the signature $\Sigma_o(R) = 2$). On the other hand, there are only 6 responses providing the signature $\Sigma_o = 1$ (or $\Sigma_o = 5$), and only one response with the signature $\Sigma_o = 0$ (or $\Sigma_o = 6$).

Let us assume that the circuit under test is combinational and the detection probability of the fault is p. If $p = 0.5$, all the responses have the same probability i.e., $1/2^6$; $\Pr[A = 0] = \Pr[A = 6] = 1/2^6 = 0.016$, $\Pr[A = 1] = \Pr[A = 5] = 6/2^6 = 0.094$, $\Pr[A = 2] = \Pr[A = 4] = 15/2^6 = 0.23$, and $\Pr[A = 3] = 20/2^6 = 0.31$. The signatures have very different probabilities, i.e., the condition in Property 12.4 is not satisfied.

12.2.2 Signature by linear feedback shift register

The behavior of an LFSR was presented in Section 10.3. Let us recall that the feedback is characterized by a *polynomial* and that a *companion matrix* is associated with the polynomial. This matrix allows calculation of the state of the register at time l from the state at time $l - 1$.

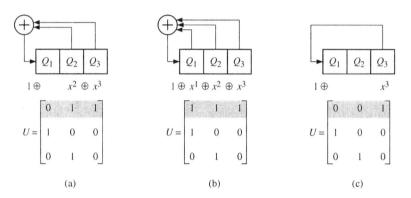

Figure 12.5 Three 3-stage LFSR with their characteristic polynomials and companion matrices.

This is illustrated in Figure 12.5 with three examples of 3-stage LFSR. In Figure 12.5a, the feedback is the sum modulo 2 of Q_2 and Q_3; the corresponding poynomial is $1 \oplus x^2 \oplus x^3$. The first row of the corresponding matrix U corresponds to this feedback, i.e., $U_{1,2} = U_{1,3} = 1$, while $U_{1,1} = 0$ because the term x^1 is not in the polynomial. The other lines of the matrix correspond to the shifting. The next state is obtained from the present state by

$$\begin{bmatrix} Q_1(l) \\ Q_2(l) \\ Q_3(l) \end{bmatrix} = \begin{bmatrix} 0 & 1 & 1 \\ 1 & 0 & 0 \\ 0 & 1 & 0 \end{bmatrix} \cdot \begin{bmatrix} Q_1(l-1) \\ Q_2(l-1) \\ Q_3(l-1) \end{bmatrix}, \text{ for } l > 0, \quad (12.10)$$

where the product is the usual Boolean product and the sum is modulo 2. Equation (12.10) may be written

$$Q(l) = U \cdot Q(l-1), \quad l > 0, \quad (12.11)$$

from which

$$Q(l) = U^l \cdot Q(0), \quad l \geq 0, \quad (12.12)$$

is obtained. Equation (12.11) and (12.12) are true for any LFSR (for example those in Figure 12.5b and c) with the corresponding U matrices.

Figure 12.6 Principle of signature analysis by LFSR.

The principle of signature analysis by LFSR is illustrated in Figure 12.6 for a single output circuit under test. The register is made up of k stages and the kth *stage is always fed back* (otherwise the polynomial would be of a lesser degree than k). Its behavior is modified by the input R, the response of the CUT. The state of the register after the test sequence of length L has been applied, i.e., $(Q_1(L), Q_2(L), ..., Q_k(L))$ is the signature of the circuit.

In the signature analyser of Figure 12.6, the bit entering the register is modified if $R(l) = 1$. For example, if the register in Figure 12.5a is used, equation (12.10) becomes:

$$\begin{bmatrix} Q_1(l) \\ Q_2(l) \\ Q_3(l) \end{bmatrix} = \begin{bmatrix} 0 & 1 & 1 \\ 1 & 0 & 0 \\ 0 & 1 & 0 \end{bmatrix} \cdot \begin{bmatrix} Q_1(l-1) \\ Q_2(l-1) \\ Q_3(l-1) \end{bmatrix} \oplus \begin{bmatrix} R(l) \\ 0 \\ 0 \end{bmatrix}, \quad l > 0. \tag{12.13}$$

As will be shown, the signature can be obtained from the **error sequence** E which is the bitwise EXOR of the response R and the faulty response R_0, i.e.,

$$E = R \oplus R_0, \qquad E(l) = R(l) \oplus R_0(l); \tag{12.14}$$

the error bit $E(l) = 1$ if and only if the output bit $R(l)$ is faulty.

Because of the linearity of the device presented in Figure 12.6, i.e., the feedback is linear and the input to the LFSR is made through an EXOR, the following property is obtained.

Property 12.5 The signature $A(R \oplus R_0)$, given $Q(0)$, contains the same information as $A(R)$ given $A(R_0)$. In other words, the schemes in Figures 12.7a and 12.7b (were $Q(0) = (0, ..., 0)$ may be used) are equivalent as far as revelation of a fault is concerned.

❑

Let us first introduce some notations useful for the proof.

Notation 12.2 (See Figure 12.7) The symbol $R_v(l)$ denotes the column vector of size k such that the first term is $R(l)$ and the others are 0 (and similarly for $R_{0v}(l)$ and $E_v(l)$), i.e.,

$$R_v(l) = \begin{bmatrix} R(l) \\ 0 \\ \cdot \\ \cdot \\ \cdot \\ 0 \end{bmatrix}, \quad R_{0v}(l) = \begin{bmatrix} R_0(l) \\ 0 \\ \cdot \\ \cdot \\ \cdot \\ 0 \end{bmatrix}, \quad E_v(l) = R_v(l) \oplus R_{0v}(l) = \begin{bmatrix} R(l) \oplus R_0(l) \\ 0 \\ \cdot \\ \cdot \\ \cdot \\ 0 \end{bmatrix} \quad (12.15)$$

□

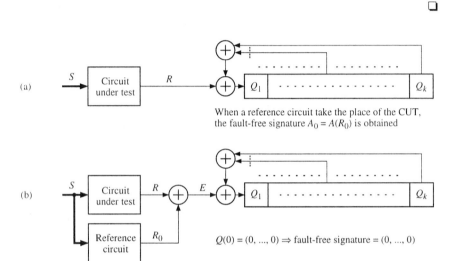

Figure 12.7 Two equivalent schemes.

Proof of Property 12.5 Using Notation 12.2, Equation (12.13) may be generalized by

$$Q(l) = U \cdot Q(l-1) \oplus R_v(l), \quad l > 0. \quad (12.16)$$

Let us develop this equation from the initial state:

$$Q(1) = U \cdot Q(0) \oplus R_v(1),$$

$$Q(2) = U \cdot [U \cdot Q(0) \oplus R_v(1)] \oplus R_v(2),$$

which can be rewritten as

$$Q(2) = U^2 \cdot Q(0) \oplus U \cdot R_v(1) \oplus R_v(2).$$

Then, in general[3]:

---

[3] Σ represents the sum modulo 2.

Signature Analysis 311

$$Q(l) = U^l \cdot Q(0) \oplus \sum_{i=1}^{l} U^{l-i} \cdot R_v(i). \qquad (12.17)$$

According to (12.15), $R_v(i) = R_{0v}(i) \oplus E_v(i)$. Hence, (12.17) may be written as

$$Q(l) = U^l \cdot Q(0) \oplus \sum_{i=1}^{l} U^{l-i} \cdot (R_{0v}(i) \oplus E_v(i)), \text{ or}$$

$$Q(l) = \left[U^l \cdot Q(0) \oplus \sum_{i=1}^{l} U^{l-i} \cdot R_{0v}(i) \right] \oplus \left[\sum_{i=1}^{l} U^{l-i} \cdot E_v(i) \right]. \qquad (12.18)$$

When the complete test sequence has been applied, i.e., at $l = L$, by definition:

$$Q(L) = A(R) \qquad (12.19)$$

is the signature (state of the register in Figure 12.7a), and

$$U^L \cdot Q(0) \oplus \sum_{i=1}^{L} U^{L-i} \cdot R_{0v}(i) = A(R_0) \qquad (12.20)$$

is the fault-free signature according to (12.17). From (12.18) where $l = L$, (12.19), and (12.20):

$$\sum_{i=1}^{L} U^{L-i} \cdot E_v(i) = A(R) \oplus A(R_0). \qquad (12.21)$$

This expression also corresponds to the content of the register in Figure 12.7b, i.e., $A(R \oplus R_0)$ given the initial state of this register is $(0, ..., 0)$.

❏

Since Figures 12.7a and b give the same information about revelation of faults, in the sequel the *calculations will be made assuming the scheme in* Figure 12.7b: the signature A is obtained from the *error vector E* and the initial state is $(0, ..., 0)$.

$$R_0 = 0\ 0\ 0\ 0\ 1\ 1\ 0\ 0\ 1\ 1\ 0\ 0\ 0\ 1$$

$$R_1 = \mathbf{1}\ 0\ 0\ \mathbf{1}\ 1\ 1\ 0\ \mathbf{1}\ 0\ 1\ 0\ 0\ \mathbf{1}\ 1$$
$$E_1 = 1\ 0\ 0\ 1\ 0\ 0\ 0\ 1\ 1\ 0\ 0\ 0\ 1\ 0$$
$$1\ \oplus\ x^3\ \oplus\ x^7 \oplus x^8\ \oplus\ x^{12}$$

$$R_2 = 0\ 0\ 0\ 0\ 1\ \mathbf{0}\ 0\ \mathbf{1}\ 0\ \mathbf{1}\ 0\ 0\ 0\ 1$$
$$E_2 = 0\ 0\ 0\ 0\ 0\ 1\ 0\ 1\ 1\ 0\ 0\ 0\ 0\ 0$$
$$x^5 \oplus x^7 \oplus x^8$$

Figure 12.8 Example of responses, with the corresponding error vectors and polynomials.

Two examples of faulty responses are presented in Figure 12.8, R_1 and R_2 (R_0 is the fault-free response). The error vector $E_1 = R_1 \oplus R_0$ (bitwise sum modulo 2) is associated with the faulty response R_1. The polynomial associated with E_1 is $1 \oplus x^3 \oplus x^7 \oplus x^8 \oplus x^{12}$ since the 1st, 4th, 8th, 9th, and 13th bits of R_1 are faulty. Similarly $E_2 = R_2 \oplus R_0$ and the corresponding polynomial is $x^5 \oplus x^7 \oplus x^8$.

According to Section 10.3.1, a polynomial may be divided by another. The signature in Figure 12.7b is different from (0, ...,0) if and only if the division of the polynomial associated with E by the polynomial associated with the LFSR is not exact (i.e., the rest is not 0); the formal proof is obtained from the theory of cyclic codes [Pe 61]. Assume, for example, that the polynomial of the LFSR is $1 \oplus x^2 \oplus x^3$ (LFSR in Figure 12.5a). For the error vectors in Figure 12.8, the following results are obtained:

$$1 \oplus x^3 \oplus x^7 \oplus x^8 \oplus x^{12} = (1 \oplus x^2 \oplus x^3)(1 \oplus x^7 \oplus x^8 \oplus x^9) \oplus x^2;$$

$$x^5 \oplus x^7 \oplus x^8 = (1 \oplus x^2 \oplus x^3) x^5.$$

Hence, there is no aliasing for E_1 since the remainder of the division[4] is $x^2 \neq 0$ while there is aliasing for E_2 since the division of $x^5 \oplus x^7 \oplus x^8$ by $1 \oplus x^2 \oplus x^3$ is exact.

12.2.3 Properties of SISR

This section is related to single input signature analysers based on LFSR, usually called **SISR**, short for *single input signature registers.*

The performance analysis of an SISR is based on the following hypothesis.

Hypothesis 12.1 Each bit in the response is faulty with a probability p, i.e.,

$$\Pr[E(l) = 1] = p, \quad 0 < p < 1, \quad l > 0. \tag{12.22}$$

and $\Pr[E(l) = 1]$ is independent from the presence of faulty bits before time l.

❑

This hypothesis exactly corresponds to the response of a single output circuit under test affected by a combinational fault whose detection probability is p, when the test sequence is random.

It is an approximation for a sequential circuit because of the memory effect (Section 5.4).

Figure 12.9a presents a 3-stage SISR (i.e., $k = 3$) whose polynomial is $1 \oplus x^2 \oplus x^3$. Its behavior can be modeled[5] by the Markov chain in

---

[4] The polynomial corresponding to the signature is not the remainder of the division but there is a one-to-one correspondence between them [Ya 90]. There is another way to built an LFSR producing the same polynomial: the value Q_k is an input of EXORs placed between two successives cells [Sm 80], [BaMcSa 87], [Ya 90]. For this kind of LFSRs, not usually implemented, the polynomial of the signature is the remainder.

[5] This example is taken from [Wi et al. 88].

Figure 12.9b. The black arcs correspond to the autonomous behavior of the LFSR, i.e., if the input E does not exist or if $E(l) = 0$ for all l; since the polynomial of the LFSR is primitive (Section 10.3.1) there is a cycle of maximum length, i.e., $2^3 - 1 = 7$ (states **2** to **8**) and a cycle of length one (state **1**). The grey arcs correspond to the "disturbance" provoked by error bits $E(l) = 1$. For example, assume the state of the register is $Q(l-1) = (0, 1, 0)$ denoted by **3** in Figure 12.9b; if $E(l) = 0$, $Q(l) = (1, 0, 1)$ is reached since the 1st and 2nd bits are shifted, while the entering bit is $Q_2(l-1) \oplus Q_3(l-1) \oplus 0 = 1$; if $E(l) = 1$, $Q(l) = (0, 0, 1)$ is reached since the entering bit is $Q_2(l-1) \oplus Q_3(l-1) \oplus 1 = 0$. Hence, the probabilities $1-p$ and p are associated with arcs $\mathbf{3} \rightarrow \mathbf{4}$ and $\mathbf{3} \rightarrow \mathbf{8}$ respectively. The corresponding transition matrix, V, is given in Figure 12.9c (in general V is a $2^k \times 2^k$ stochastic matrix).

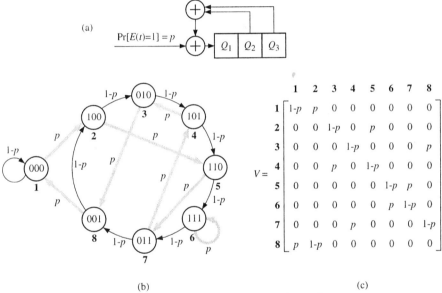

Figure 12.9 (a) SISR whose polynomial is $1 \oplus x^2 \oplus x^3$. (b) Markov chain corresponding to its behavior[6]. (c) Corresponding transition matrix.

Before giving an important property, let us observe that any state of the signature register can be reached from any other state[7] in k steps; in other words the transition matrix V is *irreducible* (Appendix C). Consider for example the SISR in Figure 12.9a. Assume the present state is $Q(l)$ and the state $Q(l+3) = \alpha\beta\gamma$ is sought after. The value $Q_3(l+3) = \gamma$ is the value of $Q_1(l+1)$, obtained after two shiftings. Since $Q_1(l+1) = Q_2(l) \oplus Q_3(l) \oplus E(l+1) = \gamma$ is

---
[6] Figure 12.9b is reprinted from [Wi *et al*. 88] (© 1988 IEEE).
[7] This is formally shown in [Wi *et al*. 88].

314   Chapter 12

wanted, it can be obtained if $E(l+1) = Q_2(l) \oplus Q_3(l) \oplus \gamma$. The values $E(l+2)$ and $E(l+3)$ can be obtained in a similar way. For example $Q(l+3) = 001$ can be reached from $Q(l) = 101$ if $E(l+1)E(l+2)E(l+3) = 010$ (see Figure 12.9b).

Property 12.6 For any SISR of length k, and for any value of p (different of 0 and 1), the probability to be in any of the 2^k states tends to $1/2^k$, i.e., for any state Q:

$$\lim_{l \to \infty} \Pr[Q(l) = Q] = \frac{1}{2^k}. \tag{12.23}$$

Proof We have already seen that the transition matrix is *irreducible*. Since all the states can be reached in exactly k steps, there is no cycle and the matrix V is also *regular* (Appendix C).

According to (C.2) in Appendix C, since V is irreducible and regular:

$$\lim_{l \to \infty} \pi(0) \cdot V^l = \pi(\infty), \tag{12.24}$$

where $\pi(\infty)$ is the limit probability vector (which is independent of $\pi(0)$). Let $\pi_i(\infty)$ be the limit probability of the state i, $i = 1, 2, \ldots 2^k$.

In the SISR, the $(k-1)$ first bits are shifted. Since the kth bit is fed back, each state of the SISR has two predecessors depending on $Q_k(l-1)$, one corresponding to $E(l) = 0$ and the other one to $E(l) = 1$. Hence, each column of V sums to 1 (the transition matrix is doubly stochastic). Because of the symmetry, $\pi_i(\infty)$ has the same value for all i (it is obviously a solution, and we know this solution is unique). Since the sum of the 2^k probabilities $\pi_i(\infty)$ is 1, then $\pi_i(\infty) = 1/2^k$ for each $i = 1, \ldots, 2^k$.

❏

Before drawing the consequences of Property 12.6, let us recall and specify some definitions and properties given in Section 12.1.1. Assume that the circuit under test is affected by a combinational fault whose detection probability is p. As in Figure 12.9b, the fault-free signature $(0, \ldots, 0)$ is given the state number **1** in the Markov chain. Let $\pi(l)$ be the 2^k-component row vector denoting the probability vector associated with the 2^k states of the SISR.

1) $\Pr[A = A_0]$ is the *probability of non-revelation* (Definition 12.2 in Section 12.1.1).

If the scheme in Figure 12.7b is considered, then

$$\Pr[A = A_0] = \Pr[Q(L) = Q(0)], \tag{12.25}$$

where $Q(0) = (0, \ldots, 0)$.

Let us denote $\eta = (1, 0, \ldots, 0)$ a 2^k-component row vector in which the first component is one and the others are 0. Since $Q(0) = (0, \ldots, 0)$, the initial probability vector is $\pi(0) = \eta$, and

$$\pi(l) = \eta \cdot V^l. \tag{12.26}$$

Since the fault-free signature corresponds to state **1**:

$$\Pr[Q(L) = Q(0)] = \pi_1(L) = \pi(L) \cdot \eta^T, \tag{12.27}$$

and from (12.25), (12.27), and (12.26),

$$\Pr[A = A_0] = \eta \cdot V^L \cdot \eta^T \tag{12.28}$$

is obtained[8].

2) $\Pr[R \neq R_0]$ is the *probability of testing* the fault (i.e., that the response is faulty). According to (6.8) in Section 6.1.1.2,

$$\Pr[R = R_0] = (1 - p)^L, \tag{12.29}$$

$$\Pr[R \neq R_0] = 1 - (1 - p)^L. \tag{12.30}$$

3) AL is the *probability of aliasing*. According to Property 12.1 in Section 12.1.1, and using (12.28), (12.29), and (12.30),

$$AL = \frac{\eta \cdot V^L \cdot \eta^T - (1-p)^L}{1 - (1-p)^L} \tag{12.31}$$

is obtained.

❏

Now let us consider two direct consequences of Property 12.6.

First: according to Property 12.6, all the signatures are asymptotically equally likely. Hence, according to Property 12.4 (Section 12.1.2), the *probability of non-revelation* is, asymptotically, the *minimum which can be obtained from any kind of k-bit signature analyzer* (under Hypothesis 12.1).

Second: since $\eta \cdot V^l \cdot \eta^T$ tends towards $1/2^k$ for large values of L, according to (12.31) the aliasing probability, AL, also tends towards $1/2^k$ because $(1-p)^L$ tends towards 0.

The main features we now want to exhibit are illustrated with examples. The SISR whose polynomial $1 \oplus x^2 \oplus x^3$ is taken as an example. Figure 12.10a illustrates $AL = \Pr[\text{aliasing}]$ (Equation (12.31)) as a function of the test length l for $p = 0.1, 0.2, 0.8,$ and 0.9. Two phenomena can be observed:

Observation 12.1 Let us compare the *probability of aliasing* for two complementary values of p, for example $p = 0.1$ and $p = 0.9$. For $p = 0.1$, AL increases apparently in a monoton way from the value 0 to the limit value $1/2^3 = 0.125$. For $p = 0.9$, AL is *greater* than for $p = 0.1$ (at least for the small values of l), and presents some peaks with a relatively great aliasing probability before converging to 0.125. A similar phenomena is observed by comparison of $p = 0.2$ and $p = 0.8$.

---

[8] The matrix X^T denotes the matrix transposed from X.

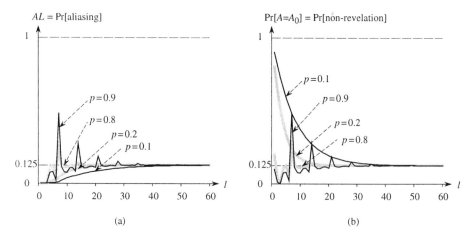

Figure 12.10 Convergence to the limit value for the polynomial $1 \oplus x^2 \oplus x^3$. (a) Pr[aliasing]. (b) Pr[non-revelation].

Observation 12.2 Let us compare the *probability of aliasing* for two values of p less than 0.5. The probability of aliasing for $p = 0.1$ is *less* than the probability of aliasing for $p = 0.2$.

❑

The *phenomenon of peaks* (observed for $p = 0.9$ and 0.8) was observed and studied by various researchers. From a theoretical point of view this is interesting, but is it an important phenomenon from a *practical point of view*?

According to (12.8) in Section 12.1.1,

$$\Pr[A = A_0] = \Pr[R = R_0] + \Pr[R \neq R_0] \cdot AL. \tag{12.32}$$

This means that the fault is not revealed ($\Pr[A = A_0]$, according to Definition 12.2 in Section 12.1.1) if

either *the fault is not detected* ($\Pr[R = R_0]$),

or *there is aliasing* ($\Pr[R \neq R_0]) \cdot AL$).

It is important to show that aliasing is *one of* the phenomena to take into account, but it is *not the only one*: we must take into account *both* aliasing and non-detection, the two components of non-revelation.

Figure 12.10b presents $\Pr[A = A_0]$ (Equation (12.28)) for the same examples. Two observations[9] (which will be compared to the preceding ones) can be made.

---

[9] Observations 12.3 and 12.4 correspond to properties due to the particular structure of the transition matrix V of the Markov chain.

Signature Analysis 317

Observation 12.3 Let us compare the *probabilities of non-revelation* for two complementary values of p. For $p = 0.1$, $\Pr[A = A_0]$ decreases in a monoton way from the value $1 - p = 0.9$ to the limit value 0.125. For $p = 0.9$, $\Pr[A = A_0]$ presents some peaks but is *smaller* (or equal) than for $p = 0.1$.

Observation 12.4 Let us compare the *probabilities of non-revelation* for two values of p less than 0.5. It is clear that $\Pr[A = A_0]$ is *greater* for $p = 0.1$ than for $p = 0.2$.

❏

Comparison of Observations 12.1 and 12.3 shows that if $a \leq 0.5$, the *probability of aliasing is greater* for $p = 1 - a$ than for $p = a$, while the *probability of non-revelation is smaller*[10] for $p = 1 - a$ than for $p = a$. Let us illustrate why with an example.

Consider the values for $l = 5$ in the examples in Figure 12.10 (obtained from (12.28) to (12.31)).

For $p = 0.1$ and $l = 5$: $\Pr[A = A_0] = 0.59220$, $\Pr[R = R_0] = 0.59049$, $\Pr[R \neq R_0] = 0.40951$, and $AL = 0.00418$. According to (12.32):

$$\mathbf{0.59220} = 0.59049 + 0.40951 \times \mathbf{0.00418}. \tag{12.33}$$

For $p = 0.9$ and $l = 5$: $\Pr[A = A_0] = 0.08020$, $\Pr[R = R_0] = 0.00001$, $\Pr[R \neq R_0] = 0.99999$, and $AL = 0.08019$. According to (12.32):

$$\mathbf{0.08020} = 0.00001 + 0.99999 \times \mathbf{0.08019}. \tag{12.34}$$

In (12.33), $AL = 0.00418$ is *low*. However, the *fault is difficult to detect* ($p = 0.1$) so that $\Pr[R = R_0] = 0.59049$. It follows that the probability of non-revelation, 0.59220, is close to $\Pr[R = R_0]$.

In (12.34), $AL = 0.08019$ is about *twenty times higher* than in the preceding case. However, *the fault is easy to detect* ($p = 0.9$) so that $\Pr[R = R_0] = 0.00001$. It follows that the probability of non-revelation, 0.08020, is close to AL (one can observe by comparing Figures 12.10a and b that AL and $\Pr[A = A_0]$ are practically similar for $p = 0.9$, except for very small values of l). It follows that the probability of non-revelation is less in (12.34) than in (12.33).

Conclusion 12.1 For $a < 0.5$, the *probability of non-revelation is greater* for $p = a$ than for $p = 1 - a$, although the *probability of aliasing is lower* (and the probability of non-revelation is the important measurement from a practical point of view).

❏

Now, for $p = 0.2$ and $l = 5$: $\Pr[A = A_0] = 0.33920$, $\Pr[R = R_0] = 0.32768$, $\Pr[R \neq R_0] = 0.67232$, and $AL = 0.01713$. According to (12.32):

$$\mathbf{0.33920} = 0.32768 + 0.67232 \times \mathbf{0.01713}. \tag{12.35}$$

---

[10] Several authors have mistaken one for the other concepts, including the author of this book in [Da 86] who specified in [Da 90].

318   *Chapter 12*

The comparison between (12.33) and (12.35) shows that the probability of *non-revelation* is lower in the second case ($p = 0.2$) than in the first one ($p = 0.1$).

Conclusion 12.2   For $a < b \leq 0.5$, the *probability of non-revelation is greater* for $p = a$ than for $p = b$, although the probability of aliasing is lower.
❑

Two features can usually be observed in combinational circuits. They are illustrated in Figure 12.11 for the arithmetic and logic unit (Section 6.3) and three circuits[11] of the ISCAS Benchmark [BrFu 85]. The first feature is that the number of faults whose detectability is greater than 0.5 is relatively small. The second feature is that, for a circuit, the detection probability of the most resistant fault, p_{min}, is closer to 0 than the detection probability of the easiest fault, p_{max}, is close to 1.

A fault with $p = p_{max}$ is easier to reveal[12] than a fault with $p = 1 - p_{max}$ (Conclusion 12.1). Since $1 - p_{max} > p_{min}$ (Figure 12.11), a fault with $p = 1 - p_{max}$ is easier to reveal than a fault with $p = p_{min}$ (from Conclusion 12.2). It follows that *the fault which is the most difficult to reveal is the fault which has the lower detection probability* p_{min} ($p_{min} \leq 0.5$, and $p_{min} < 0.5$ for any non-trivial circuit).

	Number of faults			Detection probabilities		
	$p_i \leq 0.5$	$p_i > 0.5$	Total	p_{min}	p_{max}	$\dfrac{1 - p_{max}}{p_{min}}$
ALU	204	23	227	$5.86 \cdot 10^{-3}$	0.9453	9
C 499	1018	0	1018	$1.83 \cdot 10^{-3}$	0.500	273
C 1355	1866	48	1914	$1.89 \cdot 10^{-3}$	0.501	264
C 6288	7260	610	7870	$1.18 \cdot 10^{-2}$	0.813	16

Figure 12.11   Number of faults and bounds of detection probability for some circuits.

Hence, *the probability of aliasing is important for the faults with a low detection probability*, but, practically, not for those with a high detection probability (those which present peaks), because these faults are easy to test and to reveal in spite of the peaks.

Let us now concentrate on values of $p < 0.5$. Since all the polynomials lead to the same limit $1/2^k$ for both AL and $\Pr[A = A_0]$, what are the polynomials which provide the "best" convergence to this limit? As we have shown, the most significant measurement, as far as random testing is concerned, is the probability

---

[11] For the ALU, the single stuck-at faults are considered and the detection probabilities are exact. For the three other circuits, the BB-fault model (Appendix D) was used. The values of detection probabilities, obtained by simulation, are approximate.

[12] For the stuck-at fault model, one can easily show that, for any combinational circuit, $p_{max} \geq 0.5$ [PrDa 96].

of non-revelation $\Pr[A = A_0]$. For $l = 1$, $\Pr[A = A_0] = 1 - p > 0.5$. Since $1/2^k \leq 0.5$ for any k, $\Pr[A = A_0] > 1/2^k$ at time $l = 1$. It follows that the "best" convergence is the *fastest convergence* of $\Pr[A = A_0]$ to $1/2^k$.

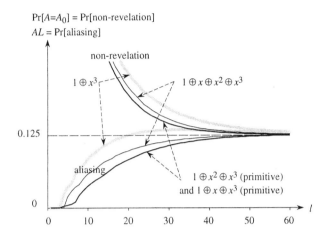

Figure 12.12 Comparison of the various polynomials of degree 3 ($p = 0.1$).

Figure 12.12 presents the behavior of all the polynomials of degree 3, for $p = 0.1$. It is clear that the two primitive polynomials, $1 \oplus x^2 \oplus x^3$ and $1 \oplus x \oplus x^3$, provide the lower value of the probability of aliasing (except for some very low values of l), hence, for the probability of non-revelation, according to (12.32). The simplest polynomial $1 \oplus x^3$ ($1 \oplus x^k$ in the general case) provides the highest probabilities.

Property 12.7 Given k and $p < 0.5$, there is no polynomial such that the convergence to $1/2^k$ of the non-revelation, $\Pr[A = A_0]$, is faster than for a primitive polynomial.

Proof The transition matrix V of the Markov chain (presented at the beginning of this section, example in Figure 12.9c) has 2^k eigenvalues: one with a value of 1 and $2^k - 1$ (qualified non-trivial) whose magnitude is less than 1.

The convergence of $\Pr[A = A_0]$ to the limit $1/2^k$ when l tends to infinity depends on the maximum magnitude of the non-trivial eigenvalues (the convergence is faster if this magnitude is lower).

According to [Da *et al.* 89], all the non-trivial eigenvalues of all the Markov chains corresponding to primitive polynomials have the same magnitude, say λ_m, and for any non-primitive polynomials, the magnitude of the biggest eigenvalue is at least equal to λ_m.

❑

This property is ilustrated in Figure 12.12: one can observe that $\Pr[A = A_0]$ converges faster to the limit for the polynomials $1 \oplus x \oplus x^3$ and $1 \oplus x^2 \oplus x^3$, the two primitive polynomials of degree 3 (same values of $\Pr[A = A_0]$ for both).

Property 12.8 Given k, and $p < 0.5$, $\Pr[A = A_0]$ for the polynomial $1 \oplus x^k$ is greater than or equal to $\Pr[A = A_0]$ for any other k-degree polynomial.

Proof In [IvAg 89], it is shown that an upper bound of the probability of aliasing for a primitive polynomial of degree k is given by the probability of aliasing for $1 \oplus x^k$. In fact, this upper bound holds not only for primitive polynomials but for all polynomials (specified in [IvPi 90], shown in [Iv 88]).

The proof in [Iv 88] and [IvAg 89] is based on the probability of aliasing $AL*$ in Remark 12.1 (Section 12.1.1). The proof that a polynomial providing an upper bound for $AL*$ provides also an upper bound for $\Pr[A = A_0]$ is straightforward (Equation (12.8) in Section 12.1.1).

Property 12.9 The probability of non-revelation for a fault whose detection probability is p by a random test sequence of length L and a SISR of degree k is bounded by

$$\Pr[A = A_0] \leq \frac{1}{2^k} \cdot \left(1 + |1 - 2p|^a\right)^b \cdot \left(1 + |1 - 2p|^{a+1}\right)^{k-b}, \quad (12.36)$$

where $a = \lfloor L/k \rfloor$ and $b = k \lceil L/k \rceil - L$.

Proof Assume the polynomial $1 \oplus x^3$ and the error vector $E = E(1)E(2)E(3)E(4)E(5)$. Because of the simple feedback, at the end of the sequence, $Q_1(5)$ is the EXOR of the last error bit $E(5)$ and of the error bit occurring 3 time units before (and so on for a longer test sequence).

Then, $Q_1(5) = E(5) \oplus E(2)$, $Q_2(5) = E(4) \oplus E(1)$, $Q_3(5) = E(3)$. Let r_i denote the number of bits in the expression of $Q_i(t)$; in the preceding example, $r_1 = r_2 = 2$ and $r_3 = 1$.

Generally for $1 \oplus x^k$ and a length L,

$$\left. \begin{array}{l} r_i = \lceil L/k \rceil \quad \text{for } i = 1, \ldots, (L - k\lfloor L/k \rfloor), \\ r_i = \lfloor L/k \rfloor \quad \text{for } i = (L - k\lfloor L/k \rfloor + 1), \ldots, k. \end{array} \right\} \quad (12.37)$$

In [Da 78 & 80] it was shown that

$$\Pr[A = A_0] = \frac{1}{2^k} \cdot \prod_{i=1}^{k}\left(1 + (1 - 2p)^{r_i}\right). \quad (12.38)$$

From (12.38) and (12.37),

$$\Pr[A = A_0] = \frac{1}{2^k} \cdot \left(1 + (1 - 2p)^a\right)^b \cdot \left(1 + (1 - 2p)^{a+1}\right)^{k-b} \quad (12.39)$$

is obtained[13] (where a and b are the values in Property 12.9).

If $p > 0.5$, according to Observation 12.3, an upper bound of $\Pr[A = A_0]$ is obtained if $|1 - 2p|$ replaces $(1 - 2p)$ in (12.39).

Now, according to Property 12.8, Property 12.9 is proven.

Remark 12.3 In Hypothesis 12.1, the input process of the SISR is assumed to be stationary, i.e., p is assumed to be constant throughout the test. This assumption is convenient for a combinational fault, but not necessarily for a *sequential fault* for which the general non-stationary case $p = p(l)$ could be more realistic.

1) If $p = p(l)$ where $0 < p(l) < 1$ for all values l, the *properties related to the limit value* $1/2^k$ *continue to be true* (Property 12.6, and also Properties 12.10 and 12.11 in the following sections where the vector p for multiple output circuits takes the place of the value p). Only the speed of convergence to the limit is affected.

2) In case of *periodic errors*, corresponding to $p(l) = 0$ for some values of l, Property 12.6 is no longer true. This case is considered in Section 12.3.4 because results on multiple output circuits are used for this case.

12.2.4 Cost of signature analysis

The cost of signature analysis is the sum of the cost of hardware (signature analyser) and the cost of storage (of the fault-free signature).

The cost of hardware corresponds to the number k of stages plus the EXOR gates. In most cell libraries the cost of two 2-input EXORs is essentially equivalent to that of a register cell [OlDaRi 93]. Let us express the *hardware cost* by αC_H where α is the cost of one stage and C_H is the normalized hardware cost:

$$C_H = \text{number of stages} + \frac{1}{2} \text{ number of 2-input EXORs} \qquad (12.40)$$

For example $C_H = 3 + \frac{1}{2} \cdot 2 = 4$ for the SISR in Figure 12.9a.

Figure 12.13 presents some primitive polynomials in the range $10 \leq k \leq 31$. The signature analyser should be in this range (Figure 12.3) since, for these values, $1/2^k$ ranges from 10^{-3} to 5.10^{-10}. For each value k, a polynomial with the minimum number of terms[14], and the corresponding normalized hardware cost is given.

The cost of storage corresponds to the number of bits of the signature. Let us express the *storage cost* by βC_S where C_S is the normalized storage cost

---

[13] A similar result was derived in [IvAg 89].
[14] Lists of polynomials for every k up to 168 and up to 300 can be found in [St 73] and [BaMeSa 87] respectively. The polynomials in Figure 12.13 are taken from these lists.

corresponding to the number of bits to be stored, and the storage cost by bit is β. Hence, the total cost is given by

$$\text{Signature analysis cost} = \alpha\, C_H + \beta\, C_S. \qquad (12.41)$$

k	Polynomial	C_H	k	Polynomial	C_H
10	$1 \oplus x^3 \oplus x^{10}$	11	21	$1 \oplus x^2 \oplus x^{21}$	22
11	$1 \oplus x^2 \oplus x^{11}$	12	22	$1 \oplus x^1 \oplus x^{22}$	23
12	$1 \oplus x^3 \oplus x^4 \oplus x^7 \oplus x^{12}$	14	23	$1 \oplus x^5 \oplus x^{23}$	24
13	$1 \oplus x^1 \oplus x^3 \oplus x^4 \oplus x^{13}$	15	24	$1 \oplus x^1 \oplus x^3 \oplus x^4 \oplus x^{24}$	26
14	$1 \oplus x^1 \oplus x^{11} \oplus x^{12} \oplus x^{14}$	16	25	$1 \oplus x^3 \oplus x^{25}$	26
15	$1 \oplus x^1 \oplus x^{15}$	16	26	$1 \oplus x^1 \oplus x^7 \oplus x^8 \oplus x^{26}$	28
16	$1 \oplus x^2 \oplus x^3 \oplus x^5 \oplus x^{16}$	18	27	$1 \oplus x^1 \oplus x^7 \oplus x^8 \oplus x^{27}$	29
17	$1 \oplus x^3 \oplus x^{17}$	18	28	$1 \oplus x^3 \oplus x^{28}$	29
18	$1 \oplus x^7 \oplus x^{18}$	19	29	$1 \oplus x^2 \oplus x^{29}$	30
19	$1 \oplus x^1 \oplus x^5 \oplus x^6 \oplus x^{19}$	21	30	$1 \oplus x^1 \oplus x^{15} \oplus x^{16} \oplus x^{30}$	32
20	$1 \oplus x^3 \oplus x^{20}$	21	31	$1 \oplus x^3 \oplus x^{31}$	32

Figure 12.13 Primitive polynomials (with minimal numbers of terms) in the range of values k of interest for signature analysis.

If $C_S = k$, since $C_H \approx k$ (C_H equals k plus 1 or 2 according to Figure 12.13), the signature analysis cost may be approximated by $(\alpha + \beta)\, k$.

Now, in some applications, several successive signatures can be performed by a single signature analyser. Assume for example that a test sequence of length $L = 100\,000$ is applied and that two signatures are performed: $A_{(50\,000)}$ when half the test sequence has been applied and $A_{(100\,000)}$ when all the test sequence has been applied. In this case, the concatenation $A = A_{(50\,000)} \cdot A_{(100\,000)}$ may be considered as the signature of the circuit. If k is the length of the register, the aliasing probability is close to $1/2^{2k}$ since the signature length is $2k$. For this example, if $k = 15$, then $C_H = 16$ (Figure 12.13) and $C_S = 15 + 15 = 30$ and the total cost is $16\,\alpha + 30\,\beta$.

Examples of multiple signatures with a single signature analyser can be found in [Be et al. 75] (see Figure 1.6 in Section 1.2.2), and [HaMc 83].

12.3 MULTIPLE INPUT SIGNATURE ANALYSERS

In Section 12.2, the SISRs were studied. They correspond to single-output circuits. In this section, multiple input signature registers (MISR), corresponding to multiple output circuits, are studied. An important feature of a multiple output circuit is that all the faults do not affect all the outputs. Then, for some fault certain bits in the response may have a probability p ($0 < p < 1$) of being faulty, while some other *bits may have a null probability of being faulty*. This observation will lead us to also consider single output circuits for which a fault provides periodical errors (more precisely, periodical fault-free bits). Section 12.3.1 illustrates the correlations existing between faulty bits. Multiple input signature analysers for $m \leq k$ and $m > k$ are presented in Sections 12.3.2 and 12.3.3 respectively. Finally, SISR with periodic errors are considered in Section 12.3.4.

12.3.1 Space dependent and time dependent errors

For a m-output circuit and a test sequence of length L, the response can be represented as an $m \times L$ matrix:

$$R = \begin{bmatrix} z_1(1) & z_1(2) & \dots & z_1(L) \\ \dots & \dots & \dots & \dots \\ z_m(1) & z_m(2) & \dots & z_m(L) \end{bmatrix}. \tag{12.42}$$

The **error matrix** is then

$$E = \begin{bmatrix} E_1(1) & E_1(2) & \dots & E_1(L) \\ \dots & \dots & \dots & \dots \\ E_m(1) & E_m(2) & \dots & E_m(L) \end{bmatrix}, \tag{12.43}$$

where $E_i(l) = z_i(l) \oplus z_{i,0}(l)$, if $z_{i,0}(l)$ represents the fault-free bit on output z_i at time l. The matrix E contains a 0 for every fault-free bit in R and a 1 for every faulty bit in R.

Let us consider the examples in Figure 12.14. A **potential error matrix** contains a zero when the corresponding bit is *surely fault-free*, while it contains an x when the corresponding bit has a non-zero probability of being faulty (the input sequence is assumed to be random). Figure 12.14a shows a fault f_1 such that the outputs z_2 and z_3 are not affected, a double fault $[f_1 \& f_2]$ and a fault f_3 such that z_2 is not affected. Let us say that they correspond to **space-dependent errors**. Figure 12.14b shows faults such that some bits are periodically fault-free. They correspond to **time-dependent** or **periodical errors**. Figure 12.14c shows a circuit made up of three parts. The test sequence $S = S_1 \cdot S_2 \cdot S_3$ is the concatenation of three sequences. Part 1 is a core which is tested by S_1. Part 2 is tested together with Part 1 by S_2, and Part 3 is tested together with Part 1 by S_3. If the fault f_4 is present in Part 2, then the response is fault-free during S_1, and during S_3 but may be faulty during S_2. The errors during this part of S are called **burst errors**. Note that these three kinds of errors are defined from the bits which are surely fault-free. Some errors may be a combination of several types.

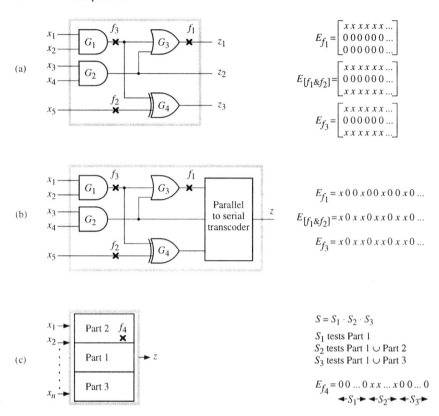

Figure 12.14 Illustration of potential error matrices[15].

Now consider the fault f_3 in Figure 12.14a. An error can appear on z_1 or on z_3 only if there is an error at the spot noted f_3. This means that $E_1(l)$ and $E_3(l)$ are not independent for this fault. The bits $E_1(l)$ and $E_3(l)$ are **correlated** error bits. In a similar way, one observes for fault f_3 in Figure 12.14b that $E(1)$ and $E(3)$ are correlated error bits, and more generally the two error bits $E(1 + 3a)$ and $E(3 + 3a)$, $a = 0, 1, 2...$, are correlated. If a potential error matrix contains only x's *without correlation* between error bits, it is called an **homogeneous error**.

12.3.2 MISR if $m \leq k$

This section applies to multiple output circuits. The first question that arises is: is it more convenient to use one or several signature analysers? The answer is illustrated using a 2-output circuit in Figure 12.15.

---

[15] This figure is reprinted from [Da 86] (© 1986 IEEE).

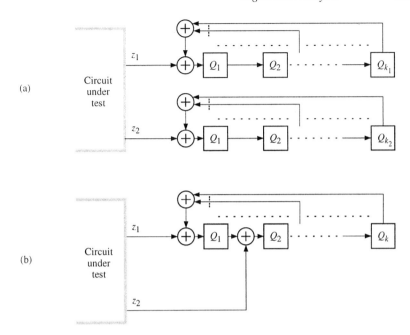

Figure 12.15 Comparison of two schemes[16].

The cost of signature analysis is approximately proportional to $k_1 + k_2$ in Figure 12.15a and to k in Figure 12.15b (Section 12.2.4). For the same cost we may have $k_1 = k_2 = k/2$. For the scheme in Figure 12.15a, there are faults whose aliasing probability is roughly $1/2^{k_1}$ (faults such that z_2 is fault-free) or $1/2^{k_2}$ (faults such that z_1 is fault-free). For the scheme in Figure 12.15b, the k bits of the signature analyser may become faulty even for a fault affecting a single output, and the aliasing probability is about $1/2^k$ which is much smaller. In other words, the equal likelihood of signatures cannot be obtained for all the faults in the first case while it may be obtained in the second case; then, according to Property 12.4 in Section 12.1.2, the scheme in Figure 12.15b is better.

The general case of multiple input shift register (MISR) is shown in Figure 12.16a for any $m \leq k$. Let us recall that the number of stages k is obtained from the required aliasing probability (Section 12.1.3) but is not necessarily linked to the number m of outputs[17].

Before giving the important Property 12.10, let us specify the hypotheses. It is assumed that k is equal to the number of outputs of the CUT; however, it will be seen that this is not a restriction.

---

[16] This figure is reprinted from [Da 86] (© 1986 IEEE).
[17] Some authors consider that $k = m$. In this case, k would be equal to 1 for a single output circuit!

(a)

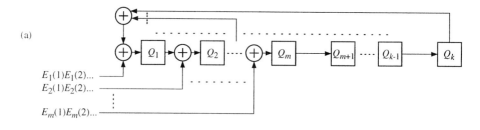

(b)

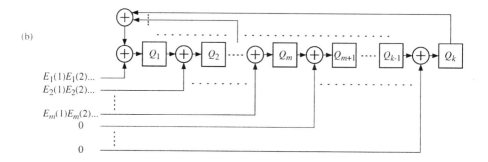

Figure 12.16 (a) k-bit MISR for a m-output circuit such that $m \leq k$. (b) Equivalent scheme.

Hypothesis 12.2 Consider the MISR in Figure 12.16a and assume that $m = k$. The error matrix (12.43) may be rewritten as

$$E = \begin{bmatrix} E_1(1) & E_1(2) & ... & E_1(L) \\ ... & ... & ... & ... \\ E_k(1) & E_k(2) & ... & E_k(L) \end{bmatrix}. \quad (12.44)$$

A column $[E_1(l), ..., E_k(l)]^T$ may have 2^k different values which may be represented by the corresponding integer $i = 2^{(k-1)} E_1(l) + ... + 2^0 E_k(l)$. Let p_i denote the probability that the vector corresponds to the value i.

It is assumed that the successive input vectors are independent and that the stationary vector $p = (p_0, p_1, ..., p_{2^k-1})$ describes the stochastic process at the MISR input.

❑

The vector p depends on the fault considered. The component p_0 corresponds to the probability of the correct response, and $1 - p_0$ corresponds to the detection probability (for a combinational circuit).

If $m < k$, the scheme in Figure 12.16b may be considered (instead of Figure 12.16a with $m = k$). This is consistent with Hypothesis 12.2: the error vectors such that $E_j(l) = 1$, $m < j \leq k$, have simply a null probability.

Property 12.10 Assume $m \leq k$. The non-revelation probability $\Pr[A = A_0]$ tends to $1/2^k$ for all the faults fulfilling Hypothesis 12.2, if and only if the polynomial of the MISR is primitive.

Proof The proof is given in [DaOlRi 91]; in this paper, the proof concerns the aliasing probability, and moving to the non-revelation probability is straightforward (Equation (12.8) in Section 12.1.1).
❏

In [DaOlRi 91], the proof of this property is followed by an example corresponding to a realistic fault and a non-primitive polynomial. It is shown that only 2^d signatures can be obtained (with $d < k$), and that the limit of aliasing probability is $1/2^d$ for this fault. This result may be compared to Property 12.4 (Section 12.1.2): the signatures are not equally likely (since 2^{k-d} of them have a zero probability) and the maximum distinction potential cannot be obtained.

12.3.3 MISR if $m > k$

According to Section 12.1.3 the length k of the MISR should be chosen as a function of the required aliasing probability ($\approx 1/2^k$). What can we do if the number of outputs of the CUT is greater than the value k chosen? If the difference is small, a few stages can be added to the MISR. However, if the difference is great, the price of the signature analysis would be too high, without necessity. Assume a 100-output circuit, a 100-stage MISR would be expensive while an aliasing probability as low as $1/2^{100} \approx 10^{-30}$ is not required. A solution remains which consists of feeding some (or all) stages of the MISR with the EXORs of two or more outputs of CUT. In this case, the outputs which are put together to feed a stage may be chosen in such a way that they are not (or not over) correlated. Let us illustrate this idea with simple examples.

Examples where $k = 3$ and $m = 4$ are presented in Figure 12.17. For the CUT in Figure 12.17a, there is no common part upstream of both z_2 and z_4. Since not any single stuck-at fault can affect both z_2 and z_4, feeding a stage of the MISR by $z_2 \oplus z_4$ is a good solution.

In Figure 12.17b, the CUT has no pair of outputs with no common part in their upstreams. Assume that $z_1 \oplus z_2$ has been chosen to feed stage 1 and that the fault $f = x_2/0$ is present (Figure 12.17b). In this case, the errors on z_1 and z_2 are correlated.

Let us generalize the notation in Section 12.3.2 to take into account the fact that $m > k$. Let E_j denote the presence of an error at the input of stage j of the MISR. With this notation, for the example in Figure 12.17b,

$$E_1(l) = (z_1(l) \oplus z_{1,0}(l)) \oplus (z_2(l) \oplus z_{2,0}(l)),$$
$$E_2(l) = z_3(l) \oplus z_{3,0}(l), \quad (12.45)$$
$$E_3(l) = z_4(l) \oplus z_{4,0}(l).$$

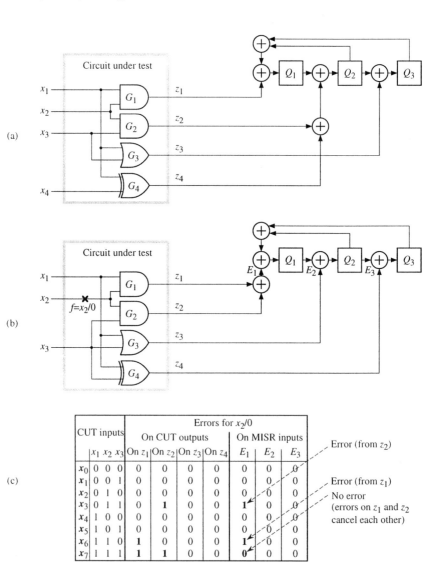

Figure 12.17 Examples where $m = 4$ and $k = 3$. (a) Errors on z_2 and z_4 are not correlated. (b) Errors on z_1 and z_2 are correlated for $f = x_2/0$. (c) Errors as a function of the input vector for fault f.

See Figure 12.17c. With the new notation, the vector of probabilities defined in hypothesis 12.2 becomes $p = (0.75, 0, 0, 0, 0.25, 0, 0, 0)$ since $E_1 E_2 E_3 = 000$ with a 0.75 probability and $E_1 E_2 E_3 = 100$ with a 0.25 probability. One can observe in Figure 12.17c that, for the input vector x_7, $z_1 \oplus z_2$ is error free while there are errors on both z_1 and z_2. However, the fault may be revealed because $p_0 \neq 1$.

Property 12.11 Assume $m > k$ and several CUT outputs are gathered by EXOR gates feeding the MISR (as explained in this section). The non-revelation probability $\Pr[A = A_0]$ tends to $1/2^k$ for all faults fulfilling Hypothesis 12.2 (where $E_j(l)$ now denote the error bit on the jth stage of the MISR), if and only if the polynomial of the MISR is primitive and $p_0 \neq 1$.

Proof This property is similar to Property 12.10. The only difference is that the condition $p_0 \neq 1$ is added (this condition was verified for any testable fault in Property 12.10).
□

Assume that $E_j(l) = \sum_{i \in Z(j)} z_i(l) \oplus z_{i,0}(l)$, where $Z(j)$ is the set of CUT outputs gathered on stage j. The condition $p_0 = 1$ would be satisfied if and only if, for every j and for every l, the number of faulty z_i in $Z(j)$ were even: only very particular cases of circuits, faults and gatherings could lead to $p_0 = 1$. Hence, $p_0 \neq 1$ is not a restriction from a *practical* point of view.

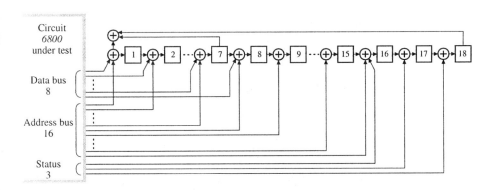

Figure 12.18 Example for the *6800* microprocessor[18].

If errors on several CUT outputs could be correlated, it may be convenient to dispatch them on several stages of the MISR. This could avoid a partial cancellation of some errors by others, and allow a faster convergence to $1/2^k$ [Da 86].

---

[18] This figure is reprinted from [Da 86] (© 1986 IEEE).

Figure 12.18 illustrates a more significant case: the CUT is the Motorola *6800* microprocessor. Let us assume that a 18-stage MISR has been chosen (polynomial $1 \oplus x^7 \oplus x^{18}$ may be chosen, according to Figure 12.13). Since the circuit has 27 outputs[19], some of them must be gathered. Figure 12.18 presents one of the possible solutions. We may assume that errors on the wires of the same bus (data or address) could be correlated, and that errors on the data bus are not correlated with errors on the address bus. In Figure 12.18, the 16 lines of the address bus are dispatched on 16 different stages, and the 8 lines of the data bus are dispatched on 8 different stages; however, an address line and a data line may be gathered on the same stage.

12.3.4 SISR for periodic errors

In section 12.2, it is assumed that the single output of the CUT may be faulty at any time, i.e., all the bits of the response have a non-zero probability of being faulty. However, for some sequential circuits, periodical errors could occur (Figure 12.14b). This case is considered in this Section 12.3 because it is linked to properties which were developed for MISR, as it will appear in the sequel.

Figure 12.19a presents a 2-stage SISR with an homogeneous error. According to Section 12.2.3, the Markov chain and the transition matrix in Figures 12.19b and c are obtained. The non-revelation probability, $\Pr[A = A_0] = \pi_1(l)$, tends towards $1/2^k = 1/2^2 = 0.25$ when l tends to infinity.

Now consider Figure 12.19d. The error sequence $E = 0 \, x \, 0 \, x \, 0 \, x...$ means that the odd bits in the response are error-free, while the even bits of this response may or may not be faulty. It is assumed that each even bit has the probability p of being faulty.

The behavior of the SISR may be modeled by the Markov chain in Figure 12.19e. Each state of the SISR is represented by two states of the Markov chain. For example the state $Q_1Q_2 = 10$ corresponds to $(10, e)$ and $(10, o)$. In $(10, e)$, the first part means that $Q_1Q_2 = 10$, and e means that an even number of test vectors has been applied to the CUT, while o means that this number is odd. As previously, there are transitions with probabilities p or $1-p$. In addition, there are transitions with a 1 probability; for example if the state is $(10, e)$, the next error bit (whose number is odd) will be error-free: then, the state$(11, o)$ will surely be reached.

The corresponding transition matrix is presented in Figure 12.19f. This matrix is cyclic: if the present state is in {1, 2, 3, 4}, the next one will be in {5, 6, 7, 8}, and vice-versa. The bottom-left part of the matrix corresponds to the matrix in Figure 12.19c, and the top-right part corresponds to the same matrix where p has the value 0.

---

[19] If the tri-state outputs are transformed to outputs with a constant value 0 or 1 (Section 11.4.2).

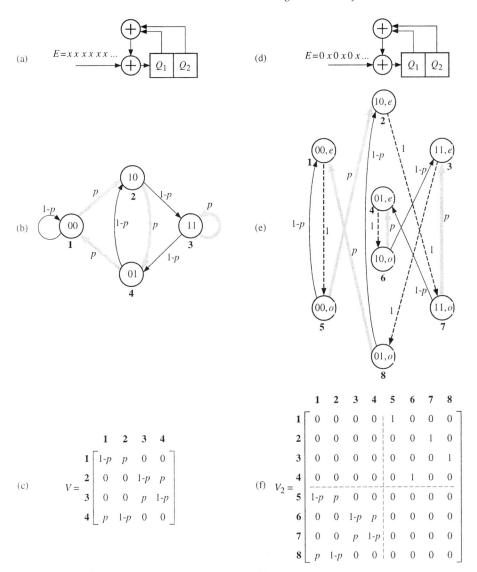

Figure 12.19 Behavior of a 2-stage SISR. (a) Homogeneous error. (b) and (c) Corresponding Markov chain and transition matrix. (d) Periodical error. (e) and (f) Corresponding Markov chain and transition matrix.

One can verify that the matrix V_2 in Figure 12.19f is irreducible. Since it is cyclic the limit probability of each state is not the same for all the times l: there is a limit for the even values of l and another one for the odd values of l, denoted by $\pi_i(e)$ and $\pi_i(o)$ respectively. The following values are found:

$$\left.\begin{aligned}
\pi_1(e) &= 0.25, & \pi_1(o) &= 0, \\
\pi_2(e) &= 0.25, & \pi_2(o) &= 0, \\
\pi_3(e) &= 0.25, & \pi_3(o) &= 0, \\
\pi_4(e) &= 0.25, & \pi_4(o) &= 0, \\
\pi_5(e) &= 0, & \pi_5(o) &= 0.25, \\
\pi_6(e) &= 0, & \pi_6(o) &= 0.25, \\
\pi_7(e) &= 0, & \pi_7(o) &= 0.25, \\
\pi_8(e) &= 0, & \pi_8(o) &= 0.25.
\end{aligned}\right\} \quad (12.46)$$

Now, $\Pr[A = A_0] = \Pr[Q = 00] = \pi_1 + \pi_5$. When the test length tends to infinity, $\pi_1(e) + \pi_5(e) = \pi_1(o) + \pi_5(o) = 0.25$ according to (12.46). Hence, the same limit $1/2^k$ is obtained. Which conditions must be satisfied in order to reach this limit?

In a general case, the Markov chain contains $d \cdot 2^k$ states if the period of the error is d. Let $\pi_{i,av}$ be the limit value, when l tends to infinity, of the average probability of state i over a period d. *If the matrix is irreducible, $\pi_{i,av}$ is the same value for all i,* because the sum of probabilities of arcs entering a state i is 1 as the sum of probabilities of the outgoing arcs. However, is the matrix irreducible? The matrix V_2 in Figure 12.19f, corresponding to a primitive polynomial of the SISR, is irreducible. The reader may verify that a matrix obtained in a similar way from the polynomial $1 \oplus x^2$ is not irreducible; then some states of the SISR have a zero probability and the limit of aliasing probability is 0.5 instead of 0.25.

Conjecture 12.1 Consider a k-stage SISR and an error sequence of period d. If the polynomial is primitive, and d does not share a common factor with $2^k - 1$, then the probability of non-revelation of a fault tends to $1/2^k$.
❑

The idea leading to this conjecture is as follows. For a k-stage LFSR, a new LFSR can be built corresponding to $\sigma = d$ shiftings (Section 10.3.2.1). Because of the Decimation Property (Section 10.3.1), the sequence produced by the new LFSR is an M-sequence if and only if d is relatively prime to $2^k - 1$. If the sequence of errors $E(j + \alpha d)$ is sent to stage j through an EXOR, a MISR is obtained and Property 12.10 (Section 12.3.2) may be applied to it. The proof is not complete because the behavior of the MISR is equivalent to the initial SISR only for the values of l (related to the initial SISR) which are multiples of d, i.e., $l = 0, d, 2d, \ldots$.

For example, consider the SISR represented in Figure 12.20a, whose polynomial is $1 \oplus x^2 \oplus x^3$. The transformation explained above is represented in Figure 12.20b for $\sigma = 2$. After reordering (Figure 12.20c), it is clear that the new circuit is still a *shift* register (loop $Q_2 \rightarrow Q_1 \rightarrow Q_3$), a MISR.

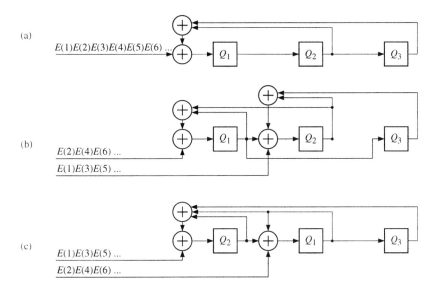

Figure 12.20 Illustration of Conjecture 12.1. (a) Initial SISR. (b) Transformation with $\sigma = 2$. (c) Reordering.

k	Factorization of 2^k-1	k	Factorization of 2^k-1
10	3·11·31	21	7·7·127·337
11	23·89	22	3·23·89·683
12	3·3·5·7·13	23	47·178481
13	8191	24	3·3·5·7·13·17·241
14	3·43·127	25	31·601·1801
15	7·31·151	26	3·2731·8191
16	3·5·17·257	27	7·73·262657
17	131071	28	3·5·29·43·113·127
18	3·3·3·7·19·73	29	233·1103·2089
19	524287	30	3·3·7·11·31·151·331
20	3·5·5·11·31·41	31	2147483647

Figure 12.21 Factorization of $2^k - 1$ for the polynomials in Figure 12.13 (Section 12.2.4).

In order to know wether d and $2^k - 1$ share a common factor, Figure 12.21 may be consulted[20].

---

[20] This information is drawn from [Br et al. 88].

Remark 12.4
a) According to Conjecture 12.1, *if* the CUT is affected a fault provoking an error of *any periodicity* which is not a multiple of 7 (i.e., 2^3-1), the probability of non-revelation is 2^3-1. We means that we are not trying to solve an aliasing problem for a specific periodicity (although those faults can be considered as somewhat realistic); we want to say that, if the conditions in Conjecture 12.1 are verified, the SISR works correctly for any fault *periodic or not* (if the period is not a multiple of 2^k-1). The author believe that the conjecture is also true for a MISR.

b) The condition in Conjecture 12.1 seems to be sufficient but not necessary. For $k = 4$, 2^k-1 = 15 = 3 ×5, one can verify that the probability of non-revelation is $1/2^4$ for $d = 3$ but not for $d = 5$.

NOTES and REFERENCES

The concept of signature analysis using linear feedback shift registers was introduced in [Be *et al.* 75], and the expression "signature analysis" appeared in [Fr 77] and [Na 77]. The notion and the word "aliasing" are introduced in [Wi *et al.* 86]; its expression in Property 12.1 (Section 12.1.1) is introduced in [Da 90]. The "non-revelation" concept is introduced in this book to the best of the author's knowledge. Property 12.4 (Section 12.1.2), related to equal likelihood of signatures, is given[21] in [Da 80].

Deterministic transition counting for combinational circuits was introduced in [Ha 75b]; a possible test sequence S_α is ordered in such a way that the response is an alternance of 0's and 1's; another possible test sequence S_γ is ordered in such a way that the response is a string of 0's followed by a string of 1's; it is shown that these test sequences present no aliasing for most faults. Some improvements were proposed in [Se 77], [Re 77].

Check-sum tests were presented in [Ha 76a]. Multiple counting was introduced in [Se 77], then developed in detail in [FuKi 78]. Other counting methods, requiring an exhaustive test (application of the 2^n input vectors), were developed: testing syndrome techniques [Sa 80] and the ubiquitous parity bit [Ca 82] are based on counting of ones; verifying Walsh coefficient is another counting method [Su 81].

Transition counting for sequential circuits by modifying the initial circuit, is proposed in [VeSa 80].

For deterministic testing, the test sequence depends usually of the counting method. For pseudorandom testing, the test sequence is independent of the

---

[21] In [Da 80], the property was more general (concerning the distinction of a number of faults which may be greater than two), but the proof was less accurate than in this book (the implicit assumption was $\Pr[A_j|f] = \Pr[A_j]$).

counting method; the fault-free signature is unique because the test sequence is pseudorandom, hence can be repeated.

In [Fr 77] it was assumed that every bit in the response has a probability 0.5 of being faulty. This is obviously far from the reality for most faults (if this hypothesis were true, there could be no need for signature analysis since a k-bit response, i.e., $L = k$, would provide a $1/2^k$ probability of non-detection). Effectiveness of fault signature analysis was studied in [Sm 80] assuming some particular classes of faulty responses (for example burst errors such that a bits, at most, are in error within N consecutive bit positions).

The general assumption of statistical independence of successive input vectors to the SISR (probability p for each response bit to be faulty) was introduced in [Da 78 & 80]. This assumption was extensively used in [Wi et al. 86 & 88] who use it to introduce modeling of the behavior of an LFSR as a Markov chain. This assumption was extended to multiple input signature analysers (Hypothesis 12.2) in [DaOlRi 91].

The properties of signature analysis with characteristic polynomial $1 \oplus x^k$ are investigated in [Da 80] for SISR and in [Da 86] for MISR. In [Wi et al. 88] it was suggested and partially shown that primitive polynomials would provide a better convergence to the aliasing probability limit $1/2^k$. In [Da et al. 89], an analytical approach was presented for the analysis of the Markov chain equations, and the proof that primitive polynomials provide better performances for SISR was given. In [DaOlRi 91] it was shown for MISR that the aliasing probability limit $1/2^k$ can be reached if and only if the polynomial is primitive.

Various authors have proposed bounds for the aliasing probability, for example [IvAg 89], [CaPiVe 91], [SaFrMc 92].

Many works on signature analysis have been published. Some examples are: analysis based in algebraic codes [PrGuKa 90]; aliasing for a three state data serial stream [Hl 86]; particular case of CUT [IwAr 90]; variants in design of MISRs [Hl 92]; use of linear automata as signature analysers [DaWiWa 90]; count-based compaction [IvZo 92]. Even if some research is still made for solving some specific problems, it may be emphasized that the general principles, presented in this chapter, have been found during about two decades (only).

An approach combining the signature analyser and pseudorandom generation (the generator is disturbed by the response of the CUT) was proposed in [EiInYa 80].

13

Design For Random Testability

The methods of design for testability which have been developped for any test sequence can obviously be used for random testing. They are briefly reviewed in Section 13.1. However, if the test sequence is random, it may be possible to take advantage of some specific features.

The random pattern generator may be simplified if a circuit is completely specified; Section 13.2 is concerned by this topic. Section 13.3 shows how some variables may be decorrelated to allow better random testability (an undetectable fault may become detectable and a detectable fault may require a shorter test length). Finally some design methods for combinational circuits, allowing shorter test lengths than circuits obtained by usual methods, are presented in Section 13.4.

13.1 DESIGN FOR TESTABILITY IN GENERAL

There are two main kinds of approaches, the *ad hoc* approaches and the *structured* approaches.

Solving a test problem for a given design is an ad hoc approach. In this case, a functional testing is generally considered. The circuit to be tested may be partitioned in order to obtain smaller parts which are easier to test[1]; this approach is called "divide and conquer". Then, test points may be added to improve the controllability (additional inputs) and observability (additional outputs). These techniques of partitioning and adding test points can obviously be used for random testing. However, due to the randomness of the inputs, additional inputs may be devised in a different way (Section 13.3).

---

[1] As already stated in Chapter 1, the computer run time required to perform test generation and fault simulation is approximately proportional to the number of logic gates to the power 3 [Go 80].

The structured approaches are based on the scan design (Section 1.1.2), including LSSD (level sensitive scan design). Their goal is to reduce sequential complexity for easier test generation and observation; in full scan situations, all the memories are interconnected into a scan path and the test sequence is applied to the remaining combinational circuit. In random testing situations, better controllability and observability may be useful, but the test sequence generation does not need this approach since it is easy without it.

For complex structures such as microprocessors, the general architecture may be considered for good testability. For such a circuit, an important general rule consists of articulating the circuit to be tested on a single bus. This bus is therefore the only means to transfer information between the different elements; testability is improved by connecting both the test pattern generator (good controllability) and the test evaluator (good observability) on the bus. A data path, designed in order to increase the number of accesses on the single bus, improves testability of the whole circuit.

Bit sliced structures are welcome; in such a structure, a b-bit block is made up of b similar 1-bit components (i.e., slices). The sliced structures present a natural partitioning which simplifies evaluation of testability because of the repetitiveness of the components in a block.

13.2 EXTENDED SPECIFICATION

The concept of *extended specification* was introduced in Section 5.2.5. Since, according to Section 5.2, *only the specified cases should be tested*, a random pattern generator able to produce a test sequence for all the specified cases and only the specified cases may be complex (Section 10.3.4 for example). An *extended specification transforms an incompletely specified circuit into a completely specified one*. Then, *the generator is easier to build* since all the cases are specified. This idea is illustrated with a few examples.

First example: combinational circuit

Figure 13.1a presents the truth table of an uncompletely specified 4-input circuit. In order to apply *only* test vectors for which the output is specified, a weighted test vector generator should be used. An example of such a generator is presented in Figure 13.1b: each pattern $Q_1Q_2Q_3Q_4$ which does not correspond to a pattern $x_1x_2x_3x_4$ for which the output is specified, is mapped into a pattern for which it is specified. For example $Q_1Q_2Q_3Q_4 = 1111$ is mapped into $x_1x_2x_3x_4 = 0011$. There are obviously many solutions and one of the simplest ones has been chosen in Figure 13.1b (the weights of the vectors depend on the chosen solution; for example, $\psi(x_0) = 1/16$ and $\psi(x_2) = 3/16$ in Figure 13.1b).

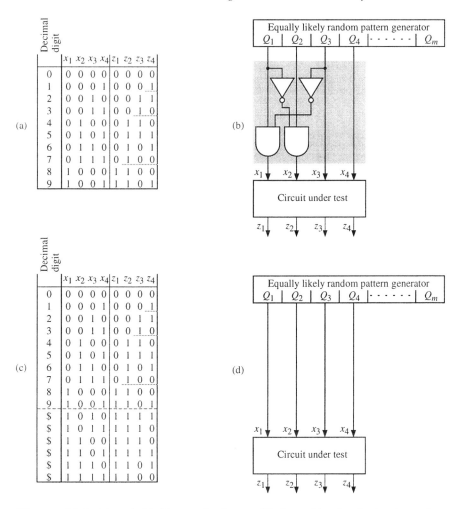

Figure 13.1 (a) and (b) Uncompletely specified combinational circuit, and a possible generator. (c) and (d) Extended specification and corresponding generator.

An extended specification of Figure 13.1a is shown in Figure 13.1c. The last six lines of the truth table have been chosen in order to obtain the simplest circuit (the Karnaugh maps for this example can be found in Figure 5.2, Section 5.2.1). In practice, the designer always needs to assign output values to the unspecified cases when the circuit is to be implemented, but two designers could make different choices. What we suggest is to *make explicit a choice in the specification*. All the implementations would thus be similar and the completely specified truth table known by everybody, including the user. It follows that the generator is simplified as illustrated in Figure 13.1d.

Second example: sequential circuit

Figure 13.2a represents an uncompletely specified sequential circuit. This machine may correspond to an embedded circuit such that:

1) When the state q_1 is reached, the input vector x_3 cannot be applied;

2) The input vector x_1 can never be applied;

3) When the state q_3 is reached, the output z_2 is not useful, i.e., it may have any value.

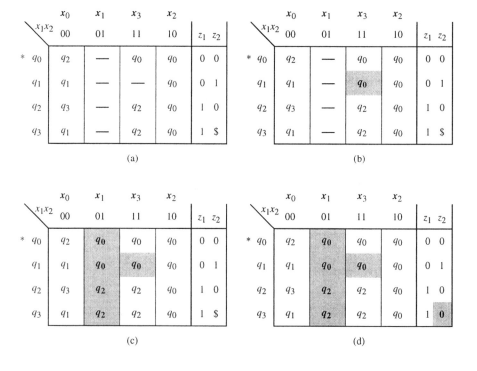

Figure 13.2 (a) Uncompletely specified sequential machine; (b) Machine for which general distribution is not needed. (c) Machine for which a weighted vector generator is not needed. (d) Machine for which the test evaluator is a simple comparison.

Obviously, once the circuit has been built, all the unspecified cases have a fixed behavior[2]. However, if the test designer does not know it, i.e., if he knows only the specification in Figure 13.2a, he has to build a complicated random tester. We shall see that this random tester becomes very simple if an *extended specification* of the machine is provided.

1) The machine in Figure 13.2a requires a *generalized distribution* (Appendix A); a *sequential circuit* between the output Q of the random pattern generator and the input x of the CUT is necessary in order to know when the state q_1 is reached; at this time the test vector x_3 should be assigned a zero probability. If $\delta(q_1, x_3)$ is specified (Figure 13.2b), a *constant distribution* generator may be used.

2) For the machine in Figure 13.2b, the constant distribution should be weighted since the test vector x_1 should have a zero probability (i.e., a *combinational circuit* between Q and x, corresponding to a weighted random vector generator, is necessary). If the next state is defined for all the entries corresponding to x_1 (Figure 13.2c), the random tester does not need any weighting (i.e., a generator similar to Figure 13.1d, using two values $x_1 = Q_1$ and $x_2 = Q_2$ may be used).

3) For the machine in Figure 13.2c, an unspecified output remains. The test *evaluator* should be able to know when the circuit is in the state q_3, in order to know that either $z_2 = 0$ or $z_2 = 1$ are admitted as correct outputs at this time. This information requires a *sequential circuit in the evaluator*. If the output is specified for this case (Figure 13.2d), the test evaluation is a simple comparison of outputs.

Third example: operation codes in a microprocessor

For the *6800* microprocessor, 197 of the $2^8 = 256$ values on the data bus correspond to valid operation codes (Appendix H). Then, 59 codes are *invalid operation codes*: their behavior is unspecified and the generator should allow for the fact that none of the 59 invalid codes is applied when the microprocessor is waiting for an operation code (Section 10.3.4). An *extended specification* would assign an instruction to each invalid operation code (then some instructions would correspond to two or more operation codes). In this way, all the 8-bit data would be *valid operation codes* and the generator would be simpler. In addition, the instructions assigned to the 59 codes in excess can be chosen in order to increase the detectability of the most difficult faults (increasing the probability of instruction *CLI* is interesting according to Section 9.5.2.2).

This approach was taken into account in the design of a microprocessor designed by the CNET (Centre National d'Etudes des Télécommunications): the operation codes are encoded with 6 bits, and each of the 64 codes is assigned an instruction [Th *et al.* 87].

---

[2] This behavior could possibly be random for the unspecified cases: two circuits with the same implementation could have a different behavior because of hazards, delays... since the designer had no constraints for these cases.

13.3 DECORRELATION BY EXOR GATES

Adding outputs to a circuit for better *observability is similar* for deterministic and random testing. Adding inputs for better *controllability is different*; in deterministic testing, a Boolean value 0 or 1 may be assigned to an additional input while a random value is to be assigned in random testing.

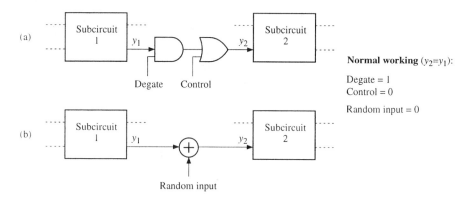

Figure 13.3 Separating subcircuits. (a) Deterministic approach. (b) Random approach.

Separating subcircuits, to simplify testing of each subcircuit is illustrated in Figure 13.3, where, in normal operation, the output y_1 of subcircuit 1 is connected to input y_2 of subcircuit 2 (the dotted lines correspond to other input and output lines). The deterministic approach is illustrated in Figure 13.3a: a *degating line*, when a 0 value is assigned to it, isolates subcircuit 2 from subcircuit 1; the value required on y_2 is equal to the Boolean value of the *control line*. The random approach is illustrated in Figure 13.3b: if the random input feeding the EXOR gate between y_1 and y_2 is independent of the inputs of subcircuit 1, then y_2 is a random variable as will be shown in the sequel (i.e., the input y_2 is "controllable" as if it were directly coming from a random generator).

Note that in both cases (Figures 13.3a and b), the observability of subcircuit 1 is improved by an extra output corresponding to y_1.

Consider the case illustrated in Figure 13.4a. The circuit under consideration has two inputs y_1 and y_2; y_1 corresponds to the primary input x_1 and y_2 corresponds to *any function* ϕ of both primary inputs x_1 and x_2. This is a general case of embedded circuit, except that $x_1 = y_1$, x_2, and y_2, could be vectors instead of variables; variables are considered only to simplify the notation.

Definition 13.1 Let C be an n-input subcircuit and Y be the set of 2^n possible input vectors y_i. The input vector y is said to be **R-controllable** (for *random-controllable*) if any sequence in Y^* has a non-zero probability to be applied.

□

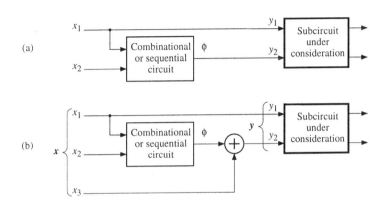

Figure 13.4 Random controllability by an additional random input. (a) Initial circuit. (b) Modified circuit.

In other words, if a vector is R-controllable, at any time each vector out of the 2^n possible vectors has a non-zero probability of being applied.

This corresponds to an *ideal random test* (Definition 5.1, Section 5.2) in case of *extended specification* (Section 13.2).

Let us show that the input vector y in Figure 13.4b is R-controllable if the EXOR gate with the random variable x_3 is added between ϕ and y_2.

Property 13.1 For the scheme in Figure 13.4b, if x is R-controllable, then y is R-controllable.

Proof Let $y(i) = (\alpha_{i1}, \alpha_{i2})$, $\alpha_{ij} \in \{0, 1\}$, denote the vector y at time i. Let $S = y(1)y(2)... y(l)$ be an arbitrary sequence of vectors y. We want to show that this sequence may be applied (i.e., has a non-zero probability).

S is applied if:

1) $x_1(1)x_1(2) ... x_1(l) = \alpha_{11}\alpha_{21} ... \alpha_{l1}$, (13.1)

and

2) $y_2(1)y_2(2) ... y_2(l) = \alpha_{12}\alpha_{22} ... \alpha_{l2}$. (13.2)

The condition (13.1) is satisfied since x is R-controllable by hypothesis. Now consider the second condition.

At time i ($1 \leq i \leq l$), the value ϕ may generally depend on all the values of x_1 and x_2 up to time i, i.e.,

$$\phi(i) = \phi(x_1(1), ..., x_1(i), x_2(1), ..., x_2(i)). \tag{13.3}$$

and

$$y_2(i) = \phi(i) \oplus x_3(i). \tag{13.4}$$

From (13.3) and (13.4), one can write that Condition (13.2) is satisfied if and only if

$$\alpha_{i2} = \phi(x_1(1), ..., x_1(i), x_2(1), ..., x_2(i)) \oplus x_3(i). \tag{13.5}$$

Then, this condition is satisfied if and only if:

$$x_3(i) = \phi(x_1(1), ..., x_1(l), x_2(1), ..., x_2(l)) \oplus \alpha_{i2}. \tag{13.6}$$

This condition (13.6) obviously has a non-zero probability of being satisfied since x is R-controllable.

❑

The proof would be similar if x_1 and x_2 were replaced by vectors instead of variables. Let us note that if y_2 were a vector instead of a variable y_2: in the general case an additional input x_j would be required for each component of y_2; if two components of y_2 depend on two disjoint sets of primary inputs, the same input variable x_j may be used for both.

Let us now present some examples[3] which were used in the design of a microprocessor with built-in self-test [PuRaTh 87]. In the first example, undetectable faults become detectable. In the second one, the test length is considerably reduced.

Consider the example in Figure 13.5. The combinational block has a bit sliced structure. The study can thus be conducted on a single slice of the block. In Figure 13.5a, the random test pattern generator sends on a single clock pulse, the same data to input y_1 of the combinational slice and to the 1-bit register. Therefore, a value x_1 generated at time t cannot be present on y_2 at time t: it will be applied at $(t + 1)$. It follows that the input vector $y = (y_1, y_2)$ is not R-controllable. For example, if $y(t) = 11$, then $y(t + 1)$ cannot be 10 or 00. Some length 2 sequences cannot be applied to the input of the combinational slice. There is a temporal correlation between both inputs of the combinational slices (some stuck-open or delay faults may not be testable, and other faults could be undetectable if the slice were sequential, according to Appendix K). An EXOR gate added as shown in Figure 13.5b allows the input vector y to become R-controllable according to Property 13.1. Because of the bit-sliced structure, the insertion of the EXOR gate is repetitive and the *same random input* x_2 may be used for all the slices.

---

[3] The examples are taken from [Th et al. 87].

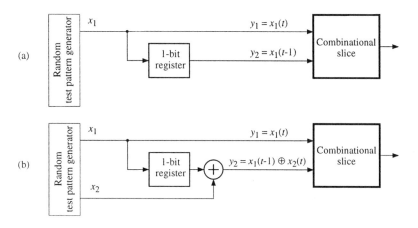

Figure 13.5 Decorrelation of the inputs of a block.

The second example is illustrated in Figure 13.6. The vector y is a 16-bit output vector of the arithmetic and logic unit, which is the result of an executed operation. A logic function detects a zero value of y, i.e., $z = 1$ if all $y_i = 0$, $i = 1, ..., 16$. The detection function is designed as a chain of 15 two-input AND gates as shown in Figure 13.6a. Let us consider the testability of this subcircuit, denoted by K.

Let us assume that the vector y is R-controllable and that the 2^{16} possible values are equally likely. Without a detailed analysis of the subcircuit K, it can easily be stated that:

1) For the set of *combinational faults* which can affect the subcircuit (this set includes all the single and multiple stuck-at faults), a lower bound of the minimum detection probability is

$$p_{min} \geq \frac{1}{2^{16}} \approx 1.5 \times 10^{-5},$$

since the circuit has 16 inputs. Let $L_{m,1}$ be an upper bound of the test length for these faults. According to (6.33) in Section 6.1.2.2, $L_{m,1}(0.001) = 452\,703$.

2) For the set of faults requiring a *subsequence of length two* (this set includes all the delay faults and most[4] of the stuck-open faults, according to Section K.1 in Appendix K), a lower bound of the minimum detection probability is

$$p_{min} \geq \frac{1}{2^{16}} \times \frac{1}{2^{16}} \approx 2.3 \times 10^{-10},$$

---

[4] Probably all the stuck open faults; the cases where a sequence of at least 3 vectors is required are very specific.

corresponding to two successive vectors whose probabilities are $1/2^{16}$. Let $L_{m,2}$ be an upper bound of the test length for these faults. The value $L_{m,2}(0.001) \approx 3 \times 10^{10}$ is obtained. For a 20 MHz clock frequency, $L_{m,2}$ corresponds to 25 minutes.

The subcircuit K may be partitioned into two subcircuits, namely K_1 and K_2, as illustrated in Figure 13.6b. The random input x_1 increases the controllability of the part in K_2 and the additional output z_1 (which may be also before the EXOR gate) increases the observability of the part in K_1. According to Property 13.1, we now have two 9-input, 1-output, subcircuits whose input vectors are R-controllable. It follows that for the new subcircuit made up of K_1 and K_2: $L_{m,1}(0.001) = 3534$, and $L_{m,2}(0.001) \approx 1.8 \times 10^6$ are obtained. For a 20 MHz clock frequency, $L_{m,2}$ corresponds to 0.1 second.

(a)
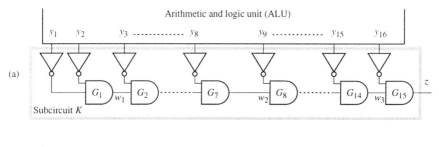

(b)
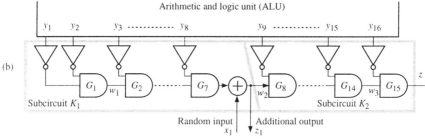

Figure 13.6 (a) Initial function $z = y'_1 y'_2 ... y'_{16}$. (b) Partitioning by a random input variable and an additional output[5].

Hence, with a small design effort, the test length has been greatly reduced. The number of faults, which can be detected with a 0.999 probability for the obtained test length, is very large. A more specific study of some particular faults (stuck-open or delay faults) is not necessary and is left as an exercise for the readers (Exercises 13.3 and 13.4).

---

[5] This figure is reprinted from [Th et al. 87] (© 1987 IEEE).

13.4 FACTORIZATION IN COMBINATIONAL FUNCTIONS

This section is based on [ToMc 94] and [ChGu 94], where the synthesis of random pattern testable combinational circuits is proposed. The basic idea is a transformation of a combinational circuit by factorization in the Boolean expression.

According to Section 6.1.2.1, the length of random testing (for a required expected fault coverage) primarily depends on the most random-resistant faults. For combinational faults, the detectability of fault f (i.e., the number of vectors detecting it, Section 6.1.1.1) is denoted by k_f, and k_{min} is the detectability of the most resistant fault in the circuit. In order to minimize the test length, the first aim is to *maximize the value* of k_{min}. Now, if the value of k_{min} cannot be increased, the second aim is to *reduce the number of faults* with this low detectability.

Two factorization methods, called *cube extracting* and *kernel extracting* are presented in the sequel. When one of these methods is applied to some part of the circuit, since the circuit is modified, a set of faults F_1 is replaced by another one F_2. Some faults in F_2 are as difficult to test as some faults in F_1, while some faults are easier to detect in F_2 than the corresponding ones in F_1. Globally, the detection of faults in F_2 is easier than the detection in F_1 (fortunately!). An important point is that *these factorization methods cannot create faults[6] more difficult to detect in F_2 than in F_1* [ToMc 94], [ChGu 94].

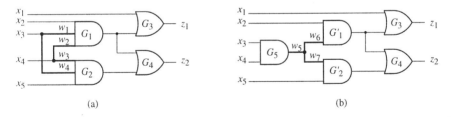

Figure 13.7 Illustration of cube extraction. (a) Initial circuit. (b) Modified circuit.

A *cube* is a product of literals (Section 2.1.1). In Figure 13.7a, G_1 and G_2 produce the sub-functions $x_2x_3x_4$ and $x_3x_4x_5$ respectively. The *cube* x_3x_4 is *common to both and may be extracted* as shown in Figure 13.7b. The new gate G_5 produces the sub-function $w_5 = w_6 = w_7 = x_3x_4$, G_1 is replaced by G'_1 producing $x_2w_6 = x_2x_3x_4$, and similarly G_2 is replaced by G'_2.

---

[6] The single stuck-at fault model is considered in this section.

The lines which have to be taken into account in the comparison of both circuits are presented in heavy lines: w_1 to w_4 in Figure 13.7a, and w_5 to w_7 in Figure 13.7b. For any other line there is no change of the corresponding detectabilities. For example, the detection functions (Section 6.2.1.1) related to the primary input x_3 are

$$T_{x_3/1} = x'_3 x_4 (x_2 + x_5),$$

$$T_{x_3/0} = x_3 x_4 (x_2 + x_5),$$

for both circuits in Figure 13.7a and b.

Fault f	Detection function T_f	Detectability k_f
$w_1/1$	$x_2 x'_3 x_4$	4
$w_2/1$	$x_2 x_3 x'_4$	4
$w_1/0, w_2/0$	$x_2 x_3 x_4 (x'_1 + x'_5)$	3
$w_3/1$	$x_3 x'_4 x_5$	4
$w_4/1$	$x'_3 x_4 x_5$	4
$w_3/0, w_4/0$	$x_3 x_4 x_5 x'_2$	2

(a)

Detectability k_f	Number of faults
2	2
3	2
4	4

(b)

Fault f	Detection function T_f	Detectability k_f
$w_5/1$	$(x'_3 + x'_4)(x_2 + x_5)$	18
$w_5/0$	$x_3 x_4 (x_2 + x_5)$	6
$w_6/1$	$(x'_3 + x'_4) x_2$	12
$w_6/0$	$x_2 x_3 x_4 (x'_1 + x'_5)$	3
$w_7/1$	$(x'_3 + x'_4) x_5$	12
$w_7/0$	$x'_2 x_3 x_4 x_5$	2

(c)

Detectability k_f	Number of faults
2	1
3	1
6	1
12	2
18	1

(d)

Figure 13.8 (a) and (b) Detectabilities of faults for the circuit in Figure 13.7a. (c) and (d) Detectabilities of faults for the circuit in Figure 13.7b.

Figure 13.8a presents the detection functions and the corresponding detectabilities for the lines w_1 to w_4 of the circuit in Figure 13.7a. It appears that there are two faults whose detectability is 2, two faults whose detectability is 3, and four faults whose detectability is 4 (Figure 13.8b). Similarly, the detectabilities of faults of the circuit in Figure 13.7b, are presented in Figure 13.8c and d. The comparison of Figures 13.8b and d, shows that the minimum value of k_f (for the set of faults under consideration) is 2 in both cases. However, for the modified circuit, there is only one fault whose detectability is 2 (Figure 13.8d); there is also a single fault whose detectability is 3; some detectabilities are up to 18 (corresponding to faults very easy to detect).

A **kernel** of an expression z is the quotient of z and a cube which is called *co-kernel*. For example if $z = ab(c + d')$, the quotient of z by the cube ab is $c + d'$; hence, $c + d'$ is a kernel and ab is the co-kernel. Factoring out a kernel affects the detection probability of faults associated with each instance of the co-kernel. In addition, if the kernel were common to several expressions, the detection probability of faults associated with each instance of the kernel would also be affected.

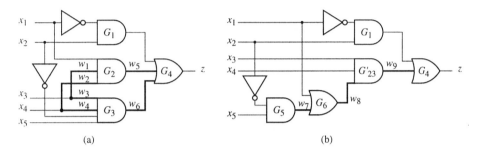

Figure 13.9 Illustration of kernel extraction. (a) Initial circuit. (b) Modified circuit.

The function performed by the circuit in Figure 13.9a is

$$z = x'_1 x_2 + x_1 x_3 x_4 + x'_2 x_3 x_4 x_5, \qquad (13.7)$$

which can be factorized as

$$z = x'_1 x_2 + x_3 x_4 (x_1 + x'_2 x_5). \qquad (13.8)$$

In (13.8), the expression $x_3 x_4 (x_1 + x'_2 x_5)$ is the product of the cube (co-kernel) $x_3 x_4$ by the kernel $(x_1 + x'_2 x_5)$. Extraction of this kernel is illustrated in Figure 13.9b. Comparing the detectabilities of the circuits in Figures 13.9a and b, requires the comparison of the heavy lines: w_1 to w_6 in Figure 13.9a on the one hand, and w_7 to w_9 in Figure 13.9b on the other hand. For any other line there is no change of the corresponding detectabilities. For example, the detection functions related to the primary input x_3 are

$$T_{x_3/1} = x'_3 x_4 (x_1 + x'_2 x_5),$$

$$T_{x_3/0} = x_3 x_4 (x_1 + x'_2 x_5),$$

for both circuits in Figure 13.9a and b.

The comparison is presented in Figure 13.10. The detection functions and detectabilities of the various faults under consideration are presented in Figure 13.10a and b for the initial circuit, in Figure 13.10c and d for the modified circuit. Comparing Figure 13.10b and d shows that three faults whose detectability is 1 remain after kernel extraction; however, there are no longer any faults whose detectability is 2, 3, or 4.

Some random-resistant faults cannot be eliminated through algebraic factoring. Their detectability can then be increased by test point insertions (additional input or additional output). The detectability related to the circuit in Figure 13.9b can be greatly improved by an additional output at the kernel output as shown in Figure 13.11a. This improvement is clear from the comparison of Figure 13.10d and 13.11b. For the set of faults under consideration, the minimum value of detectability is multiplied by 4.

Fault f	Detection function T_f	Detectability k_f
$w_1/1$	$x_1 x'_3 x_4$	4
$w_2/1$	$x_1 x_3 x'_4$	4
$w_1/0, w_2/0, w_5/0$	$x_1 x_3 x_4 (x_2 + x'_5)$	3
$w_3/1$	$x'_2 x'_3 x_4 x_5$	2
$w_4/1$	$x'_2 x_3 x'_4 x_5$	2
$w_3/0, w_4/0, w_6/0$	$x'_1 x'_2 x_3 x_4 x_5$	1
$w_5/1, w_6/1$	$x'_1 x'_2 x'_5 + (x_1 + x'_2)(x'_3 + x'_4)$	19

(a)

Detectability k_f	Number of faults
1	3
2	2
3	3
4	2
19	2

(b)

Fault f	Detection function T_f	Detectability k_f
$w_7/1, w_8/1$	$x'_1 x'_2 x_3 x_4 x'_5$	1
$w_7/0$	$x'_1 x'_2 x_3 x_4 x_5$	1
$w_8/0, w_9/0$	$x_3 x_4 (x_1 + x'_2 x_5)$	5
$w_9/1$	$x'_1 x'_2 x'_5 + (x_1 + x'_2)(x'_3 + x'_4)$	19

(c)

Detectability k_f	Number of faults
1	3
5	2
19	1

(d)

Figure 13.10 (a) and (b) Detectabilities of faults for the circuit in Figure 13.9a. (c) and (d) Detectabilities of faults for the circuit in Figure 13.9b.

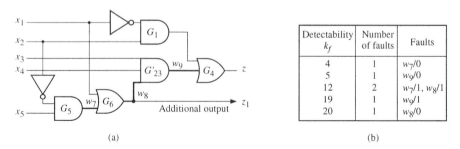

Figure 13.11 (a) Additional output to the circuit in Figure 13.9b. (b) Corresponding detectabilities.

Additional outputs (also called *observation points*) always improve the testability of a circuit. According to [ToMc 94], additional inputs (called *control points*) may improve the detectability for some faults but could lower the detectability for other faults. The author of this book believes that this observation is true for "classical" control points, i.e., additional control point to an additional OR gate after a NAND gate whose output 1-controllability is low (or vice-versa); however, if an additional input intervenes through an EXOR gate (according to Section 13.3) it could be more efficient.

Algorithms for systematic algebraic factoring (i.e., extraction of cubes or kernels), and application to some benchmark circuits are presented in [ToMc 94], [ChGu 94]. It has been observed that this factoring leads to circuits with fewer literals in most cases. On the other hand, the delays are increased since the combinational circuits have a greater number of levels (for example, the longest path corresponds to three gates in Figure 13.9a, and five gates in Figure 13.9b).

NOTES and REFERENCES

Design for testability in general is considered in many papers, including [McBo 81], [WiPa 82], [Ab 83], and [Th *et al.* 82] for microprocessors. Parts of the books [Fu 85], [Mc 86], [BaMcSa 87], and [AbBrFr 90] are also devoted to these approaches.

Test point insertion for random testing is considered in [Sa *et al.* 91]; as usual, the control points are obtained by ORing and ANDing additional inputs.

Extended specification was not widely considered in the literature. The complete specification of the field assigned to operation codes was explicitly used for a built-in self-test microprocessor [Th *et al.* 87].

The decorrelation by EXOR gates was extensively used for the BIST microprocessor cited above; generation of micro-orders and multiplexed bus access were also randomized by this means [Th *et al.* 87].

Design of random pattern testable combinational circuits is considered in [ToMc 94] and [ChGu 94], where algorithms based on algebraic factorization and application to benchmark circuits are presented. These applications to random testability are based on general results on factorization [Br *et al.* 87].

In [YoXiTa 97], a random pattern testable design, based on partial circuit duplication is proposed.

In [KrAl 91], it is shown that, contrary to common belief, a non-redundant circuit is not always easier to test than a redundant one.

Postface

As explained in Chapter 1, BIST has become an excellent way to solve some of the major problems of testing digital circuits. Furthermore, BIST leads naturally to the use of pseudorandom testing. It follows that this kind of testing is inevitable in some circumstances. However, the author's feeling is that, far from being an inescapable misfortune, the use of random testing may be a chance to perform a test whose quality may be very good, at a relatively low cost, even out of the context of BIST.

"Undoubtedly, every test engineer would like to use tests that have '100% fault coverage'. But what do such statements really mean? In deterministic testing, one postulates a fault model and then designs a test to detect all the faults in this model. In practice, it is impossible to take into account all the possible failures that may occur in a complex chip; thus some faults are certain to be missed by any test one designs, since the test models are always incomplete. The best one can hope for is to have a very good test which will detect a very large percentage of the common failures, but one has to accept the fact that once in a while a defective chip passes the test. *Given that one must accept some uncertainty about deterministic testing, one should also be willing to accept a similar uncertainty about probabilistic testing*"[1]. The uncertainty due to random testing may be as small as we want. Nevertherless, in order to achieve this end, random testing should not be performed with a method chosen at random (as already expressed in Section 1.3).

The first interesting feature of random testing is its simplicity (implying low cost) of implementation and development. There is no need for fault hypothesis to generate a random test sequence. Hypotheses on faults are necessary only for estimating the random test length required to reach some fault coverage or a lower bound of it (Chapters 6 to 9 are dedicated to these calculations for various kinds of circuits). What happens if a defective circuit is affected by a fault which is not in the hypotheses? Let us consider two cases: either the fault is a multiple fault whose components are in the hypothesis, or the fault is another non-targeted fault. The conjectures in Appendix M provide answers to the first case. Informally, these conjectures express that, if the random test length allows some "quality" of testing (minimum testing probability or fault coverage) for some model of faults, this quality is at least as high for multiple faults corresponding to the same model. It is very interesting not to have to deal with the multiple faults

---

[1] This comment is reprinted from [DaBrJü 93] (© 1993 IEEE).

since, as observed in [Wa 81], the multiple faults are far more numerous than the single faults. Now, for other non-targeted faults, a random test sequence may have a good chance of detecting them: this assertion is founded both on theoretical bases (Property 5.2 in Section 5.3.2) and experimental results (Chapter 11). Intermittent faults also have a chance of being detected (Sections 11.1.1 and 11.2, Figure 11.5 in Section 11.3.1, Remark 11.3 in Section 11.4.2).

The main weakness of random testing is the length required to reach a high defect coverage. In order to avoid a relatively long study to calculate this length, a rough upper bound is often sought [Th et al. 87], [ToMc 94]. Assume for example that a test duration of one second is admitted and that the circuit under consideration has a clock frequency of 20 MHz. Hence, a test length of 2×10^7 input vectors is acceptable. For this example, a fault requiring a test length less than 2×10^7 (for some probability of testing, for example 0.999) is qualified *random pattern testable*, while a fault requiring a test length greater than 2×10^7 is qualified *random pattern resistant*. If a circuit contains some resistant faults, there are basically two ways to tackle the problem. The first consists of shortening the test length using weighted input vectors instead of equally likely input vectors. Some examples are given in the book: for a gate in Section 10.1.1.2, for an example microprocessor in Section 9.5.1.2 (in Assumption 1), for the *6800* microprocessor in Section 9.5.2.2, and for the circuit called *AUT* in Section 11.2. Various works have been published on weighted random testing, including [Wu 87], [Wu 88], [BrGlKe 89], [Wa et al. 89], [GrSt 93], [ScLiCa 95]. The second way to reduce the test length is to modify the circuit in order that the resistant faults become random pattern testable (in this case, the generation is not weighted, hence it is simpler). Methods for reaching this goal are presented in Chapter 13. All the circuits of the ISCAS benchmark [BrFu 85] were modified to become random pattern testable[2] [Si et al. 92] by additional outputs and inputs to EXOR gates (Section 13.3).

In Chapter 11, several examples of experiments have shown that random testing is powerful. For example, the fault f_1 in Figure 11.5 related to the *6800* microprocessor (Section 11.3.1) is detected only for a particular situation: a *Reset* occurs during an operation on stack pointer. It is clear that such a fault is not in a classic fault model and may be missed by a deterministic test sequence. However, if an *ideal random test* is performed (Definition 5.1 in Section 5.2), sooner or later this particular situation occurs and the defect is detected. In [KrGa 93] it was shown that, in order to meet future test quality requirements (defect level of 1 *ppm*) for very large embedded RAMs, the use of random testing will become necessary. This is certainly true for other kinds of circuits.

---

[2] After the modification, any fault is redundant or random pattern testable in the admitted test time.

Let us recall an important result: for *any combinational or sequential fault*, the detection uncertainty related to a test length L is

$$\varepsilon(L) \approx \alpha^L, \qquad (P.1)$$

where $\alpha = 1 - p$ for a combinational fault (Equation (6.9) in Section 6.11.2) and $\alpha = \lambda_2$ for a sequential fault (Equation (7.12) in Section 7.2.2)[3]. It is remarkable that the detection uncertainty ε is exponential in L. This implies that the value of ε decreases strongly when L increases slightly. For example, assume a fault for which $\varepsilon = 10^{-3}$ for the test length L; according to (P.1), $\varepsilon = 10^{-6}$ if the test length is $2L$, $\varepsilon = 10^{-9}$ if the test length is $3L$.

In Section 8.4.2, the following observation was made about faults in RAMs: deterministic testing is much shorter than random testing for simple faults (for example stuck-at fault), and may be longer for more complicated faults (for example double toggling); when the faults are very complicated, particularly multiple faults, it becomes practically impossible to obtain a deterministic algorithm. This observation, even if it cannot be formally proven, is more general than for RAMs only. Informally, we could say that, in many cases, the faults which are *difficult to analyse* (for both deterministic and random testing) are *easy to detect* by random testing. Let us give two examples. First example: multiple faults are very difficult to analyse when the multiplicity is large, particularly if the components belong to several basic fault models. According to Conjectures M.1 and M.2, they are at least easier to test than the single faults, and, in fact, they are often much easier. Second example: the faults in the control section of a microprocessor are more difficult to analyse (because the graph model is modified) than the faults in the data processing section (Sections 9.4 and 9.3, respectively). However, the faults in the control section require shorter test lengths (Sections 9.5.1.4 and 9.5.2.1) because the behavior of the microprocessor is badly affected by such a fault.

---

[3] The value p is the detection probability of the combinational fault and λ_2 is the greatest eigenvalue (except value 1) of the transition matrix of the Markov chain corresponding to the detection process of the fault under consideration.

Appendix A

Random Pattern Sources

A random pattern source provides a random input sequence which is applied to a circuit under test and a reference circuit as shown in Figures 3.16 (Section 3.2.2.1) and 5.1 (Section 5.1), or more abstractly to an observer (Section 3.2.2).

From an alphabet X, let X^ω denote the set of infinite sequences over X. A **source** is a pair $S = (X, \psi)$ were X is an alphabet and ψ is a probability distribution over X^ω [Fe 68], [BrJü 92].

For any finite sequence $S \in X^*$, the probability of occurrence, denoted by $\psi(S)$, is $\psi(S) = \psi(SX^\omega)$, i.e., it is the probability that a random infinite sequence has S as a prefix.

In most of the rest of this book, only the special cases of memoryless and Markov sources are considered.

A.1 MEMORYLESS SOURCE: Constant Distribution

In a **memoryless source**,

$$\psi(x(1)x(2)...x(t)) = \prod_{i=1}^{t} \psi(x(i))$$

is obtained for every $t > 0$ and any $x(1)$, $x(2)$, ..., $x(t)$ in X. This means that a random pattern is drawn with a vector of probabilities which is *constant*: the distribution is called a **constant distribution**. According to the notation already used, $\psi(x_i) = \Pr[x_i]$ can be written. If $X = \{x_0, x_1, ...\}$, a constant distribution may be characterized by a vector $\psi = (\Pr[x_0], \Pr[x_1], ...)$.

Assume a 2-input circuit. There are 4 possible input patterns, namely x_0, x_1, x_2, and x_3. A *constant distribution* is a vector $\psi = (\Pr[x_0], \Pr[x_1], \Pr[x_2], \Pr[x_3])$, where the probabilities $\Pr[x_i]$, $i = 0$ to 3, are constant, and such that $\Sigma \Pr[x_i] = 1$. For example $\psi_0 = (.25, .25, .25, .25)$, $\psi_1 = (.5, .3, .1, .1)$, and $\psi_2 = (.5, .3, 0, .2)$ are three constant distributions.

Remark A.1 The language containing all the sequences which can be drawn from a constant distribution (this is also true for a generalized distribution) depends only on whether a pattern has a *zero probability or not*. For example, the same language $(x_0+x_1+x_2+x_3)^*$ corresponds to both ψ_0 and ψ_1, even if a given sequence in this language has not the same probability of occurrence for both distributions. The language corresponding to ψ_2 is $(x_0+x_1+x_3)^*$.

A.2 MARKOV SOURCE: Generalized Distribution

(a)
	x_0	x_1	x_2	x_3	
* r_0	r_0	r_0	r_0	r_0	$\psi^{r_0} = \psi_0 = (.25,.25,.25,.25)$
* r_0	r_0	r_0	r_0	r_0	$\psi^{r_0} = \psi_1 = (.5,.3,.1,.1)$
* r_0	r_0	r_0	—	r_0	$\psi^{r_0} = \psi_2 = (.5,.3,0,.2)$

(b)
	x_0	x_1	x_2	x_3	
* r_0	r_1	r_1	r_1	r_1	$\psi^{r_0} = \psi_3 = (.81,.09,.09,.01)$
r_1	r_2	r_2	r_2	r_2	$\psi^{r_1} = \psi_3 = (.81,.09,.09,.01)$
r_2	r_3	r_3	r_3	r_3	$\psi^{r_2} = \psi_3 = (.81,.09,.09,.01)$
r_3	r_3	r_3	r_3	r_3	$\psi^{r_3} = \psi_4 = (.01,.09,.09,.81)$

(c)
	x_0	x_1	x_2	x_3	
* r_0	r_1	r_1	—	—	$\psi^{r_0} = \psi_5 = (.5,.5,0,0)$
r_1	—	—	r_0	r_0	$\psi^{r_1} = \psi_6 = (0,0,.5,.5)$

(d)
	x_0	x_1	x_2	x_3	
* r_0	r_1	r_0	r_0	r_0	$\psi^{r_0} = \psi_0 = (.25,.25,.25,.25)$
r_1	r_0	—	r_0	r_0	$\psi^{r_1} = \psi_7 = (.33,0,.33,.33)$

Figure A.1 (a) Constant distributions (memoryless sources). (b)-(d) Some examples of generalized distributions (Markov sources).

A (*homogeneous*) **Markov source** may be modeled by a Moore machine $\psi = (R, X, \Psi, \delta, \mu, r_0)$ where X is the set of possible input patterns and Ψ is a set of constant distributions $\{\psi^{r_j}\}$. When a state r_j is reached, a random input pattern is drawn with the distribution ψ^{r_j}. If the pattern x is drawn, then the state $\delta(r_j,x)$ is reached, and so on.

Then,

$$\psi(x(1)x(2)...x(t)) = \psi^{r_0} \cdot \prod_{i=2}^{t} \psi^{\delta(r_0,x(1)...x(i-1))}(x(i))$$

for every $t > 0$ and any $x(1), x(2), ..., x(t)$ in X. The distribution obtained from a Markov source will be called a **generalized distribution**.

If R contains a single state r_0, then the generalized distribution reduces to a constant distribution. This is illustrated in Figure A.1a for the distributions ψ_0, ψ_1 and ψ_2 presented above.

The generalized distribution modeled in Figure A.1b is such that the 3 first vectors of the sequence are drawn with the constant distribution ψ_3 and the following ones with the constant distibution ψ_4. Since $x_0 = 00$, $x_1 = 01$, $x_2 = 10$, and $x_3 = 11$, the reader may verify that ψ_3 corresponds to a probability $\Pr[x_1 = 1]$ $= \Pr[x_2 = 1] = 0.1$; for example, $\Pr[x_2] = \Pr[x_1 x_2 = 10] = 0.1 \times 0.9 = 0.09$. Similarly, ψ_4 corresponds to a probability $\Pr[x_1 = 1] = \Pr[x_2 = 1] = 0.9$.

Figure A.1c represents a generalized distribution such that the odd vectors are either x_0 or x_1 and the even vectors are either x_2 or x_3.

The generalized distribution modeled in Figure A.1d is such that the pattern x_1 is never drawn just after the pattern x_0 (all the other combinations are possible).

The word **distribution** is used for either a constant distribution or a generalized distribution.

Remark A.2 A probability distribution over X^ω would require an infinite number of states (infinite Markov chain) in the general case. Then it could not be implemented.

Remark A.3 We assume that source S is independent of the machines it is applied to. Thus, adaptive diagnosis, where the response influences the test sequence, is not covered.

A.3 GENERATORS

Software and hardware generations of random test sequences are presented in Chapter 10. As far as hardware generation is concerned, all the generators can be based on an equally likely random pattern generator. This is illustrated in Figure A.2.

Appendix A

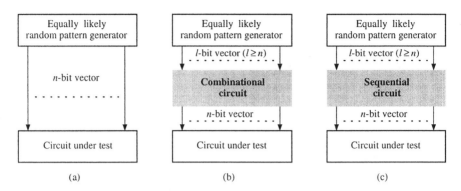

Figure A.2 Generators. (a) Equally likely distribution ψ_0. (b) Weighted constant distribution. (c) Generalized distribution.

If *equally likely* random test vectors are required (constant distribution ψ_0), an n-bit output vector of the generator is the input vector of the circuit under test. See Figure A.2a.

If **weighted test vectors** are required (i.e., a *weighted constant distribution*), they can be obtained from equally likely random vectors by a mapping through a *combinational circuit* as shown in Figure A.2b (examples are given in Section 10.3.2.2, and in Exercises 10.3 and 10.4).

A *generalized distribution* requires a *sequential circuit* between the output of the equally likely random pattern generator and the inputs of the circuit under test as shown in Figure A.2c (examples are given in Sections 10.3.3 and 10.3.4, and in Exercise 10.5).

Appendix B

Calculation of a Probability of Complete Fault Coverage

As was observed in Section 4.4.2, complete fault coverage of the stuck-at faults in Figure 4.4b, is obtained if and only if all the faults f_1, f_4, f_5 and f_6 are tested. Faults f_1, f_4 and f_5 are tested by x_7, x_3, and x_5, respectively. Fault f_6 is tested by any pattern in the set $\{x_0, x_2, x_4\}$. From this information, the observer can be built for the set of faults $\{f_1, f_4, f_5, f_6\}$.

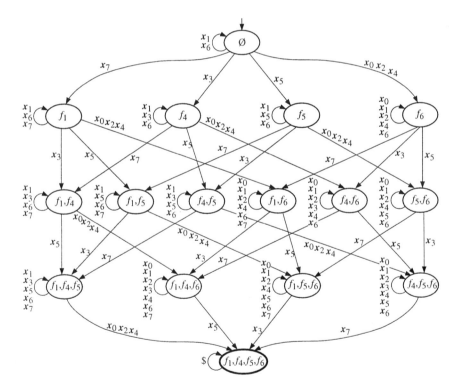

Figure B.1 Observer for the set of faults $\{f_1, f_4, f_5, f_6\}$ in Figure 4.4b (simplified representation for combinational faults).

361

Appendix B

This observer is presented in Figure B.1 using the simplified representation for combinational faults which was presented in Figure 3.18c (Section 3.2.2.2).

From the observer in Figure B.1, a discrete time Markov chain (Appendix C) can be obtained. The states of the Markov chain correspond to the states of the observer; let us denote $[f_i, ..., f_j]$ the state where the faults f_i, ..., f_j are detected. Every arc has a probability, corresponding to the probability of occurrence of the corresponding inputs. For example the probability $\Pr[x_3]$ is associated with the arc $[f_1] \rightarrow [f_1, f_4]$, the probability $\Pr[x_0] + \Pr[x_2] + \Pr[x_4]$ is associated with the arc $[f_1] \rightarrow [f_1, f_6]$. The probability 1 is associated with the self loop of the absorbing state, labelled $. The initial state corresponds to the probability 1 for the state [Ø] and 0 for the other states. The probabilities of the various states after L time units (corresponding to L successive random vectors), can be calculated by classical methods (Appendix C). The probability P_c of complete fault coverage is the probability that the absorbing state $[f_1, f_4, f_5, f_6]$ has been reached. For equally likely input vectors, i.e., $\Pr[x_i] = 0.125$ for all i, and for $L = 20$, $P_c = \Pr[\text{state } [f_1, f_4, f_5, f_6]] = 0.802$ is obtained.

Appendix C

Finite Markov Chains

Only some basic concepts that are useful in this book are presented here. For more details, the reader is referred, for example, to [KeSn 76].

We are interested in *discrete time* processes. Given a set of states, a Markov process is such that the state at time t depends only of the state at time $t - 1$ and of probabilities of transitions between states. A **finite Markov chain** is a Markov process with finitely many states whose transition probabilities do not depend on time.

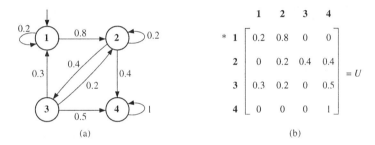

Figure C.1 Finite Markov chain. (a) Graph representation. (b) Matrix representation[1].

Figure C.1 represents a 4-state Markov chain. Part a is a graph representation. At $t = 0$, the initial state is **1**. At time $t = 1$, the probability of remaining in state **1** is 0.2 and that of reaching state **2** is 0.8, and so on. Figure C.1b is the matrix representation of that process. The array U is the **transition matrix**. The element $U_{i,j}$ is the probability of being in state j at time $t + 1$, given the system is in state i at time t. Matrix U is a stochastic matrix i.e.,

$$\sum_{j:j \text{ is a state}} U_{i,j} = 1.$$

The vector of **state probabilities** at time t is denoted $\pi(t)$. Then $\pi(0)$ is the initial **probability vector**; in our example[2] $\pi(0) = (1, 0, 0, 0)$ but, in general,

---
[1] This matrix is reprinted from [Da 97] (© 1997 IEEE).
[2] In the text, we use commas to separate the components of a vector.

363

$\pi(0)$ may be any vector such that the sum of the elements is 1. From the definitions, one has $\pi(t) = \pi(t-1) \cdot U$, for $t > 0$. Then, by iteration:

$$\pi(t) = \pi(0) \cdot U^t, \quad t \geq 0. \tag{C.1}$$

We may visualize the Markov chain as a process which moves from state to state. If at any time it is in state i, then it moves at the next "step" to state j with conditional probability $U_{i,j}$.

Consider a set W of states such that, for any pair (i, j) of states in W, Pr[reach j in one or several steps | present state is i] > 0. The corresponding subgraph is said to be **strongly connected** and the corresponding submatrix, **irreducible**. We are interested in the largest sets with that property: they correspond to equivalence classes. In our example, the set $\{2, 3\}$ is included in the set $\{1, 2, 3\}$ which corresponds to an equivalence class. Similarly the set $\{4\}$ has this property. It is always possible to split the whole set of states into equivalence classes. The set $\{1, 2, 3\}$ is a **transient set** (*transient class*) since there is no way to return to a state in that set once the chain has left that set. The set $\{4\}$ is an **ergodic set** (*ergodic class*) since there is no way to leave it once one of its states (in general, an ergodic set may contain more than one state) has been reached. The states in a transient set are **transient states** and the states in an ergodic set are **ergodic states**.

An ergodic set containing a single state, is an **absorbing state**. If every ergodic set contains a single state, the chain is an **absorbing chain**. Each equivalence class can be partitioned into cyclic classes. For example $\begin{bmatrix} 0 & U_2 \\ U_1 & 0 \end{bmatrix}$ where the 0's are square matrices contains 2 cyclic classes; that means that, if the chain is in a state of a class at time t, it will be in a state of the other class at time $t + 1$. If there is only one cyclic class, then the equivalence class is said to be **regular**, otherwise it is said to be **cyclic**.

If U is irreducible and regular,

$$\lim_{t \to \infty} \pi(0) \cdot U^t = \pi(\infty), \tag{C.2}$$

where $\pi(\infty)$ does not depend on $\pi(0)$. That means that the probability of each state q tends to a limit value $\pi_q(\infty)$ which is independent from the initial state. If U is irreducible and cyclic, an average value $\pi_q(\infty)$ may be obtained. For example, if U contains 2 cyclic classes, $\pi_q(\infty)$ may be defined as the limit of $\pi(0) \cdot (U^t + U^{t+1}) / 2$, also independent from $\pi(0)$. This property, i.e., the independence of $\pi(\infty)$ from $\pi(0)$, remains true if U is reducible but contains a single ergodic class; in that case $\pi_q(\infty) \neq 0$ if q is an ergodic state and $\pi_q(\infty) = 0$ otherwise.

Appendix D

Black-Box Fault Model

The content of this Appendix is drawn from works which were carried out in the CNET (*Centre National d'Etudes des Télécommunications*) of Grenoble [Si *et al.* 92], [Si 92].

The black-box fault model is a simple fault model which can apply to gate or higher level descriptions, and which can model realistic faults including all the single stuck-at faults.

D.1 BLACK-BOX MODEL OF A CIRCUIT

A combinational circuit can be split into black-boxes (BBs) using the following rules:

1) every line between BBs is unidirectional (so, any bidirectional wire in the initial circuit must be embedded in one BB);

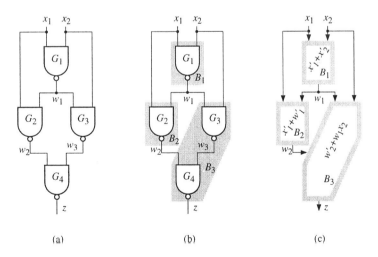

Figure D.1 Example of black-box model. (a) Initial circuit. (b) A splitting into three black-boxes. (c) The necessary and sufficient information.

2) every transistor in the initial circuit belongs to one and only one BB;
3) a BB is a v-input, 1-output, combinational function[1].

The BB model is illustrated in Figure D.1. Part (a) of the figure is the initial circuit made of 4 gates. Part (b) corresponds to a splitting into 3 BBs: the BBs B_1 and B_2 correspond to the gates G_1 and G_2, respectively, and the BB B_3 contains G_3, G_4, and the line w_3. Figure D.1c illustrates the information which is used in this model (the logic function but not the structure in the BB).

D.2 BB-FAULTS

The BB-faults are defined from the *combinations* which can be applied to BBs. In order to simplify the notation, the BB inputs have been ordered beforehand; we consider here the order from left to right or from top to bottom, depending on the drawing. A fault f which *provokes a faulty output* of the BB B_j if and only if the input **combination** a is applied to B_j is denoted

$$f = [a, B_j]. \tag{D.1}$$

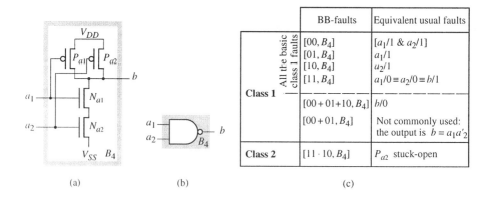

Figure D.2 Examples of BB-faults. (a) BB defined from the transistor level. (b) BB defined from the gate level. (c) Some examples of BB-faults.

---

[1] In [Si *et al.* 92] a BB may have several outputs. We restrict our attention to BBs with a *single output* since this model is easier to use.
We will consider examples of initial circuits modeled at the *gate-level*. A small example of model at the transistor level is given in Figure D.2. A more significant BB model obtained from a *transistor-level* model is shown in [Si *et al.* 92].

Some examples of BB-faults are presented in Figure D.2. For example $[01, B_4]$ is equivalent to the single stuck-at fault $a_1/1$ and $[00, B_4]$ is equivalent to a double stuck-at fault. A fault denoted by $[a_1+a_2, B_j]$ provokes a faulty output of B_j if and only if either the input combination a_1 or the combination a_2 is applied to B_j. For example, $[00 + 01 + 10, B_4]$ corresponds to $b/0$. A fault denoted by $[a_1 \cdot a_2, B_j]$ provokes a faulty output of B_j if and only if the sequence of two input combinations a_1 followed by a_2 is applied to B_j. For example, $[11 \cdot 10, B_4]$ corresponds to the stuck-open of the transistor P_{a2} (see Figure 2.15). *In the general case, a BB-fault may be denoted by* (and defined by):

$$f = [\sum S_i, B_j], \tag{D.2}$$

where $\sum S_i$ represents a set of sequences on the alphabet of input combinations of B_j, $\{a_0, ..., a_{2^r-1}\}$. In this set: 1) the length of every sequence is at least 1; 2) if S_h is in the set, there is no other sequence S_k such that S_h is a subsequence of S_k.

Three classes of faults are used.

Class 1. A class 1 fault is such that $\sum S_i$ contains at least one sequence of length 1. A **basic class 1 fault** is a class 1 fault such that $\sum S_i$ contains *exactly* one sequence.

Class 2. A class 2 fault is such that $\sum S_i$ contains no sequence of length 1 and contains at least one sequence of length 2. A **basic class 2 fault** is a class 2 fault such that $\sum S_i$ contains *exactly* one sequence.

Class 3. A class 3 fault is such that $\sum S_i$ contains no sequence of length 1 or 2 and contains at least one sequence of length 3 or more.

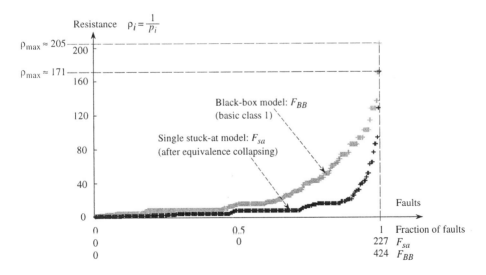

Figure D.3 Comparison of fault models (ALU circuit).

One can observe that any single or multiple stuck-at fault related to B_j, either is equivalent to a basic class 1 fault, or dominates a basic class 1 fault (the only single stuck-at fault which is not equivalent to a basic class 1 fault is $b/0$ which dominates three out of these faults).

Let us call F_{BB} the set of *single basic class 1 faults*. A test set detecting all the faults in F_{BB} detects all the single stuck-at faults (set F_{sa}), for any splitting into BBs; hence, given ε, the required test length must be larger for F_{BB} than for F_{sa}. This is illustrated in Figure D.3 for the ALU circuit (presented in Section 6.3). In this figure, the resistance profiles for the two sets of faults are represented (the BB model is such that each BB corresponds to a gate). It appears clearly that the continuous resistance profile (defined in Remark 6.4b) related to F_{sa} is under the one related to F_{BB}. Then, from Equation (6.29) in Section 6.1.2.1, the test length is larger for F_{BB} than for F_{sa}. However, the order of magnitude is the same: for $\varepsilon = 0.001$, $L(\varepsilon) = 554$ and 372 is obtained, respectively.

D.3 CONTROLLABILITY, ACTIVITY AND SOME PROPERTIES

Assume a random test sequence. When some random input vector x is applied to the circuit, there is some probability that the combination a is applied to the black-box B_j. This probability, denoted by $C(a, B_j)$ is called the **controllability**[2] of a to B_j.

Let a be an input combination of B_j. The **activity** of this combination, denoted by $A(a, B_j)$ was defined in Section 6.2.2.1. It is the probability that 1) the combination a is present at the input of B_j and 2) at least one primary output is faulty if the fault $[a, B_j]$ is present.

The following three properties are direct consequences of the definitions.

Property D.1 If an input sequence tests all the *basic* class 1 faults of B_j, it tests all the class 1 faults related to B_j.

❑

From the definitions, any class 1 fault of B_j dominates at least one basic class 1 fault of B_j. For example both $[a_1+a_2, B_j]$ and $[a_1+a_3a_4, B_j]$ dominate $[a_1, B_j]$.

Property D.2 Let $[a, B_j]$ be a basic class 1 fault. Its detection probability is

$$p_{[a,B_j]} = A(a, B_j). \tag{D.3}$$

---

[2] This notion is a generalization of the usual notion of controllability.

Property D.3 Let F_{BB} be the set of *single basic class 1 faults* in a circuit split into a set of BBs. The *detection probability of the most resistant* among the *detectable*[3] faults in F_{BB} is:

$$p_{\min} = \min_{B: B \text{ is a BB}} \left(\min_{a: a \text{ is an input combination of } B} (A(a,B)) \right). \tag{D.4}$$

❑

It appears that the BB model is an interesting fault model: it includes the single stuck-at fault model and, according to Property D.3 and thanks to the notion of *activity*, the set F_{BB} is relatively easy to manage. If v_j denotes the number of inputs of B_j, F_{BB} contains $\sum_{j: B_j \text{ is a BB}} 2^{v_j}$ faults.

D.4 SIMULATION

Estimation of the testability performed at CNET is based on the BB model [Si *et al.* 92]. The circuits are split into BBs corresponding to gates and the set of faults in F_{BB} is considered. The only objective is that the most resistant fault among the detectable faults in F_{BB} is tested with a probability 0.999 within a reasonable time, say 1 second. The following algorithm[4] is used.

Macrostep A. Simulation
  Step 1. A random vector x is generated at the primary inputs.
  Step 2. Logical values are propagated by logical simulation all over from the primary inputs to the primary outputs.
  Step 3. For each BB, the controllability counter $NC(a, B_j)$ is incremented if a is the input vector applied to B_j.
  Step 4. For each BB, the activity counter $NA(a, B_j)$ is incremented (for this step, the output value of B_j is complemented and a propagation to the primary outputs is simulated).
  Step 5. If the maximum computation time has not elapsed and if all the counters have not reached a predefined value, go to *step1*.
  Step 6. A file with the lists of counters $NC(a, B_j)$ and $NA(a, B_j)$ which are empty is supplied to the formal prover used in the following macrostep.

Macrostep B. Formal proof
  If $NC(a, B_j) = 0$, a formal prover decides if $C(a, B_j) = 0$ or if this controllability has a very low value. If $C(a, B_j) \neq 0$ and $NA(a, B_j) = 0$, the formal

---
  [3] Hence, the minimum value of $A(a, B)$ excludes the value 0.
  [4] Several circuits with random built-in-test have been designed this way, including a 16-bit microprocessor and various telecommunication circuits; this algorithm has been applied to the ISCAS benchmark circuits [Si *et al.* 92].

prover decides if $A(a, B_j) = 0$ or if this activity has a very low value. (The principle of the formal prover is given in [Si *et al.* 92].)

Macrostep C. Lower bound of p_{min}

If $NA(a, B_j) \neq 0$, $p_{[a,B_j]}$ is estimated by $NA(a, B_j) / 2^n$. If $NA(a, B_j) = 0$ and $A(a, B_j) \neq 0$, the value of $A(a, B_j)$ was obtained in *macrostep B*. If $A(a, B) = 0$, the fault $[a, B]$ is redundant, then does not change the input/output behavior of the circuit. Then p_{min} is estimated by Equation (D.4).

If p_{min} greater than a value allowing a test in less than 1 second, then END.

Macrostep D. Circuit modification

The circuit is modified in order to improve the testability for every fault such that $A(a, B)$ has a very low value other than zero. (For circuit modification, see Chapter 13.) Go to *macrostep A*.

Remark D.1 Some alternatives to the preceding algorithm may exist. When all the counters of some BB have reached a predefined value, this BB is no longer submitted to *steps 3* and *4*. If some redundant fault obviously remains redundant after circuit modification, it may be ignored in the next simulation. If the prover has not been able to decide if a fault is redundant (too time consuming), the fault is assumed to be non-redundant: the circuit will be modified to detect it more easily.

Appendix E

Exact Calculation of Activities

The content of this Appendix is drawn from works which were carried out in the CNET (*Centre National d'Etudes des Télécommunications*) of Grenoble [Si 92].

E.1 PARTIAL OBSERVABILITIES

This calculation draws inspiration from the BB-fault model. The examples consider a splitting into BBs corresponding to gates but any other splitting is possible. The main notions are presented in Section 6.2.2.1. The method for obtaining the *observability of a fanout stem* has still to be shown.

The *conditional observability* of fanout stem d, given the pattern D_i of values on the fanout stems, $O(d \mid D_i)$, was introduced in Section 6.2.2.1. From now on, all the observabilities will be conditional; then the word *conditional* is omitted and implicit. Let us introduce new notions. The **local observability** on line w is noted $O(d \mid D_i, w)$; then $O(d \mid D_i) = O(d \mid D_i, z)$ for a single output circuit. The **partial observability**, $O(d \mid D_i, w = \alpha)$, $\alpha \in \{0, 1\}$, is the observability of d given the pattern D_i on line w when the value of w is α, i.e.,

$$O(d \mid D_i, w = \alpha) = \Pr[(d \text{ is observable on } w \text{ given } D_i) \text{ AND } (w = \alpha)], \quad (E.1)$$

$$NO(d \mid D_i, w = \alpha) = \Pr[(d \text{ is not observable on } w \text{ given } D_i) \text{ AND } (w = \alpha)]. (E.1')$$

From the above definitions and the definition on conditional controllability (Section 6.2.2.1) it follows:

$$O(d \mid D_i, w) = O(d \mid D_i, w = 0) + O(d \mid D_i, w = 1), \quad (E.2)$$

$$O(d \mid D_i, w = \alpha) + NO(d \mid D_i, w = \alpha) = C(\alpha, w \mid D_i), \quad \alpha = 0, 1, \quad (E.3)$$

$$O(d \mid D_i, w = 0) + NO(d \mid D_i, w = 0) + O(d \mid D_i, w = 1) + NO(d \mid D_i, w = 1) = 1. \quad (E.4)$$

Partial observabilities of a fanout stem d can be propagated from d to the primary outputs as shown in Figure E.1. Assume the stem d, whose fault-free value is defined from D_i, is in the upstream of the gate under consideration.

372 *Appendix E*

Gate	$O(d|D_i, b=0)$		$O(d|D_i, b=1)$	
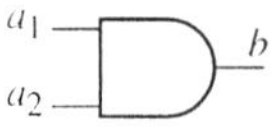 AND (a₁, a₂ → b)	$O(d|D_i, a_1=0) \cdot O(d|D_i, a_2=0)$ $+ O(d|D_i, a_1=0) \cdot NO(d|D_i, a_2=1)$ $+ NO(d|D_i, a_1=1) \cdot O(d|D_i, a_2=0)$	(E.5)	$O(d|D_i, a_1=1) \cdot O(d|D_i, a_2=1)$ $+ O(d|D_i, a_1=1) \cdot NO(d|D_i, a_2=1)$ $+ NO(d|D_i, a_1=1) \cdot O(d|D_i, a_2=1)$	(E.6)
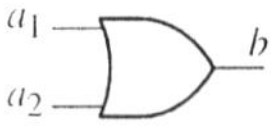 OR (a₁, a₂ → b)	$O(d|D_i, a_1=0) \cdot O(d|D_i, a_2=0)$ $+ O(d|D_i, a_1=0) \cdot NO(d|D_i, a_2=0)$ $+ NO(d|D_i, a_1=0) \cdot O(d|D_i, a_2=0)$	(E.7)	$O(d|D_i, a_1=1) \cdot O(d|D_i, a_2=1)$ $+ O(d|D_i, a_1=1) \cdot NO(d|D_i, a_2=0)$ $+ NO(d|D_i, a_1=0) \cdot O(d|D_i, a_2=1)$	(E.8)
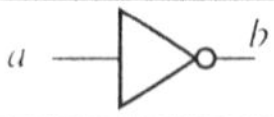 NOT (a → b)	$O(d|D_i, a=1)$	(E.9)	$O(d|D_i, a=0)$	(E.10)

Figure E.1 Propagation of partial observabilities of a fanout stem.

Consider the case $O(d \mid D_i, b=0)$ for a two input AND gate. Since this partial observability is related to the value 0 of the output b, three possible input combinations must be considered, namely $a_1 a_2 = 00$ or 01 or 10.

1) Assume $a_1 a_2 = 00$ (for the fault-free circuit); if d is faulty, the error will reach b (i.e., $b = 1$ in the faulty circuit) only if $a_1 = a_2 = 1$ in the faulty circuit, i.e., both a_1 and a_2 are faulty; that means that d is observable on b if it is observable on both a_1 and a_2 when both values should be 0. The corresponding probability is

$$O(d \mid D_i, a_1=0) \cdot O(d \mid D_i, a_2=0).$$

2) Assume now $a_1 a_2 = 01$ (for the fault-free circuit); if d is faulty, the error will reach b only if $a_1 = a_2 = 1$ in the faulty circuit; that means that d is observable on b if it is observable on a_1 (because 1 is the faulty value) but not on a_2 (because 0 is the fault-free value). The corresponding probability is

$$O(d \mid D_i, a_1=0) \cdot NO(d \mid D_i, a_2=1).$$

3) The case $a_1 a_2 = 10$ is symmetrical: the probability is

$$NO(d \mid D_i, a_1=1) \cdot O(d \mid D_i, a_2=0).$$

Equation (E.5) in Figure E.1 is obtained as the sum of the three terms. If d is not in the upstream of the AND gate, then it is observable neither on a_1 nor on a_2. That implies that $O(d \mid D_i, a_1=0) = 0$ and $O(d \mid D_i, a_2=0) = 0$, then $O(d \mid D_i, b=0) = 0$ is obtained from Equation (E.5). It follows that this equation is true whether d is in the upstream of b or not.

Equations (E.6) to (E.10) are obtained in a similar way. Equations (E.5) to (E.8) are easily generalized to v-input gates.

The method for obtaining $O(d \mid D_0)$, used in Figure 6.11 (Section 6.2.2.1), is shown in Figure E.2. The partial observabilities on the inputs of the subcircuit (w_1, w_2, and x_3) are first calculated.

Exact Calculation of Activities 373

1) *Input* w_1. Given D_0, $d = 0$. Since $w_1 = d$, $w_1 = 0$. It follows that $O(d \mid D_0, w_1 = 0) = 1$ and $O(d \mid D_0, w_1 = 1) = 0$.

Now, the conditional controllabilities are $C(1, w_1 \mid D_0) = 0$ (Figure 6.11) and $C(0, w_1 \mid D_0) = 1 - C(1, w_1 \mid D_0) = 1$. Next, the partial non-observabilities $NO(d \mid D_0, w_1 = 0) = 0$ and $NO(d \mid D_0, w_1 = 1) = 1$ are obtained from these controllabilities and Equation (E.3).

2) *Input* w_2. The calculation is quite similar.

3) *Input* x_3. Since x_3 is independent from d, the observability of d on x_3 is nul: $O(d \mid D_0, x_3 = 0) = O(d \mid D_0, x_3 = 1) = 0$.

Now, the conditional controlabilities are $C(1, x_3 \mid D_0) = 0.5$ (Figure 6.11) and $C(0, x_3 \mid D_0) = 1 - C(1, w_1 \mid D_0) = 0.5$. Next, the partial non-observabilities $NO(d \mid D_0, w_1 = 0) = 0.5$ and $NO(d \mid D_0, w_1 = 1) = 0.5$ are obtained from these controllabilities and Equation (E.3).

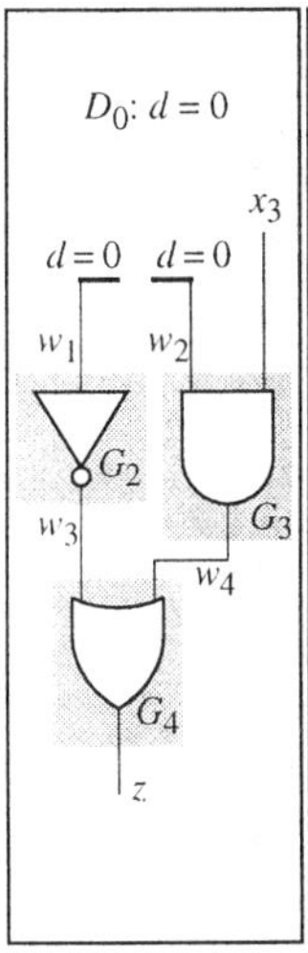

	Input partial observabilities	Obtained from (E.3) and conditional controllabilities in Fig. 6.11
w_1	$O(d \mid D_0, w_1 = 0) = 1$ $O(d \mid D_0, w_1 = 1) = 0$	$NO(d \mid D_0, w_1 = 0) = 1 - 1 = 0$ $NO(d \mid D_0, w_1 = 1) = 0 - 0 = 0$
w_2	$O(d \mid D_0, w_2 = 0) = 1$ $O(d \mid D_0, w_2 = 1) = 0$	$NO(d \mid D_0, w_2 = 0) = 1 - 1 = 0$ $NO(d \mid D_0, w_2 = 1) = 0 - 0 = 0$
x_3	$O(d \mid D_0, x_3 = 0) = 0$ $O(d \mid D_0, x_3 = 1) = 0$	$NO(d \mid D_0, x_3 = 0) = 0.5 - 0 = 0.5$ $NO(d \mid D_0, x_3 = 1) = 0.5 - 0 = 0.5$
	Propagation of partial observabilities Equations (E.5) to (E.10)	
w_3	$O(d \mid D_0, w_3 = 0) = 1$ $O(d \mid D_0, w_3 = 1) = 0$	$NO(d \mid D_0, w_3 = 0) = 0 - 0 = 0$ $NO(d \mid D_0, w_3 = 1) = 1 - 1 = 0$
w_4	$O(d \mid D_0, w_4 = 0) = 0.5$ $O(d \mid D_0, w_4 = 1) = 0$	$NO(d \mid D_0, w_4 = 0) = 1 - 0.5 = 0.5$ $NO(d \mid D_0, w_4 = 1) = 0 - 0 = 0$
z	$O(d \mid D_0, z = 0) = 0$ $O(d \mid D_0, z = 1) = 0.5$	$NO(d \mid D_0, z = 0) = 0 - 0 = 0$ $NO(d \mid D_0, z = 1) = 1 - 0.5 = 0.5$

Figure E.2 Application to the subcircuit (c) in Figure 6.11.

When the partial observabilities and partial non-observabilities have been obtained for the inputs of the subcircuit, they can be obtained for w_3 and w_4, then for z, using the propagation equations in Figure E.1.

Finally $O(d \mid D_0) = O(d \mid D_0, z = 0) + O(d \mid D_0, z = 1) = 0 + 0.5 = 0.5$. The partial observability $O(d \mid D_1)$ is obtained in a similar way.

Generalization to an m-output circuit. In this case the conditional observability of a fanout stem is obtained by the following expression where the product corresponds to the conditional probability that no output is affected by an error on d.

$$O(d\mid D_i) = 1 - \prod_{j=1}^{m}\Big(NO(d\mid D_i, z_j=0) + NO(d\mid D_i, z_j=1)\Big) \quad (E.11)$$

E.2 EXAMPLE OF LOWER BOUND p_{min}

As explained in Section 6.2.2.1, various fault models can be considered when all the conditional activities have been obtained. Let us consider the set of BB-faults F_{BB} in the initial circuit in Figure 6.11 where every gate is a BB. An activity is obtained by the weighted sum of conditional activities, as explained in Chapter 6. Values $A(00, G_3) = A(01, G_3) = A(11, G_4) = 0$ are obtained, corresponding to three redundant faults. For all the other activities, the minimum value is 0.125. Then $p_{min} = 0.125$.

Remark E.1 The fault $[a_1 \cdot a_2, B]$ is a basic class 2 fault (a sequential fault which may correspond to a stuck-open for example). If this fault is not redundant,

$$p_{[a_1 \cdot a_2, B]} = C(a_1, B) \cdot A(a_2, B).$$

A lower bound of its detection probability is

$$p_{[a_1 \cdot a_2, B]} \geq \min_{a}(C(a,B)) \cdot \min_{a}(A(a,B)),$$

such that a is an input combination of B. However, this lower bound may be very pessimistic, according to Section K.3 in Appendix K.

Appendix F

Comparing Asynchronous and Synchronous Tests

As explained in Section 7.1, a synchronous sequential circuit could be tested either by an asynchronous test or by a synchronous test.

Asynchronous test. Periodically, a single input, randomly chosen, is changed. Then the output state is compared with the fault-free one before the next input changes. This is illustrated in Figure 7.1 (Section 7.1). For a synchronous circuit, the clock is considered as any ordinary input.

Synchronous test. The input vector is changed periodically. Between two changes there is a clock pulse. This is illustrated in Figure 7.2; part of this figure is represented again in Figure F.1. The comparison of the output state is made either in the intervals (t_i, t'_i) or in the intervals (t'_i, t''_i) or in the interval (t''_i, t_{i+1}).

Extended synchronous test. Within each period there are three comparisons of the output state. See Figure F.1.

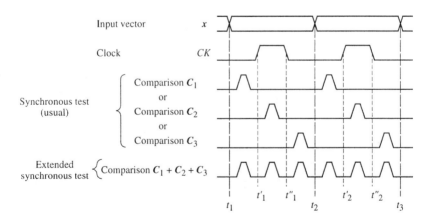

Figure F.1 Synchronous and extended synchronous test.

As shown in Figure 7.2, a primary output should have the same value for 3 consecutive intervals, then any comparison C_1, or C_2, or C_3 seems to be sufficient. However, let us assume that there is a fault such that the output is

375

faulty only for 2 intervals; then only 2 of the comparisons C_1, C_2, or C_3 will detect it. If the output is faulty only for one interval, only one kind of comparison can detect the fault.

For example if the output is faulty for (t_i, t''_i), only comparison C_1 and C_2 will detect the fault, and if the output is faulty for (t''_i, t_{i+1}), only C_3 can detect the fault. Note that the extended synchronous test can detect any fault which produces a faulty output during at least one interval.

However, does an extended synchronous test allow us to test all the specified cases? The answer generally is no: for example, if a fault produces a faulty output *only if* an input variable changes within an interval (t'_i, t''_i), this fault will never be detected by the extended synchronous test. A synchronous circuit may have asynchronous inputs, such as a reset, which could change at any time in normal operation. We have seen (Section 5.2.4) that there is no way to represent the complete behavior of a synchronous circuit with asynchronous inputs, except the asynchronous model taking the clock into account as an ordinary input. This is also true for the test: in general, there is no way to test all the specifications of this kind of circuit, except asynchronous testing taking the clock into account as an ordinary input.

Circuit	Package reference	# faults not detectable by asynchronous test	# faults detectable by asynchronous test but not detectable by:			
			Synchronous test			Extended synchronous test
			C_1	C_2	C_3	$C_1 + C_2 + C_3$
D flip-flop + S + R	474	1	7	7	7	4
JK flip-flop + R	473	1	1	1	2	0
JK flip-flop + S + R	476	2	1	1	1	0
JK flip-flop (3 J_j, 3 K_j) + S + R	472 E	2	1	1	1	0
JK flip-flop (3 J_j, 3 K_j) + S + R	472 H	3	0	2	2	0

Figure F.2 Comparing asynchronous, synchronous, and extended synchronous tests.

In Figure F.2, the performance of asynchronous, synchronous and extended synchronous tests are compared. This comparison is based on an exhaustive study of the single stuck-at faults in several circuits[1]. The D flip-flop whose package reference is *474* is the circuit in Figure F.3.

The faults which are not detectable by an asynchronous test are such that the output is always fault-free when it is stable. Such a fault may produce a transient faulty output, i.e., a glitch or hazard, according to the usual terminology (fault $w_2/1$ in Figure F.3).

---

[1] Figure F.2 is taken from [Da et al. 77]. The circuits correspond to usual TTL circuits; see [Se 75] for example.

Comparison Between Asynchronous and Synchronous Tests

Let us now concentrate only on the faults detectable by an asynchronous test. Figure F.2 shows that for any comparison C_1, C_2, or C_3 faults remain which are not detected; fault $w_1/1$ in Figure F.3 is detected only by the comparison C_2 (this fault is also considered in Section 5.2.4).

Now the extended synchronous test is as good as the asynchronous one for 4 out of the 5 circuits concerned. The fault $w_3/1$ in Figure F.3 is one of the four faults which are not detected by the extended synchronous test while they are detected by an asynchronous one. Assume that this fault is present in the circuit. The reader may verify that, when $CK = D = Q = 1$ and $S = 0$, if R takes the value 1 then returns to the value 0, the output takes the value 0 but returns to the faulty value 1 (Exercise 7.8).

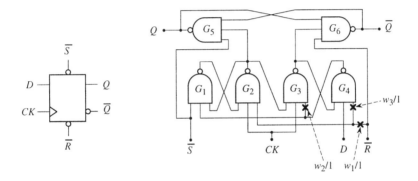

Figure F.3 D flip-flop and three faults which can be detected only in specific conditions.

Remark F.1 The analysis in this appendix is relevant *to both deterministic and random testing*. The only difference is the way in which the successive input vectors are chosen. The fact that a fault cannot be detected by a synchronous or asynchronous test is independent of the generation of the test sequence.

Appendix G

Proofs of Properties 7.1, 7.2, and 12.3

G.1 PROOF OF PROPERTY 7.1

Fefore beginning the proof[1], let us recall some elements given in Section 7.2.2.

The proof concerns commonly encountered faults whose detection process is modeled by a Markov chain illustrated in Figure G.1a. The corresponding transition matrix is shown in Figure G.1b (submatrix V is irreducible and regular).

The eigenvalues of the transition matrix U in Figure G.1b are ordered in a non-increasing order of their absolute values: λ_1, λ_2, ..., λ_m. Since U is a stochastic matrix, $|\lambda_j| \leq 1$ for every j, and $\lambda_1 = 1$. Due to the shape of the matrix U: 1) $\lambda_1 = 1$ is the eigenvalue corresponding to the submatrix associated with the absorbing state; 2) λ_2, ..., λ_m are the eigenvalues of the submatrix V and $|\lambda_j| < 1$ for $j \geq 2$ since V corresponds to a transient class.

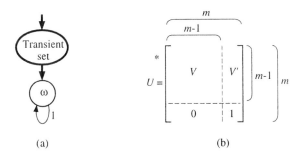

Figure G.1 Commonly encountered faults. (a) Markov chain. (b) Corresponding transition matrix.

---

[1] Some parts of the proof were presented in [Li 87]. A result of the form $\varepsilon(L) \approx - K_2 \cdot \lambda_2^L$ was obtained for a particular fault in [SaMcVe 89]. The author is grateful to B. Ycart and H. Alla for their help for finding the proof of this property. A part of this proof is reprinted from [Da 97] (© 1997 IEEE).

Proof of a) According to the Perron-Frobenius theorem (see [Ga 89] for example), since all the elements in V are non-negative and the matrix is irreducible and regular, λ_2 is a single real eigenvalue and $|\lambda_2| > |\lambda_3|$.

Proof of b) Let us use classical properties of the z-transform. The z-transform of a function ϕ of discrete time l is $\phi^*(z) = \sum_{l=0}^{\infty} \phi(l) \cdot z^{-l}$.

Equation (G.1), similar to (7.1) in Section 7.2.1, is a reccurrent expression of the probability vector at time l:

$$\pi(l+1) = \pi(l) \cdot U, \quad l \geq 0. \tag{G.1}$$

A classical result [OpSc 75] is:

$$\phi(l+1) = z\big(\phi^*(z) - \phi(0)\big). \tag{G.2}$$

Given (G.2) and the z-transform of $\pi(l) \cdot U$ is $\pi^*(z) \cdot U$, the z-transform of (G.1) is

$$z(\pi^*(z) - \pi(0)) = \pi^*(z) \cdot U. \tag{G.3}$$

From (G.3), one obtains

$$\pi^*(z) = z\, \pi(0) \cdot (zI - U)^{-1}. \tag{G.4}$$

The solutions of $\det(zI - U) = 0$ are the eigenvalues of U. After inverse transformation,

$$\pi(l) = \pi(0) \cdot \sum_{j=1}^{m} A_j \cdot \lambda_j^l \tag{G.5}$$

is obtained, where the values λ_j are the eigenvalues of U and A_j the corresponding matrices. From (G.5) and

$$\pi_\omega(l) = \pi(l) \cdot \eta = \pi(0) \cdot U^l \cdot \eta, \tag{G.6}$$

similar to (7.3) in Section 7.2.1, $\pi_\omega(l)$ is of the form

$$\pi_\omega(l) = \sum_{j=1}^{m} K_j \cdot \lambda_j^l, \tag{G.7}$$

where K_j is a constant real value if λ_j is a single real eigenvalue. If λ_j is a single complex eigenvalue, K_j is a complex value. If λ_j is a multiple eigenvalue, K_j is a polynomial in l.

Since $\pi_\omega(\infty) = 1$ and $|\lambda_j| < 1$ for $j \geq 2$, $K_1 = 1$ is obtained. Since λ_2 is a single real eigenvalue, K_2 is a real number. Then, since $|\lambda_j| < |\lambda_2|$ for $j \geq 3$,

$$\pi_\omega(l) \approx 1 + K_2 \cdot \lambda_2^l, \qquad \text{for } l \to \infty, \tag{G.8}$$

is obtained.

G.2 PROOF OF PROPERTY 7.2

The broad outline of the proof is given here. More details can be found[2] in [MoDa 97].

Let us recall some results obtained in Section 7.2.2. According to Property 7.1, for a large value of the test length l, the detection property of a sequential fault is $\pi_\omega(l) \approx 1 + K_2 \cdot \lambda_2^l$, where λ_2 is the greatest eigenvalue (except the value 1) of the $m \times m$ transition matrix corresponding to the detection process. If the initial state i of the Markov chain is unique (i.e., $\pi(0) = (0, ..., 0, 1, 0, ..., 0)$, the coefficient K_2, which may be specified by $K_2(i)$, is obtained by Equation (7.8) rewritten in (G.9):

$$K_2 = K_2(i) = \left(\frac{z - \lambda_2}{\det(zI - U)} \Gamma_{i,m}\right)_{z = \lambda_2}, \tag{G.9}$$

where $\Gamma_{i,m}$ is the co-factor of the matrix $(zI - U)^T$ corresponding to row i and column m. If the initial state is not unique, then K_2 is the weighted sum of co-factors:

$$K_2 = \pi(0) \cdot [K_2(1)\ K_2(2)\ ...\ K_2(m)]^T. \tag{G.10}$$

Property 7.2 states that, given the conditions in Property 7.1, there is an initial probability vector $\pi(0)$ such that $K_2 = -1$. The proof is based on the two following features: 1) there is at least one state i of the Markov chain such that $|K_2(i)| \leq 1$; 2) there is at least one state i of the Markov chain such that $|K_2(i)| \geq 1$.

One can easily observe that $K_2(m) = 0$, since the mth state is the absorbing state ω. It remains to show that there is a state k of the Markov chain such that $K_2(k) \leq -1$.

First, we identify the state k such that $\Gamma_{k,m}(\lambda_2)$ is maximum, then we show that the corresponding $K_2(k)$ is less than or equal to -1.

It is expected that the Markov chain state providing the lowest value of K_2 is the state whose probability $U_{k,m}$ of transition to the absorbing state is minimum. We will show that the coefficient $\Gamma_{k,m}(\lambda_2)$ is maximal for this state. We consider, without loss of generality, that the state k is unique.

Let us consider the z-matrix which is the adjoint matrix of $(zI - U)$, i.e., $\mathcal{A}_{i,j}(z) = \Gamma_{i,j}$:

$$\mathcal{A}(z) = Adj(zI - U). \tag{G.11}$$

According to [GA 89], the non-zero columns of $\mathcal{A}(\lambda_j)$ are eigenvectors of the U matrix corresponding to the eigenvalues λ_j. Particularly, the mth column of $\mathcal{A}(\lambda_2)$ (let us denote it by $\mathcal{A}_m(\lambda_2)$) is an eigenvector of U corresponding the eigenvalue λ_2.

---

[2] The author is grateful to S. Mocanu for proving this property.

Since λ_2 is the greatest eigenvalue of the substochastic matrix V (Figure G.1), from the Perron-Frobenius theorem (see [Ga 89] for example) it is obtained that all the components of $\Gamma_{i,m}(\lambda_2)$ and $\mathcal{A}_m(\lambda_2)$ have the same sign. Furthermore, since $\pi_\omega(l) \leq 1$ implies that $K_2 < 0$, from (G.9),

$$\Gamma_{i,m}(\lambda_2) \geq 0, \text{ for } i = 1, ..., m \tag{G.12}$$

is obtained.

According to the results of Gershgörin-Hadamard about the location of eigenvalues [LaTi 85], there is a state i' such that

$$\left|\lambda_2 - V_{i',i'}\right| \leq \sum_{\substack{j=1 \\ j \neq i'}}^{m-1} \left|V_{i',j}\right|. \tag{G.13}$$

From (G.13) and (G.11), given (G.12) and matrix V in non-negative,

$$\Gamma_{k,m}(\lambda_2) = \max_{i=1,...,m-1} \Gamma_{i,m}(\lambda_2) \tag{G.14}$$

is obtained after some calculation. We have identified the state k such that $\Gamma_{k,m}(\lambda_2)$ is maximum. We have now to show that the corresponding $K_2(k)$ is less than or equal to -1. From (G.9), this is equivalent to

$$\left(\frac{z-\lambda_2}{\det(zI-U)}\Gamma_{k,m}\right)_{z=\lambda_2} \leq -1. \tag{G.15}$$

Let us consider the polynomial function

$$f(z) = \frac{\det(zI-U)}{z-\lambda_2}. \tag{G.16}$$

defined on the spectrum of matrix U. The value of this function is zero for all the values of z corresponding to eigenvalues except $z = \lambda_2$.

Let us now define the function $f(U)$ of matrix U by the *fundamental formula* [Ga 89]:

$$f(U) = \frac{\mathcal{A}(\lambda_2)}{\psi(\lambda_2)} \cdot f(\lambda_2), \tag{G.17}$$

where

$$\psi(\lambda_2) = \left(\frac{\det(zI-U)}{z-\lambda_2}\right)_{z=\lambda_2} \tag{G.18}$$

According to [La 66], [Ga 89], since the values λ_i, $i = 1, ..., m$, are the eigenvalues of U, the values $f(\lambda_i)$ are the eigenvalues of $f(U)$, and the corresponding eigenvectors are the sames for both U and $f(U)$. Since $f(\lambda_i) = 0$ for any eigenvalue of $f(U)$ different of λ_2, 0 is an eigenvalue of $f(U)$ whose multiplicity is $m-1$; the other eigenvalue of $f(U)$ is $f(\lambda_2)$ which is different of 0. Thus, the eigenvector of U corresponding to $\lambda_1 = 1$, i.e., $(1, 1, ..., 1)^T$, is one of the eigenvectors of $f(U)$ corresponding to the eigenvalue 0. This property can be made explicit for the row k in the following expression:

$$\Gamma_{k,m} = -\sum_{j=1}^{m-1} \Gamma_{k,j}. \tag{G.19}$$

After some calculation, from (G.19), given (G.16), (G.17), and (G.18), one can show that:

$$\frac{\Gamma_{k,m}}{\psi(\lambda_2)} \leq -1. \tag{G.20}$$

Since the left side of (G.20) is equal to the left side of (G.15), the property is shown.

G.3 PROOF OF PROPERTY 12.3

The notations necessary to this proof[3] are introduced in Section 12.1.2. Let us recall that:

$\mathcal{A}$ is the set of possible signatures;

A_j is a signature in $\mathcal{A}$, $A(f)$ is the signature of the circuit affected by fault f;

$p_j^f = \Pr[A_j | f] = \Pr[A(f) = A_j]$;

$\mathcal{A}_v$ = subset of v signatures in $\mathcal{A}$;

$$D(\mathcal{A}_v, f, g) = \Pr[A(f) \neq A(g) | A(f), A(g) \in \mathcal{A}_v]. \tag{G.21}$$

According to the hypotheses of Property 12.3,

$$\sum_{j:A_j \in \mathcal{A}_v} p_j^f = \alpha \text{ and } \sum_{j:A_j \in \mathcal{A}_v} p_j^g = \beta, \ \alpha \text{ and } \beta \leq 1, \tag{G.22 and G.23}$$

Equation (G.21) may be expressed as

$$D(\mathcal{A}_v, f, g) = \sum_{i=0}^{v-1} p_i^f \sum_{\substack{j=0 \\ j \neq i}}^{v-1} p_j^g. \tag{G.24}$$

Using (G.23), (G.24) may be rewritten as

---
[3] The author is grateful to J. Pulou for his contribution to the proof of this property.

$$D(\mathcal{A}_v, f, g) = \sum_{i=0}^{v-1} p_i^f (\beta - p_i^g). \tag{G.25}$$

Extracting the last term of the sum in (G.25) leads to

$$D(\mathcal{A}_v, f, g) = \sum_{i=0}^{v-2} p_i^f (\beta - p_i^g) + p_{v-1}^f (\beta - p_{v-1}^g). \tag{G.26}$$

Since $p_{v-1}^f = \alpha - \sum_{i=0}^{v-2} p_i^f$ and $p_{v-1}^g = \beta - \sum_{i=0}^{v-2} p_i^g$ (from (G.22) and (G.23)), the variables p_{v-1}^f and p_{v-1}^g can be eliminated from (G.26):

$$D(\mathcal{A}_v, f, g) = \sum_{i=0}^{v-2} p_i^f (\beta - p_i^g) + \left(\alpha - \sum_{i=0}^{v-2} p_i^f\right)\left(\sum_{i=0}^{v-2} p_i^g\right). \tag{G.27}$$

The $2(v-1)$ variables p_i^f, $i = 0, 1, ..., (v-2)$, and p_j^g, $j = 0, 1, ..., (v-2)$, are independent. Then, the following partial derivatives can be obtained from (G.27):

$$\frac{\partial D(\mathcal{A}_v, f, g)}{\partial p_i^f} = \beta - p_i^g - \sum_{i=0}^{v-2} p_i^g, \quad i = 0, 1, ..., (v-2); \tag{G.28}$$

$$\frac{\partial D(\mathcal{A}_v, f, g)}{\partial p_i^g} = -p_i^f + \alpha - \sum_{i=0}^{v-2} p_i^f, \quad i = 0, 1, ..., (v-2). \tag{G.29}$$

From (G.28) and (G.29), $D(\mathcal{A}_v, f, g)$ is maximum if and only if $\beta - p_i^f - \sum_{i=0}^{v-2} p_i^g = 0$, i.e.,

$$p_i^g = \beta - \sum_{j=0}^{v-2} p_i^g = p_{v-1}^g, \quad i = 0, 1, ..., (v-2), \tag{G.30}$$

and similarly,

$$p_i^f = \alpha - \sum_{j=0}^{v-2} p_i^f = p_{v-1}^f, \quad i = 0, 1, ..., (v-2). \tag{G.31}$$

It follows, from (G.30) and (G.31), that:

$$p_i^g = \frac{\beta}{v} \quad \text{and} \quad p_i^f = \frac{\alpha}{v}, \quad i = 0, 1, ..., (v-1). \tag{G.32 and G.33}$$

Appendix H

Motorola *6800* Microprocessor

The Motorola *6800* microprocessor is used as an example in several chapters (9, 10, and 11). Its main features, useful to understand the related parts in these chapters, are gathered in this appendix.

This microprocessor was designed by the Company Motorola and manufactured by several companies [Mo 79][Th 78]. It is a 40-pin package whose inputs and outputs are illustrated in Figure H.1.

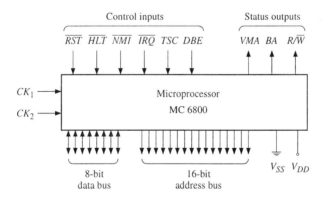

Figure H.1 Inputs and outputs of the Motorola *6800* microprocessor.

The *inputs* of this circuit are:

- An 8-bit data bus (when the microprocessor reads: bi-directional bus).

- Six control lines:
    - RST : Reset of the microprocessor.
    - HLT : Halt, stoppage of the microprocessor.
    - NMI : Non-maskable interrupt.
    - IRQ : Interrupt request.
    - DBE : Data bus enable.
    - TSC : Three-state control (high impedance of the outputs).

- Two clocks, phases opposed, CK_1 and CK_2.

The *outputs* of this circuit are:
- An 8-bit data bus (when the microprocessor writes: bi-directional bus).
- A 16-bit address bus.
- Three status outputs:
  $R/\overline{W}$: Read or write status.
  VMA : Valid memory address.
  BA : Bus available.

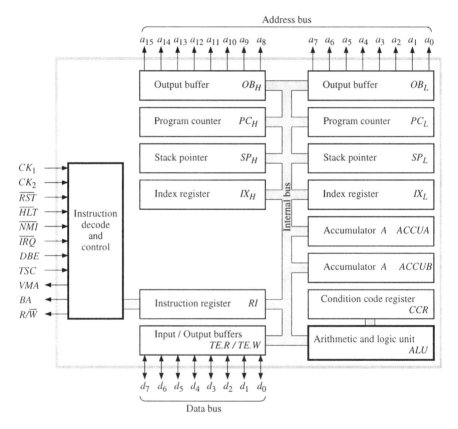

Figure H.2 Internal architecture (at the level known by the user) of the *6800* microprocessor.

The internal architecture, corresponding to the level known by the user [Mo 79], is illustrated in Figure H.2. An 8-bit internal bus links various parts:
- An instruction decoding and control unit.

Microprocessor MC 6800 **387**

- An instruction register.
- An arithmetic and logic unit.
- Two 8-bit buffers: *TE.R* (input buffer) and *TE.W* (output buffer) for connection with the bi-directional data bus.
- A 16-bit buffer: *OB* (concatenation of OB_H and OB_L, respectively high and low weights) for connection with the address bus.
- Three 8-bit registers: two accumulators (*ACCUA* and *ACCUB*) and a state register (*CCR*).
- Three 16-bit registers: program counter (concatenation of PC_H and PC_L), stack pointer (SP_H and SP_L) and an index register (IX_H and IX_L).

Code	Operation		Code	Operation			Code	Operation			Code	Operation			Code	Operation		
0 0	—		3 4	DES			6 8	ASL	IND		9 C	CPX		DIR	D 0	SUB	B	DIR
0 1	NOP		3 5	TXS			6 9	ROL	IND		9 D	—			D 1	CMP	B	DIR
0 2	—		3 6	PSH	A		6 A	DEC	IND		9 E	LDS		DIR	D 2	SBC	B	DIR
0 3	—		3 7	PSH	B		6 B	—			9 F	STS		DIR	D 3	—		
0 4	—		3 8	—			6 C	INC	IND		A 0	SUB	A	IND	D 4	AND	B	DIR
0 5	—		3 9	RTS			6 D	TST	IND		A 1	CMP	A	IND	D 5	BIT	B	DIR
0 6	TAP		3 A	—			6 E	JMP	IND		A 2	SBC	A	IND	D 6	LDA	B	DIR
0 7	TPA		3 B	RTI			6 F	CLR	IND		A 3	—			D 7	STA	B	DIR
0 8	INX		3 C	—			7 0	NEG	EXT		A 4	AND	A	IND	D 8	EOR	B	DIR
0 9	DEX		3 D	—			7 1	—			A 5	BIT	A	IND	D 9	ADC	B	DIR
0 A	CLV		3 E	WAI			7 2	—			A 6	LDA	A	IND	D A	ORA	B	DIR
0 B	SEV		3 F	SWI			7 3	COM	EXT		A 7	STA	A	IND	D B	ADD	B	DIR
0 C	CLC		4 0	NEG	A		7 4	LSR	EXT		A 8	EOR	A	IND	D C	—		
0 D	SEC		4 1	—			7 5	—			A 9	ADC	A	IND	D D	—		
0 E	CLI		4 2	—			7 6	ROR	EXT		A A	ORA	A	IND	D E	LDX		DIR
0 F	SEI		4 3	COM	A		7 7	ASR	EXT		A B	ADD	A	IND	D F	STX		DIR
1 0	SBA		4 4	LSR	A		7 8	ASL	EXT		A C	CPX		IND	E 0	SUB	B	IND
1 1	CBA		4 5	—			7 9	ROL	EXT		A D	JSR		IND	E 1	CMP	B	IND
1 2	—		4 6	ROR	A		7 A	DEC	EXT		A E	LDS		IND	E 2	SBC	B	IND
1 3	—		4 7	ASR	A		7 B	—			A F	STS		IND	E 3	—		
1 4	—		4 8	ASL	A		7 C	INC	EXT		B 0	SUB	A	EXT	E 4	AND	B	IND
1 5	—		4 9	ROL	A		7 D	TST	EXT		B 1	CMP	A	EXT	E 5	BIT	B	IND
1 6	TAB		4 A	DEC	A		7 E	JMP	EXT		B 2	SBC	A	EXT	E 6	LDA	B	IND
1 7	TBA		4 B	—			7 F	CLR	EXT		B 3	—			E 7	STA	B	IND
1 8	—		4 C	INC	A		8 0	SUB	A	IMM	B 4	AND	A	EXT	E 8	EOR	B	IND
1 9	DAA		4 D	TST	A		8 1	CMP	A	IMM	B 5	BIT	A	EXT	E 9	ADC	B	IND
1 A	—		4 E	—			8 2	SBC	A	IMM	B 6	LDA	A	EXT	E A	ORA	B	IND
1 B	ABA		4 F	CLR	A		8 3	—			B 7	STA	A	EXT	E B	ADD	B	IND
1 C	—		5 0	NEG	B		8 4	AND	A	IMM	B 8	EOR	A	EXT	E C	—		
1 D	—		5 1	—			8 5	ADC	A	IMM	B 9	ADC	A	EXT	E D	—		
1 E	—		5 2	—			8 6	LDA	A	IMM	B A	ORA	A	EXT	E E	LDX		IND
1 F	—		5 3	COM	B		8 7	—			B B	ADD	A	EXT	E F	STX		IND
2 0	BRA	REL	5 4	LSR	B		8 8	EOR	A	IMM	B C	CPX		EXT	F 0	SUB	B	EXT
2 1	—		5 5	—			8 9	ADC	A	IMM	B D	JSR		EXT	F 1	CMP	B	EXT
2 2	BNI	REL	5 6	ROR	B		8 A	ORA	A	IMM	B E	LDS		EXT	F 2	SBC	B	EXT
2 3	BLS	REL	5 7	ASR	B		8 B	ADD	A	IMM	B F	STS		EXT	F 3	—		
2 4	BCC	REL	5 8	ASL	B		8 C	CPX		IMM	C 0	SUB	B	IMM	F 4	AND	B	EXT
2 5	BCS	REL	5 9	ROL	B		8 D	BSR		REL	C 1	CMP	B	IMM	F 5	BIT	B	EXT
2 6	BNE	REL	5 A	DEC	B		8 E	LSD		IMM	C 2	SBC	B	IMM	F 6	LDA	B	EXT
2 7	BEO	REL	5 B	—			8 F	—			C 3	—			F 7	STA	B	EXT
2 8	BVC	REL	5 C	INC	B		9 0	SUB	A	DIR	C 4	AND	B	IMM	F 8	EOR	B	EXT
2 9	BVS	REL	5 D	TST	B		9 1	CMP	A	DIR	C 5	BIT	B	IMM	F 9	ADC	B	EXT
2 A	BPL	REL	5 E	—			9 2	SBC	A	DIR	C 6	LDA	B	IMM	F A	ORA	B	EXT
2 B	BMI	REL	5 F	CLR	B		9 3	—			C 7	—			F B	ADD	B	EXT
2 C	BGE	REL	6 0	NEG	IND		9 4	AND	A	DIR	C 8	EOR	B	IMM	F C	—		
2 D	BLT	REL	6 1	—			9 5	BIT	A	DIR	C 9	ADC	B	IMM	F D	—		
2 E	BGT	REL	6 2	—			9 6	LDA	A	DIR	C A	ORA	B	IMM	F E	LDX		EXT
2 F	BLE	REL	6 3	COM	IND		9 7	STA	A	DIR	C B	ADD	B	IMM	F F	STX		EXT
3 0	TSX		6 4	LSR	IND		9 8	EOR	A	DIR	C C	—						
3 1	INS		6 5	—			9 9	ADC	A	DIR	C D	—						
3 2	PUL	A	6 6	ROR	IND		9 A	ORA	A	DIR	C E	LDX		IMM				
3 3	PUL	B	6 7	ASR	IND		9 B	ADD	A	DIR	C F	—						

Figure H.3 Operation codes of the *6800* microprocessor.

Remark H.1 The register *CCR* is special. It does not contain a single value but a set of bits c_i with various meanings: c_0, carry (when an arithmetic operation is performed); c_1, overflow (when an arithmetic operation is performed); c_2, indicates a zero result; c_3, indicates a negative result; c_4, masks a hardware interrupt (*IRQ*); c_5, half-carry; $c_6 = c_7 = 1$ are not used. These flags are used in branch instructions except Non-maskable interrupt.

❑

A set of 197 operation codes is obtained from combining 72 basic instructions with the five addressing modes: *IMM* (immediate), *DIR* (direct), *REL* (relative), *IND* (indexed), *EXT* (extended). The hexadecimal codes of the operation codes are shown in Figure H.3. There are 59 hexadecimal codes (i.e., 2^8 - 197) which are invalid operation codes (for example *00* or *6B*).

For simplicity's shake, a *valid operation code* will be called an **instruction**.

The invalid operation codes can lead to some "instruction-like" processing [Ne 79] but their behavior depends heavily of the circuits manufacturer

Instruction and addressing mode	Cycles	Cycle #	Line VMA	Address bus	Line R/W	Data bus
ADD A DIR (code 9B)	3	1	1	Ad. of operation code	1	Operation code
		2	1	Ad. of operation code + 1	1	Ad. of operand
		3	1	Ad. of operand	1	Operand
JSR IND (code AD)	8	1	1	Ad. of operation code	1	Operation code
		2	1	Ad. of operation code + 1	1	Moving
		3	0	Index register	1	Non-valid data
		4	1	Stack pointer	0	Ad. of return (low weights)
		5	1	Stack pointer - 1	0	Ad. of return (high weights)
		6	0	Stack pointer - 2	1	Non-valid data
		7	0	Index register	1	Non-valid data
		8	0	Index register + Moving (no carry)	1	Non-valid data

Figure H.4 Summary of execution of two instructions.

Execution of an instruction requires several cycles (on average around 4 cycles). At the first cycle, the data bus always contains the operation code. The 3 cycles of the addition instruction *ADD A DIR* and the 8 cycles of the jump instruction *JSR IND* (respectively codes *9B* and *AD* according to Figure H.3) are shown in Figure H.4. The binary code of these instructions are 10011011 and 10101101 since the hexadecimal values *A*, *B*, and *D* correspond respectively to 1010, 1011, and 1101 as usual.

Appendix I

Pseudorandom Testing

Consider a m-bit pseudorandom source (such as an m-bit LFSR producing an M-sequence, Section 10.3.1) used to test an n-bit combinational circuit (see Figure 10.2 in Section 10.3.2.1). It is assumed that the generator produces all patterns, i.e., that the LFSR has been modified to include the all-zero pattern (Remark 10.6 in Section 10.3.1).

The pure pseudorandom testing corresponds to $m = n$. In this appendix we analyse the influence of the number $u = m - n$ of unused bits and of the detectability k (Section 6.1.1.1) of the fault in the circuit under test.

I.1 CASE $m = n$

The results in this section were given in [WaChMc 87]. In pseudorandom testing, each vector is *randomly drawn, without replacement*, with equal probability out of a set initially containing 2^n different vectors. Assume a fault f of detectability k is present, i.e., k of the 2^n vectors detect the fault.

Let P_t be the probability of first detecting the fault when the tth test vector is applied. If P_t is known for all values $t = 1, ..., L$, the probability of testing the fault f by the pseudorandom test of length L is (according to the notation in Definition 4.1 in Section 4.4):

$$P(f, L) = \sum_{t=1}^{L} P_t \tag{I.1}$$

The probability that the fault is detected by the first vector is $P_1 = \dfrac{k}{2^n}$. The probability of first detecting the fault when the 2nd vector is applied is $P_2 = \dfrac{2^n - k}{2^n} \cdot \dfrac{k}{2^n - 1}$ since the second vector is drawn for a remaining set of $2^n - 1$ vectors. The general expression is

$$P_t = \frac{2^n - k}{2^n} \cdot \frac{2^n - 1 - k}{2^n - 1} \cdot \frac{2^n - 2 - k}{2^n - 2} \cdot \ldots \cdot \frac{2^n - (t-2) - k}{2^n - (t-2)} \cdot \frac{k}{2^n - (t-1)}, \tag{I.2}$$

where the jth term ($1 \leq j \leq t - 1$) of the product expression is the probability that the jth test vector did not detect the fault.

Let us recall the combinational notation defined as

$$\binom{a}{b} = \frac{a!}{(a-b)!(b)!} \quad \text{and} \quad \binom{a}{b} = 0 \text{ for } b > a, \quad \binom{a}{0} = 1 \text{ for } a \geq 0.$$

The product of numerators and the product of denominators in (I.2) are respectively equal to

$$\frac{(2^n - k)!}{(2^n - (t-1) - k)!} k \quad \text{and} \quad \frac{2^n!}{(2^n - t)!}. \tag{I.3}$$

From (I.2) and (I.3),

$$P_t = \frac{(2^n - k)!(2^n - t)!k}{(2^n - (t-1) - k)!2^n!} = \frac{\binom{2^n - t}{k - 1}}{\binom{2^n}{k}} \tag{I.4}$$

can be obtained; the last transformation is verified by developing the last expression of P_t. From (I.1) and (I.4),

$$P(f, L) = 1 - \frac{\binom{2^n - L}{k}}{\binom{2^n}{k}} \tag{I.5}$$

can be obtained. The proof is given in [WaChMc 87].

Remark I.1 According to the hypothesis leading to the result (I.5), this result is related to any generator producing each test vector exactly once. It applies, for example, to an n-bit binary counter (which produces an exhaustive test sequence whose length is 2^n).

I.2 GENERAL CASE $m \geq n$

The results in this section were presented in [DaWa 90]. The case $m < n$ is not considered since, obviously, all the input vectors cannot be applied and the fault coverage may not be high (Remark 10.5b in Section 10.3.1).

The generator has $m = n + u$ outputs. It is called a *u-order pseudorandom generator* since it has u unattached outputs. Then the 0-order pseudorandom test corresponds to pseudorandom testing in [MaYa 84], [WaChMc 87], and the ∞-order pseudorandom generator corresponds to random testing.

To find $P(f, L)$, we use the notation $P_t(m, n, k)$ for denoting the probability of first detecting the fault when the tth test vector is applied, given the generator has m outputs, the circuit under test n inputs, and the detectability of fault f is k.

Let $a = 2^u = 2^{m-n}$. Out of the set of 2^n vectors, k detect the fault and $2^n - k$ do not. If the pseudorandom generator has u unused outputs, it will produce $a = 2^u$ copies of each possible input vector during its period, where all the output vectors are generated. Vectors are thus sampled from this set of available vectors with limited replacement. There are $a \cdot k$ vectors which detect the fault and $a \cdot (2^n - k)$ which do not detect it in the set of $a \cdot 2^n = 2^m$ vectors produced by the generator. This is equivalent to a 0-order generator of size m testing for a fault of detectability $a \cdot k$ in a m-input circuit. This property can be written as

$$P_t(m, n, k) = P_t(m, m, a \cdot k). \tag{I.6}$$

It follows that the values of $P_t(m, n, k)$ and $P(f, L)$ can be obtained from (I.4) and (I.5) by replacing 2^n with $a \cdot 2^n$ and k with $a \cdot k$. In particular,

$$P(f, L) = 1 - \frac{\binom{a \cdot 2^n - L}{a \cdot k}}{\binom{a \cdot 2^n}{a \cdot k}}. \tag{I.7}$$

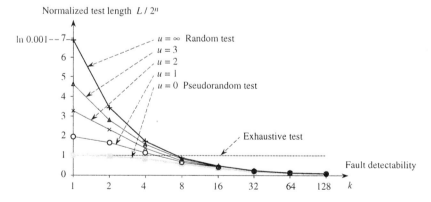

Figure I.1 Test length versus fault detectability[1] for $P(f, L) = 0.999$.

---

[1] This figure is reprinted from [DaWa 90] (© 1990 IEEE).

From (I.7), the results presented in Figure I.1 can be obtained. Given a required detection probability (0.999 in our example) and u, the normalized test length $L/2^n$ is a function only on k. The following can be observed.

1) The 0-order *pseudorandom test requires test lengths approaching those of exhaustive testing for small values of detectability* k: if $k = 1$, 99.9% of the test vectors must be applied for pseudorandom testing, which is close to the 100% of the exhaustive testing.

2) *The pseudorandom test requires length approaching those of random tests for large values of detectability* k: if $k = 100$, for example, only a small part of the possible test vectors is applied to reach a probability of detection 0.999 (then drawing with or without replacement is almost similar). In this case random testing is shorter than exhaustive testing.

3) The u-order pseudorandom test behaves approximately like random test when u is greater than 6 (i.e., the normalized test lengths for $u = 7$ and $u = \infty$ are very close to each other).

Appendix J

Random Testing of Delay Faults

Traditionally, testing of a digital system has ben concentrated on the detection of steady-state logic malfunctions. With increasing of systems speed, detection of timing failures has become important. Such defects are modeled by **delay faults**. The maximum allowable path delay T_C in a computer is determined by its clock rate. If the delay on a path exceeds T_C, incorrect output values may be latched.

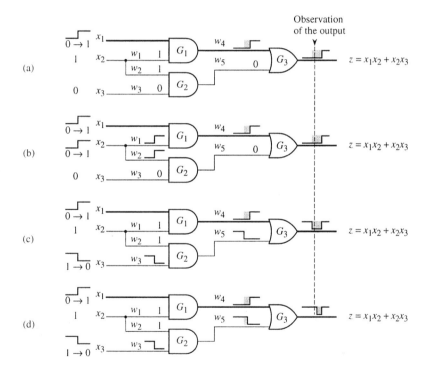

Figure J.1 Test of a delay fault: rising edge on the path from x_1 to z (a) A robust test. (b) Another robust test. (c) Detection if there is no delay fault on the path $x_3 w_3 G_2 w_5 G_3 z$. (d) No detection if there is a delay fault on this path.

393

System timing failures may either occur due to lumped delay faults caused by manufacturing defects or due to the accumulation of distributed delays along the active signal path. So, two different fault models, called the **gate delay fault model** and the **path delay fault model** respectively, have been proposed. In the *path delay fault model* [LiRe 87], a delay fault is said to have occurred if the propagation time of a transition along a path in the circuit is greater than the specified clock interval. Both lumped and distributed delays are covered by the path delay fault model which is known to be the most convenient.

Delay faults are not sequential faults since they are not logic faults. However, the detection of such a fault always requires a subsequence of two input vectors [Wa et al. 87].

An example is presented in Figure J.1. Consider the fault f illustrated in Figure J.1a: when x_1 change from 0 to 1, the rising edge of output z occurs too late (i.e., the path $x_1 G_1 w_4 G_3 z$ is too slow). This fault is detected if the following three conditions are met:

1) Input x_1 changes from 0 to 1;
2) Input $x_2 = 1$, then $w_1 = 1$, allows the propagation of the change to w_4;
3) Input $x_3 = 0$, then $w_5 = 0$, allows the propagation of the change from w_4 to z.

Thus, the subsequence of input vectors $x_2 x_6$ allows the detection of the fault ($x_2 = 010$ and $x_6 = 110$). This **test** (made up of the two successive vectors x_2 and x_6) is said to be **robust** because it detects the fault whatever the other delay faults in the circuit [PaMe 87], [LiRe 87].

Do other subsequences detect the fault f? It is clear that the values 1 on w_1 and 0 on w_3 are required *at least after the transition* (the reader may verify that, otherwise, the error could not be propagated). Now, let us observe what happens if x_2 changes from 0 to 1 or if x_3 changes from 1 to 0 simultaneously with the x_1 change.

Figure J.1b illustrates the behavior for the subsequence $x_0 x_6$ (input change of both x_1 and x_2). Even if the rising edge of x_2 were over delayed (on the path $x_2 w_1 G_1 w_4$), the rising edge at w_4 could not occur sooner than in the case of subsequence $x_2 x_6$. Thus $x_0 x_6$ is also a robust test.

Figure J.1c illustrates the behavior for the subsequence $x_3 x_6$ (input change of both x_1 and x_3) if there is no delay on the path $x_3 w_3 G_2 w_5 G_3 z$: the fault f is detected. However, according to Figure J.1d, the fault is not detected if there is another delay fault on the falling edge of the path $x_3 w_3 G_2 w_5 G_3 z$ (fault g). Hence, the subsequence $x_3 x_6$ is a test for the fault f, but it is *not robust*.

It follows that fault f is detected by an input sequence S if

$$[(x_0 x_6 \text{ in } S) \text{ OR } (x_2 x_6 \text{ in } S) \text{ OR } ((x_3 x_6 \text{ in } S) \text{ AND (fault } g \text{ not present}))] \quad (J.1)$$

Similarly one can find that the subsequences $x_3 x_2$ and $x_7 x_2$ are robust tests for the fault g (there are also non-robust tests which are not taken into account in order to simplify the presentation). The test x_3x_6 is said to be **validated**[1] by x_3x_2 or x_7x_2 [DeKe 92]. Hence, from (J.1):

$$\Pr[S \text{ tests } f] > \Pr[x_0x_6 \text{ OR } x_2x_6 \text{ OR } (x_3x_6 \text{ AND } (x_3x_2 \text{ OR } x_7x_2)) \text{ in } S] \quad (J.2)$$

The observer in Figure J.2 is such that, given the initial state b_0, the absorbing state ω has been reached if $[x_0x_6 \text{ OR } x_2x_6 \text{ OR } (x_3x_6 \text{ AND } (x_3x_2 \text{ OR } x_7x_2)) \text{ in } S]$. That observer is built from this expression (it is not built from a faulty automaton as in Section 3.2.2 because a delay fault is not a logic fault). Let us point out that: 1) if a prefix S_1 of S leads to state b_4 or b_5, x_3x_2 OR x_7x_2 is in S_1 (then it is sufficient to apply a subsequence in $(x_0 + x_2 + x_3)x_6$ to reach ω); 2) if a prefix S_1 of S leads to state b_6, b_7, or b_8, x_3x_6 is in S_1 (then it is sufficient to apply a subsequence in $(x_0 + x_2)x_6 + (x_3 + x_7)x_2$ to reach ω).

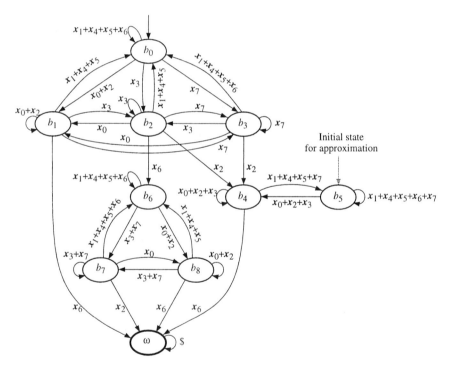

Figure J.2 Observer corresponding to the right part of (J.2).

Given a distribution of the input probabilities (Appendix A), a Markov chain is obtained from the observer in Figure J.2 and a test length can be obtained from

---

[1] A fault f, may have only *non-robust* tests which *cannot be validated*. Such a fault is said to be **weakly verifiable**.

this chain (Section 5.3.1). For an equal likelihood distribution of the input vectors, $L(0.001) = 141$ is obtained.

Let us observe that

$$\Pr[S \text{ tests } f] \approx \Pr[(x_0 x_6 \text{ OR } x_2 x_6 \text{ OR } x_3 x_6) \text{ in } S] \qquad (J.3)$$

is a good approximation for $\Pr[S \text{ tests } f]$ close to 1 (value slightly greater than the value in (J.2)). The test length defined from (J.3) is obtained from the observer (and the subsequent Markov chain) in Figure J.2 with b_5 as the initial state (only the states b_5, b_4, and ω are reachable). The test length $L(0.001) \approx 138$ is obtained. This is close to 141. The reason can be explained from a generic case (the numerical results in the sequel are easily obtained from Markov chains, assuming equal likelihood distribution of the input vectors).

Assume that $x_k x_j$ is a non-robust test validated by $x_k x_l$. If the length L_1 of S is such that $\Pr[x_k x_j \text{ in } S] = 0.999$, then $\Pr[x_k x_j \text{ AND } x_k x_l \text{ in } S] = 0.998$. This last probability would be 0.999 if the test length were $1.1 \times L_1$. Hence the necessary *validation has not a great influence on the test length*. In [CrJaDa 96] an example of two faults f_1 and f_2 in a circuit is given such that: there is one robust test for f_1 and two non-robust tests for f_2, and the required test length is much shorter for f_2.

Another approximation can be obtained from (J.3). Since the successive vectors are independent,

$$\Pr[x_0 x_6 \text{ has just been applied at time } t] = \Pr[x(t-1) = x_0] \times \Pr[x(t) = x_6]$$

$$= \left(\frac{1}{8}\right)^2 = \frac{1}{64}.$$

It follows that $\Pr[f \text{ is detected at } t] \approx \dfrac{3}{64} = 0.047$ since the probabilities of detections by the three tests in (J.3) can be added. Hence, using $p_f \approx 0.047$, $L(0.001) \approx 144$ is obtained from (6.14) in Section 6.1.1.2.

Remark J.1 If $x(t-1)$ and $x(t)$ are independent, $\Pr[x_k x_j \text{ has just been applied at } t] = \dfrac{1}{2^n} \times \dfrac{1}{2^n}$. If $x(t)$ is adjacent to $x(t-1)$ (i.e., only one input-variable changes) and x_j is adjacent to x_i, $\Pr[x_k x_j \text{ has just been applied at } t] = \dfrac{1}{2^n} \times \dfrac{1}{n}$, which is much greater than $\dfrac{1}{2^n} \times \dfrac{1}{2^n}$ for large values of n. A test sequence made up of adjacent vectors (see Section 10.3.3) leads to shorter test lengths for delay faults when n is large. This observation was made both in the deterministic test context [GlMe 89] and in the random test context [CrJaDa 96].

Appendix K

Subsequences of Required Lengths

As explained in Section 10.1.2, the order of application of test vectors is important for sequential faults. In this appendix, examples of faults in combinational circuits and in sequential circuits requiring subsequences of length at least two are presented.

K.1 FAULTS IN COMBINATIONAL CIRCUITS

A sequential behavior can be obtained from a *stuck-open* fault. An example is given in Figure 2.15 (Section 2.3.2): the subsequence $x_3 x_2$ is required to detect the fault. Examples requiring subsequences of length 3 or more can be found [DaRaRa 90].

Delay faults also require subsequences of length two, although they are not sequential faults (since they are not logic faults). See Appendix J. A subsequence of length greater than two is never required for the test of a delay fault since such a fault is associated with a transition between two input vectors.

A combinational circuit may become sequential if a *feedback bridging fault* is present in the circuit. However, a subsequence of length two is not necessarily required. This is illustrated in Figure K.1.

The circuit in Figure K.1a corresponds to the output function (when it is fault-free).

$$z = x_1 x_2 + x'_3. \tag{K.1}$$

In presence of the AND bridging fault, the circuit behaves like the circuit in Figure K.1b (Section 2.2.3.1), i.e.,

$$w = x_1 x_2 w + x'_3, \tag{K.2}$$

$$z = x_1 w. \tag{K.3}$$

From (K.2) and (K.3), given q_0 and q_1 correspond to $w = 0$ and 1 respectively, the state table in Figure K.1c is obtained. It is clear that the faulty circuit is sequential since there are two stable states for the input vector x_7. This fault can be detected by the subsequence $x_5 x_7$ because, according to (K.1), z should have the value 1 for x_7. However, this fault can also be detected by a single input vector x_0 or x_2. As a matter of fact, for these two input vectors the value of z is 0 in the faulty circuit, instead of 1 according to (K.1).

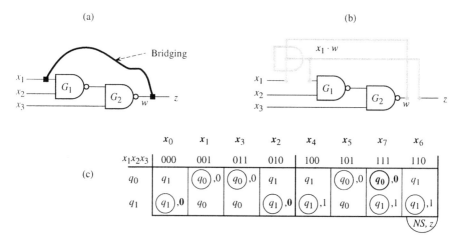

Figure K.1 (a) Combinational circuit with a feedback bridging. (b) Equivalent circuit (wired AND). (c) State table corresponding to the faulty circuit.

In [Da 75] a bridging fault in a combinational circuit requiring a subsequence of two input vectors is presented. However, this circuit is redundant. The author does not know any example of bridging fault in a non-redundant combinational circuit which cannot be tested by at least one single input vector.

K.2 FAULTS IN SEQUENTIAL CIRCUITS

Naturally, one can find faults in sequential circuits requiring subsequences of length two or more.

From the observer in Figure 7.3d (Section 7.2.1), one can observe that the two shortest test sequences (from the initial state (q_0, q'_0) to ω) are $x_1 x_1 x_0$ and $x_1 x_1 x_1$. After the first x_1, any subsequence in $(x_0 x_1)^*$ can be applied without returning to the initial state. Hence, all the input sequences leading from the initial state to ω without repassing by the initial state are in

$$A = x_1 (x_0 x_1)^* x_1 (x_0 + x_1).$$ (K.4)

Every sequence in A contains a subsequence x_1x_1. Hence, the detection of the corresponding fault requires this subsequence (this is a necessary but not sufficient condition).

The input sequence of Figure 10.3b in Section 10.3.2.1 ($\sigma = 1$) does not contain the subsequence x_1x_1; thus it is not able to detect the fault in Figure 7.3. On the other hand, the input sequence in Figure 10.3c ($\sigma = 2$) contains all the subsequences of length two (Section 10.3.2.1), for example $x(2)x(3) = x_1x_1$.

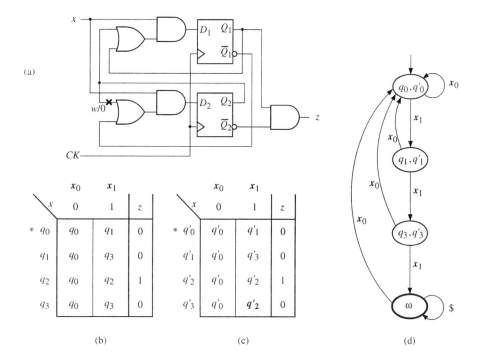

Figure K.2 Example of a fault requiring a subsequence of length three. (a) Circuit. (b) and (c) Fault-free and faulty state tables. (d) Observer.

Figure K.2 presents a sequential circuit and a fault w/0. The fault-free and faulty state tables are presented in Figures K.2b and c respectively. From the observer in Figure K.2d, it is clear that the subsequence $x_1x_1x_1$ is required to detect the fault. One can easily build faults requiring longer subsequences.

K.3 ARBITRARY SUBSEQUENCE AMONG A SET

A required subsequence may concern only a part of the circuit. It follows that only one of a set of subsequences should be applied. This is illustrated in Figure K.3.

The stuck-open illustrated in Figure K.3a requires the subsequence $x_1x_2 = 11$ then 10 to be tested (Section 2.3.3).

According to Figure K.3b representing the same circuit at the gate level, $x_3 = 1$ is required in order to propagate the fault to z_1, but x_4 may have any value.

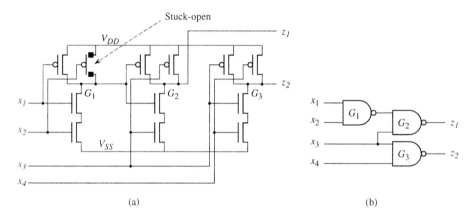

Figure K.3 (a) Stuck open in a CMOS circuit. (b) Corresponding gate circuit.

Thus $x_1x_2x_3x_4 = 111\$$ then $101\$$ is required i.e., only one of the subsequences $x_{14}x_{10}$, $x_{14}x_{11}$, $x_{15}x_{10}$, or $x_{15}x_{11}$ is necessary to test the fault.

Hence, generally, a fault requiring a sequence of two vectors for being detected, has a greater probability of being tested than the probability that two random n-input vectors are applied (in our example, this probability is four times greater). Some faults in the subcircuit of Figure 13.6 (Section 13.3) may be concerned by this observation. See Remark E.1 in Appendix E.

Appendix L
Diagnosis from Random Testing

If a circuit is faulty, the aim of diagnosis is to determine what is the fault (in case of single fault) or to specify *one of the faults* if several faults are present in the circuit. This diagnosis may be either behavioral or structural. The diagnosis from random testing is made up of two phases. The first phase consists of finding a fault which is coherent with the observation: an *hypothesis* of fault is made. The second phase is a *verification* that the hypothesis is correct: if it is correct, a fault has been diagnosed, otherwise another hypothesis is to be made, then verified.

For an integrated circuit, diagnosis may be useful either to improve the design (*manufacturer*'s point of view) or to prove to the manufacturer that a circuit which has passed his test is faulty (*user*'s point of view). The diagnosis could be only behavorial in the second case.

Two main approaches are considered. In the first one, a faulty response is analyzed. In the second one, a partially random testing is performed (some primary inputs are constant) and the detection/non-detection result is used to restrict the set of possible faults.

L.1 OBSERVATION OF THE FAULTY RESPONSE

Given an initial state, observation of the input sequence (random) and of the corresponding response (with a logic analyzer) determines when the output is faulty and proposes some fault hypotheses. For this purpose, the length to first detection should not be too large: this length can be shortened by changing the probability of each input. Let us illustrate this approach with an example.

Consider the fault f_1 in Figure 11.5 in Section 11.3.1. Its diagnosis was performed using the test machine described in Figures 10.7 (Section 10.3.4.1) and 10.9 (Section 10.3.4.3) [Fé 84]. According to Figure 11.6 (Section 11.3.1), the length to first detection may be about 10 000 instructions: this is too long. Various trials may be carried out with the test machine: if $\Pr[Reset] = 0$, we observe that the fault is not detected; hence, the *Reset* is involved in this fault and giving a high probability to this control input allows the length to first detection to be shortened. After some other trials on the probabilities of the instructions, it is possible to obtain a very short length to first detection (say about 10 or 20).

From the analysis of the corresponding response, the *hypothesis* corresponding to the behavioral description in Figure 11.5 can be made. Some other short input sequences detecting the fault have then to be applied in order to *verify* this hypothesis.

Remark L.1 Assume two circuits, C_1 and C_2, have the same length to first detection (for the same input sequence). If this length is large, for example $l_\omega = 2\,187$, the *hypothesis* that the same fault affects both circuits may be made (this hypothesis has a high probability of being true: it then has to be verified). If this length is very short, the probability that the fault is the same is not so great; see for example faults f_8 and f_9 in Figures 11.6 ($l_\omega = 1$ for both).

❑

All the faults presented in Figure 11.5 have been diagnosed in this way [Fé 84].

A diagnosis has been performed for all the circuits detected as faulty in [LaGéHe 89] (Section 11.4.2). The faults hypotheses were made using the random test machine presented in Section 10.3.4; the verifications were performed using short deterministic test sequences on a Genrad GR 125 tester [HuDa 92].

L.2 PARTIALLY RANDOM TESTING EXPERIMENTS

This approach proceeds by dichotomy: the result of an experiment (detection or non-detection of a fault during this experiment) implies a set of possible faults; successive experiments allow the fault to be diagnosed.

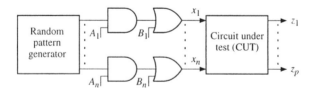

Figure L.1 Device for partially random experiments[1].

The experiments are partially random. Each input of the circuit under test is either 0, or 1, or random; these experiments can be performed according to the scheme in Figure L.1. Assume a circuit under test (combinational or sequential)

---

[1] This figure is reprinted from [DaFu 90] (© 1990 IEEE).

whose behavior is specified for any input sequence; the random pattern generator produces a sequence of random vectors (weighted or not). If $A_iB_i = 10$ for $i = 1, ..., n$ in Figure L.1, this sequence is applied to the CUT; the observation of the output sequence is performed by one of the possible means presented in Figure 5.1 (reference circuit or signature analyser). Now, $x_i = 1$ if $B_i = 1$, and $x_i = 0$ if $A_iB_i = 00$.

A *partially random testing experiment* is such that each x_i may be either 0 ($A_iB_i = 00$), or 1 ($A_iB_i = \$1$), or random ($A_iB_i = 10$).

It is assumed that, when a circuit is faulty, the fault f is in a prescribed set of faults F. Let E_i be an experiment. The main result of this experiment is either 1) a fault is detected by this experiment (that is denoted by $Det(E_i) = 1$), or 2) no fault is detected by the experiment (that is denoted by $Det(E_i) = 0$ or $NDet(E_i) = 1$). If $Det(E_i) = 1$, one can define F_1, $F_1 \subseteq F$, such that f is in F_1. If $NDet(E_i) = 1$, then $f \in F_2$, $F_2 \subseteq F$, and $F_1 \cup F_2 = F$.

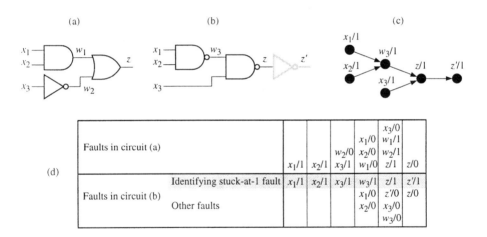

Figure L.2 (a) Initial circuit. (b) NAND-gate circuit with the same classes of faults. (c) Graph of the classes of faults. (d) Classes of faults.

Let us illustrate the diagnosis approach using a simple combinational tree circuit. The circuit under consideration is in Figure L.2a. The transformation of any circuit into a circuit made up of NAND gates preserving the structure, keeps all the information for the fault diagnosis [Ha 71]; this transformation of the circuit in Figure L.2a leads to the circuit in Figure L.2b *without the grey part* (inverter). If this inverter is added, the classes of faults are the same and, furthermore, *each class of faults corresponds to the stuck-at-1* of a line in the circuit [Li 87]. This is illustrated in Figure L.2d. Now, the classes of faults can be partially ordered according to the graph in Figure L.2c (same structure as the circuit in Figure L.2b). From now on, since each class is identified by a particular stuck-at-1, we shall use the term "fault" instead of a "class of faults."

The diagnosis process is illustrated in Figure L.3 for a tree circuit: a fault has been detected in the circuit and we want to diagnose which fault in the set shown in Figure L.2 is present in the circuit (a single fault is assumed).

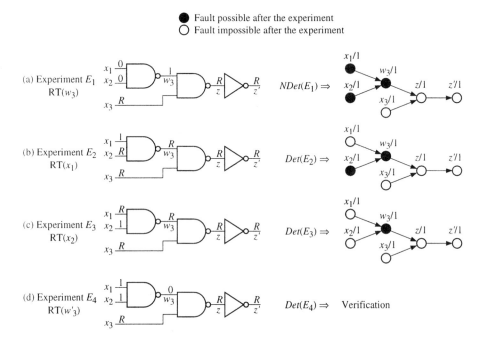

Figure L.3 (a) (b) and (c) Three successive experiments. (d) Verification that the fault is $w_3/1$.

Experiment E_1 is denoted by $RT(w_3)$. This means that all the inputs upstream of w_3 have a specified value such that $w_3 = 1$ and all the other inputs are random. The line w_3 is **strongly forced** to value 1, i.e., $x_1 = x_2 = 0$ in our example. "Strongly" means that both lines x_1 and x_2 have the value 0 while only one is enough to force w_3 to value 1. Strongly forcing reveals two interesting features: 1) every primary input in the upstream cone of the line under consideration has a specified value (0 or 1); 2) if a fault is detected during the experiment, it cannot be a fault in this upstream cone [Li 87]. All the inputs which are not upstream of w_3 are random (denoted by R in Figure L.3a).

Assume that no fault is detected, i.e., $NDet(E_1) = 1$. That means that all the lines which are randomly tested, i.e., w_3, z, and z' in Figure L.3a, behave correctly[2]. It follows that the fault is in $F_1 = \{x_1/1, x_2/1, w_3/1\}$.

Let us note that the choice of the first experiment was arbitrary. It is interesting that an experiment roughly divides by 2 the number of possible faults in order to minimize the number of successive experiments (if a fault were detected in experiment E_1, the set of possible faults would be $\{x_3/1, z/1, z'/1\}$).

Now the second experiment must split the set F_1 into a subset of possible faults and a subset of impossible faults. Experiment $RT(x_1)$ is chosen (Figure L.3b). If a fault is detected it is in $\{x_2/1, w_3/1\}$, since only x_2 and w_3 are random among the three lines x_1, x_2, and w_3.

The third experiment $RT(x_2)$ splits the remaining set of possible faults. If a fault is detected, then the fault is $w_3/1$ (Figure L.3c).

The conclusion of the set of experiments $\{E_1, E_2, E_3\}$ is that $w_3/1$ is present in the circuit, *assuming a single stuck-at fault* is present in the circuit. This assumption is *verified* by experiment $E_4 = RT(w'_3)$ illustrated in Figure L.3d. From the result of experiment E_4 we are sure that $w_3/1$ is present in the circuit.

Now, what should be the conclusion if $NDet(E_4) = 1$? It should be that the hypothesis of single stuck-at fault is not verified. Then, in a general case we would research in a similar way for a *double* stuck-at fault in the circuit, both being upstream of w_3 (in our simple example this is not possible since the double fault $[x_1/1 \ \& \ x_2/1]$ is equivalent to $z/1$).

This diagnosis method was developed for any combinational circuit (tree circuit or not) in [Li 87]. The experiments are more complicated if the circuit is not a tree. When a double stuck-at fault is assumed and the verification does not work, a triple stuck-at fault is assumed and so on. For a multiple fault, the verification requires several experiments.

A similar approach was proposed for faults in RAMs [DaFu 90]. Each experiment corresponds to a random test of a part of the memory (column, row, or cell), which is defined by the address lines whose values are not random. Starting from a set of possible faults, each experiment allows this set to be split and hence the number of possible faults to be reduced down to diagnosis of one fault.

---

[2] The test length is long enough. It is assumed that no fault would be detected even if the test length were longer.

Appendix M
Conjectures on Multiple Faults

Informally, the conjectures in this appendix are as follows: if a random test sequence tests all the single faults corresponding to a model of faults with a given level of confidence, then all the multiple faults corresponding to the same model are tested with at least the same level of confidence. There are two conjectures corresponding to two measurements of confidence: the first corresponds to the *worst case fault*, and the second to the *expected fault coverage* (P_m and P according to the notations in Sections 4.4.2 and 3.1.4.4). It is assumed that the circuits are completely specified.

Let us recall some notations and introduce new ones. According to Notation 4.1 in Section 4.2, a fault model is denoted by $\mathcal{M}_i$ (for example the stuck-at model is denoted by $\mathcal{M}_{sa}$); the set of single faults corresponding to $\mathcal{M}_i$ is denoted by F_i; the set of multiple faults whose components are in F_i is denoted by F_i^*.

Let us introduce a new notation. Let $F_{i,d}$ and $F_{i,d}^*$ denote the subsets of *detectable* faults in F_i and F_i^*.

Conjecture M.1 For any fault model $\mathcal{M}_i$, any multiple fault f in $F_{i,d}^*$ *R-dominates* at least one single fault g in $F_{i,d}$ for the constant distribution ψ_0, i.e.,

$$P(f, \psi_0, L) \geq P(g, \psi_0, L), \text{ for any } L, \tag{M.1}$$

according to Definition 4.2 in Section 4.4.3.2.

❑

For combinational faults, according to Section 4.4.3.2, (M.1) is verified if and only if the *detectability* of fault f is at least equal to the detectability of fault g, i.e.,

$$k_f = |T_f| \geq k_g = |T_g|.$$

A consequence of Conjecture M.1 is as follows. Assume a random test length, L, ensures a *minimum detection probability*[1] $P_{\min}$ for the set of faults $F_{i,d}$; for any subset F_α of faults in $F_{i,d}^*$, the *minimum detection probability* is at least $P_{\min}$ for the same length L. Generally, the faults in $F_{i,d}$ and in $F_{i,d}^*$ may be combinational

---
[1] Section 6.1.2.2

or sequential. If all the faults considered are combinational, then $k_{\min}(F_a) \geq k_{\min}(F_{i,d})$.

Let us consider an example illustrating this conjecture. The fault model is the stuck-at model. Consider the double fault $[w_1/1 \ \& \ w_2/1]$ in the circuit in Figure M.1a. According to Figure M.1b, this fault may be detected by any test vector corresponding to the detection function $x_1 x_2 (x'_3 + x'_4) + (x'_1 + x'_2) x_3 x_4$; the corresponding detectability is 6. This double fault is more difficult to detect than each of its components, since $k_{[w_1/1 \ \& \ w_2/1]} = 6 < 12 = k_{w_1/1} = k_{w_2/1}$. Nevertheless, there is at least one single fault, *corresponding to the same model*, which is more difficult to detect than the double fault considered, for example $x_1/0$ since $k_{x_1/0} = 4 < 6$.

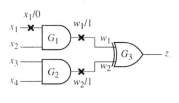

Fault f	Detection function T_f	Detectability k_f
$[w_1/1 \ \& \ w_2/1]$	$x_1 x_2(x'_3+x'_4)+(x'_1+x'_2)x_3 x_4$	6
$w_1/1$	$x'_3+x'_4$	12
$w_2/1$	$x'_1+x'_2$	12
$x_1/0$	$x_1 x_2$	4

(a) (b)

Figure M.1 Illustration of Conjecture M.1. (a) A combinational circuit. (b) Detection function and detectability of some faults in this circuit.

Conjecture 8.1 on multiple faults in RAMs (Section 8.3.2), appears to be a particular case of Conjecture M.1, in which the single fault g is a component of the multiple fault f.

The second conjecture is related to fault coverages. Additional notations are needed to write this conjecture.

Let $P(F, \psi_0, L)$ denote the fault coverage related to the set F of faults, expected for a test sequence of L equally likely vectors. Let the subset of $F^*_{i,d}$ containing all the detectable faults whose multiplicity is at most a be denoted by $F^{(a)}_{i,d}$.

Conjecture M.2 For any fault model $\mathcal{M}_i$,

$$P(F^{(a)}_{i,d}, \psi_0, L) \geq P(F_{i,d}, \psi_0, L),$$

for any $a \geq 1$ and for any $L \geq 1$.

❑

Let us illustrate the meaning of this conjecture with an example. Consider a circuit C and the stuck-at model $\mathcal{M}_{sa}$. Assume that the test length L is such that the coverage of single detectable faults is 0.999, i.e., $P(F_{sa,d}, \psi_0, L) = 0.999$. Consider now the set of all detectable single and double stuck-at faults in this circuit, i.e., $F_{sa,d}^{(2)}$. According to Conjecture M.2, $P(F_{sa,d}^{(2)}, \psi_0, L) \geq 0.999$.

The usefulness of the two conjectures is that we need only to consider single faults (for any useful model). If the length for the *worst case fault* is considered, then, according to Conjecture M.1, any additional multiple fault corresponding to the same model does not change the required test length. If the *fault coverage* is considered, according to Conjecture M.2, the test length required for the set of single faults is an upper bound of the test length required for all the multiple faults up to any multiplicity.

For example, when the faults in a microprocessor were studied (Section 9.5), it was assumed that a single fault in either the registers (set F_r), or the operators (F_o), or the register decoding function (F_d), or instruction decoding and control function (F_i) were present. According to Section 4.2, the union of several fault models is a fault model. For our example, the union of the four basic models is a model, $\mathcal{M}$, corresponding to the set of single faults $F = F_r \cup F_o \cup F_d \cup F_i$. Assume that the test length is such that the most difficult fault in F is detected with a probability $P_{min} = 0.999$. According to Conjecture M.1, any detectable multiple fault whose components are in F is detected with a probability at least equal to 0.999.

Exercises

Chapter 1

1.1 A board contains 80 circuits. What are the expected percentages of boards containing at least one faulty circuit for the three following cases (let D denote the percentages):
a) The defect level is 5 *ppm* for all circuits;
b) The defect level is 200 *ppm* for all circuits;
c) The defect level is 5 *ppm* for 40 circuits and 200 *ppm* for 40 circuits.
Comment on the results.

1.2 A 25-input, 15-output, 20-internal variable circuit is full-scan designed. How many clock cycles are necessary to apply a test sequence of 10^5 test vectors?

1.3 What is the difference between on-line BIST and off-line BIST?

1.4 What are the advantages of BIST?

Chapter 2

2.1 Give the output fonction of the circuit in the figure when it is fault-free. What is the function obtained for the bridging fault in the figure in the two following cases: a) wired OR; b) wired AND?

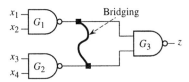

Figure E 2.1

2.2 Consider the sequential circuit in the figure, and assume the initial state corresponds to $Q = 0$. Give the state tables of the fault-free circuit and of the circuit affected by the fault $w/1$.

411

Exercises

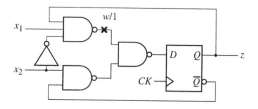

Figure E 2.2

2.3 Consider a 4-input circuit. The input vector is in a 2-out-of-4 code. Give the input language of the circuit.

2.4 Consider the JK flip-flop in the figure. What is its output language, given its input language is $\mathcal{P}((x_1 + x_2)(x_0 + x_2))^*$ (See Remark 2.4, Section 2.1.3.1)?

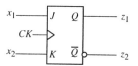

Figure E 2.4

Chapter 3

3.1 Give all the test vectors testing each single stuck-at fault in the figure.

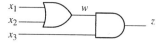

Figure E 3.1

3.2 Give the fault coverage of the test set $T = \{x_1, x_3, x_4\}$ for the circuit in Figure E 3.1: 1) before fault collapsing; 2) after equivalence and dominance fault collapsing.

3.3 Give a set of vectors detecting all the single and multiple stuck-at faults of the circuit in the figure.

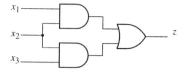

Figure E 3.3

3.4 Give the checkpoints of the circuit in Figure E 3.3 and a test vector for each single stuck-at fault affecting them.

3.5 Build the observer for the fault in Figure S 2.2b (the fault-free machine is in Figure S 2.2a) given the input language X^*. Give a sequence detecting the fault. Give the language corresponding to all the sequences detecting the fault. Do sequences $S_1 = x_0 x_1 x_1 x_3 x_2 x_0 x_1$ and $S_2 = x_2 x_3 x_0 x_1 x_1 x_2 x_3$ detect the fault?

3.6 Consider the fault-free automaton A and the fauty automaton A' in figures a and b and assume that every state of each automaton can be the starting state. Give the OR-observers for the two following cases: 1) the synchronizing sequence is x_0; 2) the synchronizing sequence is $x_0 x_1$.

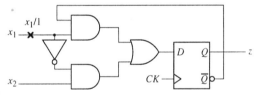

Figure E 3.6

3.7 Consider the circuit in the figure and the fault $x_1/1$. Is this fault detectable, partially detectable, or undetectable, given that the synchronizing sequence is x_0, and that $Q = z$ may have any Boolean value when the circuit is powered up?

Figure[1] E 3.7

3.8 Consider again the circuit in Figure E 3.7 and the fault $x_1/1$, and assume that $Q = z$ may have any Boolean value when the circuit is powered up.

a) What are the synchronizing sequences of the circuit?

b) Is the fault detectable, partially detectable, or undetectable, given that the synchronizing sequence used is unknown?

c) Is there an input sequence detecting the fault for any synchronizing sequence?

Chapter 4

4.1 Consider all the single stuck-at faults corresponding to the circuit in Figure E 3.1, and a random test sequence (constant distribution ψ_0) of length $L = 12$.

---

[1] This example is taken and the figure is reprinted from [AbPa 92](© 1992 IEEE).

414   Exercises

The probability of testing a fault f which can be detected by k_f input vectors is $P(f) = 1 - (1 - k_f/8)^{12}$ (this will be explained in Chapter 6).
a) Give the expected fault coverage P.
b) Give the minimum testing probability P_m.

4.2 Consider again the circuit in Figure E 3.1. According to Solution 3.1, there are 6 classes of equivalent faults (each class may be represented by one of its components). If F is the set of all these classes, $\Pr[f \mid F]$ is the conditional occurrence probability of class f, given the circuit is faulty.

Give the weighted fault coverage P_w for the same distribution and test length as in Exercise 4.1 for the two following cases.

a) The conditional occurrence probabilities of the fault classes are: 0.3 for $x_1/0$, $x_2/0$, and $x_3/0$; 0.1 for $x_1/1$, 0 for $x_3/1$ and $z/1$.

b) The conditional occurrence probabilities are: 0.1 for $x_1/0$, $x_2/0$, $x_3/0$, and $x_1/1$; 0.3 for $x_3/1$ and $z/1$.

c) Compare the results with P and P_m in Solution 4.1.

4.3 Consider again the fault classes in Exercise 4.2. Give the weighted minimum testing probability given the partition $\rho = \{\rho_1, \rho_2\}$ with $\rho_1 = \{x_1/0, x_2/0, x_3/0, x_1/1\}$, $\rho_2 = \{x_3/1, z_1/1\}$, and the following occurrence probabilities: 0.6 for ρ_1 and 0.4 for ρ_2.

4.4 Give examples of R-dominance and R-equidetectability for the distribution ψ_0, among the single stuck-at faults of the circuit in Figure E 3.1.

Chapter 5

5.1 The specification of the combinational circuit whose Karnaugh map is in figure a is incomplete because the input vectors $x_1x_2x_3 = 001$ and 111 cannot be applied in normal operation.
a) Give the input language of this circuit.

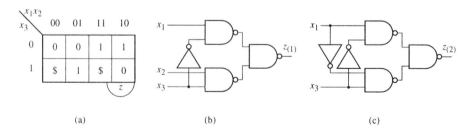

Figure E 5.1

b) Figures b and c present two solutions, denoted by $z_{(1)}$ and $z_{(2)}$, implementing the corresponding function. Is the input sequence $S_1 = x_3x_0x_0x_1x_6x_4x_0$ in the language obtained in a? What happens if this sequence compares the behaviors of the circuits in figures b and c.

5.2 Given the initial state of the asynchronous machine presented in the figure is q_0, what is the input language of this machine?

x_1x_2	x_0 00	x_1 01	x_3 11	x_2 10	z
*	(q₀)	q₁	—	q₂	1
	q₀	(q₁)	q₃	—	0
	q₀	—	q₄	(q₂)	0
	—	q₁	(q₃)	q₂	0
	—	q₁	(q₄)	q₂	1

Figure E 5.2

5.3 Give the Markov chain corresponding to the observer in Figure S 3.5, given the constant distribution ψ_0 (give two representations: graph and matrices).

5.4 Consider the fault-free machine in figure a whose input language is $I = \mathcal{P}(x_0 + x_1x_0)^*$. Give the observer corresponding to the fault in figure b, given that:
 a) the random source corresponds to the general distribution in figure c;
 b) the starting state of the faulty machine is either q'_0 (probability 0.8) or q'_1 (probability 0.2), and the synchronization sequence is x_0.

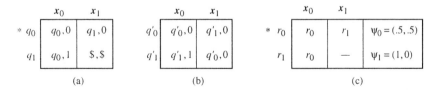

Figure E 5.4

5.5 Consider the fault-free circuit and the faulty circuit in figures a and b. Give the Markov chain modeling the detection process of this combinational fault, given the general distribution in figure c.

416   Exercises

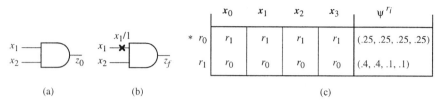

(a)   (b)   (c)

Figure E 5.5

5.6 Consider the circuit in the figure.
a) Give the state table of the fault-free circuit, given the initialization sequence is x_4x_1.
b) Give the state table of the faulty circuit (fault $x_1/0$), given its starting state corresponds to the output $z = 1$.
c) Give the sequence of detection/non-detection for the random test sequence (or another randomly generated by the reader):
$S = x_2x_1x_2x_2x_3x_2x_2x_6x_1x_4x_2x_5x_6x_4x_1x_4x_5x_1x_4x_7x_2x_3x_6x_7x_1x_1x_6x_1x_2x_3$.
Bundles of detections are observed. Give an intuitive explanation of their origin for this example.

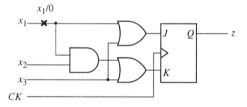

Figure E 5.6

Chapter 6

6.1 Consider the circuit in Figure E 3.1. What are the probabilities of testing faults $x_1/1$, $x_1/0$, $x_3/0$, and $z/1$, for a random test length $L = 20$ in the two following cases. Comment on the results.
   a) Constant distribution ψ_0.
   b) Constant distribution $\psi_1 = (.04, .3, .14, .2, .04, .2, .04, .04)$.

6.2 Consider again the circuit in Figure E 3.1. Give the minimum testing probability for the set of single stuck-at faults, for $L = 20$, in both cases of distributions defined in Exercise 6.1, i.e.:
   a) Constant distribution ψ_0.
   b) Constant distribution $\psi_1 = (.04, .3, .14, .2, .04, .2, .04, .04)$.

6.3 Consider again the circuit in Figure E 3.1. Give the expected fault coverage for $L = 20$ and the distribution ψ_0 in the two following cases.

a) Set F_1 of single stuck-at faults without equivalence fault collapsing.
b) Set F_2 of single stuck-at faults after equivalence fault collapsing.

6.4 For the fault $f = w_1/0$ in the figure, build an auxiliary circuit whose output corresponds to the detection function. Give the detectability of this fault.

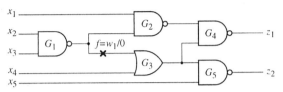

Figure E 6.4

6.5 Consider the circuit in the figure and assume that $\Pr[x_i] = 0.5$ for each i.
a) Give the controllabilities $C(0, w_1)$ and $C(1, w_1)$.
b) Give the observability $O(w_1)$.
c) Give the activity $A(11, G_2)$.

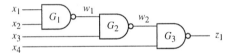

Figure E 6.5

6.6 Consider the set of all single and multiple stuck-at faults for the circuit in the figure.
a) Give a lower bound of the detectability.
b) Give an upper bound of the test length ensuring a 0.999 minimum testing probability.

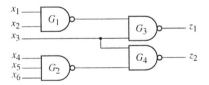

Figure E 6.6

Chapter 7

7.1 Consider the observer in Figure S 3.5. Assume the constant distribution ψ_0. For the three cases $\varepsilon = 10^{-1}$, $\varepsilon = 10^{-3}$, and $\varepsilon = 10^{-6}$:
a) Give the exact random test length which is required.
b) Give an approximate value of the random test length obtained from Equation (7.13) in Section 7.2.2 (give the integer greater than this approximate value).

7.2 Consider again the observer in Figure S 3.5. Assume now the constant distribution $\psi_1 = (.1, .1, .7, .1)$. Give the exact random test length required for $\varepsilon = 10^{-3}$, and the approximate value obtained from (7.13).

7.3 Consider the fault f in Figure S 2.2.
a) Give the detection set associated with this fault.
b) Give the approximate value of the required test length obtained from this detection set (assuming ψ_0).

7.4 Same questions as in Exercise 7.3 for the fault g in Figure S 5.6.

7.5 Consider the circuit in Figure E 2.2 whose state table is in Figure S 2.2a (don't take into account the fault $w/1$) and let F be the set of single transition faults (Section 7.3.2). Assume the distribution ψ_0.
a) Give the minimum detection probability p_{min} for the set F.
b) Give $L_m(10^{-3})$ for the set F.

7.6 Consider the sequential fault (in a combinational circuit) presented in Figure 2.15 (Section 2.3.3). The corresponding observer is in Figure 7.7 (Section 7.2.3). Assume the distribution ψ_0.
a) Give the test length $L(10^{-3})$ for this fault.
b) What would be the detection uncertainty for this fault if the test length was based on the worst case combinational fault?

7.7 The Markov chain in Figure 7.7c (Section 7.2.3) assumes that the input vector is x_0 when the circuit is powered-up.
a) Give the Markov chain if all the input vectors are equally likely when the circuit is powered-up (and the internal states are equally likely when both are possible). Comment on the construction.
b) Give the test length $L(10^{-3})$ for this case. Comment on the result.

7.8 Consider the fault $w_3/1$ in Figure F.3. Show that, when $CK = D = Q = 1$ and $S = 0$, if R takes the value 1 then returns to the value 0, the output takes the value 0 but returns to the faulty value 1.

Chapter 8

8.1 Given the following faults (where a, b, c, and d are cells of a n-word by 1-bit RAM):
$f_1 = (\uparrow a \Rightarrow \downarrow b$ when $d')$,
$f_2 = (\uparrow b \Rightarrow \uparrow a)$,
$f_3 = (c$ stuck-at-1$)$,
$f_4 = [f_1 \& f_2 \& f_3]$,
Show that $P(f_4, \psi, L) \geq P(f_i, \psi, L)$, for at least one i in $\{1, 2, 3\}$ and for any ψ and any L.

8.2 Same question as in Exercise 8.1 for
$f_1 = (\uparrow c \Rightarrow \uparrow d)$,
$f_2 = (\downarrow c \Rightarrow \updownarrow a)$,
$f_3 = (\downarrow a \Rightarrow \downarrow b$ when $cd' = 1)$,
$f_4 = [f_1 \& f_2 \& f_3]$.

8.3 Consider a 2048-word by 1-bit RAM and a random test sequence of length $L = 10^5$. If a toggling fault is present in the RAM, what is the probability of detecting it?

8.4 Consider the double fault $f = [i/0 \& j/0]$ where i and j are two cells of a 1024-word by 1-bit RAM.
a) Give the observer for this fault.
b) Give the required test length for $\varepsilon = 10^{-3}$, given the distribution ψ_0 and that the initial state is unknown. What is the length coefficient for this fault?

8.5 Assume a batch of 1 Mbit RAM (1Mword by 1 bit), and a random test sequence of length $L = 50 \cdot 10^6$. What is the expected defect level given the following information.
80% of the RAM are fault-free.
8% of the RAMs are affected by a single stuck-at cell (*).
8% "   "   by at least two stuck-at cells (*).
1% "   "   by an idempotent coupling (*).
1% "   "   by a toggling (*).
1% "   "   by a double toggling (*).
1% "   "   by unmodeled faults at least as easy to detect as an active PSF with $V = 3$.
(*) means that other faults can be present, in addition.
The length coefficients for $\varepsilon = 10^{-3}$ are found in Solution 8.4 for a double stuck-at and in Figure 8.14 for the other fault models.

Chapter 9

9.1 Consider the microprocessor whose graph model is in Figure 9.1. Assume the instructions are equally likely and the data are equally likely.
a) Give the detection set associated with the stuck-at fault of a cell in the register R_4 (fault f).
b) What is the approximate test length for this fault, for $\varepsilon = 10^{-3}$?

9.2 Same question as in Exercise 9.1 for the fault g defined as follows. Two bits of the transfer operator O_{10} are stuck-at: $b_i/0$ and $b_j/1$.

9.3 Consider a 16-bit adder. The words A and B are two 16-bit inputs, $A = a_{15}...a_0$ and $B = b_{15}...b_0$, and the output $Z = (A + B)$ mod 2^{16} is a 16-bit output. Assume the set of faults F including all the single stuck-at faults of the inputs, of the outputs, and of the carries. Give the required random test length for the worst case fault in F, given the equal likelihood of the possible values of the random words A and B.

9.4 Consider the shift register in the figure whose serial input is x and whose output is $z = Q_4$. If $r = 0$ when $\uparrow CK$ occurs, x shifts to Q_1 and Q_i to Q_{i+1}. If $r = 1$ when $\uparrow CK$ occurs, all the cells are reset.

Give a graph model containing two kinds of nodes like in Figure 9.1 (Section 9.1.1). For this purpose, the following approach may be used.

1) Consider 1-bit registers corresponding to the primary input line (R_{in}) and the various cells.

2) Define fictitious instructions corresponding to the various workings of the system.

3) Consider two kinds of fictitious operators: transfer from one register to the next, and reset of registers.

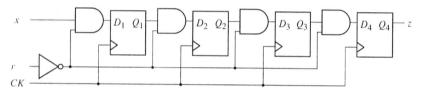

Figure E 9.4

9.5 Consider again the circuit in Figure E 9.4. Given $\Pr[x = 1] = \Pr[r = 1] = 0.5$ and $\varepsilon = 10^{-3}$, give the approximate test lengths for the following faults (using the model in Figure S 9.4).
  a) $f = Q_1$ stuck-at-0.
  b) $g = Q_2$ stuck-at-1.
  c) $h = r$ stuck-at-1.

9.6 Same questions as in Exercise 9.5, now assuming that the reset has a low probability: $\Pr[r = 1] = 0.01$. Comment on the results.

Chapter 10

10.1 Consider the LFSR in the figure.
  a) Give its characteristic polynomial.
  b) Give the sequence $Q_4(1)Q_4(2)...Q_4(15)$ produced by this LFSR, given $Q_1(1)Q_2(1)Q_3(1)Q_4(1) = 1111$. Is the characteristic polynomial primitive?
  c) Let B be the M-sequence obtained in b. Consider the sequences $B_{(2)}$ and $B_{(3)}$ (notation used at the end of Section 10.3.1) obtained by decimation. Comment on the results.

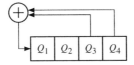

Figure E 10.1

10.2 Assume a 3-input circuit.
a) How can we build an LFSR generating all the sequences of length two, i.e., all the sequences $x_i x_j$ such that $x_i, x_j \in \{x_0, ..., x_7\}$.
b) Give the simplest solution to this problem.

10.3 Given an equally likely generator whose outputs are $Q_1, Q_2, Q_3, Q_4, ...$, give a combinational circuit producing the weighted distribution $\psi_1 = (0, .25, .5, .25)$ for a 2-output circuit whose inputs are x_1 and x_2.

10.4 Given an equally likely generator whose outputs are $Q_1, Q_2, Q_3, Q_4, ...$, give a combinational circuit producing the weighted distribution $\psi_2 = (.25, .5, .125, .125)$ for a 2-output circuit whose inputs are x_1 and x_2.

10.5 Given an equally likely generator whose outputs are $Q_1, Q_2, Q_3, Q_4, ...$, give a sequential circuit producing the generalized distribution presented in the figure for a 2-output circuit whose inputs are x_1 and x_2. Circuits in Figures S 10.3 and S 10.4 may be used for this purpose.

	x_0	x_1	x_2	x_3	
$x_1 x_2$	00	01	10	11	
* r_0	—	r_0	r_1	r_1	$\psi_1 = (0, .25, .5, .25)$
r_1	r_1	r_1	r_0	r_0	$\psi_2 = (.25, .5, .125, .125)$

Figure E 10.5

Chapter 11

11.1 Compare the length to detection l_ω to the average length between detections Av(l_t) for the experimental result in Figure 11.2b.

11.2 Consider the fault f_2 in Figure 11.5. How can we diagnose or verify the presence of this fault in a circuit?

11.3 In Figure 11.6, Min(l_ω) has the same value 103 for five classes of circuits. Explain why.

11.4 Let us say that the memory effect is *favorable* to the detection if the average length to detection is shorter than the average length between detections. Consider the results in Figure 11.8.

422   Exercises

a) For what classes of circuits is the memory effect favorable?
b) Comment on about the most unfavorable cases.

Chapter 12

12.1 Assume a type of circuit under test and a random test sequence such that the expected fault coverage is P. What length k of signature analyser would you suggest in the following two cases?
  a) $P = 0.99$.
  b) $P = 0.999999$.
  c) Comment on your choices.

12.2 Why is it useful to use LFSRs with primitive polynomials for signature analysers?

12.3 In some cases, the convergence of the aliasing probability AL to the limit value may present peaks whose values are much greater than the limit value. Explain why these peaks are not important from a practical point of view.

12.4 Consider the 3-bit SISR whose polynomial is $1 \oplus x^2 \oplus x^3$. Give the probability of aliasing, AL, and the probability of non-revelation, $\Pr[A = A_0]$, for the test lengths $l = 8$ and $l = 40$, in the following two cases.
  a) The detection probability of the fault is $p = 0.2$.
  b) The detection probability of the fault is $p = 0.8$.
  c) Comment on the results.

12.5 Compare the costs and the limit probabilities of non-revelation of SISRs of length $k = 16$ and $k = 17$.

12.6 Consider a 3-output circuit under test. A signature analysis whose probability of non-revelation is close to 10^{-6} is required for all the faults which could affect this circuit.
  Compare the costs of the two following solutions: a single three-input MISR and three SISRs (one for each output).

12.7 Assume a circuit under test which can produce periodical errors such that the period is a power of 2. What polynomials would you suggest for the signature analyser?

Chapter 13

13.1 What is the usefulness of an extended specification?

13.2 Figures 13.3a and b illustrate, respectively, a deterministic and a random approach for separating subcircuits in order to improve controllability. Compare the resulting observabilities.

13.3 Give the detection probabilities of the following faults related to Figure 13.6 in Section 13.3, given $\Pr[x_i=1] = \Pr[y_i=1] = 0.5$ for all indices i and all these variables are independent of each other.
a) $f = w_1/0$ in subcircuit K and in subcircuit K_1.
b) $g = w_2/0$ in subcircuit K and in subcircuit K_2 (w_2 is the input of G_8; it is not detected on z_1).

13.4 Consider again Figure 13.6 and assume a CMOS technology. In this technology, each AND gate can be made up of a NAND gate and an inverter as shown in the figure below.

Figure E 13.4

Give the average detection probabilities of the stuck-open faults for the PMOS transistors (see Figures 2.9 and 2.13a) specified below.
a) Fault f = stuck-open of the PMOS transistor in G'_1 associated with y_2 in subcircuit K.
b) Fault f in subcircuit K_1.
c) Fault g = stuck-open of the PMOS transistor in G'_{15} associated with y_{16} in subcircuit K.
d) Fault g in subcircuit K_2.

13.5 a) Modify the circuit in the figure by cube extraction.
b) Give the detection probabilities for the stuck-at faults affecting the lines w_1 to w_6 in the figure.
c) Give the detection probabilities for the stuck-at faults affecting the lines created by the modification.

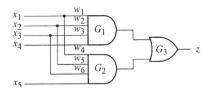

Figure E 13.5

Solutions to Exercises

Chapter 1

1.1 a) Pr[not any faulty circuit] = $(0.999995)^{80} \approx 0.99960$, hence $D \approx 0.04\%$.
b) Pr[not any faulty circuit] = $(0.999800)^{80} \approx 0.984$, hence $D \approx 1.6\%$.
c) Pr[not any faulty circuit] = $(0.999995)^{40} \cdot (0.999800)^{40} \approx 0.99183$, hence $D \approx 0.82\%$.

Comparison between cases a and b shows that D is practically proportional to the defect level: $5/200 = 0.04/1.6$.

In the case c where there is a big difference between the defect levels, D depends essentially on the circuits with a high defect level (40 circuits with a 200 *ppm* defect level $\Rightarrow D \approx 0.80\%$).

1.2 For one test vector, 21 cycles are necessary: 20 for setting the internal variables, plus 1 for application of the test vector. After the last test vector, 19 clock pulses are required to shift the last result in the scan path: $N = 21 \times 10^5 + 19 \approx 2.1 \times 10^6$ clock pulses are required.

1.3 The first error on the primary outputs, appearing in *normal operation*, is detected by on-line BIST. For off-line BIST, a test sequence can be applied only when the circuit is not working.

1.4 See "Conclusion on BIST" at the end of Section 1.2.

Chapter 2

2.1 Fault-free: $z = x_1x_2 + x_3x_4$.
a) Wired OR (Figure a): $z_f = x_1x_2x_3x_4$.
b) Wired AND (Figure b): $z_g = x_1x_2 + x_3x_4$. The output function is not changed: the fault is undetectable.

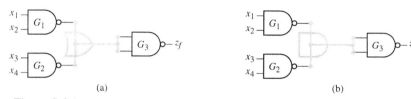

Figure S 2.1

2.2 According to Figure E 2.2, $D = Qx_1x'_2 + Q'x_2$ for the fault-free circuit and $D = Q'x_2$ for the faulty one. In both cases $z = Q$. The states tables are given in the figure: fault-free in figure a and faulty in figure b.

x_1x_2	x_0 00	x_1 01	x_3 11	x_2 10	z
* q_0	q_0	q_1	q_1	q_0	0
q_1	q_0	q_0	q_0	q_1	1

(a)

x_1x_2	x_0 00	x_1 01	x_3 11	x_2 10	z
* q'_0	q'_0	q'_1	q'_1	q'_0	0
q'_1	q'_0	q'_0	q'_0	q'_0	1

(b)

Figure S 2.2

2.3 $I = (x_3 + x_5 + x_6 + x_9 + x_{10} + x_{12})^*$

2.4 The input language may be rewritten as $\mathcal{P}(x_1x_0 + x_1x_2 + x_2x_0 + x_2x_2)^*$. Hence, the corresponding output language is $\mathcal{P}(z_1z_1 + z_1z_2 + z_2z_2 + z_2z_2)^*$ = $\mathcal{P}(z_1z_1 + z_1z_2 + z_2z_2)^*$.

Chapter 3

3.1 $T_{x_1/0} = \{x_5\}$, $T_{x_2/0} = \{x_3\}$, $T_{x_3/0} = T_{w/0} = T_{z/0} = \{x_3, x_5, x_7\}$,
$T_{x_1/1} = T_{x_2/1} = T_{w/1} = \{x_1\}$,
$T_{x_3/1} = \{x_2, x_4, x_6\}$, $T_{z/1} = \{x_0, x_1, x_2, x_4, x_6\}$.

3.2 The answer to this exercise is based on the solution to Exercise 3.1 above. $T = \{x_1, x_3, x_4\}$ tests all the single stuck-at faults except $x_1/0$.
1) The fault coverage before fault collapsing is 9/10 i.e., 90%;
2) Fault $x_3/0$ and equivalent faults dominate $x_2/0$, and $z/1$ dominates $x_3/1$. Then after equivalence and dominance fault collapsing, 4 faults have to be considered: $x_1/0$, $x_2/0$, $x_1/1$, and $x_3/1$. It follows that the fault coverage is 3/4 i.e., 75%.

3.3 The algebraic method may be used. The test set $T = \{x_2, x_3, x_5, x_6\}$ is a solution according to the figure (minimum number of test vectors).

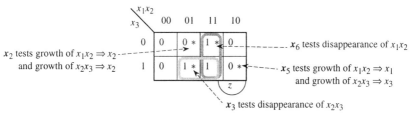

Figure S 3.3

3.4   The checkpoints are x_1, x_2, x_3, w_1, w_2 shown in the figure.

Faults $x_1/0$, $x_2/0$, and $w_1/0$ are tested by the vector $\boldsymbol{x}_6$. Faults $x_3/0$ and $w_2/0$ are tested by the vector $\boldsymbol{x}_3$. Faults $x_1/1$ and $x_3/1$ are tested by the vector $\boldsymbol{x}_2$. Faults $x_2/1$, $w_1/1$, and $w_2/1$ are tested by the vector $\boldsymbol{x}_5$.

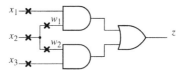

Figure S 3.4

3.5   The observer is given in the figure.

The sequence $S = \boldsymbol{x}_1\boldsymbol{x}_2$ tests the faults. The set of sequences detecting the fault corresponds to the language

$$L = \big((\boldsymbol{x}_0+\boldsymbol{x}_2)^* + ((\boldsymbol{x}_1+\boldsymbol{x}_3)(\boldsymbol{x}_0+\boldsymbol{x}_1+\boldsymbol{x}_3))^*\big)^* (\boldsymbol{x}_1+\boldsymbol{x}_3)\boldsymbol{x}_2(\boldsymbol{x}_0+\boldsymbol{x}_1+\boldsymbol{x}_2+\boldsymbol{x}_3)^*.$$

There are other solutions for expressing the same language. Note that the set of sequences in $\big((\boldsymbol{x}_0+\boldsymbol{x}_2)^* + ((\boldsymbol{x}_1+\boldsymbol{x}_3)(\boldsymbol{x}_0+\boldsymbol{x}_1+\boldsymbol{x}_3))^*\big)^*$ corresponds to all the sequences leading from (q_0, q'_0) to itself.

Sequence S_1 is in L and, thus, detects the fault (at the 5th vector). Sequence S_2 is not in L and, thus, does not detect the fault.

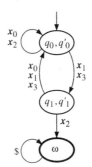

Figure S 3.5

3.6   If the synchronizing sequence is $\boldsymbol{x}_0$, then the initial state of A is q_0 and the initial state of A' is in $Q'_0 = \{q'_0, q'_1\}$; the OR-observer in figure a is obtained.

If the synchronizing sequence is $x_0 x_1$, then the initial state of A is q_1 and the initial state of A' is in q'_1; the observer in figure b is obtained.

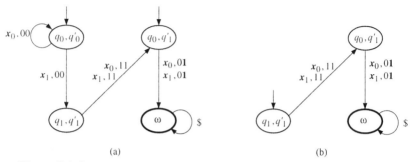

Figure S 3.6

3.7 The fault-free automaton A and the faulty automaton A' are given in figures a and b respectively. Since the synchronizing sequence is x_0, the initial state of A is q_0 and the initial state of A' is in $Q'_0 = \{q'_0, q'_1\}$.

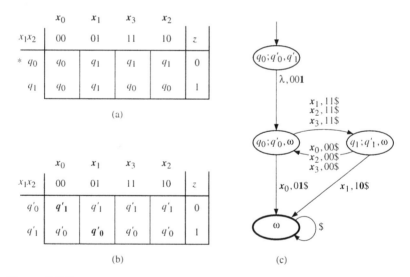

Figure S 3.7

The AND-observer is presented in figure c. Although the Moore representation is used for the automata, the Mealy model is used for the observer (to avoid introducing an additional notation for observers). Note that the fault is *immediately* detected if the initial state of A' is q'_1 since the output state is faulty (in a Moore model, the output is defined at time $t = 0$); in the AND-observer, this is represented by the transition labelled "$\lambda, 001$", where λ is the sequence of length zero.

Since there is a sequence $S = \lambda \cdot x_0 = x_0$ leading from the initial state $(q_0; q'_0, q'_1)$ to the final state ω in the AND-observer, the fault is detectable (Definition 3.17 and Remark 3.10).

3.8 a) The shortest synchronizing sequences are x_0 (initial state q_0) and x_1 (initial state q_1). The set of all synchronizing sequences correspond to the sequences in the language $\mathcal{L} = (x_0 + x_1 + x_2 + x_3)^*(x_0 + x_1)(x_0 + x_1 + x_2 + x_3)^*$.

b) For any synchronizing sequence such that the initial state of A is q_0, the the initial state of A' is in $Q'_0 = (q'_0, q'_1)$: in this case the fault is detectable according to Solution 3.7.

For any synchronizing sequence such that the initial state of A is q_1, the initial state of A' is in $Q'_1 = (q'_0, q'_1)$: in this case the fault is detectable since there is a sequence $S = \lambda \cdot x_1 = x_1$ leading from the initial state $(q_1; q'_0, q'_1)$ to the final state ω in the AND-observer in Figure S 3.8.

Hence, according to Definition 3.18, the fault is detectable.

c) After any synchronizing sequence, the input sequence $x_0 x_0$, for example, detects the fault (see the AND-observers in Figures S 3.7c and S 3.8).

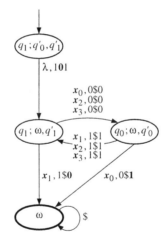

Figure S 3.8

Chapter 4

4.1 According to Solution 3.1:
$P(x_1/0) = P(x_2/0) = P(x_1/1) = P(x_2/1) = P(w/1) = 0.799$;
$P(x_3/0) = P(w/0) = P(z/0) = P(x_3/1) = 0.996$;
$P(z/1) \approx 1$.
a) $P = (0.799 \times 5 + 0.966 \times 4 + 1) / 10 = 0.898$.
b) $P_m = 0.799$.

4.2 a) $P_w = ((0.799 + 0.799 + 0.996) \times 0.3 + 0.799 \times 0.1) = 0.858$.
b) $P_w = ((0.799 + 0.799 + 0.996 + 0.799) \times 0.1 + (0.996 + 1) \times 0.3)$
$= 0.938$.

c) Obviously, in both cases P_w is greater than P_m. In case a: $P_w < P$. In case b: $P_w > P$.

4.3 $P_m(\rho_1) = 0.799$, $P_m(\rho_2) = 0.996$.
$P_{wm} = 0.799 \times 0.6 + 0.996 \times 0.4 = 0.878$.

4.4 According to Solution 3.1: $x_3/0$ dominates $x_1/0$; $z/1$ dominates $x_3/1$ and $x_1/1$. Faults $x_1/0$ and $x_2/0$ are equidetectable; faults $x_3/0$ and $x_3/1$ are equidetectable.

Chapter 5

5.1 a) $I = (x_0 + x_2 + x_3 + x_4 + x_5 + x_6)*$
b) The output sequences are $R_{(1)}(S) = 1000110$ and $R_{(2)}(S) = 1001110$. They differ at the 4th test vector corresponding to x_1 although both circuits correspond to the (incomplete) specification. The input sequence $x_3 x_0 x_0 x_1$ is not in I. Two different values have been assigned to this input vector in figures b and c.

5.2 $I = \mathcal{P}((x_1 + x_2)(x_0 + x_3))*$ (See Remark 2.4 about the prefixes).

5.3 The graph representation is in figure a and the matrix representation (containing both the initial state $\pi(0)$ and the transition matrix U) is in figure b.

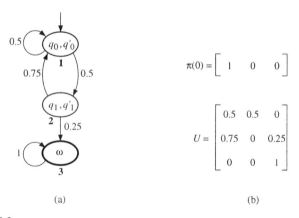

(a)   (b)

Figure S 5.3

5.4 Since the synchronization sequence is x_0, the initial state is q'_0 if the starting state is q'_0, and the initial state is q'_1 if the starting state is q'_1. The Markov chain is presented in the figure.

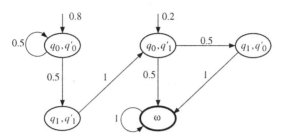

Figure S 5.4

5.5 The fault in Figure E 5.5b is detected by the input vector x_1 (i.e., $x_1 x_2 = 01$). Since the fault is combinational, the fault-free and the faulty circuits have a single state, then their product is also a single state, here denoted by q_0.

The observer is presented in figure a. From the initial state (q_0, r_0), if x_1 is applied the fault is detected; else, if x_0 or x_2, or x_3 is applied, the state (q_0, r_1) is reached according to Figure E 5.5c, and so on.

The Markov chain is given in figure b. When the generator is in state r_0, $\psi(x_1) = 0.25$ and $\psi(x_0 + x_2 + x_3) = 0.75$. When it is in state r_1, $\psi(x_1) = 0.4$ and $\psi(x_0 + x_2 + x_3) = 0.6$.

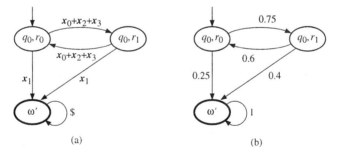

Figure S 5.5

5.6 a) The state table of the fault-free machine is in figure a. The initial state q_0 is reached (consequence of the initialization sequence $x_4 x_1$).

b) The state table of the faulty machine is in figure b. From the starting state q'_1, the initialization sequence $x_4 x_1$ leads to initial state q'_0. Note that only the next state function is faulty.

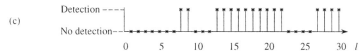

Figure S 5.6

c) See figure c. The fault-free values J_0 and K_0 and the faulty values J_f and K_f are not very different: $J_0 = x_1 + x_3$, $K_0 = x_1 x_2 + x_3$, $J_f = x_3$, and $K_f = x_3$. It follows that many input vectors are such that $J_0 = K_0 = J_f = K_f$. If $J_0 = K_0 = J_f = K_f = 0$ at time t, the state does not change: there is a detection at t if and only if there is a detection at $t-1$. If $J_0 = K_0 = J_f = K_f = 1$ at time t, the state changes in both machines: there is a detection at t if and only if there is a detection at $t-1$ (since the output z corresponds to the state).

Chapter 6

6.1 The test sets for the various faults are found in Solution 3.1.

a) Probabilities of testing: 0.93 for $x_1/1$, 0.93 for $x_1/0$, $(1 - 8 \cdot 10^{-5})$ for $x_3/0$, and $(1 - 3 \cdot 10^{-9})$ for $z/1$.

The faults with a great detectability (3 for $x_3/0$ and 5 for $z/1$) are detected with a probability very close to 1.

b) Probabilities of testing: $(1 - 8 \cdot 10^{-4})$ for $x_1/1$, 0.99 for $x_1/0$, $(1 - 9 \cdot 10^{-6})$ for $x_3/0$, and $(1 - 7 \cdot 10^{-8})$ for $z/1$.

Faults $x_1/1$ and $x_1/0$ which were R-equidetectable for ψ_0 are no longer R-equidetectable for the distribution ψ_1.

6.2 a) For ψ_0. According to Solution 3.1: $k_{min} = 1$, then $p_{min} = 1/8 = 0.125$. It follows $P_m = 0.93$.

b) For ψ_1. From the results in Solution 3.1 and ψ_1:
$p_{x_1/0} = 0.2$, $p_{x_2/0} = 0.2$, $p_{x_3/0} = p_{w/0} = p_{z/0} = 0.44$,

$p_{x_1/1} = p_{x_2/1} = p_{w/1} = 0.3$, $p_{x_3/1} = 0.22$, $p_{z/1} = 0.56$.
Then $p_{min} = 0.2$. It follows $P_m = 0.99$.

The weighted distribution ψ_1 increases the probabilities of detections for the faults which have a low detectability, hence the minimum testing probability.

6.3 a) $P(x_1/0) = P(x_2/0) = P(x_1/1) = P(x_2/1) = P(w/1) = 0.93$;
$P(x_3/0) = P(w/0) = P(z/0) = P(x_3/1) = 1 - 8 \cdot 10^{-5}$ (i.e., ≈ 1);
$P(z/1) = 1 - 3 \cdot 10^{-9}$ (i.e., ≈ 1).
$P \approx \dfrac{1}{10}(5 \times 0.93 + 4 + 1) = 0.965$, for the set F_1.

b) The set F_2 is obtained by deleting $w/0$, $z/0$, $x_2/1$, and $w/1$ from F_1. Then
$P \approx \dfrac{1}{6}(3 \times 0.93 + 2 + 1) = 0.965$, for the set F_2.

6.4 Since there is no reconverging fanout in the downstream cone of the line w_1, the exact detection function can be built: see the figure. The fault is detected if three conditions are met:
1) $w_1/1$ (provoking an error at w_1);
2) $w_2 = 0$ (propagation through G_3);
3) $w_3 = 1$ OR $w_4 = 1$ (propagation through G_4 or G_5).

The detection function is then $T_f = (x'_2 + x'_3) x'_4 (x'_1 + x_5)$. Since $T_f = 1$ for 9 minterms, $k_f = 9$ is obtained.

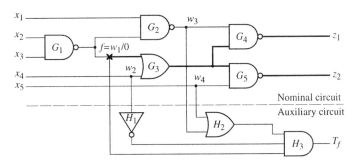

Figure S 6.4

6.5 a) $C(0, w_1) = 0.25$ and $C(1, w_1) = 0.75$.
b) $O(w_1) = 0.25$.
c) $A(11, G_2) = 0.1875$.

6.6 a) Since z_1 does not depend on three input variables (namely x_4, x_5, and x_6), $k_{1,min} \geq 2^3 = 8$. Similarly $k_{2,min} \geq 2^2 = 4$. Then, according to Property 6.5, $k_{min} \geq 4$.
b) Since $k_{min} \geq 4$, $p_{min} \geq 4/2^6 = 0.0625$. Hence, $L_m(0.001) \leq 108$.

Chapter 7

7.1 a) The corresponding Markov chain is given in Figure S 5.3.
$L(10^{-1}) = 26$, $L(10^{-3}) = 76$, $L(10^{-6}) = 150$.
b) For the transition matrix U in Figure S 5.3b, the value $\lambda_2 = 0.9114$ is found. It follows: $L(10^{-1}) \approx 25$, $L(10^{-3}) \approx 75$, $L(10^{-6}) \approx 149$.

7.2 For ψ_1, the transition matrix U in Figure S 5.3 is modified as follows: $U_{1,1} = 0.8$, $U_{1,2} = 0.2$, $U_{2,1} = 0.3$, $U_{2,3} = 0.7$.
Exact value: $L(10^{-3}) = 51$.
Approximate value: $\lambda_2 = 0.8690 \Rightarrow L(10^{-3}) \approx 50$.

7.3 a) Comparison of Figures S 2.2a and S 2.2b shows that: 1) the output function is fault-free; 2) there is a single faulty entry in the next state function, namely $\delta(q_1, x_2)$. It follows that $D_f = q_1 \cdot x_2$.
b) $\Pr[D_f] = \Pr[q_1] \cdot \Pr[x_2]$. We know that $\Pr[x_2] = 0.25$. The value $\Pr[q_1]$ can be obtained from the following set of equations:
$\Pr[q_1] = \Pr[q_0] \cdot \psi_0(x_1 + x_3) + \Pr[q_1] \cdot \psi_0(x_2)$,
$\Pr[q_0] + \Pr[q_1] = 1$.
The values $\Pr[q_1] = 0.4$ and $\Pr[D_f] = 0.4 \times 0.25 = 0.1$ are obtained. From (7.18), $L(10^{-3}) \approx 66$ is obtained (according to Solution 7.1, the exact value is $L(10^{-3}) = 76$).

7.4 a) $D_g = q_0 \cdot (x_4 + x_6) + q_1 \cdot x_6$.
b) $\Pr[q_0] = 0.45$ and $\Pr[q_1] = 0.55$. Hence: $\Pr[D_g] = 0.18$ and $L(10^{-3}) \approx 35$.

7.5 a) In Solution 7.3, it was found that $\Pr[q_0] = 0.6$ and $\Pr[q_1] = 0.4$.
$\Pr[q_0, x_i, q_j] = 0.6 \times 0.25 = 0.15$ and $\Pr[q_1, x_i, q_j] = 0.4 \times 0.25 = 0.1$, for all i and j such that the transition (q_0, x_i, q_j) or (q_1, x_i, q_j) exist.
It follows that $p_{min} = 0.1$.
b) $L_m(10^{-3}) \approx \ln 10^{-3}/\ln(1 - p_{min}) = 66$.

7.6 a) From the Markov chain in Figure 7.7c, $L(10^{-3}) = 101$ is obtained.
b) Any detectable combinational fault is detectable by at least one test vector. Hence, $p_{min} = 0.25$. It follows $L_m(10^{-3}) = 25$. From the Markov chain in Figure 7.7c, $\varepsilon(25) = 0.19$ is obtained.

7.7 a) A Mealy machine is more concise than a Moore machine. However, the output state is not defined at time 0 for a Mealy machine. Figures a and b represent the Moore machines corresponding to the Mealy machines in Figure 7.7a and b. From figures a and b, the Markov chain in figure c can be obtained. The starting/initial state is:
(q_1, q'_1) if the input vector is x_0 or x_1, or x_2 and the internal state is q'_1 (probability 0.625);
(q_0, q'_0) if the input vector is x_3 (probability 0.25);

$(q_1, q'_0) = \omega$ if the input vector is x_2 and the internal state is q'_0 (probability 0.125).

This is an example where the fault can be detected at time 0.

b) $L(10^{-3}) = 98$ is obtained. One can observe that, even if $\pi_\omega(0) = 0.125$ and $\pi_{(q_0,q'_0)} = 0.25$ (instead of zero for both in Figure 7.7c), the difference on the test length is small (98 instead of 101). The initial state has a slight influence (consistent with Property 7.3 in Section 7.2.2), and this is a justification for using Mealy models.

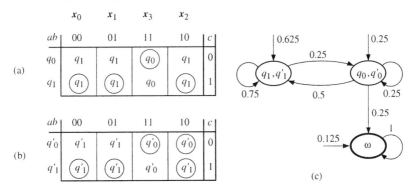

Figure S 7.7

7.8 The behavior is illustrated in the figure. The logical values corresponding to the lines in the circuit are shown in standard characters for the state where $CK = D = Q = 1$ and $S = R = 0$ (i.e., $\overline{S} = \overline{R} = 1$).

When R takes the value 1, the values presented with bold characters are obtained (due to the fault $w_3/1$, some lines have a faulty value). The output Q takes the correct value 0.

When R returns to the value 0, the values presented with italic characters are obtained. The output returns to the faulty value 1.

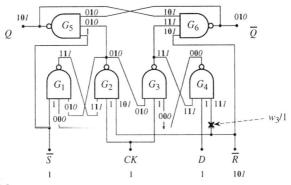

Figure S 7.8

Chapter 8

8.1 Since f_3 is a stuck-at fault, the proof is obtained from Property 8.5 in Section 8.3.2.

8.2 According to the notation in Section 8.3.2, $I_1 = \{c\}$, $O_1 = \{d\}$, $O_2 = \{a\}$, $O_3 = \{b\}$, then $\{I_1 \cup O_1\} \cap \{O_2 \cup O_3\} = \emptyset$. Hence, the proof follows from Property 8.6 in Section 8.3.2.

8.3 The average number of accesses to every cell is $10^5/2048 = 48.8$. According to Figure 8.11a, the average number of accesses to every cell allowing a detection with a probability $1 - 10^{-3}$ is $H = 100$ for a toggling fault.

According to (8.23) in Property 8.9 (Section 8.4.1), $L(\varepsilon)$ is proportional to $\ln \varepsilon$. Thus, the detection uncertainty for the considered test length is $10^{-\beta}$ such that $\dfrac{\beta}{3} = \dfrac{48.8}{100}$ i.e., $\beta = 3 \times (48.8/100) = 1.46$. Hence, the probability of detecting a toggling fault is $1 - 10^{-1.46} = 0.965$.

8.4 a) The observer is in figure a, where each state different than ω represents (state of i, state of j).

b) The Markov chain is in figure b. The test length $L(10^{-3}) = 23\,041$ is obtained from this Markov chain. The length coefficient is $H = 23\,041 / 1\,024 = 22.5$. Note that this length coefficient is approximately half the value of the length coefficient for a single stuck-at cell.

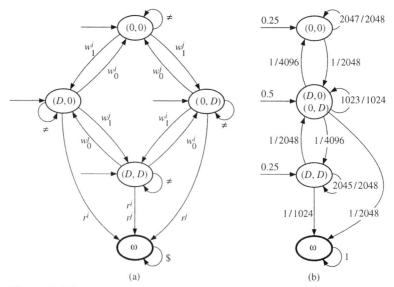

Figure S 8.4

8.5 Let us note F_1 to F_6, the six subsets of faults in the order they appear in the list. The number of words is $n = 1\,048\,576$, then the test length is $L \approx 47.7n$.

According to (8.23) in Property 8.9, and using the results in Figure 8.11, for $\varepsilon = 10^{-3}$, one can conclude that a single stuck-at cell is detected with a probability close to $1 - 10^{-\beta_1}$, where $\beta_1 = 3 \times (47.7/46) \approx 3$ (see Solution 8.3 where a similar calculation is explained).

All the faults in F_1 are at least as easy to detect as a single stuck-at cell, then $P_m(F_1) \approx 1 - 10^{-\beta_1}$.

Similarly, for the other sets of faults, $P_m(F_i) \approx 1 - 10^{-\beta_i}$, where $\beta_2 \approx 6$ (the test length for a double stuck-at is approximately half the test length for a single stuck-at, as found in Solution 8.4), $\beta_3 \approx 3 \times (47.7/219) = 0.65$, $\beta_4 \approx 3 \times (47.7/100) = 1.4$, $\beta_5 \approx 3 \times (47.7/66) = 2.2$, and $\beta_6 \approx 3 \times (47.7/447) = 0.32$.

Then, we obtain $P_m(F_1) = 0.999$, $P_m(F_2) = 0.999999$, $P_m(F_3) = 0.776$, $P_m(F_4) = 0.96$, $P_m(F_5) = 0.994$, and $P_m(F_6) = 0.52$.

The subsets F_i define a partition $\rho = \{F_1, F_2, F_3, F_4, F_5, F_6\}$ of the faults in the batch. The conditional occurrence probabilities of the subsets F_i are $\Pr[F_1|F] = \Pr[F_2|F] = 8/20 = 0.4$ and $\Pr[F_3|F] = \Pr[F_4|F] = \Pr[F_5|F] = \Pr[F_6|F] = 1/20 = 0.05$.

From Equation (4.27) in Section 4.4.2, we obtain $P_{wm}(\rho) = \Sigma P_m(F_i) \cdot \Pr[F_i|F] = 0.962$.

Since $P_{wm}(\rho)$ is a lower bound of P_w (Property 4.5 in Section 4.4.3.1) and $P_u = P_w$ because all the faults have been considered (the unmodeled faults are in the subset F_6), we have $P_u \geq 0.962$.

From Equation (4.6) in Section 4.1, given $Y = 0.8$, and $P_u \geq 0.962$, one obtains $DL \leq 0.0094$, i.e., $DL \leq 0.94\%$.

Chapter 9

9.1 a) Let Q_f denote the set of states such that there is an error in R_4. Then $D_f = Q_f \cdot (I_{16} + I_{18})$.

b) $\Pr[Q_f] = 0.5$, $\Pr[I_{16}] = \Pr[I_{18}] = 1/21$. Hence, $\Pr[D_f] = 0.5 \times (2/21) = 0.048$. It follows that $L(10^{-3}) \approx \ln 10^{-3} / \ln(1 - 0.048) \approx 141$.

9.2 a) Let Q_g denote the set of states such that an error will be provoked in R_2 if instruction I_3 is executed. Specifically, Q_g is such that the value of the ith bit in R_1 is 1 OR the value of the jth bit in R_1 is 0.

According to the notation in Section 9.3.2: $K_2 = (I_1 + I_5 + I_6 + I_7 + I_9 + I_{10} + I_{11} + I_{12} + I_{14} + I_{15} + I_{16} + I_{18} + I_{20} + I_{21})$, and $D_g \supset Q_g \cdot I_3 K_2^* (I_8 + I_{17})$.

b) $\Pr[Q_g] = 0.75$, and $\Pr[D_g] > 0.75 \times \dfrac{1}{21} \times \dfrac{1}{1-\dfrac{14}{21}} \times \dfrac{2}{21} = 0.010$. It follows that $L(10^{-3}) \approx \ln 10^{-3} / \ln(1 - 0.010) = 688$.

9.3 According to part f in Figure 9.9 (Section 9.3.2), the fault which is the most resistant is the stuck-at-0 of the right-most carry. The detection probability of this fault is $p_f = \Pr[a_0 = 1 \text{ AND } b_0 = 1] = 0.25$. It follows that $L(10^{-3}) = 25$.

9.4 The graph model is presented in the figure. The system under consideration is assimilated to a structure performing computations (with 1-bit registers and 1-bit buses). As R_{in} represents the input bus in a microprocessor, it represents the input line here.

At each cycle (i.e., at each occurrence of $\uparrow CK$), two operations can be performed: either the system is reset if $r = 0$, or a shifting is performed if $r = 1$. Then we define two fictitious instructions: $I_1 = r \cdot \uparrow CK$ (reset) and $I_2 = r' \cdot \uparrow CK$ (shifting). Four operators (O_1 to O_4) are activated simultaneously by instruction I_2, and the operator O_5 is activated by instruction I_1. Let us observe that operator O_5 has no input register (i.e., no operand) and has four output registers.

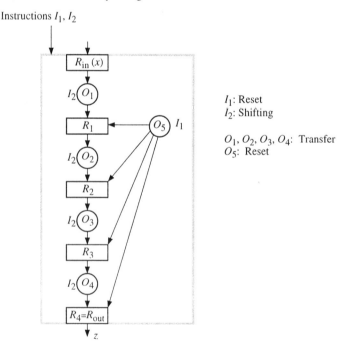

Figure S 9.4

9.5 a) $D_f = Q_f \cdot I_2 I_2 I_2$, where Q_f represents the set of states such that there is an error on Q_1 (Figure E 9.4) i.e., an error in R_1 (Figure S 9.4).

There is an error in R_1 if 1) the last instruction executed was instruction I_2 (probability 0.5) and 2) when this instruction was executed, the value in R_{in} was 1, i.e., $x = 1$ (probability 0.5). Hence, $\Pr[Q_f] = 0.5 \times 0.5 = 0.25$.

Thus, $\Pr[D_f] = \Pr[Q_f] \cdot (\Pr[I_2])^3 = 0.25 \times 0.5^3 = 0.031$. From this value, $L(10^{-3}) \approx 220$ is obtained.

b) Similarly, $D_g = Q_g \cdot I_2 I_2$ where Q_g represents the set of states such that there is an error in R_2.

$\Pr[Q_g] = \Pr[\text{the value in } R_2 \text{ is } 0] = 1 - \Pr[\text{the value in } R_2 \text{ is } 1]$
$= 1 - (0.5 \times 0.5^2) = 0.875$.

$\Pr[D_g] = \Pr[Q_g] \cdot (\Pr[I_2])^2 = 0.875 \times 0.5^2 = 0.22$. From this value, $L(10^{-3}) \approx 28$ is obtained.

c) If $r/1$, instruction I_1 is always executed. The value 0 is always present in R_1, R_2, R_3, and $R_4 = R_{out}$. Hence, this fault is equivalent to Q_4 stuck-at-0. $\Pr[D_h] = \Pr[x = 1] \cdot (\Pr[I_2])^4 = 0.031$, and $L(10^{-3}) \approx 220$ is obtained again.

9.6 $\Pr[I_1] = 0.01$ and $\Pr[I_2] = 0.99$.
a) $\Pr[D_f] = 0.5 \times (0.99)^4 = 0.48$, from which $L(10^{-3}) \approx 11$ is obtained.
b) $\Pr[Q_g] = 1 - (0.5 \times 0.99^2) = 0.51$. Then, $\Pr[D_g] = 0.51 \times (0.99)^2 = 0.50$, from which $L(10^{-3}) \approx 10$ is obtained.
c) $\Pr[D_h] = 0.5 \times (0.99)^4 = 0.48$, from which $L(10^{-3}) \approx 11$ is obtained.

A 0.5 probability of the reset leads to a low probability of propagation of an error up to the output. The weighted patterns considered in this exercise lead to a shorter test length (11 test vectors instead of 220 for the most difficult out of the faults considered).

Chapter 10

10.1 a) $1 \oplus x^3 \oplus x^4$.

b) $Q_4(1)...Q_4(15)... = 111100010011010...$ The characteristic polynomial is primitive since the period of the sequence obtained is $15 = 2^4 - 1$ (necessary and sufficient condition).

c) $B_{(2)} = 110101111000100...$; $B_{(3)} = 100101001010010...$; $B_{(2)}$ is an M-sequence, while $B_{(3)}$ is not an M-sequence (its period is 5). The reason is that 2 and 15 do not share a common factor, while 3 and 15 share the common factor 3.

10.2 a) According to Property 10.3, this can be done with an LFSR (having a primitive polynomial) whose length m is greater than $3 \times 2 = 6$, and with a shifting $\sigma = 3$ such that σ and $2^m - 1$ do not share a common factor.

440 *Solutions to Exercises*

b) See the figure. According to Figure 10.4, the polynomial $1 \oplus x \oplus x^7$ can be used. For simplicity of the construction, the reciprocal polynomial (Section 10.3.1) $1 \oplus x^6 \oplus x^7$ is chosen here.

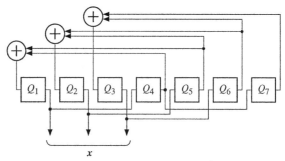

Figure S 10.2

10.3 The weighted distribution can be written as $\psi_1 = (0, 1/4, 2/4, 1/4)$. It follows that this distribution can be obtained from two variables Q_i. A solution is given in the figure.

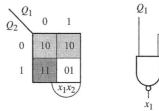

Figure S 10.3

10.4 The weighted distribution can be written as $\psi_2 = (2/8, 4/8, 1/8, 1/8)$. It follows that this distribution can be obtained from three variables Q_i. A solution is given in the figure.

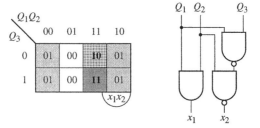

Figure S 10.4

10.5 When the generator of Figure E 10.5 is in state r_0, the distribution ψ_1 can be obtained from the circuit in Figure S 10.3; let us denote by $x_1^{(0)}$ and $x_2^{(0)}$ the values x_1 and x_2 in this case.

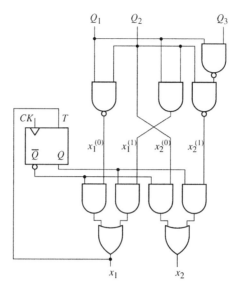

Figure S 10.5

When the generator of Figure E 10.5 is in state r_1, the distribution ψ_2 can be obtained from the circuit in Figure S 10.4; let us denote by $x_1^{(1)}$ and $x_2^{(1)}$ the values x_1 and x_2 in this case.

An internal variable is needed for coding the states of the generator in Figure E 10.5. Let the values $Q = 0$ and $Q = 1$ be assigned to r_0 and r_1 respectively. It follows that $x_1 = Q' \cdot x_1^{(0)} + Q \cdot x_1^{(1)}$ and $x_2 = Q' \cdot x_2^{(0)} + Q \cdot x_2^{(1)}$.

According to Figure E 10.5, the internal state changes if $x_1 = 1$. The sequential part can then be obtained using a T flip-flop, as shown in the figure.

Chapter 11

11.1 The length to detection l_ω corresponds approximately to 75 test vectors. In this experiment, the 30th detection occurs approximately at the 450th test vector, thus $\mathrm{Av}(l_\tau) \approx 450/30 = 15$.

The length to detection is approximately five times the average time between detections. The memory effect is quite clear in Figure 11.2b.

11.2 Two kinds of experiments may be considered: random or deterministic.

Random experiment: give a high probability to the control input *RST* and to instruction *WAI*, and low probabilities to the other controls and instructions. The fault has a high probability to be detected by a short test sequence; the output can then be observed from the initial state up to detection and compared with the expected output sequence corresponding to the specification.

Deterministic experiment: build a short test sequence such that the fault under consideration produces an error (a reset is imposed during execution of a *WAI* instruction). The error is observed when the reset is completed.

11.3 The results in Figure 11.6 correspond to 15 differents experiments. Let $S = \{S_1, ..., S_{15}\}$ denote the set of test sequences which have been applied.

According to the results for class 1, there is a test sequence S_a in S which detects the fault f_1 at the 103th instruction (i.e., $l_\omega = 103$), while all other sequences in S are such that $l_\omega \geq 103$.

For classes 2, 3, 4, and 6, the first fault detected by S_a is f_1; furthermore, there is no sequence in S detecting f_2, f_3, f_4, f_{11}, f_{14}, or f_{15}, with a length shorter than 103.

Note that the circuit in class 5 is affected by at least one fault in addition to f_1, f_3, f_4, and f_{14}; this fault, which has not been diagnosed, is represented by "?".

11.4 a) The memory effect is favorable if $\text{Av}(l_\omega) < \text{Av}(l_r)$, i.e., for classes 8 and 15. It is neither favorable nor unfavorable for classes 11 (difference less than 1%), 16, and 17. It is unfavorable for all the other classes.

b) The greater ratio $\text{Av}(l_\omega) / \text{Av}(l_r)$ is found for class 10: this ratio is 23. The ratios are 11 and 13 for classes 1 and 2; this is more important in practice because the corresponding faults are more difficult to detect.

Chapter 12

12.1 a) For $P = 0.99$, i.e. $\varepsilon = 0.01$, $k = 12$ to 15 would be convenient.
For $k = 12$, $AL \approx 2^{-12} \approx 0.00024$; according to Property 12.2,
$\Pr[A = A_0] \approx 0.01 + 0.99 \times 0.00024 \approx 0.0102$.
For $k = 15$, $AL \approx 2^{-15} \approx 0.00003$;
$\Pr[A = A_0] \approx 0.01 + 0.99 \times 0.00003 \approx 0.01003$.
b) For $P = 0.999999$, i.e. $\varepsilon = 10^{-6}$, $k = 25$ to 28 would be convenient.
For $k = 25$, $AL \approx 2^{-25} \approx 3 \cdot 10^{-8}$;
$\Pr[A = A_0] \approx 10^{-6} + (1-10^{-6}) \times 3 \cdot 10^{-8} \approx 1.03 \cdot 10^{-6}$.
For $k = 28$, $AL \approx 2^{-28} \approx 4 \cdot 10^{-9}$;
$\Pr[A = A_0] \approx 10^{-6} + (1-10^{-6}) \times 4 \cdot 10^{-9} \approx 1.004 \cdot 10^{-6}$.
c) We have chosen values of k such that the aliasing probability may be neglected. However values such as $k = 10$ or 11 for case a, and $k = 23$ or 24 for case b, would be also admissible since the probability of aliasing would remain low in comparison with the uncoverage probability.

12.2 Because a primitive polynomial:
   1) allows faster convergence of the non-revelation probability to the limit $1/2^k$, in the case of SISR with an homogeneous error;
   2) is necessary (and sufficient) to ensure convergence of the non-revelation probability to the limit $1/2^k$, in the case of MISR with either $m \leq k$ or $m > k$;
   3) is necessary to ensure convergence of the non-revelation probability to the limit $1/2^k$, in the case of SISR with periodic errors.
   More details are given in Properties 12.7, 12.10, 12.11, and Conjecture 12.1.

12.3 The peaks on AL occur for faults whose detection probability p is greater than 0.5, and these faults are easy to detect. Let f_1 be such a fault. A fault f_2 whose detection probability is $1-p$ has a greater probability of non-revelation than f_1, even if its aliasing probability is lower, because it is more difficult to detect.
   In practice, if a fault whose detection probability $p > 0.5$ exists in a circuit (easy to detect), there is a fault (difficult to detect) whose detection probability is less than or equal to $1-p$ in the circuit.

12.4 The values of AL and $\Pr[A = A_0]$ are obtained from Equations (12.31) and (12.28) respectively. The corresponding matrix V is in Figure 12.9.
   a) Detection probability $p = 0.2$.
   For $l = 8$, $AL = 0.05372$ and $\Pr[A = A_0] = 0.21248$.
   For $l = 40$, $AL = 0.12489$ and $\Pr[A = A_0] = 0.125$.
   b) Detection probability $p = 0.8$.
   For $l = 8$, $AL = 0.13472$ and $\Pr[A = A_0] = 0.13472$.
   For $l = 40$, $AL = 0.125$ and $\Pr[A = A_0] = 0.125$.
   c) For $l = 8$, the probability of aliasing is greater for $p = 0.8$, but the probability of non-revelation is greater for $p = 0.2$. For $l = 40$, the limit value $1/2^3 = 0.125$ is practically reached for both AL and $\Pr[A = A_0]$ for $p = 0.2$ and $p = 0.8$.

12.5 1) The hardware cost is the same, i.e., 18α, for both because a 3-term primitive polynomial can be used for $k = 17$, while there is no primitive polynomial having less than 5 terms for $k = 16$ (Figure 12.13).
   2) The storage cost is 16β for $k = 16$ and 17β for $k = 17$.
   3) The probability of non-revelation is close to $15 \cdot 10^{-6}$ for $k = 16$ and $8 \cdot 10^{-6}$ for $k = 17$.

12.6 A probability of non-revelation close to 10^{-6} can be obtained with a 20-bit signature since $1/2^{20} = 0.95 \cdot 10^{-6}$.
   For a three-input MISR, the hardware cost is 22α (21α for a single-input LFSR according to Figure 12.13, plus α for two additional EXORs) and the storage cost is 20β.

For a fault affecting the three outputs, we could use three 7-stage SISRs. But for a fault affecting only one output, the probability of non-revelation would be $1/2^7 = 8 \cdot 10^{-3}$: the requirement is not satisfied. Hence, the three SISRs should have 20 stages each. Then, the hardware cost is $63\,\alpha$ and the storage cost is $60\,\beta$.

12.7 A primitive polynomial is necessary. According to Conjecture 12.1, the period and the length of the M-sequence should not share a common factor. Since $2^k - 1$ is odd, it cannot share a common factor with a power of 2. Hence any primitive polynomial could be used.

Chapter 13

13.1 An extended specification is such that the behavior of the circuit is specified for any test sequence. Hence, the random test sequence generator may be easy to implement.

13.2 In Figures 13.3a, if $Degate = 0$, the output y_1 is no longer observable on y_2 (or on a line of its downstream cone if there is no other path from y_2 to this line).

In Figures 13.3b, if the random input has effectively a random value, the output y_1 remains observable on y_2 or on any line of its downstream cone; given $r(t)$, this random value at time t is $y_1(t) = y_2(t) \oplus r(t)$.

13.3 a) In subcircuit K, $f = w_1/0$ is detected on output z if $y_1 = \ldots = y_{16} = 0$. Hence $p_f = 1/2^{16} = 1.5 \cdot 10^{-5}$.

In subcircuit K_1, $f = w_1/0$ is detected on output z_1 if $y_1 = \ldots = y_8 = 0$ (for any value of x_1). Hence $p_f = 1/2^8 = 4 \cdot 10^{-3}$.

b) In subcircuit K, $g = w_2/0$ is detected on output z if $y_1 = \ldots = y_{16} = 0$. Hence $p_g = 1/2^{16} = 1.5 \cdot 10^{-5}$.

In subcircuit K_2, $g = w_2/0$ is detected on output z if $w_2 = 1$ and $y_9 = \ldots = y_{16} = 0$. Now, $\Pr[w_2 = 1] = 0.5$ since $w_2 = 1$ if $((y_1 = \ldots = y_8 = 0$ AND $x_1 = 0)$ OR (at least one $y_j = 1$ in $\{y_1, \ldots, y_8\}$ AND $x_1 = 1))$. Hence $p_g = 1/2^9 = 2 \cdot 10^{-3}$.

13.4 a) In circuit K, fault f is detected at time t if: 1) $y_1 = y_2 = 0$ at $t-1$, and 2) $y_1 = 0$, $y_2 = 1$, and the fault is propagated to z at t; the fault is propagated to z if $y_3 = \ldots = y_{16} = 0$.

Since $\Pr[y_1 = y_2 = 0$ at $t-1] = 1/4$ and $\Pr[y_1 = 0, y_2 = 1$, and the fault is propagated to z at $t] = 1/2^{16}$, $p_{av,f} = 1/2^{18} = 4 \cdot 10^{-6}$ is obtained.

b) In circuit K_1, fault f is detected at time t if: 1) $y_1 = y_2 = 0$ at $t-1$, and 2) $y_1 = 0$, $y_2 = 1$, and the fault is propagated to z_1 at t; the fault is propagated to z_1 if $y_3 = \ldots = y_8 = 0$.

$p_{av,f} = 1/2^{10} = 10^{-3}$ is obtained.

c) In circuit K, fault g is detected at time t if: 1) $w_3 = 1$ and $y_{16} = 0$ at $t-1$, and 2) $w_3 = 1$ and $y_{16} = 1$ at t; $w_3 = 1$ if $y_1 = \ldots = y_{15} = 0$.
$P_{av.g} = 1/2^{32} = 2 \cdot 10^{-10}$ is obtained.

d) In circuit K_2, fault g is also detected at time t if: 1) $w_3 = 1$ and $y_{16} = 0$ at $t-1$, and 2) $w_3 = 1$ and $y_{16} = 1$ at t; $w_3 = 1$ if $w_2 = 1$ and $y_9 = \ldots = y_{15} = 0$.
$P_{av.g} = 1/2^{18} = 4 \cdot 10^{-6}$ is obtained.

13.5 a) The circuit is given in the figure.
b) The detection probability of $w_i/0$ is $1/2^5 = 0.03$ for $i = 1, \ldots, 6$.
The detection probability of $w_i/1$ is $2/2^5 = 0.06$ for $i = 1, \ldots, 6$.
c) Detection probabilities for the lines created: $3/2^5 = 0.09$ for $w_7/0$; $21/2^5 = 0.66$ for $w_7/1$; $1/2^5 = 0.03$ for $w_8/0$ and $w_9/0$; $7/2^5 = 0.22$ for $w_8/1$ and $w_9/1$.

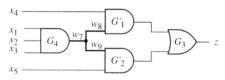

Figure S 13.5

Bibliography

[AbTh 88] Z. Abazi and P. Thévenod-Fosse, "Test aléatoire de cartes à microprocesseur : étude théorique basée sur des modèles markoviens," *Technique et Science Informatiques*, vol. 7, n° 5, pp. 477-492, 1988.

[Ab 83] J. A. Abraham, "Design for Testability," in *Proc. of IEEE Custom Integrated Circuit Conference*, Rochester, NY, pp. 278-283, May 1983.

[AbBrFr 90] M. Abramovici, M. A. Breuer, and A. D. Friedman, *Digital Systems Testing and Testable Design*, Computer Science Press, New York, 1990.

[Ab et al. 86] M. Abramovici J. J. Kulikowski, P. R. Menon, and D. T. Miller, "SMART and FAST: Test Generation of VLSI Scan-Design Circuits," *IEEE Design & Test of Computers*, vol. 3, n° 4, pp. 43-54, August 1986.

[AbPa 92] M. Abramovici and P. S. Parikh, "Warning: 100% Fault Coverage May Be Misleading!!," in *Proc. International Test Conference*, IEEE CS Press, pp. 662-668, 1992.

[AgKiSa 93] V. K. Agarwal, C. R. Kime, and K. K. Saluja, "A Tutorial on Built-InSelf-Test", Part 1: Principles", *IEEE Design & Test of Computers*, pp. 73-81, March 1993.

[AgAg 72] V. D. Agrawal and P. Agrawal, "An Automatic Test Generation System for Illiac IV Logic Boards", *IEEE Transactions on Computers*, vol. C-21, n° 9, pp. 1015-17, September 1972.

[AgAg 75] P. Agrawal and V. D. Agrawal, "Probalistic Analysis of Random Test Generation Method for Irredundant Combinational Logic Networks", *IEEE Transactions on Computers*, vol. C-24, n° 5, pp. 573-578, May 1975.

[AgSeAg 82] V. D. Agrawal, S. C. Seth, and P. Agrawal, " Fault Coverage Requirements in Production Testing of LSI Circuits," *IEEE Journal of Solid-State Circuits*, vol. SC-17, n° 1, pp. 57-61, February 1982.

[AhSeUl 86] A. V. Aho, R. Sethi, and J. D. Ullman, *Compilers: Principles, Techniques, and Tools*, Addison-Wesley, Reading, MA, 1986.

[An 71] D. A. Anderson, *"Design of Self-Checking Digital Networks Using Coding Techniques,"* Urbana, CSL, University of Illinois, Report 527, September 1971.

[Ba 78] A. Bader, *"Test aléatoire synchrone de circuits séquentiels,"* Student Project, Laboratoire d'Automatique de Grenoble, INPG, 1978.

[BaMc 82] P. Bardell and W. McAnney, "Self-Testing of Multichip Logic Modules," in *Proc. International Test Conference*, IEEE CS Press, pp. 200-204, November 1982.

[BaMcSa 87] P. H. Bardell, W. H. McAnney, and J. Savir, *Built-in Test for VLSI Pseudorandom Techniques*, John Wiley & Sons, New York, 1987.

[BeMa 87] F. Beenker and C. Maunder, "Boundary-Scan: A Framework for Structured Design-For-Test," in *Proc. International Test Conference*, IEEE CS Press, pp. 724-729, September 1987.

[BeVe 84] C. Bellon and R. Velazco, "Hardware and Software Tools for Microprocessor Functional Test," in *Proc. International Test Conference*, IEEE CS Press, pp. 804-810, October 1984.

[Be 87] P. G. Belomorski, "Pseudorandom Self-Testing of Microprocessors," *Microprocessing and Microprogramming 19*, North-Holland, pp. 37-47, 1987.

447

Bibliography

[BeMaRo 81] R. G. Bennetts, C. M. Maunder, and G. D. Robinson, "CAMELOT : A Computer Aided Measure for Logic Testability", *IEE Proceedings*, vol. 158-E, pp. 177-189, September 1981.

[Be et al. 75] N. Benowitz, D. F. Calhoun, G. E. Alderson, J. E. Bauer, and C. T. Joeckel, "An Advanced Fault Isolation System for Digital Logic," *IEEE Transactions on Computers*, vol. C-24, n° 5, May 1975.

[BoHo 71] D. C. Bossen and S. J. Hong, "Cause-Effect Analysis for Multiple Fault Detection in Combinational Networks, " *IEEE Transactions on Computers*, vol. C-20, n° 11, pp. 1252-1257, November 1971.

[Br et al. 87] R. K. Brayton, R. Rudell, A. Sangiovanni-Vicentelli, and A. Wang, "MIS: a Multiple-Level Logic Optimization System," *IEEE Transactions on Computer-Aided Design*, vol. 6, n° 11, pp. 1062-1081, November 1987.

[BrFr 76] M. A. Breuer and A. D. Friedman, *Diagnosis and Reliable Design of Digital Systems*, Computer Science Press, Rockville, 1976

[BrFu 85] F. Brglez and H. Fujiwara, "A Neutral Netlist of 10 Combinational Benchmark Circuits and a Target Translator in Fortran", in *Proc. IEEE Symposium on Circuits and Systems*, IEEE CS Press, pp. 663-698, 1985.

[BrGlKe 89] F. Brglez, C. Gloster, and G. Kedem, "Hardware-Based Weighted Random Pattern Generation", in *Proc. International Test Conference*, IEEE CS Press, pp. 1-11, 1989.

[Br et al. 88] J. Brillhart, D. H. Lehmer, J. L. Selfridge, B. Tuckerman, and S. S. Wagstaff Jr., *Factorizations of $b^n \pm 1$, $b = 2, 3, 5, 6, 7, 10, 11, 12$ up to high power*, American Mathematical Society, Providence, Rhode island, Second Edition, 1988.

[BrJü 92] J. A. Brzozowski and H. Jürgensen, "A model for Sequential Machine Testing and Diagnosis," *Journal of Electronic Testing: Theory and Application*, n° 3, pp.219-234, March 1992.

[BrJü 96] J. A. Brzozowski and H. Jürgensen, "An Algebra of Multiple Faults in RAMs," *Journal of Electronic Testing: Theory and Application*, n° 8, pp. 129-142, 1996.

[BrSe 95] J. A. Brzozowski and C.-J. Seger, *Asynchronous Circuits*, Springer-Verlag, New York, 1995.

[Ca 82] W. C. Carter, "The Ubiquitous Parity bit," in *Proc. Int. Fault-Tolerant Computing Symposium*, IEEE CS Press, Portland, USA, June 1982.

[CaSc 68] W. C. Carter and P. R. Schneider, "Design of Dynamically Checked Computers," in *Proc. IFIP Congress*, Edinburg (GB), North-Holland Publisher, pp. 878-883, 1968.

[Ca 88] P. Caspi, personal communication, 1988.

[CaPiVe 91] P. Caspi, J. Piotrowski, and R. Velazco, "An *A Priori* Approach to the Evaluation of Signature Analysis Efficiency," *IEEE Transactions on Computers*, vol. 40, n° 9, pp. 1068-1071, September 1991.

[ChLi 92] S. Chakravarty and M. Liu, "Algorithms for IDDQ Measurements Based Diagnosis of Bridgering Faults," *Journal of Electronic Testing: Theory and Application*, vol. 3, n° 4, pp. 377-386, December 1992.

[ChGu 96] C. Chen and S. K. Gupta, "BIST Test Pattern Generators for Two-Pattern Testing — Theory and Design Algorithms," *IEEE Transactions on Computers*, vol. C-45, n° 3, pp. 257-269, March 1996.

[ChJo 92] K.-T. Cheng and J.-Y. Jou, "A Functional Fault Model for Sequential Machines," *IEEE Transactions on Computer-Aided Design*, vol. 11, n° 9, pp. 1065-1073, September 1992.

[ChGu 94] C.-H. Chiang and S. K. Gupta, "Random Pattern Testable Logic Synthesis," in *Proc. International Conference on Computer-Aided Design*, San-José, CA, pp. 125-128, 1994.

Bibliography 449

[Ch 78]   T.S. Chow, "Testing Software Design Modeled by Finite State Machines," *IEEE Transactions on Software Engineering*, vol. SE-4, pp. 178-187, March 1978.

[Co 94]   B. F. Cockburn, "Deterministic Tests for Detecting Single V-Coupling Faults in RAM's," *Journal of Electronic Testing: Theory and Application*, 5, pp. 91-113, 1994.

[CoBr 92]   B. F. Cockburn and J. A. Brzozowski, "Near-Optimal Tests for Classes of Write-Triggered Coupling Faults in RAM's," *Journal of Electronic Testing: Theory and Application*, 3, pp. 251-264, 1992.

[CoCz 96]   N.L. Cooray and E.W. Czeck, "Guaranteed Faut Detection Sequences for Single Transition Faults in Finite State Machine Models Using Concurrent Faults Simulation," *Journal of Electronic Testing: Theory and Application*, n° 8, pp. 261-273, August 1996.

[CoMaWi 93]   F. Corsi, S. Martino, and T. W. Williams, "Defect Level as a Function of Fault Coverage and Yield," in *Proc. European Test Conference*, Rotterdam, NL, IEEE CS Press, pp. 507-508, April 1993.

[CoReCa 92]   C. Counil, M. Renovell, G. Cambon, "BIST for Mixed Signal Analog/Digital Circuits," in *Proc. International Symposium on Signals Systems and Electronics*, IEEE CS Press, pp. 958-961, 1992.

[CrJaDa 94]   S. Crépaux, M. Jacomino, and R. David, "On Robustness of Required Random Test Length with Regard to Fault Occurrence Hypotheses," in *Proc. IEEE VLSI Test Symposium*, Cherry Hill (NJ), USA, IEEE CS Press, pp. 348-355, April 1994.

[CrJaDa 96]   S. Crépaux-Motte, M. Jacomino, and R. David, " An Algebraic Method for Delay Fault Testing," in *Proc. IEEE VLSI Test Symposium*, IEEE CS Press, pp. 308-315, April 1996.

[DaWiWa 90]   W. Daehn, T. W. Williams, and K. Wagner, "Aliasing Errors in Linear Automata Used as Multiple-Input Signature Analyzers," *IBM Journal of Research and Development*, vol. 34, n° 2/3, March/May 1990.

[Da *et al.* 89] M. Damiani, P. Olivo, M. Favalli, and B. Riccò, "An Analytical Model for the Aliasing Probability in Signature Analysis Testing," *IEEE Transactions on Computer-Aided Design*, vol. 8, n° 11, pp. 1133-1144, November 1989.

[DaOlRi 91]   M. Damiani, P. Olivo, and B. Riccò, "Analysis and Design of Linear Finite State Machines for Signature Analysis Testing," *IEEE Transactions on Computers*, vol. 40, n° 9, pp. 1034-1045, September 1991.

[Da *et al.* 90] D. V. Das, S. C. Seth, P. T. Wagner, J. C. Anderson, and V. D. Agrawal, "An Experimental Study on Reject Ratio Prediction for VLSI Circuits: Kokomo Revisited," in *Proc. International Test Conference*, IEEE CS Press, pp. 712-720, 1990.

[Da 75]   R. David, "Paradoxe du test des circuits combinatoires," *Digital Processes*, n° 1, pp. 333-336, 1975.

[Da 78]   R. David, "Feedback Shift Register Testing," in *Proc. Int. Fault-Tolerant Computing Symposium*, IEEE CS Press, Toulouse, France, pp. 103-107, June 1978.

[Da 80]   R. David, "Testing by Feedback Shift Register," *IEEE Transactions on Computers*, vol. C-29, n° 7, pp. 668-673, July 1980.

[Da 84]   R. David, "Signature Analysis of Multi-output Circuits," in *Proc. Int. Fault-Tolerant Computing Symposium*, IEEE CS Press, Orlando, USA, pp. 366-371, June 1984.

[Da 86]   R. David, "Signature Analysis for Multiple-Output Circuits," *IEEE Transactions on Computers*, vol. C-35, n° 9, pp. 830-837, September 1986.

[Da 90]   R. David, "Comments on Signature Analysis for Multiple Output Circuits," *IEEE Transactions on Computers*, vol. 39, n° 2, pp. 287-288, February 1990.

[DaBl 76]　　R. David and G. Blanchet, "About Random Fault Detection of Combinational Networks," *IEEE Transactions on Computers*, vol.C-25, n° 6 , pp. 659-664, June 1976.

[DaBrJü 92]　R. David, J. A. Brzozowski, and H. Jürgensen, *Testing for Bounded Faults in RAM's*, Department of Computer Science, University of Waterloo, Research Report CS.92-30, June 1992.

[DaBrJü 93]　R. David, J. A. Brzozowski, and H. Jürgensen, "Random Test Length for Bounded Faults in RAMs," in *Proc. European Test Conference*, Rotterdam, NL, IEEE CS Press, pp.149-158, 1993.

[DaBrJü 97]　R. David, J. A. Brzozowski, and H. Jürgensen, " Testing for Bounded Faults in RAMs," *Journal of Electronic Testing: Theory and Application*. vol. 10, pp. 197-214, 1997.

[Da et al. 77]　R. David, P. Deschizeaux, G. Blanchet, P. Fosse, F. Martin, *Test aléatoire des circuits logiques et informatiques*, Final Report, Contract IRIA 74-147, n° LAG 77-09, April 1977.

[DaFé 83]　　R. David and X. Fédi, "*Dispositif de test aléatoire pour circuits logiques, notamment microprocesseurs*" French Patent n° 83 16285, October 13, 1983. International Extension PCT-FR-84/00 229, October 12, 1984 under the title "*Dispositif de transformation de la probabilité d'apparition de vecteurs logiques et de génération de séquences de vecteurs à probabilités variables dans le temps*".

[DaFé 85]　　R. David and X. Fédi, "Résultats améliorés sur le test aléatoire des mémoires," *Revue d'Automatique, Informatique et Recherche Opérationnelle / APII*, vol. 19, n° 6, pp. 553-560, 1985.

[DaFoTe 75]　R. David, P. Fosse, and R. Tellez-Giron, "*Expériences de test aléatoire*," Report LAG 75-05, Workshop "Prévention des pannes dans les systèmes logiques," Paris, December 1975.

[DaFu 90]　　R. David and A. Fuentes, "Fault Diagnosis in RAM's from Random Testing Experiments," *IEEE Transactions on Computers*, vol. C-39, n° 2, pp. 220-229, February 1990.

[DaFuCo 89]　R. David, A. Fuentes, and B. Courtois, "Random Pattern Testing Versus Deterministic Testing of RAM's," *IEEE Transactions on Computers*, vol. 38, n° 5, pp. 637-650, May 1989.

[DaRaRa 90]　R. David, S. Rahal, and J.-L. Rainard, "Some Relationships Between Delay Testing and Stuck-Open Testing in CMOS Circuits," in *Proc. European Design Automation Conference*, Glasgow, GB, IEEE CS Press, pp. 339-343, June 1990.

[DaTe 79]　　R. David and R. Tellez-Giron, "Comments on The Error Latency in a Sequential Digital Circuit," *IEEE Transactions on Computers*, vol. C-27, n° 1, pp. 85-86, January 1979.

[DaTh 81]　　R. David and P. Thévenod, "Random Testing of Integrated Circuits," *IEEE Transactions on Instrumentation and Measurement*, vol. IM-30, n° 1, pp. 20-25, March 1981.

[DaTh 78]　　R. David and P. Thévenod-Fosse, "Design of Totally Self-Checking Asynchronous Modular Circuits," *Journal of Design Automation and Fault Tolerant Computing*, pp. 271-287, October 1978.

[DaTh 80a]　R. David and P. Thévenod-Fosse, "Minimal Detecting Transition Sequences: Application to Random Testing", *IEEE Transactions on Computers*, vol. C-29, n° 6, pp. 514-518, June 1980.

[DaTh 80b]　R. David and P. Thévenod-Fosse, "Random Testing of Intermittent Faults in Digital Circuits," in *Proc. Int. Fault-Tolerant Computing Symposium*, Kyoto, Japan, IEEE CS Press, pp. 182-184, October 1980.

[DaWa 86]　　R. David and K. Wagner, "*Application of the Cutting Algorithm to Analysis of Pseudorandom Testing*," CRC Report 86-16, Center for Reliable Computing, Stanford University, November 1986.

Bibliography 451

[DaWa 90]   R. David and K. Wagner, "Analysis of Detection Probability and some Applications," *IEEE Transactions on Computers*, vol. C-39, n° 10, pp. 1284-1291, October 1990.

[DeThBe 84]   H. Deneux, P. Thévenod-Fosse, and L. Beghin, "Test aléatoire de circuits développés par le CNET/CNS," *International Conference on Reliability and Maintenability*, Perros-Guirec, France, May 1984.

[De et al. 76] P. Deschizeaux, M. Silva-Suarez, G. Nicoud, and F. Martin, "Statistical Fault Location in Logical Circuits," in *Proc. Int. Fault-Tolerant Computing Symposium*, Pittsburgh (PA), USA, IEEE CS Press, pp. 88-92, June 1976.

[DeKe 92]   S. Devadas and K. Keutzer, "Validatable Non Robust Delay-Fault Testable Circuits Via Logic Synthesis," *IEEE Transactions on Computer-Aided Design*, vol. 11, n° 12, pp. 1559-1573, December 1992.

[DuZo 97]   C. Dufaza and Y. Zorian, "On the Generation of Pseudo-Deterministic Two-Pattern Test Sequence with LFSRs," in *Proc. European Design & Test Conference*, IEEE CS Press, Paris, pp. 69-76, March 1997.

[DuVr 75]   L. Dugard and B. Vray, *Test aléatoire des mémoires MOS*, Student Report, Laboratoire d'Automatique de Grenoble, Institut National Polytechnique de Grenoble, June 1975.

[EiLi 83]   E. B. Eichelberger and E. Lindbloom, "Random-Pattern Coverage Enhancement and Diagnosis for LSSD Logic Self-Test", *IBM Journal of Research and Development*, vol. 27, n° 3, pp.265-272, May 1983.

[EiWi 78]   E. B. Eichelberger and T. W. Williams, "A Logic Design Structure for LSI Testability", *J. Design Automation and Fault Tolerant Computing*, vol. 2, n° 2, pp. 165-178, 1978.

[EiInYa 80]   H. Eiki, K. Inagaki, S. Yajima, "Autonomous Testing and its Applications to Testable Design of Logic Circuits," in *Proc. Int. Fault-Tolerant Computing Symposium*, IEEE CS Press, Kyoto, Japan, pp. 173-178, November 1980.

[El 59]   R. D. Eldred, "Test routines Based on Symbolic Logical Statements,"*Journal of Association of Computer Machines*, vol. 6, n° 1, pp. 33-36, 1959.

[Fé 84]   X. Fédi, *"Contribution à l'étude expérimentale du test aléatoire des microprocesseurs,"* Thèse de Doctorat, INP Grenoble, France, February 1984.

[FéDa 86]   X. Fédi and R. David, "Some Experimental Results from Random Testing of Microprocessors," *IEEE Transactions on Instrumentation and Measurement*, vol. IM-35, n° 1, pp. 78-86, March 1986.

[Fe 68]   W. Feller, *An Introduction to Probability Theory and Its Applications*, John Wiley & Sons, New York, 1968 (3rd edition).

[Fl 74]   M. J. Flynn, "Trends and Problems in Computer Organization," in *IFIP Proc.*, Amsterdam (NL), North-Holland, pp. 3-10, 1974.

[FrMe 71]   A. D. Friedman and P. R. Merron, *Fault Detection in Digital Circuits,* Prentice Hall, Englewood Cliffs, New Jersey, 1971.

[Fr 77]   R. A. Frohwerk, "Signature Analysis: A New Digital Field Service Method," *Hewlett-Packard Journal*, pp. 2-8, May 1977.

[FuAb 84]   K. Fuchs and J. A. Abraham, "A Unified Approach to Concurrent Error Detection in Highly Structured Logic Arrays," in *Proc. Int. Fault-Tolerant Computing Symposium*, IEEE CS Press, Kissemmee, FL, June 1984.

[Fu 84]   A. Fuentes, *Réalisation d'un testeur aléatoire pour microprocesseurs MC 6800*, Student Project, Laboratoire d'Automatique de Grenoble, INPG, France, June 1984.

[Fu 86a]   A. Fuentes, unpublished manuscript, 1986.

Bibliography

[Fu 86b] A. Fuentes, *Contribution à l'étude du test aléatoire des mémoires RAM's*, Thèse de Doctorat, INP Grenoble, France, December 1986.

[FuDaCo 86] A. Fuentes, R. David, and B. Courtois, "Random Testing versus Deterministic Testing of RAM's," in *Proc. Int. Symposium on Fault Tolerant Computing Systems*, Vienne, Austria, pp. 266-271, June 1986.

[Fu 85] H. Fujiwara, *Logic Testing and Design for Testability*, The MIT Press, Cambridge, 1985.

[FuKi 78] H. Fujiwara and K. Kinoshita, "Testing Logic Circuits with Compressed Data," in *Proc. Int. Fault-Tolerant Computing Symposium*, IEEE CS Press, Toulouse, France, pp. 108-113, June 1978.

[FuWaAr 75] S. Funatsu, N. Wakatsuki, and T. Arima, "Test Generation Systems in Japan," in *Proc. Design Automation Symposium*, IEEE CS Press, pp. 114-122, June 1975.

[FuMc 91] K. Furuya and E. J. McCluskey, "Two-Pattern Test Capabilities of Autonomous TPG Circuits," in *Proc. International Test Conference*, IEEE CS Press, pp. 704-711, October 1991.

[Ga 89] F. R. Gantmacher, *The Theory of Matrices*, 3rd edition, Chelsea, New-York, 1989.

[GéRaLa 83] G. Gérard, M. Raeth, and A. Laviron, *Sensibilité des microprocesseurs aux rayonnements. Importance de la méthode de test*, Commissariat à l'Energie Atomique, Report SRSC n°173, 1983.

[GlBr 89] C. S. Gloster and F. Brglez, "Boundary Scan with Built-In Self-Test," *IEEE Design & Test of Computers*, pp. 36-44, February 1989.

[GlMe 89] C. T. Glover and M. R. Mercer, "A Deterministic Approach to Adjacency Testing for Delay Faults", in *Proc. Design Automation Conference*, IEEE CS Press, pp. 351-356, 1989.

[Go 80] P. Goel, "Test Generation Cost Analysis and Projections," presented at 17th Design Automation Conference, Minneapolis, MN, 1980.

[Go 79] L. H. Goldstein, "Controllability/Observability Analysis Program of Digital Circuits", *IEEE Transactions on Circuits and Systems*, vol. CAS-26, n° 9, pp. 685-693, September 1979.

[Go 82] S. W. Golomb, *Shift Register Sequences*, Aegean Park Press, Laguna Hills, CA, 1982.

[Go 91] A. J. van de Goor, *Testing Semiconductor Memories, Theory and Practice*, John Wiley & Sons, Chichester, 1991.

[GöGr 93] M. Gössel and S. Graf, *Error Detection Circuits*, Mc Graw-Hill Book Company, London, 1993.

[Gr 79] J. Grason, "TMEAS, A Testability Measurement Program", in *Proc. Design Automation Conference*, IEEE CS Press, San Diego (CA), USA, pp. 156-161, June 1979.

[GrSt 93] M. Gruetzner and C. W. Starke, "Experience with Biased Random Pattern Generation to Meet the Demand for a High Quality BIST," in *Proc. European Test Conference*, Rotterdam,NL, IEEE CS Press, pp. 408-417, 1993.

[GsMc 75] H. W. Gschwind and E. J. Mc Cluskey, *Design of Digital Computers*, Springer-Verlag, New York, 1975.

[GuHa 93] R. K. Gulati and C. E. Harvkins (Editors), "I_{DDQ} Testing of VLSI circuits," Kluwer Academic Publishers, Assinippi Park, Norwell, MA, 1993.

[HaBo 84] M. P. Halbert and S. M. Bose, "Design Approach for a VLSI Self-Checking MIL-STD-1750 Microprocessor," in *Proc. Int. Fault-Tolerant Computing Symposium*, IEEE CS Press, Kissemmee, FL, June 1984.

[HaSt 66] J. Hartmanis and R. E. Stearns, *Algebraic Structure Theory of Sequential Machines*, Prentice Hall, Englewood Cliffs, 1966.

[HaMc 83] S. Z. Hassan and E. J. McCluskey, "Testing PLAs using Multiple Parallel Signature Analysers," in *Proc. Int. Fault-Tolerant Computing Symposium*, IEEE CS Press, Milano, Italy, pp. 422-425, June 1983.

[Ha et al. 89] C. Hawkins, J. Soden, R. Frietzemeier, and L. Harning, "Quescient Power Supply Current Measurement for CMOS IC Defect Detection," *IEEE Transactions on Industrial Electronics*, vol. 36, n° 2, pp. 211-218, May 1989.

[Ha 71] J. P. Hayes, "A NAND Model for Fault Diagnosis in Combinational Logic networks," *IEEE Transactions on Computers*, vol. C-20, n° 12, pp. 1496-1505, December 1971.

[Ha 75a] J. P. Hayes, "Detection of Pattern Sensitive Faults in Random Access Memories," *IEEE Transactions on Computers*, vol. C-24, pp. 150-157, February 1975.

[Ha 75b] J. P. Hayes, "Testing Logic Circuits by Transition Counting," in *Proc. Int. Fault-Tolerant Computing Symposium*, IEEE CS Press, Paris, France, pp. 215-219, June 1975.

[Ha 76a] J. P. Hayes, "Check Sum Test Methods," in *Proc. Int. Fault-Tolerant Computing Symposium*, IEEE CS Press, Pittsburgh, USA, pp. 114-119, June 1976.

[Ha 76b] J. P. Hayes, "On the Properties of Irredundant Logic Networks," *IEEE Transactions on Computers*, vol. C-25, n° 9, pp. 884-892, September 1976.

[He 64] F. C. Hennie, "Fault Detecting Experiments for Sequential Circuits," in *Proc. 5th Ann. Symposium on Switching Circuit Theory and Logical Design*, pp. 95-110, 1964.

[Hl 86] A. Hlawiczka, "Compression of Three-State Data Serial Streams by Means of a Parallel LFSR Signature Analyser," *IEEE Transactions on Computers*, vol. C-35, n° 8, pp. 732-741, August 1986.

[Hl 92] A. Hlawiczka, "Parallel Signature Analyzers Using Hybrid Design of Their Linear Feedbacks," *IEEE Transactions on Computers*, vol. 41, n° 12, pp. 1562-1571, December 1992.

[Ho 91] G. J. Holzmann, *Design and Validation of Computer Protocols*, Prentice-Hall, Englewood Cliffs, NJ, 1991.

[HoMcCa 89] P. D. Hortensius, R. D. McLeod, and H. C. Card, "Parallel Pseudorandom Number Generation for VLSI Systems using Cellular Automata," *IEEE Transactions on Computers*, vol. C-38, n° 10, pp. 1466-1473, October 1989.

[HuDa 92] J. W. Huang and R. David, "Fault Diagnosis of a Batch of Microprocessors," *Proceedings of Safety of Computer Control Systems 1992 (IFAC Symp. SAFECOMP '92)*, Zurich (CH), pp. 197-201, October 1992.

[Hu 93] L. M. Huisman, "Fault Coverage and Yield Predictions : Do we need more than 100% Coverage," *Proc. European Test Conference*, Rotterdam (NL), IEEE CS Press, pp.180-187, 1993.

[In 87] Intel Corp., *Components Quality/Reliability Handbook*, 1987.

[IsJaDa 93] W. Issa, M. Jacomino, and R. David, "Caractérisation des fautes dans un lot de circuits à tester", *1st Report on contract CNET-INPG N° 931B055*, Laboratoire d'Automatique de Grenoble, December 1993.

[Iv 88] A. Ivanov, *BIST Signature Analysis: Analytical Techniques for Computing the Probability of Aliasing*, Ph. D. Thesis, McGill University, Montréal (CA), 1988.

[IvAg 89] A. Ivanov and V. K. Agrawal, "An Analysis of the Probabilistic Behavior of Linear Feedback Signature Register," *IEEE Transactions on Computer-Aided Design*, vol. 8, n° 10, pp. 1074-1088, October 1989.

[IvPi 90] A. Ivanov and S. Pilarksi, *Signature Analysis for VLSI Built-In Self-Test: A Survey*, Report TR 90-3, Centre for Integrated Computer Systems Research, The University of British Columbia, March 1990.

454 Bibliography

[IvZo 92]   A. Ivanov and Y. Zorian, "Count-Based BIST Compaction Schemes and Aliasing Probability Computation," *IEEE Transactions on Computer-Aided Design*, vol. 11, n° 6, June 1992.

[IwAr 90]   K. Iwasaki and F. Arakawa, "An Analysis of the Aliasing Probability of Multiple-Input Signature Registers in the Case of a 2^m-ary Symmetric Channel," *IEEE Transactions on Computer-Aided Design*, vol. 9, n° 4, pp. 427-438, April 1990.

[Ja 89]   M. Jacomino, "*Sur la théorie du test des circuits digitaux : Mesures de la confiance*," Thèse de Doctorat, INP Grenoble, France, February 1989.

[JaDa 89a]   M. Jacomino and R. David, "Sur les mesures de la confiance dans un test," *Technique et Science Informatiques*, vol. 8, n° 5, pp. 451-469, 1989.

[JaDa 89b]   M. Jacomino and R. David, "A New Approach of Test Confidence Estimation," in *Proc. Int. Fault-Tolerant Computing Symposium*, Chicago (IL), USA, IEEE CS Press, pp. 307-314, June 1989.

[JaRaDa 89]   M. Jacomino, J.-L. Rainard, and R. David, "Fault Detection in CMOS Circuits by Comsumption Measurement," *IEEE Transactions on Instrumentation and Measurement*, vol. IM-38, n° 3, pp. 773-778, June 1989.

[JaAg 85]   S. K. Jain and V. D. Agrawal, "Statistical Fault Analysis", *IEEE Design & Test of Computers*, vol. 2, n° 1, pp. 38-44, February 1985.

[Jh 93]   N. K. Jha and S. J. Wang, "Design and Synthesis of Self-Checking VLSI Circuits", *IEEE Transactions on Computer-Aided Design*, vol. 12, n° 6, pp. 878-887, June 1993.

[JTAG 88]   Joint Test Action Group, *Boundary-Scan Architecture Standard Proposal, Version 2.0*, (available by writing to AT&T Eng. Res. Ctr., PO900, Princeton, N.J. 08540), March 1988.

[Jo 86]   M. Journeau, "A Note on Restricted Range Cutting Algorithm," *IEEE Transactions on Computers*, vol. C-35, p. 73, January 1986.

[KeMe 94]   W. Ke and P.R. Menon, "Synthesis of Delay-Verifiable Two-Level Circuits," in *Proc. European Design & Test Conference*, IEEE CS Press, pp. 297-301. February 1994.

[KeSn 76]   J. G. Kemeny, J. L. Snell, *Finite Markov Chains*, Springer Verlag, Berlin, 1976.

[Kl 88]   H.-P. Klug, "Microprocessor Testing by Instruction Sequences Derived from Random Patterns," in *Proc. International Test Conference*, IEEE CS Press, pp. 73-80, 1988.

[Kn 69]   D. E. Knuth, *The Art of Computer Programming*, vol. 2, *Seminumerical Algorithms*, Addison-Wesley, Reading (MA), 1969.

[Ko 78]   Z. Kohavi, *Switching and Finite Automata Theory*, 2nd edition, Mc Graw-Hill, New Dehli, 1978.

[Ko 82]   D. Komonytsky, "LSI Self-Test Using Level-Sensitive Scan Design and Signature Analysis," in *Proc. International Test Conference*, IEEE CS Press, pp. 414-424, November 1982.

[KöMuZw 79]   B. Könemann, J. Mucha, and G. Zwiehoff, "Built-In Logic Block Observation Technique," in *Digest of Papers Test Conference*, pp. 37-41, October 1979.

[KöMuZw 80]   B. Könemann, J. Mucha, and G. Zwiehoff, "Built-In Test for Complex Digital Integrated Circuits," *IEEE Journal of Solid State Circuits*, vol. SC-15, n° 3, pp. 315-318, June 1980.

[Ko *et al.* 87]   S. Kong *et al.*, "Design Methodology of a VLSI Multiprocessor Workstation," *VLSI Systems Design*, vol. 8, n° 2, pp. 44-54, February 1984.

[KrAl 91]   A. Krasniewski and A. Albicki, "Random Testability of Redundant Circuit," in *Proc. IEEE International Conference on Compter Design*, pp. 424-427, 1991.

[KrGa 93]	A. Krasniewski and K. Gaj, "Is There Any Future for Deterministic Self-Test of Embedded RAMS ?," in *Proc. European Test Conference*, Rotterdam, NL, IEEE CS Press, pp.159-168, April 1993.
[La 66]	P. Lancaster, *Lambda Matrices and Vibrating Systems*, Pergamon Press, Oxford, England, 1966.
[LaTi 85]	P. Lancaster and M. Tismenetsky, *The Theory of Matrices*, Academic Press Inc., Orlando, FL, 1985.
[LaGéHe 89]	A. Laviron, G. Gérard, and T. Y. Heury, "Effects of Low Irradiation Dose Rates on Microprocessors to Simulate Operation in Nuclear Installation. A Safety Approach," in *Proc. Int. Conference on Operability of Nuclear Systems in Normal and Adverse Environment*, Lyon, France, September 1989.
[Le 84]	J. LeBlanc, "LOCST: A Built-In Self-Test Technique," *IEEE Design & Test of Computers*, vol. 1, n° 4, December 1984.
[LeYa 96]	D. Lee and M. Yannakakis, "Principles and Methods of Testing Finite State Machines - A Survey," *Proceedings of the IEEE*, vol. 84, n° 8, pp. 1090-1123, August 1996.
[Le 51]	D. H. Lehmer, in *Proc. Symposium on Large-Scale Digital Computing Machinery*, Harward University Press, Cambridge (MA), USA, pp. 142-145, 1951.
[LiRePa 87]	C. Lin, S. M. Reddy, and S. Patil, "An Automatic Test Pattern Generator for the Detection of Path Delay Faults," *Proc. of Int. Conf. on Computer-Aided Design*, pp. 284-287, 1987.
[Li 87]	J. M. Liu, *Test aléatoire de microprocesseur : calcul de la longueur du test*, Student Project, Laboratoire d'Automatique de Grenoble, June 1987.
[Lo 77]	J. Losq, "Efficiency of Compact Testing for Sequential Circuits," in *Proc. Int. Fault-Tolerant Computing Symposium*, IEEE CS Press, pp. 168-174, June 1977.
[Lu 76]	W. Luciw, "Can a User test LSI Microprocessors Effectively?," *IEEE Transactions on Manufacturing Technology*, vol. MFT-5, n° 1, pp. 21-23, March 1976.
[MaYa 84]	Y. K. Malaiya and S. Yang, "The Coverage Problem for Random Testing," in *Proc. International Test Conference*, IEEE CS Press, pp.237-242, 1984.
[Ma 80]	M. Marinescu, "Test fonctionnel de mémoire vive à grande couverture de pannes," in *Proc. Int. Conf. Reliability & Maintainability*, Perros-Guirrec Tregastel, France, September 1980.
[Ma 95]	G. Massebœuf, "*Méthode hiérarchique de calcul de la longueur de test aléatoire de circuits VLSI, et analyse de testabilité*," Thèse de Doctorat, INP Grenoble, France, April 1995.
[MaPu 96]	G. Massebœuf and J. Pulou, "*Hierarchical test analysis of VLSI circuits for random BIST*," unpublished manuscript, 1996.
[MaAi 93]	P. C. Maxwell and R. C. Aitken, "Test Sets and Reject Rates: All Fault Coverages are not Created Equal," *IEEE Design & Test of Computers*, pp. 42-51, March 1993.
[Mc 84]	E. J. McCluskey, "Verification Testing - A Pseudoexhaustive Test Technique," *IEEE Transactions on Computers*, vol. C-33, n° 6, pp.541-546, June 1984.
[Mc 86]	E. J. McCluskey, *Logic Design Principles with Emphasis on Testable Semicustom Circuits*, Prentice-Hall, Englewood Cliffs, 1986.
[McBo 81]	E. J. McCluskey and S. Bozorgui-Nesbat, "Design for Autonomous Test," *IEEE Transactions on Circuits and Systems*, vol. CAS-28, n° 11, pp. 1070-1079, November 1981.
[McBu 89]	E. J. McCluskey and F. Buelow, "IC Quality and Test Transparency," *IEEE Transactions on Industrial Electronics*, vol 36, n° 2, pp 197-202, May 1989.

[Mc 87]   R. J. McEliece, *Finite Fields for Computer Scientists and Engineers*, Kluwer Academic Publishers, Assinippi Park, Norwell, MA, 1987.

[Me 74]   K. C. Y. Mei, "Bridging and Stuck-at Faults", *IEEE Transactions on Computers*, vol. C-23, n° 7, pp.129-153, July 1974.

[Me 97]   Mentor Graphics, *BISTArchitect Reference Manual, Software version 8.5-5*, Mentor Graphics Corporation, P.O. Box 5050, Wilsonville, Oregon, 97070-5050 USA

[MiJa 87] M. Milan and M. Jannin, *Microprocessor board for testing irradiated 6800*," Research Report, University of Dijon, France, June 1987.

[MoDa 97] S. Mocanu, R. David, *Sur l'évaluation de la longueur de test aléatoire pour les machines séquentielles*, Research Report 97-089, Laboratoire d'Automatique de Grenoble, France, July 1997.

[Mo 56]   E. F. Moore, "Gedanken Experiments on Sequential Machines", in *Automata Studies*, pp. 129-153, Princeton University Press, Princeton, New Jersey, 1956.

[Mo 79]   Motorola, "*Microcomputer Components. Systems on Silicon*," Data book, Motorola, Austin (TX), USA, 1979.

[Na 77]   H. J. Nadig, "Signature Analysis — Concepts, Examples, and Guidelines," *Hewlett-Packard Journal*, pp. 15-21, May 1977.

[NaKaLe 91] P. Nagvajara, M. G. Karpovsky, and L. B. Levitin, "Pseudorandom Testing for Boundary-Scan Design with Built-In Self-Test," *IEEE Design & Test of Computers*, pp. 58-65, September 1991.

[NaThAb 78] R. Nair, S. M. Thatte, and J. A. Abraham, "Efficient Algorithms for Testing Semiconductor Random Access Memories," *IEEE Transactions on Computers*, vol. C-27, pp. 572-576, June 1978.

[NaCh 96] S. Nandi and P. P. Chauduri, "Analysis of Periodic and Intermediate Boundary 90/150 Cellular Automata, *IEEE Transactions on Computers*, vol. C-45, n° 1, pp. 1-12, January 1996.

[NaKa 85] T. Nanya and T. Kawamura, "Error Secure / Propagating Concept and its Application to the Design of Strongly Fault Secure Microprocessors," in *Proc. Int. Fault-Tolerant Computing Symposium*, IEEE CS Press, Ann Arbor, MI, pp. 396-401, June 1985.

[NSC 81]   National Semiconductor Corporation, *CMOS Data Book*, Santa Clara, CA, 1981.

[Ne 79]   M. Nemmour, "Etude des codes invalides des microprocesseur *6800*," Research Report n° 165, IMAG, Grenoble, France, June 1979.

[Ni 85]   M. Nicolaïdis, "Evaluation of a Self-Checking Version of the MC 68000 Microprocessor," in *Proc. Int. Fault-Tolerant Computing Symposium*, IEEE CS Press, Ann Arbor, MI, June 1985.

[Ni 89]   M. Nicolaïdis, "Self-Exercising Checkers for Unified Built-In Self-Test (UBIST)," *IEEE Transactions on Computer-Aided Design*, vol. 8, n° 3, pp. 203-218, 1989.

[NiCo 85] M. Nicolaïdis and B. Courtois, "Layout Rules for the Design of Self-Checking Circuits," in *Proc. IEEE VLSI Test Symposium*, IEEE CS Press, Tokyo, Japan, August 1985.

[NiJaCo 84] M. Nicolaïdis, I. Jansch, and B. Courtois, "Strongly Code Disjoint Checkers," in *Proc. Int. Fault-Tolerant Computing Symposium*, IEEE CS Press, Kissemmee, FL, June 1984.

[OlDaRi 93] P. Olivo, M. Damiani, and B. Riccò, "Aliasing Minimization of Signature Analysis Testing," in *Proc. European Test Conference*, IEEE CS Press, Rotterdam, NL, pp. 451-456, March 1993.

[OpSc 75] A. V. Oppenhein and R. W. Schaefer, *Digital Signal Processing*, Prentice-Hall, Englewood Cliffs, NJ, 1975.

Bibliography

[Pa 90] A. Papoulis, *Probability and Statistics*, Prentice Hall, Englewood Cliffs, 1990.

[PaSa 85] C. Papachristou and N. Sahgal, "An Improved Method for Detecting Functionnal Faults in Semiconductor Random Access Memories," *IEEE Transactions on Computers*, pp. 110-116, February 1985.

[PaMe 87] E. S. Park and M. R. Mercer, "Robust and Nonrobust Tests for Path Delay Faults in a Combinational Circuit," in *Proc. International Test Conference*, IEEE CS Press, pp. 1027-1034, September 1987.

[Pa 76] K. P. Parker, "Compact Testing: Testing with Compressed Data," in *Proc. Int. Fault-Tolerant Computing Symposium*, IEEE CS Press, Pittsburg, USA, pp. 93-98, June 1976.

[PaMc 75a] K. P. Parker and E. J. McCluskey, "Analysis of Logic Circuits with Faults Using Input Signal Probablity", *IEEE Transactions on Computers*, vol. C-24, n° 6, pp. 668-670, June 1975.

[PaMc 75b] K. P. Parker and E. J. McCluskey, "Probabilistic Treatment of General Combinational Networks", *IEEE Transactions on Computers*, vol. C-24, n° 6, pp. 668-670, June 1975.

[Pa 71] A. Paz, *Introduction to Probabilistic Automata*, Academic Press, New York, 1971.

[Pe 61] W. W. Peterson, *Error correcting codes*, John Wiley and Sons, New-York, 1961.

[PiStZa 90] C. Piguet, A. Stauffer, and J. Zahnd, *Conception des circuits ASIC numériques CMOS*, Dunod, Paris, 1990.

[PoRe 93] I. Pomeranz and S. M. Reddy, "Classification of Faults in Synchronous Sequential Circuits," *IEEE Transactions on Computers*, vol. 42, n° 9, pp. 1066-1077, September 1993.

[Pr 86] D. K. Pradhan, *Fault-Tolerant Computing, Theory and Techniques*, Prentice-Hall, Englewood Cliffs, 1986.

[PrGuKa 90] D. K. Pradhan, S. K. Gupta, and M. G. Karpovsky, "Aliasing Probability for Multiple Input Signature Analyzer," *IEEE Transactions on Computers*, vol. 39, n° 4, April 1990.

[Pr 97] V. Prépin, "*Test aléatoire de circuits combinatoires. Nouvelle mesure de testabilité*," Thèse de Doctorat, INP Grenoble, July 1997.

[PrDa 96] V. Prépin and R. David, "Testability Measure for Combinational Circuits when Random Testing", in *Proc. European Test Workshop*, Montpellier, June 1996.

[PrDa 97] V. Prépin and R. David, "Fault Coverage of a Long Random Test Sequence Estimated from a Short Simulation", in *Proc. IEEE VLSI Test Symposium*, IEEE CS Press, Monterrey (CA), USA, April 1997.

[PuRaTh 87] J. Pulou, J.-L. Rainard, and P. Thorel, "*Microprocesseur à test intégré (MTI). Description fonctionnelle et architecture*," Technical Report CNET, NT/CNS/CCI/59, January 1987.

[RaTy 96] J. Rajski and J. Tyszer, "Multiplicative Window Generators of Pseudo-Random Test Vectors, " in *Proc. European Design & Test Conference*, IEEE CS Press, pp. 42-49, March 1996.

[Ra 71] J.-C. Rault, "A Graph Theoritical and Probabilistic Approach to the Fault Detection of Digital Circuits," in *Proc. Int. Fault-Tolerant Computing Symposium*, Pasadena (CA), USA, IEEE CS Press, pp. 26-29, June 1971.

[Re 77] S. M. Reddy, "A Note on Testing Logic Circuits by Transition Counting," *IEEE Transactions on Computers*, pp. 313-314, March 1977.

[RoFi 97] R. Rodriguez-Montañès and J. Figueras, "Bridges in Sequential CMOS Circuits: Current-Voltage Signature," in *Proc. IEEE VLSI Test Symposium*, IEEE CS Press, pp. 68-73, April 1997.

[RoOrPa 83] C. Rose, G. Ordy, and F. Parke, "N.mpc: A Retrospective," in *Proc. Design Automation Conference*, IEEE CS Press, pp. 497-505, 1983.

[Ro 66]   J. P. Roth, "Dynamic of Automata Failures : A Calculus and a Method," *IBM Journal of Research and Development*, vol. 10, n° 4, pp.278-291, July 1966.

[Sa *et al.* 91] Y. Savaria, M. Youssef, B. Kaminska, and M. Koudil, "Automatic Test Point Insertion for Pseudo-Random Testing", in *Proc. IEEE Symposium on Circuits and Systems*, IEEE CS Press, pp. 1960-1963, 1991.

[Sa 80]   J. Savir, "Syndrome-testable Design of Combinational Circuits," *IEEE Transactions on Computers*, vol. C-29, n° 6, pp. 442-451, June 1980.

[Sa 83]   J. Savir, "Good Controllability and Observability Do Not Guarantee Good Testability", *IEEE Transactions on Computers*, vol. C-32, n° 12, pp. 1198-1200, December 1983.

[SaBa 84]   J. Savir and P. H. Bardell, "On Random Pattern Test Length," *IEEE Transactions on Computers*, C-33, n° 6, pp. 467-474, June 1984.

[SaDiBa 84] J. Savir, G. S. Ditlow, and P. H. Bardell, "Random Pattern Testability," *IEEE Transactions on Computers*, vol. C-33, n° 1, pp. 79-90, January 1984.

[SaMcVe 89] J. Savir, W. H. McAnney, and S. R. Vecchio, "Testing for Coupled Cells in Random Access Memories," in *Proc. Int. Test Conference*, IEEE CS Press, pp. 439-451, 1989.

[SaFrMc 92] N. R. Saxena, P. Franco, and E. J. McCluskey, "Simple Bounds on Serial Signature Analysis Aliasing for Random Testing," *IEEE Transactions on Computers*, vol. 41, n° 5, pp. 638-645, May 1992.

[ScLiCa 75] H. D. Schnurmann, E. Lindbloom, and R. Carpenter, "The Weigthed Random Test Pattern Generator," *IEEE Transactions on Computers*, vol. C-24, n° 7, pp. 695-700, July 1975.

[Se 85]   R. M. Sedmark, "Built-In Self-Test: Pass or Fail?," *IEEE Design & Test of Computers*, (Guest Editor's Introduction of a Special issue on BIST), pp. 17-19, April 1985.

[Se *et al.* 90] M. Serra, T. Slater, J. C. Muzio, and D. M. Miller, "The Analysis of One-Dimensional Linear Cellular Automata and their Aliasing Properties," *IEEE Transactions on Computer-Aided Design*, vol. CAD-9, n° 7, pp. 767-778, July 1990.

[Se 75]   Sescosem (Thomson-CSF, Semiconductor Division), *"Logic TTL Integrated Circuits"*, 50 rue Jean-Pierre Timbaud, Courbevoie, France, 1975.

[Se 65]   S. Seshu, "On an Improved Diagnosis Program*"*, *IEEE Transactions on Electronic Computers*, vol. EC-12, n° 2, pp. 76-79, February 1965.

[Se 77]   S. C. Seth, "Data Compression Techniques in Logic Testing: an Extension of Transition Counts," *J. Design Automation and Fault Tolerant Computing*, pp. 99-114, February 1977.

[SeAg 89]   S. C. Seth and V. D. Agrawal, "On the Probability of Fault Occurrence," in *Proc. International Workshop on Defects and Fault Tolerance in VLSI Systems*, I. Koren, Ed., Plenum, New-York, pp. 47-52, 1989.

[SeAgFa 90] S. C. Seth, V. D. Agrawal, and H. Farhat, "A Statistical Theory of Digital Circuit Testability," *IEEE Transactions on Computers*, vol. 39, n° 4, pp.582-586, April 1990.

[SePaAg 85] S. C. Seth, L. Pan, and V. D. Agrawal, "PREDICT : Probalistic Estimation of Digital Circuit Testability", in *Proc. Int. Fault-Tolerant Computing Symposium*, IEEE CS Press, Ann Arbor, USA, pp. 220-225, June 1985.

[ShMc 75]   J. J. Shedlestsky and E. J. McClusky, "The Error Latency of a Fault in a Combinational Digital Circuit," in *Proc. Int. Fault-Tolerant Computing Symposium*, Paris, France, pp. 210-214, June 1975.

[ShMc 76]   J. J. Shedletsky and E. J. McCluskey, "The Error Latency of a Fault in a Sequential Digital Circuit," *IEEE Transactions on Computers*, vol. C-25, n° 6, pp. 655-659, June 1976.

[Si 92] E. Simeu, *Test aléatoire : évaluation de la testabilté des circuits combinatoires*, Thèse de Doctorat, INP Grenoble, France, July 1992.

[Si et al. 92] E. Simeu, A. Puissochet, J.-L. Rainard, A.-M. Tagant, and M. Poize, "A New Tool for Random Testability Evaluation Using Simulation and Formal Proof", in *Proc. IEEE VLSI Test Symposium*, IEEE CS Press, pp. 321-326, April 1992.

[Sm 80] J. E. Smith, "Measures of the Effectiveness of Fault Signature Analysis," *IEEE Transactions on Computers*, vol. C-29, n° 6, pp. 510-514, June 1980.

[SmMe 78] J. E. Smith and G. Metze, "Strongly Fault-Secure Logic Networks," *IEEE Transactions on Computers*, vol. C-27, n° 6, June 1978.

[So et al. 92] J. Soden, C. Hawkins, R. Gulati, and W. Mao, "I_{DDQ} Testing: A Review," *Journal of Electronic Testing: Theory and Application*, vol. 3, n° 4, pp. 291-303, December 1992.

[SoKu 96] J. Sosnowski and A. Kusmierczyk, "Pseudorandom Testing of Microprocessors at Instruction/Data Flow Level," in *Proc. Dependable Computing - EDCC2*, A. Hlawiczka, J. G. Silva, and L. Simoncini (Eds), Springer, pp. 246-263, October 1996.

[SRC 85] Semiconductor Research Corporation, *Guidelines for Research Proposals*, August 1985.

[St 73] W. Stahnke, "Primitive Binary Polynomials," *Mathematics of Computation*, vol. 6, n° 124, pp. 977-980, 1973.

[SuRe 80] D. S. Suk and S. M. Reddy, "Test Procedures for a Class of Pattern Sensitive Faults in Semiconductor Random Access Memories," *IEEE Transactions on Computers*, vol. C-29, pp. 419-429, June 1980.

[Su 81] A. K. Suskind, "Testing by Verifying Walsh Coefficients," in *Proc. Int. Fault-Tolerant Computing Symposium*, IEEE CS Press, Madison, USA, pp. 206-208, June 1981.

[Sy 97] Synopsis, *Test Computer Reference Manual*, V 1997.01, Mountain View, CA, 1997.

[SzFl 71] S. A. Szygenda and M. J. Flynn, "Failure Analysis of Memory Organization for Utilization of Self-Repair Memory System," *IEEE Transactions on Reliability*, vol. R-20, n° 2, pp. 64-70, May 1971.

[Te 74] R. Tellez-Giron, *Contribution à l'étude du test aléatoire des systèmes logiques*, Thèse de Doctorat, INP Grenoble, France, March 1974.

[TeDa 74] R. Tellez and R. David, "Random Fault Detection in Logical Networks," in *Proc. Int. Symposium on Discrete Systems*, Zinatne Ed., Riga, Lettonia, pp. 232-241, September 1974.

[TeEr 87] R. Tellez-Giron Lopez and K. Ergang, *Testability Analysis, a Survey on Methods and Applications*, Verlag TUV Rheinland GmbH, Koln, 1987.

[ThAb 80] S. M. Thatte and J. A. Abraham, "Test Generation for Microprocessors," *IEEE Transactions on Computers*, vol. C-29, n° 6, pp. 429-441, June 1980.

[Th et al. 82] S. M. Thatte, D. S. Ho, H. T. Yuan, T. Sridhar, and T. J. Powell, "An Architecture for Testable VLSI Microprocessors," in *Proc. of IEEE Test Conference*, vol. 16.5, pp. 484-492, 1982.

[Th 83] P. Thévenod-Fosse, "*Test aléatoire de microprocesseurs 8-bits. Application au Motorola 6800*," Doctorat d'Etat, Grenoble University/INP Grenoble, France, October 1983.

[ThDa 78a] P. Thévenod-Fosse and R. David, "Test aléatoire des mémoires," *Revue d'Automatique, Informatique et Recherche Opérationnelle*, vol. 12, n° 1, pp. 43-61, 1978.

[ThDa 78b] P. Thévenod-Fosse and R. David, "A Method to Analyse Random Testing of Sequential Circuits," *Digital Processes*, vol. 4, pp. 313-332, March-April 1978.

[ThDa 81]   P. Thévenod-Fosse and R. David, "Random Testing of the Data Processing Section of a Microprocessor," in *Proc. Int. Fault-Tolerant Computing Symposium*, Portland, USA, IEEE CS Press pp. 275-280, June 1981.

[ThDa 83]   P. Thévenod-Fosse and R. David, "Random Testing of the Control Section of a Microprocessor," in *Proc. Int. Fault-Tolerant Computing Symposium*, Milano, Italy, IEEE CS Press, pp. 366-373, June 1983.

[ThDaJo 81] P. Thévenod-Fosse, R. David, and D. Jourdan, "*Etude du test aléatoire des microprocesseurs,*" Final Report, Contract IRIA 78-195 (Projet Pilote SURF), May 1981.

[Th 78]   Thomson-CSF, "*Microprocesseur SF-F 96800 et circuits associés,*" Sescosem, 50 rue Jean-Pierre-Timbaud, Courbevoie, France, 1978.

[Th 87]   P. Thorel, "*Contribution au test autonome des circuits VLSI : un microprocesseur à test aléatoire intégré,*" Thèse de Doctorat, INP Grenoble, France, July 1987.

[Th et al. 87] P. Thorel, R. David, J. Pulou, and J.-L. Rainard, "Design for Random Testability," in *Proc. International Test Conference*, IEEE CS Press, Washington, USA, September 1987.

[Th et al. 91] P. Thorel, J.-L. Rainard, A. Botta, A. Chemarin, and J. Majos, "Implementing Boundary-Scan and Pseudo-Random BIST in Asynchronous Transfer Mode Switch," in *Proc. International Test Conference*, IEEE CS Press, September 1991.

[ThDa 75]   J. Thuel and R. David, "Propriétés des réseaux logiques en présence de pannes multiples. Application à la détection," *Revue d'Automatique, Informatique et Recherche Opérationnelle*, n° J-1, pp. 71-102, February 1975.

[Ti et al. 83] C. Timoc, M. Buehler, T. Griswold, C. Pina, F. Scott, and L.Hess, "Logical Models of Physical Failures," in *Proc. International Test Conference,* IEEE CS Press, pp. 546-553, October 1983.

[ToMc 94]   N. A. Touba and E. J. McCluskey, "Automated Logic Synthesis of Random Pattern Testable Circuits," in *Proc. International Test Conference*, IEEE CS Press, Washington, USA, pp. 174-183, 1994.

[Un 69]   S. H. Unger, *Asynchronous Sequential Switching Circuits*, John Wiley & Sons, New York, 1969.

[VeBeMa 90] R. Velazco, C. Bellon, and B. Martinet, "Fault Coverage of Functional Test Methods: A Comparative Experimental Evaluation," in *Proc. International Test Conference*, IEEE CS Press, pp. 91-97, September 1990.

[VeSa 80]   C. S. Venkatraman and K. K. Saluja, "Transition Count Testing of Sequential Machines," in *Proc. Int. Fault-Tolerant Computing Symposium*, IEEE CS Press, Kyoto, Japan, pp. 167-172, October 1980.

[ViDa 80]   J. Viaud and R. David, "Sequentially Self-Checking Circuits," in *Proc. Int. Fault-Tolerant Computing Symposium*, IEEE CS Press, Kyoto, Japan, pp. 263-268, October 1980.

[Vi 71]   G. H. de Visme, *Binary Sequences*, The English University Press Ltd., London, 1971.

[Wa 78]   R. L. Wadsack, "Fault Coverage in Digital Integrated Circuits," *Bell Systems Technical Journal*, vol. 57, pp. 1475-1488, June 1978.

[Wa 81]   R. L. Wadsack, "VLSI : How Much Fault Coverage is Enough ?," in *Proc. International Test Conference*, IEEE CS Press, pp.547-554, 1981.

[WaChMc 87] K. Wagner, C. Chin, and E. J. McCluskey, "Pseudorandom Testing," *IEEE Transactions on Computers*, vol. C-36, n° 3, pp. 332-343, March 1987.

[Wa et al. 89] J. A. Waicukauski, E. Lindbloom, E. B. Eichelberger, and O.P. Forlanza, "WRP: A method for Generating Weighted Random Test Patterns," *IBM Journal of Research and Development*, vol. 33, n° 2, pp.149-161, March 1989.

Bibliography

[Wa et al. 86] J. A. Waicukauski, E. Lindbloom, B. K. Rosen, and V. S. Iyengar, "Transition Fault Simulation by Parallel Pattern Single Fault Propagation," in *Proc. International Test Conference*, IEEE CS Press, pp. 542-549, September 1986.

[WaMeWi 95] L. C. Wang, M. R. Mercer, and T. W. Williams, "Enhanced Testing Performance via Unbiased Test Sets," in *Proc. European Design & Test Conference*, IEEE CS Press, Paris, France, pp. 294-302, March 1995.

[WiBr 81] T. W. Williams and N. C. Brown, "Defect Level as a Function of Fault Coverage," *IEEE Transactions on Computers*, vol. C-30, n° 12, pp. 987-988, December 1981.

[Wi et al. 86] T. W. Williams, W. Daehn, M. Gruetzner, and C. W. Starke, "Comparison of Aliasing errors for primitive and non-primitive polynomials," in *Proc. International Test Conference*, IEEE CS Press, pp. 282-288, September 1986.

[Wi et al. 88] T. W. Williams, W. Daehn, M. Gruetzner, and C. W. Starke, "Bounds and Analysis of Aliasing Errors in Linear Feedback Shift Registers," *IEEE Transactions on Computer-Aided Design*, vol. CAD-7, n° 1, pp. 75-83, January 1988.

[WiPa 82] T. W. Williams and K. P. Parker, "Design for Testability — A Survey," *IEEE Transactions on Computers*, vol. C-31, n° 1, pp. 2-15, January 1982.

[Wo 87] D. Wood, *Theory of Computation*, Harper & Row, New York, 1987.

[WoGiKa 90] D. A. Wood, G. A. Gibson, and R. H. Katz, "Verifying a Multiprocessor Cache Controller using Random Test Generation," *IEEE Design & Test of Computers*, pp. 13-25, August 1990.

[Wu 87] H.-J. Wunderlich, "Self-Test Using Unequiprobable Random Patterns," in *Proc. Int. Fault-Tolerant Computing Symposium*, IEEE CS Press, Pittsburg, USA, June 1987.

[Wu 88] H.-J. Wunderlich, "Multiple Distributions for Biased Random Test Patterns," in *Proc. International Test Conference*, IEEE CS Press, 1988.

[Ya 90] V. N. Yarmolik, *Fault Diagnosis of Digital Circuits*, John Wiley & Sons, Chichester, 1990.

[YaDe 88] V. N. Yarmolik and S. N. Demidenko, *Generation and Application of Pseudorandom Sequences for Random Testing*, John Wiley and Sons, New-York, 1988.

[YoTaNa 97] H. Yokoyama, H. Tamamoto, and Y. Narita, "Built-in Current Testing for CMOS Logic Circuits Using Random Patterns," *Systems and Computers in Japan*, vol. 25, n° 11, pp. 1-10, 1994.

[YoXiTa 97] H. Yokoyama, W. Xiaoqing, and H. Tamamoto, "Random Pattern Testable Design with Partial Circuit Duplication," in *Proc. of Asian Test Symposium*, IEEE CS Press, Akita, Japan, November 1997.

Index

Absorbing
   chain 120, 364
   state 120, 123, 170, 214, 364
Accepting state 22
Acceptor 22, 63
Activate (a fault) 42
Activity 154, 157, 368-70, 371-4
   conditional 156, 157
Adjacent vectors 113, 253
   generation 268
Algebraic method 46, 164
Algorithms
   hardware generation (µprocessor) 272, 273
   software generation 254, 255
   test length for sequential fault 177
Aliasing 300-2
   bounds 320, 335
   peaks 316, 318
   probability of 301, 315-21, 325, 327
Alphabet 22
   input 23
   output 24
AND-coupling in a RAM 210
AND-observer 67, 75-6, 78
Approximate methods 178-88
Approximate value 174-6, 190, 217, 232, 246
   accurate 171-6
Asynchronous
   circuit 113, 167
   partly 116

   sequential machine 19-21
Asynchronous test 168, 190, 253, 280, 375-7
ATPG 3, 41, 81
Automaton 21, 197
   finite 17
   initialized 21, 185
Auxiliary circuit 149, 150
Average
   initialization 199
   probability 182

Backtracking 46
Basically AND 44
Basically OR 44
Batch of circuits 219, 419
BB-fault 365-70
Benchmark ISCAS 163, 318, 354
BILBO 11
Bipolar 29
BIST 7-12, 353
   off-line 9
   on-line 8
   properties 12
Black-box fault (*see* BB-fault)
Black-box model 365-6
Blocks (partition) 95
Board 2, 5, 247, 411
Boolean 15-6, 20, 264, 269, 271, 300, 308, 342, 347
Bound
   aliasing 320, 335

Index

[Bound]
　detectability 159
　non-revelation 320
　test length 354
Bounded fault 213
　non-evasive 214, 215, 216
Bridging fault 33, 39, 398
　feedback 33
Bug 296
Built-in self-test (*see* BIST)

Cache controller (multiprocessor) 295-6
CAMELOT 164
Carry 226, 238-9, 242-3
CEA random tester 292, 294
Cell, stuck-at (*see* Stuck-at cell)
Cellular automata 276
Chain
　absorbing 120, 364
　Markov (*see* Markov chain)
Chakravarty and Hunt 165
Checker 7-8, 14
Checking sequence 38
Checkpoints 51-2
Check-sum 306, 307
Circuit
　auxiliary 149, 150
　batch of 219, 419
　combinational (*see* Combinational circuit)
　fault-free 84, 90
　faulty (*see* Faulty circuit)
　iterative 68
　nominal 9, 149
　sequential (*see* Sequential circuit)
　tree 153, 155
Circuit under test 3, 6, 8, 9, 10, 107-10
Clock 4-5, 17, 19, 22, 27-9, 116, 168-9, 184, 221, 257, 264, 267, 269, 271, 346, 375-6, 385, 393-4
CMOS 29, 110, 281

CNET 282, 341, 365, 371
Code disjoint 8
　strongly 14
Coefficient
　fault detection 128
　length 199, 200, 202-3, 205-10, 217
Co-factor 174
Collapsing (*see* Fault collapsing)
Combinational circuit 15-7, 41, 360, 397-8
　factorization 347-51
　numerical results 162
　specification 110, 338-40
　test length 135-65
Combinational fault 32, 120, 130, 175, 249-52, 345, 355
　dominance 50-1, 56
　equidetectable 50, 142, 146
　equivalent 49, 142, 145
　undetectable 54
Common factor 261, 265-6, 332-3
Comparable faults 105
Comparison
　aliasing and non-revelation 317-20
　asynchronous and synchronous test 375-7
　CUT with a reference circuit 108
　CUT with several reference circuits 108
　deterministic and random test 117-8, 218, 280, 282, 342, 353, 377
　experiments vs theory 287-9
　faults in registers and faults in operators 237
　faulty and fault-free circuits 42, 63-8
　of faults 49
　of measurements 101
　of responses 168-9, 375-7
　of signatures 108
　probability of aliasing wrt p 315, 316
　probability of non-revelation wrt p 317
　pseudorandom and random test 389-92

Index   465

[Comparison]
  signature analysers 325, 422
  stuck-at fault and BB-fault 367-8
Compatibility
  machine 181
  table 181
Compatible 25
  faults 87
  initial states 59-62, 70-8, 123, 126
  input 179
  output 179
Complement 15
Complete fault coverage 95, 98, 102, 106, 361-2
Complete specification 117, 149, 338-41
Concatenation 23
  of transition sequences 180
Conclusion on
  detection power 128
  faults in RAM 203
  specification 117
  topics neglected (approximate methods) 188
Cone 46
Confidence level 97, 127, 140, 142, 241
  measurements 90-102
Congruential linear 276
Consensus 160
Constant distribution 92, 115, 186, 213, 261-8, 267, 341, 357, 360
  hardware generation 261-9
  software generation 254
Control inputs probabilities (μprocessor) 275
Control line 342
Control point 351
Control section 224
  test length 239-41
Controllability 6, 154, 239, 342, 368-70
  conditional 156
  propagation 155

Counter 189, 390
Coupling 195-6, 207, 210, 225
  idempotent (*see* Idempotent coupling)
  multiple 236
Coverage of
  faults (*see* Fault coverage)
  faulty circuits 85, 86, 94
Cube 17, 347
  extraction 347-8
CUT (*see* Circuit under test)
Cutting Algorithm (Extended) 151-4
Cycle 221, 235, 242, 271-5, 282, 284, 287, 296, 313-4, 388

D-algorithm 81
Data (restriction of) 287
Data processing section (μprocessor) 224
  test length 233-9
Decimation property 261
Decoder and read/write logic stuck-at faults 195
Decorrelation 342-6
Defect 1, 3, 6, 13, 30-1, 290, 293-5
Defect coverage 85, 290
Defect level 2, 83-6, 88, 90, 94, 105-6, 219
Degree (of polynomial) 259, 309, 319-20
Delay fault 31, 393-6
Design for testability 4, 337-51
Detectability 72-9, 137, 142, 143, 348-51, 407
  lower bound 159-61
  most resistant fault 160
  profile 143, 144
Detectable 72-9, 154, 159-61, 168, 369, 376-7
  just 70-1
  partially (*see* Partially detectable)
  sequential fault 72, 74, 75, 77
Detecting transition sequence 178-80
Detection 31, 129

466 Index

[Detection]
  coefficient 128
  first (*see* First detection)
  function 148, 150, 348-50
  graph 181
  power 117-28
  quality 106
  set 180-5
  subset 183, 190
  surface 164
Detection probability 136-7, 318, 374
  average 131
  computation 147-61
  definition 137
  distribution 164
  minimum 146, 318, 369-70
Detection uncertainty 144-5, 355 (*see also* Uncertainty level)
Deterministic test 3, 167, 279
  vs random test 117-8, 218, 280, 282, 342, 353, 377
DFT (*see* Design for testability)
Diagnosis 31, 51, 278, 280, 285, 293, 295, 401-5
Digital circuit 21
  models 15-26
Distinction potential 302
Distinguishing sequence 38, 180
DL (*see* Defect level)
Distribution 92-3, 357-60
  constant (*see* Constant distribution)
  equally likely (*see* Equally likely distribution)
  generalized (*see* Generalized distribution)
Dominance 102, 183, 189, 211-3, 236
  combinational fault 50
  fault collapsing 50, 80
  sequential fault 70
  under a test set 51
Double fault in a RAM 211

Downstream cone 46, 151
DTS (*see* Detecting transition sequence)
Dynamic tests 2

Edge triggered 27
Eigenvalue 172, 190, 215, 355
Enable 29
Equally likely distribution 92, 137, 147, 360, 407-9
Equidetectable 103
  combinational faults 50-1, 142, 146
  sequential faults 70-1
  strongly 71
  under a test set 51
Equivalence 102
  combinational faults 49, 142, 145
  fault collapsing 50
  machines 24
  sequential faults 70, 72
  under a test set 51
Error 31, 229, 231
  correlated 324, 328, 330
  design 295
  homogeneous 324, 330
  matrix 323-4, 326
  periodic 321, 323, 330-4
  propagation 42-6, 68-9, 231, 234-5, 237, 239, 241
  sequence 309
  space-/time-dependent 323-4
Escape probability 106
Evaluator 3, 341
Evasive fault 59-62, 123, 126, 127, 128, 172 (*see also* Non-evasive fault)
  for a compatible pair of states 61
Exhaustive test 37-9
EXOR gate 257, 261, 267, 309, 312, 320, 321, 327, 329, 332
Expected fault coverage 94 (*see also* Fault coverage)
Experiments random test 277-97, 354

Index 467

[Experiments random test]
   by Laviron 290-3
   by Luciw 290
   by Velazco *et al.* 293-5
   by Wood *et al.* 295-6
Extended synchronous test 168, 375-7
External fault (operator) 225, 238

Factorization 347-51
Failed 84
Fanout 45-6, 156
   branches 45, 51, 153
   reconverging 46, 151-3
   stem 45, 153, 155, 157, 371-3
Fanout-free 46
Fault 30
   bounded (*see* Bounded fault)
   bridging (*see* Bridging fault)
   collapsing 49, 69
   combinational (*see* Combinational fault)
   comparable 105
   compatible 87
   delay 31, 393-6
   easy to detect 239, 243, 246, 355
   evasive (*see* Evasive fault)
   external (operator) 225, 238
   hardest (to detect) (*see* Most resistant fault)
   in control section 243
   in instruction decoding and control function 227-8
   in memory cell array 206
   in operators 225-6, 236-9, 242
   in register 225, 233-6, 242
   in register decoding function 226-7
   intermittent 31, 165, 191, 280, 282, 293, 354
   internal (operator) 226
   logical 31
   masking 54
   most difficult (*see* Most resistant fault)
   most resistant (*see* Most resistant fault)
   multiple (*see* Multiple fault)
   non-evasive (*see* Non-evasive fault)
   non-target (*see* Non-target fault)
   parametric 280
   passive 196
   permanent 31
   prescribed (*see* Prescribed fault)
   sequential (*see* Sequential fault)
   set of 55-7, 141, 154
   single (*see* Single fault)
   stuck-at (*see* Stuck-at fault)
   stuck-on (*see* Stuck-on)
   stuck-open (*see* Stuck-open)
   target (*see* Target fault)
   transient 31
   transition (*see* Transition fault)
   under consideration 87
   whole microprocessor 243
   worst case (*see* Most resistant fault)
Fault collapsing 87
   dominance 50
   equivalence 50
Fault coverage 55-6, 88, 94, 98, 101, 106, 141, 187, 304, 407-9
   complete (*see* Complete fault coverage)
   expected 94
   weighted 88, 94, 98
Fault model 86, 87
   basic 26-35
   black-box 365-70
   delay 394
   microprocessor 224-8
   RAM 195-7, 209-11
Fault secure 8
   strongly 14
Fault simulation 158, 369-70
   parallel 158
Fault-free circuit 84, 90

Faulty circuit 84, 85
  coverage 85, 86, 94
Favorable initial state 175
Feedback (linear) 257
Field reject rate 85
Final state 22, 59, 63, 181, 182
First detection 70, 129, 139
Flip-flop 27, 189, 376-7
Flow table 19
  primitive 113
Functional
  models 26, 35
  specifications 35
  testing 35-6
Fundamental mode 19, 113

Gate
  primitive 28
  reconverging 46
Generalized distribution 92, 112, 114, 126, 253, 268, 269-75, 341, 358-60, 421
  observer for 121
Generation
  control pattern (μprocessor) 273
  data pattern (μprocessor) 272
  hardware 257-75
  needs 249-53
  software 254-7
  test set/test sequence (*see* Test generation)
Generator 359-60, 440-1
  for Motorola *6800* microprocessor 269-75
  input-bit 251, 268
  input-vector 251, 268
  pseudorandom 1, 249, 389-92
Graph (detection) 181
Graph model 222, 229, 231, 244, 247, 355
I-compatible 179
IDDQ testing 14

Ideal random test 110, 112, 123, 126, 249, 264, 343, 354
Idempotent coupling 196, 208, 209
  PSF 196
Implicant 17, 47
  prime 17
Incomplete specification 117, 141, 148, 338-41
Inconsistencies 46
Influence
  of initialization (RAM) 200-2
  of the confidence level (RAM) 203
  of the distribution ψ (RAM) 199-200
  of the number of cells (RAM) 202-3
Initial probability vector 121, 174, 363
Initial state 18, 57, 63, 65, 181, 188, 216
  compatible 59-62, 123
  favorable 175
  known 118
  unfavorable 175
  unknown/not unique 120, 174, 434-5
Initialization 57-62, 78, 81, 108, 127, 179, 196-204, 272-5
  average 199
  explicit 59
  implicit 59
  kinds of 58
  sequence 57, 125
Input
  additional 342, 351, 354
  alphabet 23
  cone 46
  primary 31
  sequence 3, 22, 271
  states 15
  variables 15
  vector 15
Input language 22-4, 112, 115, 124
  specified 24
Input line probability 158
Input vector

[Input vector]
  generator 268
  probability 148
Instruction 221, 223, 235, 244, 386-8
  keeping 236
  probabilities 246
Instruction sequences 233
Intermittent fault 31, 165, 191, 280, 282, 293, 354
Internal
  fault (operator) 226
  state 17
Involved cells (RAM) 207, 208
Irreductible polynomial 259
ISCAS benchmark 163, 318, 354
Iteration 23
Iterative circuit 68

Just (test result) 102
Just detectable/detecting 70
Justification 43

Karnaugh map 16
Keeping instruction 236
Kernel extraction 349-50

Language 22, 357
  input (*see* Input language)
  output (*see* Output language)
Latch 27
Layout level 30
Length 22, 93
  between detections 129, 289
  coefficient 199, 208
  test (*see* Test length)
  to (first) detection 129, 139, 289
Level
  of confidence (*see* Confidence level)
  uncertainty (*see* Uncertainty level)
LFSR
  generator 261-7

  properties 257-61
  signature analyser 308-12
Linear
  congruential 256, 276
  feedback 257
Literal 17, 47
Logic tests 2
Logical fault 31
Low cost 280, 290, 353
Lower bounds
  detectability 159
  detection probability 370, 374
LSI 90, 281
LSSD 14, 338

Machine (*see* Sequential machine)
Manufacturer 2, 7, 293, 401
Markov chain 118-23, 131, 169, 216, 229, 246
  vs MDTS 228-33
Markov source 258-9
Matrix
  companion 264
  transition (*see* Transition matrix)
Masking (of fault) 54
MDTS 178-85, 233
  vs Markov chain 228-33
Mealy machine 18
  transformation 25
Measurements of confidence level 90-102
  comparison 101-2
Memory effect 132, 175, 183, 187, 280, 289
Microprocessor
  fault model 224-8
  hypothetical 222
  Intel *8080* microprocessor 290
  model 222-4
  Motorola *6800* microprocessor 244-7, 269-75, 283-9, 290, 293, 385-8
  numerical results 241-7

470   Index

[Microprocessor]
    structural analysis 247
    test length 221-47
Minimum testing probability 95, 98, 102, 106, 146-7, 353
    weighted 96, 100
Minterm 17
MISR 322-30
Model
    digital circuits 15-26
    fault (*see* Fault model)
    functional 26, 35
    gate-level 26, 28
    microprocessor 222-4
    RAMs 193-5
    register-level 26
    structural 26
    transistor-level 26
Moore machine 17
    transformation 25
MOS 29, 281
Most resistant fault 95, 146, 161, 369, 407-9
    6800 microprocessor 245, 247
    detectability 160
    in register (μprocessor) 235
    in usual operators (μprocessor) 238
Motorola (*see* Microprocessor)
M-sequence 257, 266
    properties 257-61
MSI 162, 189, 277
Multiple faults 33, 87, 88, 160, 161, 213, 236, 353-5, 407-9
    in a RAM 211-3
Multiprocessor cache controller 295-6

Needs
    generation random test 249-53
    testing 1-4
Neighborhood 196
Next state function 21

NMOS 29
Nominal
    circuit 9, 149
    speed 6
Non-evasive fault 61, 79 (*see also* Evasive fault)
    bounded fault 214, 215, 216
    given an initial state 61
    given an initialization sequence 61
Non-revelation 300-2
    bound 320
    convergence 319
    probability of 302, 304, 314-21, 327-34
Non-target faults 87, 88, 90-4, 118, 127, 277, 284-5, 353-4
Numerical results
    combinational circuit 162
    microprocessor 241-7
    RAM 208, 218
    sequential circuit 189

Observability 6, 154, 239, 342, 371-4, 423
    conditional 156
    propagation 155
Observation point 351
Observer 63-8, 81, 118-23, 132, 169, 229, 231, 361, 395
    AND-observer 67, 75-6, 78
    definition 64
    for generalized distribution 121
    OR-observer 67, 75-6, 78
Occurrence probability 91, 96, 98
O-compatible 179
Open 32
Operation codes
    specification 341
    valid 253, 270-4, 285-7, 341, 387-8
Operator 222, 223, 244
OR-coupling in a RAM 210
OR-observer 67, 75-6, 78

Output
   additional 346, 351, 354
   alphabet 24
   cone 46
   primary 31
   sequence 3, 22, 24
   states 16
   variables 16
   vector 16
Output function 21, 23
Output language 22-4

Parametric tests 2, 280, 281, 292, 296-7
Parity 46
Partially detectable 74-9, 81
Partially random test 402-5
Partition 95, 99, 100, 246, 437
Passed 84
Passive
   fault 196
   PSF 196
Path 42-6, 149-54, 231-5, 244, 393-4
   reconverging 46, 149, 153-4
   scan 4
   sensitized 42, 149-50
Pattern 195 (*see also* Vector)
Pattern sensitive fault (*see* PSF)
Period/Periodic 221, 256-66, 276, 292, 321, 323, 330-4, 375
Permanent fault 31
PMOS 29
Polynomial 258
   degree 259, 309, 319-20
   irreducible 259
   primitive (*see* Primitive polynomials)
   reciprocal 260
Power up/on 57-60, 62, 73, 75, 77-8, 124, 127, 177, 196, 413
ppm 2, 354, 411
PREDICT 165
Prefix 23, 24, 58, 70, 128-9, 253

Prescribed fault 87, 284-6 (*see also* Target fault)
Preset sequence 57 (*see also* Initialization sequence)
Primary
   inputs 31
   outputs 31
Primitive
   flow table 113
   gates 28
Primitive polynomials 259, 266, 319, 320, 322, 327, 329, 332, 335
Probability 88
   aliasing 301, 315-21, 325, 327
   average 182
   complete fault coverage 95
   control input (μprocessor) 275
   data (μprocessor) 275
   detection (*see* Detection probability)
   detection set 182
   e-detection 175
   input lines 158
   input vector 148, 158
   instruction 246, 275
   non-revelation 302, 304, 314-21, 327-34
   occurrence 91, 96, 98
   of state (*see* State probability)
   propagation of 152
   R/W input (RAM) 199
   theory of 133
   transition 185
   transition sequence 182
   zero, non-zero 124, 264, 357
Probability of testing 93, 121, 124, 137, 315
Probability vector 119-23, 170, 363
   initial 121, 174, 363
Product of literals 17, 47
Product of sums 48
   complete 48

Propagation
    of controllability 155
    of error 42-6, 68-9, 231, 234-5, 237, 239, 241
    of observability 155
    of probabilities 152
    of probability bounds 153
Properties of LFSR 257-61
    decimation 261
    maximum length 257, 259
    shift-and-add 261
    window 260
Provoking an error 42-3, 68-9, 149, 154, 156, 158, 239
Pseudo-equivalence 25
Pseudorandom
    generators 1, 12, 249-76
    test 389-92
    test sequence 12, 262
PSF 196, 208, 209
    idempotent coupling 196
    passive 196
    toggling 196

Quality level 2, 14

RAM
    fault models 195-7, 209-11
    model 193-5
    numerical results 208, 218
    test length 193-220
Random
    defect 293-5
    generation 80-1
    test sequence 3, 12, 92-106, 109, 136-8, 146, 167, 170, 249, 288-93, 320, 337, 407-8
Random access memory (*see* RAM)
Random pattern resistant 354
Random pattern source 108, 357-60
Random pattern testable 354

Random test 167, 279, 353-5
    experiments 277-97, 402-5
    ideal (*see* Ideal random test)
    implementation 108
    limits of 123
    principle 107-33
    vs deterministic test 117-8, 218, 280, 282, 342, 353, 377
Random tester 270-5, 277-8, 295
    CEA 292, 294
R-controllable 342-5
R-dominance 103-5, 211, 213, 236, 407
Reconverging
    fanout 46
    gate 46
    paths 46
Redundant
    circuit 52-3, 398
    fault (*see* Undetectable)
Register 189, 222, 272, 275
    shift (*see* Shift register)
Regular expression 22-3
R-equidetectable 103-5, 196-7
Reset 27, 79, 124, 128, 184, 271-3, 286-7
    external 58-62, 78
    power-up 58-62, 78
Resistance 145
    profile 144, 145, 163, 367
Response 3, 22, 401-2
Restricted-range 152
Revelation 302 (*see also* Non-revelation)
R/W input probability (RAM) 199

Scan
    boundary 5
    design 4-6, 135, 338, 411
    path 4
SCOAP 164
Self-testing 8
Sequence
    checking (*see* Checking sequence)

[Sequence]
    distinguishing (*see* Distinguishing sequence)
    homing 59
    initialization (*see* Initialization)
    input 3, 22
    maximum length (*see* *M*-sequence)
    of instructions (*see* Instruction sequence)
    of transitions (*see* Transition sequence)
    output 3, 22, 24
    synchronizing (*see* Synchronizing sequence)
    test (*see* Test sequence)
Sequential circuit 17-22, 57, 360, 398-9
    numerical results 189
    specification 113, 115, 340-1
    test length 167-91
Sequential fault 33, 69, 34, 131, 175, 252, 355
    detectable 72, 74, 75, 77
    dominance 70
    equidetectable 70
    equivalence 70, 72
    in a combinational circuit 135, 136, 165, 177, 397-8
    intermittent 191
    undetectable 72, 74, 77, 78
Sequential machine
    asynchronous 19-21
    equivalent 24
    finite state 17
    synchronous 17-9
Set (input) 27
Set of bits 225
Set of subsequences 252, 263-7, 400
Set of vectors 249-52, 261-2
Shift register 257-8 (*see also* LFSR)
Shift-and-add property 261
Shifting (number of) 263, 264, 265, 266
Short 32

Signature 292, 304
    multiple 322
Signature analysis 3, 299-335
    by LFSR 308-12
    cost 321-2
    counting methods 306-7
    dimensioning 304-6, 422
    multiple input (*see* MISR)
    principle 300
    single input (*see* SISR)
Simeu 165
Simulation (*see* Fault simulation)
Single fault 33, 86
SISR 306-22
    properties 312-21
Slice 338, 344
Software generation 254-7
    comments 256
Specification
    basic 117
    combinational circuit 110
    complete 117, 149
    extended 117, 149, 338-41, 343
    functional 35
    incomplete 117, 141, 148, 338-41
    sequential circuit 113, 115
Specified cases 109-17, 249, 250
Specified input language 24
SSI 162, 189, 277
Stable state 19
STAFAN 165
Starting state 38, 57, 65, 73, 75, 76, 77, 78, 125, 184, 196
State 17, 363-4
    absorbing 120, 123, 170, 214, 364
    accepting 22
    current 181
    final (*see* Final state)
    initial 18 (*see* Initial state)
    input 15
    internal 17

[State]
  next 17
  output 16
  present 17
  stable 19
  starting (*see* Starting state)
  total 17
  unstable 19
State diagram 18, 20
State probability 119-21, 131, 170, 184, 188-90, 199-200, 205, 217, 230, 308, 314, 363-4
State table 18, 19, 20
Strongly code disjoint 14
Strongly connected 58-9, 61, 77, 186, 364
Strongly fault secure 14
Structural
  model 26
  testing 36-7
Stuck-at 87
Stuck-at cell 196, 197-204, 208, 209, 212
Stuck-at fault 32, 39, 160, 161
  address line (RAM) 205
  decoder and read/write logic 195, 204
  multiple 48, 52, 54
  single 36, 48, 49, 51, 52, 54
Stuck-on 34, 40, 87
Stuck-open 34, 39, 87
Subsequence 23, 345, 397-400
Suffix 23
Sum of products 47, 48
  complete 48, 160
Sure (test result) 102
Synchronizing sequence 38, 58, 59, 60, 73, 75, 76, 77, 78, 170
Synchronous
  circuit 115, 168
  sequential machine 17-9
Synchronous test 168, 190, 252, 375-7

Table
  compatibility 181
  truth 16
Target fault 80, 87, 90-4, 101, 118, 277
  (*see also* Prescribed fault)
Test
  asynchronous 168, 190, 253, 280, 375-7
  dynamic 2
  evaluator 3, 341
  exhaustive 37-9
  logic 2
  of program 167
  parametric 2
  random *see* Random test
  robust 394
  set 48
  synchronous 168, 190, 252, 375-7
  validated 395
Test generation 12, 41-81
  random test sequence 79-81, 249-76
Test length 162, 354
  6800 microprocessor 245
  approximate method 178-88
  bound 354
  combinational circuits 135-65
  control section 239-41
  data processing section 233-9
  exact calculation (sequential fault) 169-71
  microprocessors 221-47
  obtaining the 176-7
  RAMs 193-220
  reduce the 246-7, 354
  sequential circuits 167-91
  single fault (RAM) 197-209
  whole microprocessor 243
Test methods (Velazco *et al.*) 293
Test of communication protocols 167
Test sequence 2, 22, 41, 55, 57, 79, 83-106, 107-10, 212, 291, 300, 323, 390, 396, 402

Index **475**

[Test sequence]
    random 3, 12, 92-106, 109, 136-8, 146, 167, 170, 249, 288-93, 320, 337, 407-8 (*see also* Test generation)
Test vector 42, 44, 48, 80-1, 136-8, 141, 149, 306-7, 341, 408
    set of 49, 136, 249-52, 278, 389-92
    weighted 246, 267-8, 338, 360
Testers 270-5, 277-8, 293, 295
Testing
    deterministic (*see* Deterministic test)
    exhaustive 37
    functional 35
    probability (*see* Probability of testing)
    random (*see* Random test)
    structural 36
Testing quality 106
TMEAS 164
Toggling 196, 206-7, 208, 209
    PSF 196
Totally self-checking 8
TPG 3, 8
Transient fault 31
Transition fault 185-8
Transition function 21, 23
Transition matrix 172-6, 198, 214, 215, 313, 319, 363
    cyclic 330-2, 364
Transition sequence 182, 236
    detecting 178-80
    minimal detecting *see* MDTS
Tree circuit 46, 153, 155
Tri-state 271, 292
Truth table 16
TSC goal 14
TTL 162, 189, 277

Uncertainty level 97, 140, 142 (*see also* Detection uncertainty)
Undetectable/redundant
    combinational fault 54, 425-6
    for a test set 54
    sequential fault 72, 74, 77, 78
Unfavorable initial state 175
Union 22
Universal fault model 106
Universal fault set 86
Unspecified 19, 21, 197
Unstable state 19
Upstream 43, 46, 159-61, 371-2, 404-5
User 2, 7, 221, 247, 401

Variable
    input 15
    output 16
Vector (or pattern)
    adjacent 113
    input 15
    output 16
    probability (*see* Probability vector)
    test (*see* Test vector)

Weakly equidetectable 71
Weighted
    fault coverage 88, 94, 98
    minimum testing probability 96, 100
Weighted test vectors/patterns 246, 267-8, 341, 354, 360
Window Property 260, 264
Word-oriented RAM 209-11
Worst case fault (*see* Most resistant fault)

Yield 84, 88, 93